Biohistory

"The most powerful books tend to challenge common wisdom and widespread beliefs. Jim Penman does that in this volume."
—Dr Steven A. Peterson, Professor of Politics and Public Affairs and Director of the School of Public Affairs, Penn State Harrisburg

"A theory which only the unwise will ignore."
—Dr Michael T. McGuire, MD, Professor of Psychiatry and Biobehavioral Science, University of California at Los Angeles

"If a fraction of the argument presented [...] is borne out, it will shatter mainstream political science and grand history."
—Dr Frank Salter, Principal, Social Technologies Pty Ltd; Senior Fellow, International Strategic Studies Association

"Biohistory surpasses all existing efforts to explain the major patterns of human history both in originality and scientific potential"
—Dr Ricardo Duchesne, Department of Social Science, University of New Brunswick Saint John

Biohistory

By

Jim Penman

Cambridge
Scholars
Publishing

Biohistory

By Jim Penman

This book first published 2015

Cambridge Scholars Publishing

Lady Stephenson Library, Newcastle upon Tyne, NE6 2PA, UK

British Library Cataloguing in Publication Data
A catalogue record for this book is available from the British Library

ISBN (10): 1-4438-7165-6
ISBN (13): 978-1-4438-7165-5

TABLE OF CONTENTS

ACKNOWLEDGEMENTS

I offer my sincere thanks to Professor Paolini and his team, to LaTrobe University, RMIT and the Florey Institute for access to their laboratories and equipment, and to the Australian Research Council for providing matching funds. Neither the researchers nor the Institutions had any knowledge of the theory and ideas presented in this book. The conclusions and the interpretations are wholly mine.

My thanks also to Professor Ricardo Duchesne for his unstinting support and assistance with many aspects of the theory, to Dr Andrew Fear of Manchester University for his help with the section on Roman history, to Professor Michael McGuire for advice on primatology and general editing for sense and purpose, and to Dr Frank Salter for his advice and valuable connections.

A special thanks to my son Andrew for his support over the years and as editor, researcher and co-writer. Without his contributions this book would not have been written, or at least delayed for many years.

Jim Penman
Melbourne, Australia
August 2014

INTRODUCTION

In the study of human societies certain questions have exercised the minds of scholars for thousands of years. These include:

- Why are some societies so much wealthier than others?
- Why do some groups within a society suffer disadvantage?
- Why do wars occur?
- Why do economic recessions occur?
- Why do birth rates tend to decline in affluent societies?
- Why do some countries form stable democracies and others do not?
- Why do certain peoples rise to power and prominence?
- Why do civilizations fall?

Each of these questions has attracted multiple answers. For example, the rise of the West has been attributed to Protestant Christianity, the Enlightenment and the Industrial Revolution. The decline and fall of nations and empires has been explained as the result of economic stagnation, social pressures, disease and climate change. Populist historians have explained historical change as the result of great leaders such as Augustus, who brought an end to Roman civil war, Charlemagne, who established the Frankish Empire, Napoleon as conqueror of Europe, and Hitler as a crazed demagogue driving his people to destruction.

While these explanations are diverse they share one thing in common—they are not testable empirically. Even the most rigorous historical analysis can do little more than show that selected factors trend together and are possible causes of change. Historians may argue that the Second World War was launched by Hitler's personality or the Versailles Treaty or the economic crisis of the Weimar Republic, but short of repeating history *without* Hitler or the Versailles Treaty there is no way to be certain.

The theory presented in the following pages sees changes in social, political and economic behavior as reflecting changes in temperament. It accepts the prevailing view that temperament is a behavioral and emotional state that varies among individuals, is relatively stable over time and situation, is biologically based and appears early in life, but is influenced by parenting style and other environmental variables which condition how the inherited temperament is expressed.[1]

Where it differs from other approaches is in seeing the prevailing temperament of the population as the key to all those questions asked earlier. It is far more important than political and economic institutions, the decisions of leaders, or *any other factor*.

This theory also identifies the key aspects of temperament for this purpose as two separate but related biological systems which help animals adjust their attitudes and behaviors to changes in the environment.

One of the systems is triggered by relatively mild yet chronic food shortage, the other by occasional famine or predator threat. Both work via physiological signals that influence the expression of genes. In turn, behavioral change renders individuals more likely to survive and prosper. For example, mild food shortage leads animals to drive away members of their species and become more active and exploratory—behaviors that increase their chances of survival in environments with limited food.

But these systems can be triggered in other ways. Human cultures have developed codes of behavior, especially related to religion, that have the same effect as calorie restriction. These codes conflict with "natural" human inclinations but have been strengthened by competition between cultures. They change human temperament in a way that has made the rise of farming possible, along with wider political loyalties and more advanced economies—what we term 'civilization.'

These codes of behavior include fasting, religious rituals, patriarchy, and (above all) the restriction of sexual activity. By mimicking the physiological effects of hunger they help individuals and societies to survive and prosper.

---

[1] Jan Kristal, *The Temperament Perspective, Working with Children's Behavioral Styles* (Paul H. Brooks Publishing, 2005); Jerome Kagan, *Galen's Prophecy: Temperament in Human Nature* (Basic Books, 1994); Mary Rothbart, David Evans & Stephan Ahadi, "Temperament and Personality: Origins and Outcomes," *Journal of Personality and Social Psychology* 78 (1) (2000); This view is distinct from the ancient Greek concepts of four bodily "humours" (choleric, melancholic, phlegmatic and sanguine) widely accepted up to the eighteenth century; Robert Stelmack & Anastasios Stalikas, "Galen and the Humour Theory of Temperament," *Personality and Individual Differences* 12 (3) (1991).

To understand how this works, contrast the survival strategies of hunter-gatherers with those that drive success in civilization. Hunter-gatherers normally need only a few hours a day to find enough food, and the work is varied and interesting (especially for the men as hunters). They spend much of their time socializing, which develops bonds that aid group defense. Individuals who work harder but are less sociable would probably have fewer surviving children than others.

In a civilized society the state tends to handle defense, and socializing detracts from the crucial work of making a living. Thus, an individual who works harder and socializes less is likely to have *more* surviving children. Success in different social environments thus requires a different form of temperament.

Civilized societies also require physical technologies such as agriculture, metalworking, writing and trade. Yet, on their own, these technologies fail to explain why some civilizations rise and why others fail.

The changes in human behavior and temperament that are induced by cultural strategies and practices are rooted in epigenetics. This means that environmental influences, especially in early life, alter the level of activity and expression of key genes. These in turn affect behavior and temperament, including attitudes toward authority, capacity for work, economic and mechanical skills, and creativity.

The development of civilization thus depends on developing not only physical technologies but cultural technologies, especially religions. For example, if a religion induces behavioral change such that the society farms more productively, has more surviving children, and organizes itself into a large state, it is likely to expand and conquer its neighbors.

A weak point about these cultural systems is that they are vulnerable to the effects of abundance and population density. Wealthy urban societies with plentiful food tend to abandon ascetic behaviors, such as restrictions on sexual activity, which mimic the effects of food shortage. This in turn leads to society-wide change in temperament and behavior which undermine success. In effect, the greater the wealth and density of a society's population, the harder it is to maintain the cultural strategies responsible for the society's rise. In the chapters to come we propose that the collapse of civilizations, along with their replacement by people from less-developed societies, can be understood in this way.

This is a theory of mammalian social behavior which offers a novel and robust understanding of human history. Because it is a historical theory based on biology, it is given the name "biohistory."

While initially developed from the study of human societies, the biological basis of the theory makes it possible to generate hypotheses that may be tested in both animal and human populations. To continue the example given earlier, there are a number of theories for the origin of the Second World War that are broadly consistent with the evidence, including the personality of Adolf Hitler and resentment against the Versailles Treaty. Biohistory proposes a different reason—that it was (in part) the result of anxiety transmitted to infants born at the close of the 1914–18 war, which caused a permanent epigenetic change that made them more aggressive. When these young men reached their early twenties they brought about a more militaristic tone to society which helped launch another war. So far this is a standard historical theory, broadly consistent with the evidence, but no more.

The key difference is that this particular theory can be tested quite rigorously by *testing the men born in 1917–18* for a specific epigenetic signature associated with aggression in rats *and* people, which should be more prominent in this cohort than in those born earlier or later. If too few subjects are available, the same could be done for Chinese born in 1948–49 or Europeans born in 1944–45. To the extent that this pattern is *not* found, the theory is weakened or must be modified; to the extent that it is, the theory is confirmed.

There are hundreds of other potential tests that could be done, some of which are detailed in the final chapter. Clearly, not every application of biohistory can be tested in this way since most of it relates to the distant past, but that is no different from any scientific theory. Physicists assume the laws of gravity apply to distant galaxies as they do on Earth, even though there is no way to test them directly. The measure of any scientific theory is not that every application must be tested but that it gives rise to testable hypotheses through which it *can* be tested, in the sense that it may be falsified or confirmed. In this sense, as a theory of history, biohistory is unique.

Chapter one reviews the family and social patterns that are present in civilized societies, including nuclear monogamous families, control of children and restriction of sexual activity. Studies of non-human primates in their natural environments, principally baboons and gibbons, show that many of these behaviors are associated with food-restricted environments. Animals adapt to such environments by delaying breeding, reducing group size, and moving from promiscuous mating towards nuclear monogamous families. It is a physiological response which allows populations to adapt quickly to food shortage, but to abandon such behavior when food is once more plentiful.

Chapter two presents laboratory studies on the effects of mild food restriction on rats. Among other affects it improves maternal care, reduces sexual activity and increases exploration. These studies show how food restriction during infancy changes the activity levels of genes which affect the behavior and the biochemistry of animals during later life. These changes may be inherited by an individual's offspring, at least partly through changes in parental behavior.

Chapter three examines how human cultural norms, especially control of sexual activity, have the same effect as calorie restriction. Religion drives the development of civilization through its influence on behavior and temperament, at least as much as technologies such as metalworking and trade. We refer to this "civilized" temperament as "C."

Chapter four uses zoological and ethnographic evidence to introduce a second set of characteristics that are distinct from those related to C. These include vigorous aggression, intolerance of crowding, hierarchical cooperative social organization, male domination of females, and a switch from indulgence and protection of infants to rejection of juveniles after weaning. This behavior is labeled "V" (for vigor) and is an adaption to environments where food is generally plentiful but with occasional famines. Such environments require mutual defense and fast population growth. V is triggered by occasional but severe stresses, such as famine or predator attack.

Chapter five further develops the concepts of C and V. It focuses on how cultural practices which promote C and V have very different effects depending on the age of exposure. Control of children during early childhood increases "Infant C," which renders individuals open to change and being skilled with machines. "Child V," which results mainly from experience of authority and punishment during late childhood, renders people more traditional and accepting of authority. Punishment of children also raises the level of stress in the society. Ethnographic case studies are reviewed to further test the concepts of C and V. The chapter shows that cultural norms and childrearing patterns influence adult temperament, which in turn determines political systems and levels of economic success.

Chapter six traces the rise of C in England over more than five centuries. Changes in age of puberty, family patterns, attitudes towards sex, work habits and increased control of children, especially infants, are documented. The striking success of the Industrial Revolution during the nineteenth century is explained as the result of an unprecedented peak of C, especially Infant C.

Chapter seven further develops the biohistory model by addressing events in England, Europe and Japan. The analysis indicates that a rise in C is driven by high V resulting in a high level of stress, which reached a peak in the sixteenth century. The subsequent decline in V and stress eventually allows C to fall. This is called the "civilization cycle."

Chapter eight explores how population fluctuations in species such as lemmings and muskrats can be explained by changes in C and V. These "lemming cycles" are used to explain patterns such as the decline of Chinese dynasties, the virulence of the Black Death in Medieval Europe, the timing of the Renaissance, and conflicts such as the Wars of the Roses.

Chapter nine looks at wars and revolutions that follow peaks of population growth and/or the end of previous wars. These are explained as a consequence of larger families and anxious mothers causing an increase in V in an age cohort. When males reach their early twenties their greater aggressiveness has a disproportionate influence on society and makes it easier for governments to engage in war. Findings are used to explain both the cause and timing of the French Revolution, the First and Second World Wars in Europe, the Japanese attack on Pearl Harbor, the Chinese Cultural Revolution, and the more aggressive attitudes of Iranians in recent years.

Chapter ten addresses the decline and fall of civilizations. C promotes larger states and more advanced economies, while V promotes higher birthrates and martial vigor. Civilizations collapse because declining V causes military weakness, and declining C causes political weakness and economic decay.

Chapter eleven analyses the rise and fall of Rome. Roman civilization was the result of an unprecedented rise in Infant C, driven by cultural systems imported from the Middle East. V and stress among the Romans appear to have peaked in the sixth century BC, while C peaked around 250 BC and was followed by a prolonged decline in both C and V. The fall of C explains the change from Republic to Empire and the subsequent collapse.

Chapters twelve and thirteen focus on why some societies are less vulnerable to the loss of V and C and thus become more durable. This is attributed to the presence of a stability factor known as "S" which leads people to indulge infants yet be stricter with older children. S reduces the unstable infant C and increases child V, making the society more conservative. Increased S is likely due to genetic change that arises from the experience of civilization, especially civilization collapse, which confers a demographic advantage on people with higher S and renders future collapse less serious and prolonged. The rise of S is traced in China and India.

Chapter fourteen traces the rise of S in the Middle East, from the low S Sumerians to the higher S empires of later times. Civilized societies developed high C cultural systems, while their "barbarian" neighbors had higher V because of harsh living conditions. The civilized peoples transmitted higher C cultures to the barbarians, who transmitted higher V to the settled lands by immigration and conquest. A gradual rise in both C and V culminated in the Arab conquest and the rise of Islam, seen as an especially powerful and durable cultural system which promotes long-term success at the expense of economic progress. A clear implication is that the Muslim populations of the Middle East will spread and gradually assimilate most of the world into their own faith and culture, beginning with Europe. Patriarchy and purdah, not liberal democracy, will be the true "end of history."

Chapter fifteen approaches the most pressing issue in this book—the decline of Western civilization. Changes in Western countries over the past 150 years, and especially since the 1960s, are the result of a dramatic fall in both C and V. Evidence of falling C includes the declining age of puberty, increased sexual freedom, reduced control of children, declining work ethic, and economic stagnation. Signs of falling V include female emancipation, plunging birth rates, and a reduced enthusiasm for war. Though not all of these are negative, the end result must be economic decline and political collapse. Knowledge of the underlying biology indicates that no conventional social or political policy can reverse the process.

The model in this book presents our current understanding of the topics addressed. There will of course be additions and amendments as a result of further research and testing. Selected proposals are described in a final section.

This book is intended as a companion volume to *Biohistory: Decline and Fall of the West.*[2] It contains fuller evidence and the references not included in what is intended to be a shorter, more popular work. So that readers may easily cross from one version to another, the chapter structure remains the same. For example, anyone wanting more information about war than contained in chapter nine of the popular work may open chapter nine of this version and find further examples and longer descriptions. The shorter version also contains substantial material not included in this version, including different quotations and illustrations, but either book will give the reader a comprehensive understanding of biohistory.

---

[2] J. Penman, *Biohistory: Decline and Fall of the West* (Newcastle: Cambridge Scholars Publishing, 2014).

CHAPTER ONE

OF SCIENCE AND TEMPERAMENT

A key purpose of this book is to explain why some human societies have developed civilizations and others have not. For example, what differentiates humans who practice agriculture and form large-scale political and economic systems from those that live as hunter-gatherers in small bands? Why do the power and affluence of civilizations change so much and so quickly, with large, stable societies dissolving into anarchy, and others growing rapidly in influence and wealth?

Historians, archaeologists and economists have sought to understand these changes by looking at economic pressures, population growth, warfare, environmental change and the actions of charismatic leaders. Biohistory uses a different approach. It starts with the premise that humans are biological beings and therefore influenced by the same basic principles that affect our close non-human relatives. Genetically speaking we are very similar to other mammals. We share 95–98% of our genes with chimpanzees and, according to the latest results from Celera Genomics, about 85% with mice.[1]

There are many forms of behavior unique to humans. Apart from ants and termites, no other species unites thousands of individuals to work together and fight against outsiders. No other species develops market economies, uses money, builds machinery, establishes religions and formal codes of morality, or wears clothing by choice.

But in terms of family and social patterns, human behavior is less distinct. We control or punish our offspring, neglect them or provide intensive care. We can be monogamous, polygynous or promiscuous, and our levels of sexual activity vary enormously. We may mate immediately after puberty

---

[1] The more commonly used figure is 98%, but recent analysis suggests 95% might be more meaningful. R. J. Britten, "Divergence between Samples of Chimpanzee and Human DNA Sequences is 5%, Counting Indels," *Proceedings of the National Academy of Science U.S.A.* 99 (21) (2002): 13633–5; E. R. Winstead, "Humans and Mice Together at Last," *Genome News Network* (May 31, 2002).

or delay breeding for a decade or more. Males can be dominant over females or vice versa. Societies can be egalitarian or hierarchical. We can work hard, even in the absence of real need, or lie back and take it easy. We can be aggressive or peaceful, angry or affectionate, suspicious or trusting. Every one of these behaviors has a direct equivalent in terms of animal behavior, as described later in this chapter.

Cross-cultural evidence

But the interesting point is that certain family and social behaviors are more often found in large-scale civilizations with complex economies. As detailed in the following chapters, ethnographic studies indicate that people living in long-civilized societies are more likely than those in pre-literate societies to restrict sexual activity, marry later, form monogamous nuclear families, and control their children's behavior. The extremes of such behavior can be found in Northern Europe during the nineteenth century, when children were rigorously controlled from infancy and sexual behavior was strictly limited—especially for women. But similar patterns can be seen in other parts of the world.[2]

Biohistory proposes that there is an underlying *temperamental* difference between civilized and non-civilized societies, expressed to some extent in these family and personal behaviors but also in attitudes towards political authority and the market. In this and future chapters we will see that this temperamental difference can be explained in physiological terms. It will also be proposed that it is the development of this "civilized" temperament that makes civilization possible, whereas the loss of it is followed by a weakening and eventual collapse of the society affected. The best-known example is the decline and fall of the Roman Empire. Consider some of this evidence.

First, as long ago as the 1930s, J. D. Unwin found that civilized societies are more likely to control sexual behavior, a finding confirmed by later

---

[2] Historical studies provide a more mixed picture, with extended families and very early marriage evident in areas such as Eastern Europe and Asia. But since this evidence largely deals with elite families it is not clear how typical they are of the general population. Recent ethnographic studies, dealing with non-elite families and using precise quantitative measures, tend to show the differences indicated.

cross-cultural studies.[3] Related studies have found that societies with severe obedience training are more politically complex and more likely to be farmers and herders than hunters or fishers.[4]

These studies are meticulous and well researched, but from the viewpoint of biohistory they have significant problems. The first is that, apart from Unwin, they tend to exclude the societies with the biggest states and most advanced economies such as those of India, China, Europe and the Middle East, which show the most extreme forms of these behaviors.

Second, in terms of parental behavior these studies focus on the *aim* of control or punishment, such as to promote obedience or sharing. This is thinking in rational terms—that someone taught to be obedient will more likely obey others as he or she grows up. But in biological terms what is far more important is the *level* of control or punishment. For example, a severe punishment increases the level of stress hormones such as cortisol, whether the punishment is for disobedience, breaking cultural taboos or merely because the parent is bad-tempered. Similarly, we will see that it is the level of control that matters more than the purpose of the control. Both punishment and control have profound effects on hormones, on epigenetics (the way in which certain genes are switched on and off), and thus on adult temperament and behavior.[5]

Third, most studies fail to distinguish between control and punishment, using terms such as "severity of obedience socialization" which include both. In biochemical and behavioral terms, control and punishment have very different effects.

Finally, they do not always distinguish the age at which training applies. In later chapters evidence is presented from rat and monkey experiments showing that the same influence at different ages can have different or even opposite effects. The key distinction is between infancy, when

---

[3] J. D. Unwin, *Sex and Culture* (Oxford: Oxford University Press, 1934), 13–14, 27–9, 315–7, 321; W. N. Stephens, *The Family in Cross-cultural Perspective* (New York: Holt, Rinehart & Winston, 1963), 256–258.

[4] F. B. Aberle, "Culture and Socialization," in *Psychological Anthropology*, edited by F. L. K. Hsu, 386 (Homewood, Illinois: Dorsey Press, 1961); R. Barry, I. Child & M. K. Bacon, "Relation of Child Training to Subsistence Economy," in *American Anthropologist* 61 (1959): 56–60.

[5] Guy Riddehough & Laura M. Zahn, "What is Epigenetics" *Science* 330, (October 2010), http://www.sciencemag.org/content/330/6004.toc (accessed 13 September 2014).

mammals are nursed by their mother, and the juvenile period before puberty. In humans these ages are roughly 0–2 and 6–12.

The Cross-cultural survey

To better understand the family and social patterns linked to civilization, a study of 67 societies was made, ranging from hunter-gatherers to the long-civilized peoples of Europe and Asia. As with other cross-cultural studies, information from ethnographic studies provided quantitative scores for political, economic, family and childrearing variables. The ethnographies chosen were those with relatively detailed information on childrearing patterns. Full details of the study are given at www.biohistory.org.

The first point to note is that the findings of earlier studies linking controls on sexual behavior to measures of political and economic complexity are verified (see Table 1.1 below).

Table 1.1. Correlations between political and economic complexity and limits on sexual behavior.[6]Societies which restrict sexual behavior are more likely to be politically complex and with advanced economies. For example, the size of political units is positively correlated with restrictions on premarital sex.

	Premarital sex restricted	Adultery restricted	Divorce restricted
Size of political unit	.53**	.51**	.35*
Hereditary status	.39**	.43**	.41
Market economy	.45*	.54*	
Status from wealth versus generosity	.46*	.47*	.30
Routine work	.42**	.48**	
Deities enforce morality	.30*	.30*	
Modesty in dress	.51**	.53**	
significance	**.001	* .01	Others: .05

[6] A full account of the coding system, codes for each society and significant correlations are given at www.biohistory.org.

Societies which insist on premarital chastity, sanction adultery and restrict divorce are more likely to form large political units, have market economies, work at routine jobs such as farming, and strive to achieve individual wealth. Their religious systems more often include moral codes.

Table 1.2 shows that societies which are politically and economically complex also have distinct family patterns. Compared with small-scale societies they are more likely to form nuclear monogamous families, marry late and control their children's behavior.

Table 1.2. Correlations of political and economic variables with family and social variables.[7] Societies which are politically and economically complex are more likely to form nuclear monogamous families, marry late and control their children's behavior.

	Premarital sex restricted	Monogamy versus polygyny	Nuclear family	Marry late	Control children	Children wanted
Size of political unit	.53**	.52**	.46**	.51**	.56**	.28*
Hereditary status	.39**	.42**		.55**	.50**	.31
Market economy	.45*	.43**	.45**	.39*	.59**	.28
Status from wealth vs generosity	.46*	.41**	.65**	.42*	.56**	
Routine work	.42**	.51**	.35*	.41**	.58**	
Deities enforce morality	.30*		.42**	.30	.51**	.38
Modesty in dress	.51**	.62**	.54**	.54**	.68**	.25
significance	**.001	* .01	Others:	.05		

[7] See www.biohistory.org

These variables also correlate strongly with each other, as shown in Table 1.3 below.

Table 1.3. Significant Correlations among family and social variables linked to civilization.[8] Patterns of family behavior found in more complex societies, such as monogamous nuclear families, late marriage and control of children, correlate independently with each other.

	Monogamy versus polygyny	Nuclear family	Marry late	Control children	Children wanted
Premarital sex restricted	.48**	.30*	.43**	.38**	.32*
Monogamy vs polygyny		.30	.50**	.36*	
Nuclear family			.29	.33*	
Marry late				.54**	
Control children					.31*
significance	**.001	* .01	Others:	.05	

Table 1.4 below shows that measures of political and economic complexity correlate strongly. Note the link to modesty in dress, which has no obvious connection with other features of civilization.

[8] See www.biohistory.org.

Table 1.4. Correlations among measures of political and economic complexity.[9] Politically complex societies tend to have advanced economies and to be modest in dress.

	Hereditary status	Market economy	Status wealth	Routine work	Deities moral	Modest dress
Size of political unit	.58**	.62**	.61**	.47**	.47**	.69**
Hereditary status		.39**	.45**	.34*	.28	.48**
Market economy			.49**	.55*	.41**	.62**
Status from wealth				.38**	.39**	.69**
Routine work					.22	.57**
Deities enforce morality						.39*
significance	**.001	* .01	Others:	.05		

It is no surprise that politically complex societies should have more advanced economies. What is interesting is that they are just as likely to insist on premarital chastity and control children. In other words, limits on sexual behavior and control of children are just as distinctive a feature of larger states as markets. It is logical that larger states should have market economies, if only because political union makes trade easier, but why should they limit sexual behavior or control their children?

This link is not without exceptions, of course. There are societies combining advanced political and economic systems with liberal standards of sexual behavior. The modern West and late Republican Rome are two obvious examples, which will become highly significant once we understand *why* civilized societies restrict sexual behavior. These apparent exceptions will then help us to understand why civilizations fall.

---

[9] www.biohistory.org.

The temperamental basis of civilization

What is needed is an explanation of why certain behaviors such as sexual restraint are more prevalent in civilized societies. Biohistory proposes that they represent underlying biological systems that adjust people's temperaments to the needs of civilization. This makes them more accepting of wider political authority, more inclined to perform routine work, better suited to a market economy, and more accepting of impersonal moral codes such as those taught by religious systems. They also change their behavior in other ways, such as increasing modesty in dress and reducing tolerance of premarital sex.

The next step is to explore the implications of these behaviors. A core theme of biohistory is that humans are in many ways similar to animals, sharing up to 95% of our genes with other species. Thus it is that the family and personal behavior associated with civilized societies can also be observed in animals. Primates in particular can delay breeding, the equivalent of premarital chastity and late marriage. They can form nuclear monogamous families, and they can control the behavior and movements of their offspring. But which species are more likely to show such behaviors, and in what circumstances?

Gibbons and baboons

The study begins with two primate groups: gibbons and savannah baboons, which display extremes of these behaviors. Gibbons, a tree dwelling ape living in the forests of South-East Asia, act more like civilized peoples. Baboons, a largely terrestrial monkey living in the open grasslands of Africa, act more like the people of small-scale societies.

For example, gibbons are far less sexually active than baboons. They are also less sociable. Studies of three different populations show that they spend only 6%, 4% and 1.3% of the day in social activities.[10] By comparison, studies of 18 baboon populations show that they spent an average of 11.9% of their day in social activities, ranging from a low of 4.5% to a high of 22.7%.[11]

---

[10] S. M. Cheyne, "Behavioural Ecology of Gibbons (Hylobates albibarbis) in a Degraded Peat-Swamp Forest," *Indonesian Primates: Developments in Primatology: Progress and Reports* (2010): 121–56.

[11] D. L. Cheney & R. M. Seyfarth, *Baboon Metaphysics: The Evolution of a Social Mind* (Chicago: University of Chicago Press, 2007).

Gibbons are not only less social but also less tolerant of each other than baboons, and indeed of most primates. Males usually drive away other adult males, and females will not tolerate other females, including their own offspring after they reach puberty. Mated pairs defend territories, which represents an extreme in social intolerance. They do not totally avoid other gibbons, as they have been observed grooming and playing with individuals from neighboring territories.[12] But the only instance in which individuals of the same gender *share* a territory is when two males bond with a single female, a pattern rare in primate societies but forming 15% of families in one well-studied population.[13] In other cases, gibbons typically form nuclear monogamous families.

Among baboons, by comparison, no monogamous or polyandrous population has ever been found. Baboon troops normally consist of multiple males and females, although troops in some areas may have a single male with multiple females. One study comparing 23 populations found an average group size of 67, ranging from a low of 19 to a high of 247.[14] In these troops, the dominant male tends to monopolize females in estrus and sire most of the young, with other males having access only when the dominant male is distracted.[15] In human terms, baboons are promiscuous or in some cases polygynous, but never monogamous.

Gibbon populations tend to be limited in size by restricted breeding.

---

[12] T. Q. Bartlett, "Intragroup and Intergroup Social Interactions in White-Handed Gibbons," *International Journal of Primatology* 24 (2) (2003): 239–59; U. Reichhard & V. Sommer, "Group Encounters in Wild Gibbons (Hylobates lar): Agonism, Affiliation, and the Concept of Infanticide," *Behaviour* (1997): 1135–74.

[13] W. Y. Brockelman, U. Reichhard, U. Treesucon & J. J. Raemaekers, "Dispersal, Pair Formation and Social Structure in Gibbons (Hylobates lar)," *Behavioral Ecology and Sociobiology* 42 (5) (1998): 329–39; T. Savini, C. Boesch & U. H. Reichhard, "Varying Ecological Quality Influences the Probability of Polyandry in White-Handed Gibbons (Hylobates lar) in Thailand," *Biotropica* 41(4) (2009): 503–13; V Sommer & U. Reichhard, "Rethinking Monogamy: The Gibbon Case," in *Primate males: Causes and Consequences of Variation in Group Composition*, edited by P. M. Kappeler (Cambridge: Cambridge University Press, 2000); U. Reichhard, "The Social Organization and Mating System of Khao Yai White-Handed Gibbons: 1992–2006, " *The Gibbons* (2009): 347–84.

[14] D. L. Cheney & R. M. Seyfarth, *Baboon Metaphysics: The Evolution of a Social Mind* (Chicago, University of Chicago Press, 2007).

[15] S. C. Alberts, J. C. Buchan & J. Altmann, "Sexual Selection in Wild Baboons: From Mating Opportunities to Paternity Success," *Animal Behaviour* 72 (5) (2006): 1177–96.

Mating is delayed until after animals establish a territory. One study found that this did not happen until around the age of 10, two years after gibbons reach full adult size and several years after sexual maturity.[16] For this and other reasons, gibbons reproduce below their potential reproductive rate. For example, a study of 7 gibbon females over a 6-year period found that only one gave birth more than once, two failed to breed at all, one remained unpaired, and the seventh female lost her mate to another female. This is the non-human equivalent of late marriage and limited sexual activity.

By contrast, baboon populations appear to be limited mainly by predation and infanticide. Females are sexually active soon after puberty and males whenever they can be.

A related difference is that gibbons are highly discriminating in their choice of mates. While different species are fertile with each other and their ranges often overlap, only occasionally do they interbreed in the wild. They can distinguish even closely related species living in the same area by the sound of their calls, which are thought to be important in the maintenance of pair bonds.[17] Efforts to breed gibbons in captivity result in only half the presented mates being accepted, even when no other mate is available.[18]

Baboons, on the other hand, seem to be far less picky. A male baboon will mate with any fully mature female in estrus. Females also appear to be less discriminating because they solicit mating from around ten days before ovulation, even though only juveniles and adolescents are interested at such times.[19]

There are no direct observations of parental control among gibbons, but parents stay in close and continuous contact with their young until the age of puberty. Baboon mothers, by contrast, wean their young early and cease

---

[16] Brockelman et al., "Dispersal, pair formation and social structure in gibbons," 329–39; T. Geissmann, "Reassessment of Age of Sexual Maturity in Gibbons (Hylobates spp.)," *American Journal of Primatology* 23 (1) (1991): 11–22.

[17] J. C. Mitani, "Species Discrimination of Male Song in Gibbons," *American Journal of Primatology* 13 (4) (1987): 413–23.

[18] A. W. Breznock, J. B. Harrold & T. G. Kawakami, "Successful Breeding of the Laboratory-Housed Gibbon (Hylobates lar)," *Laboratory Animal Science* 27 (2) (1977): 222–8.

[19] Cheney et al., *Baboon Metaphysics*.

to provide much care thereafter. Gibbons also spend far more time foraging for food than do baboons. Their food is widely scattered and generally in small quantities, so only constant movement and searching can provide enough food to survive. This is the non-human equivalent of very hard work.

Limited food as an explanation for gibbon behavior

These differences can be explained by one simple observation—gibbons are far more likely to be short of food.[20] They live in one of the most food-restricted environments on Earth—the tropical forests of south-east Asia. Typically, tropical forests are lush and productive with a wide variety of plants, including fruiting trees. If one plant species is not productive, another is likely to be. But there are major problems—many leaves and fruit contain dangerous toxins, and forest primates tend to be very selective feeders. Only the fruit or leaves of particular trees at certain stages of ripeness are suitable. Favored food plants are also widely scattered, so that forest-living primates such as gibbons must spend a great deal of their time foraging.

Another factor in gibbon behavior is that food supplies, though restricted, are relatively constant throughout the year. There are variations in the availability of certain foods, so the type of food taken will differ from season to season, but the sheer number of plant species means that there is always something to feed on however difficult it may be to find.[21]

In such conditions gibbon numbers grow to the absolute limit allowed by the forest larder. Hunger is thus likely to be a daily rather than yearly problem.[22] A further factor is that trees make gibbons less vulnerable to

---

[20] Monogamy is commonly found in species with limited food and the consequent need for females to defend an exclusive territory, though there is an alternative explanation in terms of avoiding infanticide by males. C. Gelling, "Evolution of Mammalian Monogamy Remains Mysterious," *Science News* August 2, 2013. https://www.sciencenews.org/article/evolution-mammalian-monogamy-remains-mysterious.

[21] P. Fan, Q. Ni, G. Sun, B. Huan & X. Jiang, "Gibbons under Seasonal Stress; the Diet of the Black Crested Gibbon (Nomascus concolor) on Mt. Wuliang, Central Yunnan, China," *Primates* 50 (1) (2009): 37–44.

[22] L. A. Isbell, "Predation on Primates: Ecological Patterns and Evolutionary Consequences," *Evolutionary Anthropology Issues, News and Reviews* 3 (2) (1994): 63–4.

predation, so populations are limited primarily by the carrying capacity of the environment.[23]

Other evidence is consistent with this shortage of food. In Malaysia, for example, gibbons survive only below 500 meters. This is not because of predation or competition from other species, but rather because any significant reduction in food quality or increased effort involved in traveling put their energy budget in the red. In other words, food is so hard to find that a gibbon can barely find enough to maintain itself, even with an exclusive territory. Nursing mothers have been seen in notably poor condition. Intra-group squabbling over food is common and may account for why the young are expelled from their group. Some gibbon species are found at slightly higher altitudes, but even with these there is often too little food to survive.[24] One study found significantly higher rates of juvenile mortality among gibbons where home ranges were larger and animals had to travel further in search of food.[25]

Baboons, on the other hand, tend to live in environments where food supplies are highly variable. Through much of the year and even (in some areas) for several years in a row, food may be plentiful. But during times of drought it can be very scarce, resulting in severe stress and even starvation.[26] This environment is associated with a different set of behavioral responses.

---

[23] D. F. Makin, H. F. P. Payne, G. I. H. Kerley & A. M. Shrader, "Foraging in a 3-D World: How does Predation Risk Affect Space use of Vervet Monkeys?" *Journal of Mammalogy* 93 (2) (2012): 424–7.

[24] J. O. Caldecott, "Habitat Quality and Populations of two Sympatric Gibbons (Hylobatidae) on a Mountain in Malaya," *Folia Primatologica* 33(1980): 291–309.

[25] T. Savini, C. Boesch & U. H. Reichhard, "Home-Range Characteristics and the Influence of Seasonality on Female Reproduction in White-Handed Gibbons (Hylobates lar) at Khao Yai National Park, Thailand," *American Journal of Physical Anthropology* 135 (1) (2008): 1–12.

[26] W. J. Hamilton, "Demographic Consequences of a Food and Water Shortage to Desert Chacma Baboons, Papio ursinus," *International Journal of Primatology* 6 (5) (1985): 451–62; Cheney et al., *Baboon Metaphysics*. L. R. Gesquiere, M. Khan, L. Shek, T. L. Wango, E. O. Wango, S. C. Alberts & J. Altmann, "Coping with a Challenging Environment: Effects of Season Variability and Reproductive Status on Glucocorticoid Concentrations of Female Baboons (Papio cynocephalus)," *Hormones and Behaviour* 54 (3) (2008): 410–6; S. C. Alberts & J. Altmann, "The Evolutionary Past and the Research Future: Environmental Variation and Life History Flexibility in a Primate Lineage," *Developments in Primatology: Progress and Prospects Part II* (2006): 277–303; D. K. Brockman & C. P. van Schaik, *Seasonality in Primates: Studies of Living and Extinct Human and Non-Human Primates* (Cambridge: Cambridge University Press, 2005), 159.

The advantages of different behavior in different environments

Primatologists consider troop size to be a trade-off between foraging efficiency and defense against predators. Feeding in large groups is inefficient. If food supplies are scarce or scattered in small patches, a great deal of time must be spent in searching for nutrients. A patch that would feed an individual for an hour might feed a larger group for only minutes. Socializing is an unnecessary distraction from time spent searching for food or resting, and relatively infrequent socializing is a trade mark of groups when food is scarce.

But if food is plentiful or in large patches, either because the land is productive or because there are many deaths from predation or occasional famine, then larger groups are better. More animals are available to provide warning against predators, or even to attack them.[27] Larger groups may also provide defense against competing groups of the same species. For example, in one study of three baboons baboon troops impacted by a drought, the smallest troop was driven from the most productive area so that its numbers dropped by three quarters. Meanwhile, the larger groups maintained or even increased their numbers.[28] When large groups are advantageous, socializing becomes a valuable way of cementing social ties and holding the group together. This is the same benefit we saw from socializing in hunter-gatherer bands.

An example of the trade-off between food supply and group size can be seen in a study of baboon troops in Amboseli National Park, where groups ranged in size from 8 to 44 members.[29] Baboons in the smallest troop obtained the same energy intake while spending half as much time foraging as those in the largest troop. However, they were observed spending more time near trees and were more likely to choose an elevated spot for resting, indicating a greater caution about predators. In a Botswana baboon population studied intensively over ten years, deaths from predation were greatest during the floods, when troop members were

[27] Cheney et al., *Baboon Metaphysics*.

[28] C. Brain, "Deaths in a Desert Baboon Troop," *International Journal of Primatology* 13 (6) (1992): 593–99.

[29] P. B. Stacey, "Group Size and Foraging Efficiency in Yellow Baboons," *Behavioral Ecology and Sociobiology* 18 (3) (1986): 175–87.

scattered and less able to warn other members of predators.[30]

On the other hand, gibbons are rarely taken by predators because of their agility and the time spent high in the forest canopy.[31] On average, gibbon populations contain a relatively high proportion of adults, indicating both low mortality and low birth rate.[32] In their environment, having too many young could be a disaster. Pregnancy is a highly demanding state when food is in short supply. Young that are born in less than optimal conditions are unlikely to survive, and those that do will be weak and unable to compete. Thus the best strategy is to limit breeding by delaying puberty and limiting sexual activity. Territory is also a factor. Normally, gibbons will not breed if unable to command an exclusive territory that can feed the mated pair and their young. Without such a territory, a pregnancy is unlikely to produce successful young and could put the female's life at risk. Given that death from predation is rare, it is far more sensible to wait for a suitable territory to become available.

Environmental constraints also govern reaction to predators. Gibbons are timid about predators, fleeing through the treetops at any sign of disturbance. Generally, baboons are far bolder, though the actual response depends on the predator. For lions, against which they have no defense, they can only be vigilant, giving alarm calls and hiding in trees. But for leopards:

> [If] the baboons are able to isolate a leopard in a bush, tree, or aardvark hole, they immediately surround it, screaming, alarm-calling, and lunging at it, seemingly without fear. Although male baboons, with their size and enormous canines, are much better equipped than females to fight a leopard, the mass mobbing involves baboons of every age and sex. Juveniles, adult females, even mothers with young infants join to form a huge, hostile mob that tries to corner the leopard. The attack continues

---

[30] Cheney, *Baboon Metaphysics*.

[31] U. Reichhard, "Social Monogamy in Gibbons: The Male Perspective," in *Monogamy: Mating Strategies and Partnerships in Birds, Humans and Other Mammals*, edited by U. Reichhard & C. Boesch (Cambridge: Cambridge University Press, 2003), 192; N. L. Uhde & V. Sommer, "Antipredator Behavior in Gibbons (Hylobates lar, Khao Yai/Thailand)," in *Eat or be Eaten: Predator Sensitive Foraging among Primates*, edited by L. E. Miller (Cambridge: Cambridge University Press, 2002).

[32] Mitani, "Species Discrimination of Male Song in Gibbons."; T. Geissmann, "Inheritance of Song Parameters in the Gibbon Song, Analysed in 2 Hybrid Gibbons (Hylobates pileatus X H. lar)," *Folia Primatologica* 42 (1984): 216–35.

> even after some baboons have received slashes on their arms, legs, and face that open up huge wounds.[33]

Leopards are not uncommonly killed by such attacks.

More abundant food makes animals bolder and more group-minded but does not fully account for the level of aggression found in savannah baboons. Aggression is part of a complex of social traits which will be discussed in chapter four.

Paradoxically, timidity makes sense when the risk of predation is low. Warning of a predator is dangerous and mobbing it far more so. Thus, there is less point in doing it when escape is easy. But when the predator is likely to make a kill, attempting to discourage or even kill it may be worth the risk of confrontation.

Not only are gibbons, with their arboreal habitats, less vulnerable to predators, they are also remarkably long-lived and have been known to reach 44 years in captivity.[34] Baboons are much shorter-lived. Female baboons may live to more than 20 years if not taken by a predator, but the life of male baboons could be described as "nasty, brutish and short." Fierce battles for dominance, together with predation, mean most never reach this age.[35] Even in captivity baboons rarely live beyond the age of 30.[36] This is especially striking since larger animals typically live longer, and gibbons are half or less the weight of baboons.[37]

---

[33] Cheney et al., *Baboon Metaphysics*.

[34] Uhde et al., "Antipredator Behavior in Gibbons."

[35] Cheney et al., *Baboon Metaphysics*.

[36] L. D. Chen, R. S. Kushwaha, H. C. McGill, K. S. Rice & K. D. Carey, "Effect of naturally reduced ovarian function on plasma lipoprotein and 27-hydroxycholesterol levels in baboons (Papio sp.)," *Atherosclerosis* 136 (1) (1998): 89–98;

A. G. Comuzzie, S. A. Cole, L. Martin, K. D. Carey, M. C. Mahaney, J. Blangero & J. L. vandeBerg, "The Baboon as a Nonhuman Primate Model for the Study of the Genetics of Obesity," *Obesity Research* 11 (1) (2003): 75–80.

[37] Ma, S., Y. Wang & F. E. Poirier (1988). "Taxonomy, Distribution and Status of Gibbon (Hylobates) in Southern China and Adjacent Areas," *Primates* 29 (2): 277–86; A. H. Schultz, "The Relative Weight of the Testes in Primates," *The Anatomical Record* 72 (3) (1938): 387–94; R. W. Barton, D. G. Reynolds & K. G. Swan, "Mesenteric Circulatory Responses to Hemorrhagic Shock in the Baboon," *Annals of Surgical Innovation and Research* 175 (2) (1972): 204–9; N. J. Espat, J. C. Cendan, E. A. Beierle, T. A. Auffenberg, J. Rosenberg, D. Russell, J. S.

Both longer *and* shorter lives aid survival and success. If premature death is unlikely, as for gibbons, the most successful animals are those which can maintain their bodies and wait for better times. But if death can happen at any moment, as among baboons, the best strategy is to put maximum effort into breeding fast, even if it often shortens life. In one Botswana baboon population most deaths among adult females and juveniles were due to predation, causing up to 95% of adult female deaths. In a single year, 25% of the troop's adult females disappeared from confirmed or suspected predation.[38] To maximize offspring a female must breed as fast as possible, because she may not be around next year.

This applies even more to males, since dominant males tend to sire most of the young. Fierce competition for dominance and the consequent breeding rights mean that male baboons often die from wounds sustained during fights.[39] So when predation is high the advantage shifts from long-term survival to faster breeding.

A similar argument applies to choice of mates. It makes sense that animals in food-limited environments should choose mates similar to themselves. Success in a stable, competitive environment means adapting to local conditions. An animal that does well in local conditions will reproduce most successfully with a mate that is similarly adapted, not one with variant genes that may be better suited to living somewhere else. Thus it is that gibbons have elaborate courtship rituals, so that subtle differences in behavior and appearance act to prevent interbreeding. Thus, regional populations become more distinct and eventually form a multiplicity of species, which is the case with gibbons.

By contrast, baboon environments are highly changeable so that having variant genes may be an advantage. Thus, baboons are far less discriminating and local sub-species readily interbreed, so that baboons are

---

Kenney, E. Fischer, W. Montegut, S. F. Lowry, E. M. Copeland III & L. L. Moldawer, "PEG-BP-30 Monotherapy Attenuates the Cytokine-Mediated Inflammatory Cascade in Baboon Escherichia Coli Septic Shock," *Journal of Surgical Research* 59 (1) (1995): 153–8.

[38] D. L. Cheney, R. M. Seyfarth, J. Fischer, J. Beehner, T. Bergman, S. E. Johnson, D. M. Kitchen, R. A. Palombit, D. Rendall & J. B. Silk, "Factors Affecting Reproduction and Mortality Among Baboons in the Okavango Delta, Botswana," *International Journal of Primatology* 25 (2) (2004): 401–28.

[39] Brain, C. (1992). "Deaths in a Desert Baboon Troop."

considered as a single species.[40]

Though there is no direct evidence that gibbons train or control their young, such behavior in a food-limited environment would make sense. Offspring are rare and vulnerable in a hungry world and need every care and protection if they are to survive. In particular they must learn which plants are good to eat and which are poisonous and where they can be found.

Overall, gibbon behavior is an adaptation to chronic food shortage and low mortality, while baboon behavior is an adaption to normally plentiful food and high mortality.

Changes in behavior as a response to environment

While gibbon behavior is an adaptation to an environment with limited food, it is not a *response* to limited food. In a large number of studies no clear exception has been found to the common gibbon pattern of one male to one female (monogamy), with occasional families of two males to one female (polyandry). As has been mentioned, even in captivity where food is plentiful, gibbons are picky about mates and intolerant of their same-sex adult offspring.[41] Thus, gibbon behavior seems to be set by genes.

But this is not the case for many other species, including baboons and vervets—a small monkey native to southern and eastern Africa. Both baboons and vervets are found in a variety of habitats including mountain, desert, savannah, and dense forest, sometimes in the form of different sub-species. By their adaptability to a wide range of environments they differ from gibbons, which are only found in dense forest. And just as baboons and vervets can thrive in different environments, so they show a variety of social structures to suit different environments. None of these include monogamy with males and females paired, but the number of males in a troop can vary from one to many. Troops thus vary widely in size. And just as gibbons form small troops to adjust to a food-limited environments,

---

[40] R. I. M. Dunbar, "Time: A Hidden Constraint on the Behavioural Ecology of Baboons," *Behavioral Ecology and Sociobiology* 31 (1) (1992): 35–49; S. P. Henzi & L. Barrett, "The Historical Socioecology of Savannah Baboons (Papio hamadryas)," *Journal of Zoology (London)* 265 (3) (2005): 215–26.

[41] B. L. Burns, H. M. Dooley & D. S. Judge, "Social Dynamics Modify Behavioural Development in Captive White-Cheeked (Nomascus leucogenys) and Silvery (Hylobates moloch) Gibbons," *Primates* 52 (3) (2011): 271–7.

baboons and vervets form smaller troops in marginal habitats with lower predation pressure. The link between behavior and environment is shown in Table 1.5 below, taken from a study of two populations of vervets in the wild—one in a productive swamp and the other in less productive dry woodland.[42]

Table 1.5 Ecological and demographic differences between two vervet groups[43]

	Dry Woodland	Swamp
Food quality	Lower	Higher
Water availability	Poor	Good
Group size	9–13	11–25
Average age of female at first birth (years)	5–6	4–5
Median interval between births (years)	2	1
Average infant mortality in first year	59%	57%
Predator sightings per hundred hours	4	6

The dry woodland population shows behavior associated with species in food-limited environments, such as gibbons. Troops are smaller and mothers gave birth later and less often, presumably allowing them to invest more in their young. This explains why the infant mortality rate is similar to that of the better-fed population, despite their poorer living conditions. Mothers did not spend more time with their young, but they were less likely to break contact and deny them their nipple. Predators were less of a factor, which is consistent with populations living near to the limits of its food supply.

It is not just the mother's behavior that increases care of the young in marginal habitats. A study of baboons showed that during seasons of food shortage the young were more likely to throw tantrums and thus achieve

[42] M. D. Hauser & L. A. Fairbanks, "Mother-Offspring Conflict in Vervet Monkeys: Variation in Response to Ecological Conditions," *Animal Behaviour* 36 (3) (1988): 802–13.
[43] Ibid.

more attention and care from their mothers.[44]

It is an axiom of primate social behavior that groups of the same species are smaller in areas with scarcer food. As mentioned, the theory is that larger groups protect members against predators and provide an advantage in inter-troop competition. On the other hand, feeding is less efficient which reduces female reproductive rates and increases mortality from causes such as malnutrition or disease.[45] Smaller groups are more efficient for feeding purposes but provide less protection against predators and other troops.

For example, baboon populations in a marginal mountain area had smaller troops compared to those in a more food productive national park. Mothers also provided greater levels of care to their infants, with a longer interval between births. In turn, infant survival rates were actually better than in more typical baboon habitats with plentiful food.[46] These findings are consistent with the view that group size can be explained entirely in terms of feeding strategy. Troops of wide ranging sizes allocated similar amounts of time to foraging, in that they foraged with the same efficiency and travelled approximately the same distances to do so.[47] In terms of survival and success this is the optimum strategy because it permits

---

[44] L. Barrett & S. Peter-Henzi, "Are Baboon Infants Sir Phillip Sydney's Offspring?" *Ethology* 106 (7) (2000): 645–58.

[45] C. P. van Schaik, "Why are Diurnal Primates Living in Groups?" *Behaviour* 87 (1/2) (1983): 120–44; C. A. Chapman, L. J. Chapman & R. W. Wrangham, "Ecological Constraints on Group Size: An Analysis of Spider Monkey and Chimpanzee Subgroups," *Behavioral Ecology and Sociobiology* 36 (1) (1995): 59–70; C. Janson, "Aggressive Competition and Individual Food Consumption in Wild Brown Capuchin Monkeys (Cebus apella)," *Behavioral Ecology and Sociobiology* 18 (2) (1985): 125–38; G. Ramos-Fernández, D. Boyer & V. P. Gómez, "A Complex Social Structure with Fission–Fusion Properties can Emerge from a Simple Foraging Model," *Behavioral Ecology and Sociobiology* 60 (4) (2006): 536–49; Y. Takahata, S. Suzuki, N. Okayasu, H. Sugiura, H. Takahashi, J. Yamagiwa, K. Izawa, N. Agetsuma, D. Hill & C. Saito, "Does Troop Size of Wild Japanese Macaques Influence Birth Rate and Infant Mortality in the Absence of Predators?" *Primates* 39 (2) (1998): 245–51.

[46] J. E. Lycett, S. P. Henzi & L. Barrett, "Maternal Investment in Mountain Baboons and the Hypothesis of Reduced Care," *Behavioral Ecology and Sociobiology* 42 (1) (1998): 49–56; S. P. Henzi, J. E. Lycett & S. E. Piper, "Fission and Troop Size in a Mountain Baboon Population," *Animal Behaviour* 53 (3) (1997): 525–35.

[47] Lycett et al., "Maternal Investment in Mountain Baboons"; Henzi et al., "Fission and Troop Size."

animals to balance the advantages of smaller groups for effective foraging with the advantages of larger groups for mutual support and defense. On balance, group size among both baboons and vervets is a function of foraging efficiency, food availability and predator density.

Another difference is that, though animals in food-limited environments spend about the same time moving as animals in a more food abundant environment, they tend to travel a great deal faster. Baboon troops in areas with less than 700 mm of rain per year had a mean travel speed of around 2 km/hour, while those in areas with more than 1000 mm of rain travelled at an average of 0.5 km/hour.[48] When food is scarce and widely scattered, faster movement has clear advantages in terms of locating food and thus survival.

Many other studies have shown that primates in food-limited areas form smaller groups. For example, a study of Japanese macaques found that on an island where food quality was poor, the time animals spent on feeding was 1.7 times greater than on an island with better quality food. In the poor environment, monkeys also spent significantly less time grooming each other. There were fewer males in the groups but more solitary males outside of groups—an indication of social intolerance. These macaques were also more vigilant, even though there was far less aggression between groups.[49]

A further finding from this study is that males on the less productive island were never seen to mate with females in other groups or to transfer groups, a behavior common on the more productive island. This perhaps indicates a preference for mating with more similar and familiar animals, which is a feature of behavior in food limited environments.

All of these species seem to be genetically adapted to environments with more plentiful food, forming multi-male or one-male troops but not monogamous pairs. But there are other species which vary in behavior but with a bias towards food-limited patterns. Spider monkeys, which live in

---

[48] Dunbar, "Time: A Hidden Constraint on the Behavioural Ecology of Baboons," 35–49.

[49] C. Saito, S. Sato, S. Suzuki, H. Sugiura, N. Agetsuma, Y. Takahata, C. Sasaki, H. Takahashi, T. Tanaka & J. Yamagiwa, "Aggressive Intergroup Encounters in Two Populations of Japanese Macaques (Macaca fuscata)," *Primates* 39 (3) (1998): 303–12; N. Agetsuma & N. Nakagawa, "Effects of Habitat Differences on Feeding Behaviors of Japanese Monkeys: Comparison between Yakushima and Kinkazan," *Primates* 39 (3) (1998): 275–89.

similar environments to gibbons, form gibbon-size groups when food is scarce but much larger ones in times of food abundance.[50]

The snub-nosed langur of Mentawai, a forested island off the coast of Sumatra, is one of the few primate species to be monogamous in its natural environment. Part of the island had been logged some ten years before the researchers arrived, and the forest regrowth provided ample food. The langurs in this area were more numerous than in the non-logged area and commonly formed larger troops with one male and multiple females. They were also bolder. Though hunted intensively by local people in both areas, those in the regrowth areas were noisier and less vigilant and thus more likely to be captured. People came from far afield to hunt them and rarely left without an ample catch, whereas in the untouched forest they often ended up empty handed.[51]

It should be noted here that spider monkeys and langurs in larger troops exhibit the same boldness and large group dynamics but do not show the same levels of aggression found among large troops of baboons. Again, we will return to this important topic in chapter four.

The advantages of being able to change behavior as a response to changes in food

For species in environments where food availability varies there are benefits in being able to adapt behavior to different levels of food and predator threat, especially since the change in food supplies can happen very quickly. For example, snub-nosed monkeys in Yunnan form smaller groups during the winter when food is less available, and larger ones at other times of the year.[52]

When food is scarce and predators few, the best strategy is to spread out in smaller groups for more efficient feeding, travel faster in search of food, and spend less time in social activities such as grooming. And when predators are scarce, flight is a better response than dangerous practices

---

[50] D. Robbins, C. A. Chapman & R. W. Wrangham, "Group Size and Stability: Why do Gibbons and Spider Monkeys Differ?" *Primates* 32 (3) (1991): 301–5.

[51] K. Watanabe, "Variation in Group Composition and Population Density of the Two Sympatric Mentawaian Leaf-Monkeys," *Primates* 22 (2) (1981): 145–60.

[52] B. Ren, D. Li, P. A. Garber & M. Li, "Fission–Fusion Behavior in Yunnan Snub-Nosed Monkeys (Rhinopithecus bieti) in Yunnan, China," *International Journal of Primatology* 33 (5) (2012): 1–14.

such as mobbing. Mobbing is also less likely to be effective when groups are smaller.

The same argument applies to reproduction. When food is short, breeding rapidly and too early is a danger to mothers and offspring. It is far better to delay breeding until mothers are fully mature, and even then only when sufficient food is available. Births should be spaced and greater attention given to offspring to optimize their chance of survival. Ideally, individuals in such a group would spend less energy on fast reproduction and more on body maintenance and long-term survival.

When food is plentiful and predators are common, the opposite strategy is preferable. Coming together in large groups provides better defense against other groups and predators, to provide warnings or even attack them.

For optimum genetic advantage in food-affluent environments, breeding should start early because there is ample food for even immature females to rear their young. Birth intervals may be short and maternal care less intensive but the young will likely survive. Males should compete fiercely for reproduction, even at risk to their own lives, because they are unlikely to live very long as a result of predation.

We can also appreciate why animals in a food-limited environment, such as gibbons or monogamous langurs, should be more cautious about predators. When there are few dangerous predators, population rises to match the level of the food supply, so food becomes scarce. There is less point risking your life by mobbing a predator when it is unlikely to catch you. When predation is strong, however, population stays below the level supportable by the food supply, so food becomes relatively plentiful. Predators pose an extreme danger, so group size increases to allow for mutual warning and defense. In parallel, animals become bolder which allows for active mobbing.

The preceding points explain why Mentawaian langurs apparently acted contrary to their own survival interests, showing greater boldness in the productive areas where hunting was more prevalent. Plentiful food switches on the anti-predator response of larger and bolder troops, effective against snakes and eagles but not against men armed with shotguns.

There are studies documenting that food shortage is associated with the same behavioral response in humans as in non-human primates. In the cross-cultural survey, people who live in societies that are short of food are more monogamous and modest in dress than people in societies where food is plentiful. Women in these societies are also more likely to report that they do not enjoy sex (see Table 1.6 below).

Table 1.6. Behavior related to food shortage in the cross-cultural survey[53] Behaviors associated with food shortage in non-human primates are also more common in human societies reported as short of food.

	Shortage of food
Divorce restricted	.25
Monogamy	.22
Women less interested in sex	.44
Modesty in dress	.37*
* .01 significance	Others: .05 significance

How do changed food conditions affect behavior?

Having identified the behavioral effects of food shortage, the next step is to identify how this comes about. Granted that it benefits individuals in times of food shortage to delay breeding or forming monogamous families, how could shortage of food give rise to such behavior? And why would it cause people to be modest in dress—something that has no parallel in animal behavior and yet, as we have seen, is strongly linked to other food-restricted behaviors (see Tables 1.1 and 1.2 above).

One hypothesis is that the primate groups become smaller in tougher conditions because animals have less time to service their social relationships. Another is that they split up because of competition for food.[54] Both of these ideas could be valid. Grooming plays a vital role in social cohesion, so animals that need to spend more time searching for food will have less energy for strengthening social bonds. And direct competition for food has a role in splitting groups, as we saw with

[53] See www.biohistory.org.

[54] Janson, "Aggressive Competition and Individual Food Consumption in Wild Brown Capuchin Monkeys," 125–38; Chapman et al., "Ecological constraints on group size," 59–70.

gibbons. It is also not unreasonable that groups short of food should travel faster between feeding patches. But such explanations do not easily account for the full range of behavioral changes, such as why animals in food-limited environments flee from predators rather than mob them, breed more slowly, and give more attention to their young.

Biohistory proposes that the changes in behavior noted in tables 1.5 and 1.6 stem from a direct physiological response to hunger. This in turn has an epigenetic effect—hunger changes gene expression so that genes become more or less active.

Epigenetics is a relatively new field, but it has already led to powerful insights into the ways in which individuals develop. The key point is that animals which experience limited food availability, either because of a food-limited environment or reduced predator pressure, undergo physiological changes that alter their behavior. In effect, they behave more like gibbons and less like baboons. There are of course limits to this change. For example, the social behavior of baboons is far more flexible than that of gibbons, but no baboon groups have been found with the social organization of gibbons.

Note that a limited food supply or calorie restriction, which from now on will be identified as "CR," does not mean starvation. Numerous studies have found that, short of malnutrition, CR has health benefits—limiting food intake delays aging and extends lifespan for many species, including primates. Other effects include reducing the likelihood of diabetes, cancer, cardiovascular disease and brain atrophy.[55] This reflects a shift of body resources from fast reproduction to body maintenance and longevity, responses that are optimal in environments with stable but limited food.

CR also appears to improve learning and memory—useful skills when locating scattered food resources in an environment such as a tropical forest. For example, a gibbon will be more likely to survive if it is good at remembering the location of a fruiting tree and how to reach it.

In chapter two we discuss evidence that hormone changes resulting from CR mediate the behavioral changes. Establishing this point is not easy in

---

[55] R. J. Colman, R. M. Anderson, S. C. Johnson, E. K. Kastman, K. J. Kosmatka, T. M. Beasley, D. B. Allison, C. Cruzen, H. A. Simmons, J. W. Kemnitz & R. Weindruch, "Caloric Restriction Delays Disease Onset and Mortality in Rhesus Monkeys," *Science* 325 (5937) (2009): 201–4.

the wild, but a study of four vervet populations found that leptin levels were up to four times higher in the wet season, when food was more readily available (leptin is a hormone known to increase with plentiful food). And although female leptin levels varied widely depending on breeding conditions, and the other results were not entirely consistent, male leptin was lowest in the population with lowest rainfall and smallest troop size.[56]

A recent study supplemented the diet of mice with minimal extra sugar, an amount equivalent to a human drinking three cans of soft drink a day. This is, in effect, a condition of super-abundant food, because of the high calorie content of sugar. The mice not only had a higher death rate than the mice which did not receive the supplement, but 26% fewer males were able to establish territories.[57] In other words, well-fed animals are less territorial.

The genetic and environmental components of CR behavior

The most successful species are those most well adapted in body and behavior to their environments. For example, gibbons are adapted to life in the rainforest in that their long arms are well suited to brachiating through the trees. A genetic predisposition to high CR behavior would similarly adapt their *behavior* to the tropical forest, where predation is low and population presses against the limits of the food supply. Adults would have to work hard to find food. They would have fewer young but spend more time looking after them so as to produce offspring that can flourish in their environment.

Baboons are adapted to life on the savannah and are far better than gibbons at moving along the ground. A genetic predisposition to low CR

[56] P. L. Whitten & T. R. Turner, "Ecological and Reproductive Variance in Serum Leptin in Wild Vervet Monkeys," *American Journal of Physical Anthropology* 137 (2008): 441–8; N. Dracopoli, F. L. Brett, T. R. Turner & C. J. Jolly, "Patterns of Genetic Variability in the Serum Proteins of the Kenyan Vervet Monkey (Cercopithecus aethiops)," *American Journal of Physical Anthropology* 61 (1983): 39–49.

[57] J. S. Ruff., A. L. Suchy, S. A. Hugentobler, M. M. Sosa, B. L. Schwartz, L. C. Morrison, S. H. Gient, M. K. Shigenaga & W. K. Potts, "Human-Relevant Levels of Added Sugar Consumption Increase Female Mortality and Lower Male Fitness in Mice," *Nature Communications* 4 (2013): Article 2245.

behavior would be adaptive to their environment where food is normally plentiful and breeding can occur early in the lives of females. They have no need to spread out or work overly hard to search for food. The threat of predators means they stand to gain from membership in larger troops, which provide many pairs of eyes to detect danger and contribute mutual defense. Larger groups also give them an advantage in competition with other troops. Further, if local food resources become limited, a larger group provides protection should it choose to migrate. A large, well-organized troop is better able to cope with the dangers of a strange environment than individuals or scattered nuclear families.

Such broad differences arise through the process of natural selection. When ancestral gibbons moved into the tropical forest, individuals with longer arms are likely to have found more food than those less capable of moving through the trees. Better-adapted animals would have more offspring carrying their genes which, over thousands of generations, would lead to their becoming the superb acrobats that we now observe. Similarly, in a resource-scarce environment, animals with CR genes would tend to be more reproductively successful in the sense that the survival rate of offspring would exceed that of animals such as baboons, despite producing fewer offspring in each generation. Such genes would also predominate in future generations.

Moving into an environment with more plentiful food, adaptation would work the other way. Genetically determined changes in patterns of sociability would accompany changes in sexual behavior and care of offspring. Behavioral changes would be hastened if all the CR systems operated as a single mechanism.

However, mammals have a faster (though not perfect) way of adapting to the environment than the slow process of natural selection. In the next chapter we discuss how the complex of behavior associated with rapid adaptation to changing food availability applies to other mammal species such as rats and mice. Within limits, what we as human beings have inherited is the capacity of individuals to change their level of CR-related behavior as a direct response to specific environmental conditions. The stimuli in this case are food-abundant and food-limited environments. The genetic code does not need to change for this process to work. The relevant genes can simply be switched on or off epigenetically via environmental triggers acting through hormones.

A clear implication of the preceding points is that the level of CR-related behavior is not always fixed or constant. Species adapted to a stable environment, such as gibbons, have minimal need to alter their behavior. But others, such as baboons and vervets, which inhabit multiple ecological niches, need to alter their CR-related behavior to suit. One can predict, therefore, that the greater the range of habitats, the more CR-related behaviors can change.

Such plasticity has its limits. Forest-dwelling species such as Mentawaian langurs vary between monogamous and polygamous behaviors, while more open-country species such as baboons and vervets vary between multi-male and polygynous. This suggests that each species has a genetic "set point" for behavior that fits its most typical habitat, but that, within limits, behavior can vary as an adaptation to different habitats.

"Natural" levels of CR-related behavior in humans

Based on the observation of the behavior of current hunter-gatherers, whose subsistence patterns and social structures mirror those of our distant ancestors, the human set point for CR-related behavior appears to be low. Hunter-gatherers normally live in multi-family groups which travel and hunt together. For example, camps of the Mbuti pygmies consist of at least 6–7 families, the minimum required for the Mbuti practice of hunting with nets. Family groups move periodically and also change in composition as people attach themselves to different relatives by blood or marriage. The maximum size of such groups is determined by the needs of hunting. Too many people are seen as a disadvantage.[58] This pattern is nothing like that observed among gibbons and langurs where couples or polygynous males defend an exclusive patch of land. Subsistence patterns also have more in common with baboons than gibbons. Men spend much of their time hunting, which is far less routine than the intensive foraging of a rainforest primate. In this sense they also show no trace of CR-related behavior.

For human beings, having a low set point for CR-related behavior makes sense. Humans are physically adapted to life in the open, probably more so than baboons. Our efficient striding walk allows us to cover long distances in pursuit of prey, and compared with most primates we are poor at swinging through trees or even climbing them. Unable to outrun lions or leopards and poor at climbing trees, our ancestors suffered at the hands of numerous predators, which once again would make group defense and

[58] Turnbull, C. (1963) The Forest People, London, The Reprint Society, Pp 38-9

other low CR-related behaviors advantageous. This cluster of behaviors can be still be observed in group-living hunter-gatherers.

Like baboons and vervets, humans show a relatively large range of CR-related behavior. As mentioned earlier, civilized peoples tend to have very high levels. A farmer or factory worker usually lives with their spouse and dependent young on a defined plot of land (even if only an apartment), with most of their working hours spent on routine tasks. In this sense he acts more like a gibbon than a baboon or, in general terms, more like animals in a food-restricted environment.

However, unlike many nonhuman primate species, in no known human society do breeding couples defend their plots against all intruders, with each partner driving away visitors of the same sex. On the contrary, people in most societies tend to form pair bonds but are also highly cooperative. This behavior suggests the presence of a genetic bias towards monogamy, which takes a peculiar form. Instead of defending a territory, mated pairs have a strong and enduring bond within what (in primate terms) consists of a multi-male band. This applies to hunter-gatherers as much as farmers. In this we are unlike gibbons or langurs, because it permits us to form monogamous nuclear families even when CR-related behavior is only moderately high.

Still, the level of activation of the CR systems does determine the degree to which humans are faithfully monogamous. In some societies, such as Victorian Britain, monogamous norms were immensely strong, though even then not always observed. In other societies, such as in much of Africa, polygyny and even promiscuity are widely accepted. The level of CR-related behavior is indicated by the strength of the social forces requiring monogamy or permitting polygyny and/or promiscuity. So allowing for the fact that humans are more monogamous than many other primates, when their CR-related behavior is weak they are more likely to exercise polygynous and promiscuous behavior. Fig. 1.1 below depicts their overall CR-related behavior in a number of primate species, including humans.

Fig. 1.1. Proposed set points and range of variation in CR-related behavior in various primates. Baboons and vervets are changeable but more likely to form multi-male bands. Langurs and gibbons are changeable but more likely to form single-male troops or even monogamous pairs. Gibbons are finely adapted to a food stable environment and thus less able to change behavior. Human hunter-gatherers are more like group-living baboons, with the exception of unusually strong pair bonds.

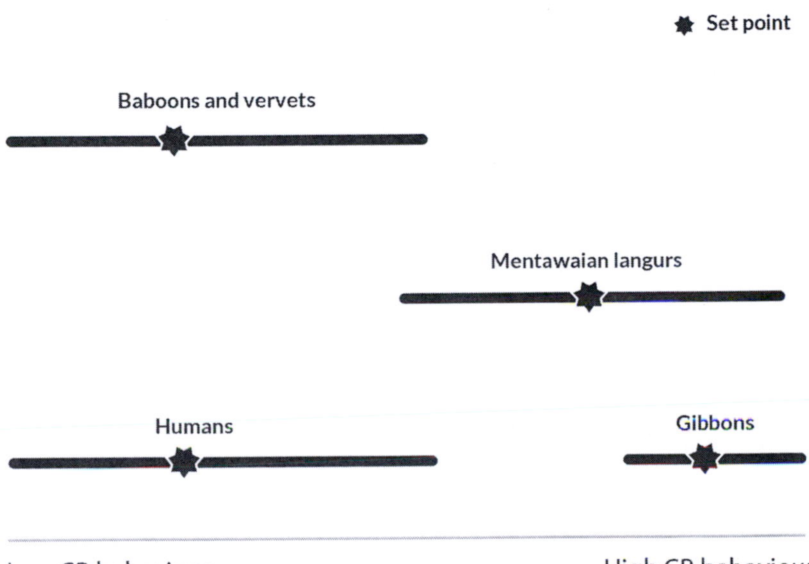

An obvious objection to this approach is that many civilized peoples today and in the recent past have relatively high levels of CR-related behavior without being short of food. For example, America in the nineteenth century had plentiful land and a fast-growing population, and yet was clearly high in CR-related behavior by human standards. This suggests that some factor other than food shortage was responsible for the behavior. To identify this factor or factors we must first study the biochemical and epigenetic effects of food shortage to gain an understanding of what CR-related behavior is really about.

CHAPTER TWO

FOOD RESTRICTION

In the previous chapter we discussed how changes in primate behavior are a direct and adaptive response to environments where food supplies are limited or abundant. This chapter continues the same topic by exploring the physiological and behavioral effects of calorie restriction or "CR" in the laboratory.

Starvation versus hunger

Starvation is highly stressful and dangerous to health. Starving rats have compromised immune systems and are more likely to die from infection.[1] Experiments in humans establish that semi-starvation results in extreme fatigue and weakness, irritability, anemia, apathy, reduced coordination, and loss of concentration.[2] And, malnutrition as a result of anorexia can cause acute liver damage.[3]

By contrast, mild CR can have beneficial effects. In the majority of studies reviewed in this chapter rats were provided only 25% less food than if unrestricted, or 50% less for three days only, similar to that of properly conducted weight-loss programs for humans. These diets contain adequate levels of protein, vitamins and minerals. The initial set of studies, conducted by the author and his associates at LaTrobe and RMIT Universities in Melbourne, addresses CR-induced changes in blood hormone levels and behavior.

---

[1] R. Faggioni, A. Moser, K. R. Feingold & C. Grunfeld, "Reduced Leptin Levels in Starvation Increase Susceptibility to Endotoxic Shock," *American Journal of Pathology* 156 (5) (2000): 1781–7.

[2] L. M. Kalm & R. D. Semba, "They Starved So That Others Be Better Fed: Remembering Ancel Keys and the Minnesota Experiment," *The Journal of Nutrition* 135 (6) (2005): 1347–1352.

[3] L. Di Pascoli, A. Lion, D. Milazzo & L. Caregaro, "Acute Liver Damage in Anorexia Nervosa," *International Journal of Eating Disorders* 36 (1) (2004): 114–17.

The effect of CR on testosterone and other hormones

Numerous previous studies establish that CR reduces blood testosterone levels.[4] We replicated these studies for 25% CR and 50% CR for three days only. Fig. 2.1 below summarizes the results.

Fig. 2.1 confirms that ongoing food restriction reduces testosterone, though not for the group that experienced restriction for only three days. To understand the significance of this finding, we will consider the attitudes and behavior associated with different levels of testosterone.

---

[4] E. A. Levay, A. H. Tammer, J. Penman, S. Kent & A. G. Paolini, "Calorie Restriction at Increasing Levels Leads to Augmented Concentrations of Corticosterone and Decreasing Concentrations of Testosterone in Rats," *Nutrition Research* 30 (5) (2010): 366–73;
A. Amorio, J. L. Montero & T. Jolin, "Chronic Food Restriction and the Circadian Rhythms of Pituitary-Adrenal Hormones, Growth Hormone and Thyroid-Stimulating Hormone," *Annals of Nutrition and Metabolism* 31 (2) (1987): 81–7; J. Stewart, M. J. Meaney, D. Aitken, L. Jensen & N. Klant, "The Effects of Acute and Life-Long Food Restriction on Basal and Stress-Induced Serum Corticosterone Levels in Young and Aged Rats," *Endocrinology* 123 (1988): 1934–41; R. J. Seeley, C. A. Matson, M. Chavez, S. C. Woods, M. F. Dallman & M. W. Schwartz, "Behavioural, Endocrine and Hypothalamic Responses to Involuntary Overfeeding," *American Journal of Physiology* 271 (3) (1966): R819–23; K. M. Heiderstadt, R. M. McLaughlin, S. E. Wrighe & C. E. Gomez-Sanchez, "The Effect of Chronic Food and Water Restriction on Open-Field Behaviour and Serum Corticosterone Levels in Rats," *Laboratory Animal Science* 34 (1) (2000): 20–8; F. Chacón, A. I. Esquifino, M. Perello, D. P. Cardinali, E. Spinedi & M. P. Alvarez, "24-Hour Changes in ACTH, Corticosterone, Growth Hormone, and Leptin Levels in Young Male Rats Subjected to Calorie Restriction," *Chronobiology International* 22 (2) (2005): 253–65; L. S. Brady, M. A. Smith, P. W. Gold & M. Herkenham, "Altered Expression of Hypothalamic Neuropeptide mRNAs in Food-Restricted and Food-Deprived Rats," *Neuroendocrinology* 52 (5) (1990): 441–7; E. S. Han, N. Levin, N. Bengani, J. L. Roberts, Y. Sun, K. Karelus & J. F. Nelson, "Hyperadrenocorticism and Food Restriction-Induced Life Extension in the Rat: Evidence for Divergent Regulation of Pituitary Proopiomelanocortin RNA and Adrenocorticotropic Hormone Biosynthesis," *Journals of Gerontology Series A: Biological and Medical Sciences* 50A (5) (1995): B288–94; J. Lindblom, T. Haitina, R. Fredriksson & H. B. Schloth, "Differential Regulation of Nuclear Receptors, Neuropeptides and Peptide Hormones in the Hypothalamus and Pituitary of Food Restricted Rats," *Molecular Brain Research* 133 (1) (2005): 37–46; J. E. Schneider, D. Zhou & R. M. Blum, "Leptin and Metabolic Control of Reproduction," *Hormones and Behavior* 37 (4) (2000): 306–26;
J. S. Flier & E. Maratos-Flier, "Obesity and the Hypothalamus: Novel Peptides for New Pathways," *Cell* 92 (4) (1998): 437–40.

Fig. 2.1. Mean serum testosterone levels of calorie-restricted adult rats.[5] Even mild food restriction reduces testosterone. The Control group was not calorie restricted. The CR 25% and CR 50% groups had 25% and 50% of their normal calorie intake restricted throughout the study, and the Acute group had 50% calorie restriction for three days.

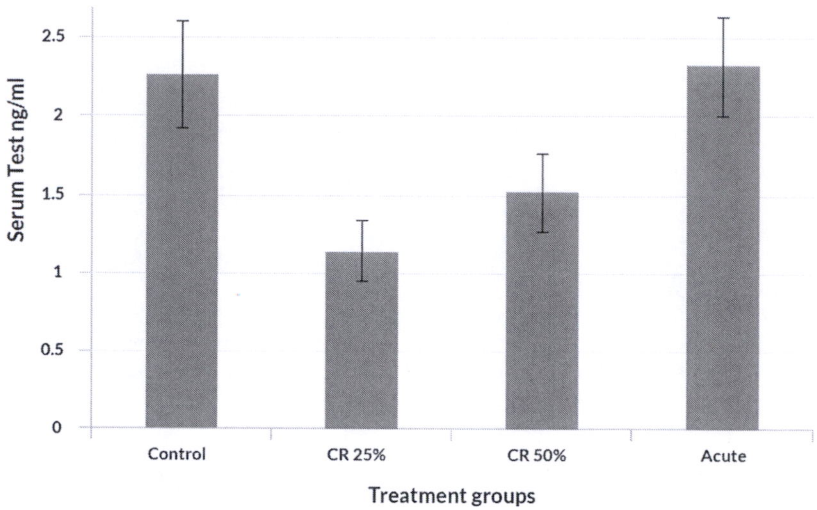

Testosterone is a major male sex hormone also present in females but to a lesser degree. Among males it is associated with a strong sex drive, aggression and striving for dominance. Men with high testosterone are less likely to be married or in a committed relationship. They spend less time with their wives and divorce more often.[6] Higher levels of testosterone in

---

[5] A. Govic, E. A. Levay, A. Hazi, J. Penman, S. Kent & A. G. Paolini, "Alterations in Male Sexual Behaviour, Attractiveness and Testosterone Levels Induced by an Adult-Onset Calorie Restriction Regimen," *Behavioural Brain Research* 190 (1) (2008): 140–6.

[6] H. Persky, H. I. Lief, D. Strauss, W. M. Miller & C. P. O'Brien, "Plasma Testosterone Level and Sexual Behavior of Couples," *Archives of Sexual Behavior* 7 (3) (1978): 157–73; T. C. Burnham, J. F. Chapman, P. B. Gray, M. H. McIntyre, S. F. Lipson & P. T. Ellison, "Men in Committed, Romantic Relationships have Lower Testosterone," *Hormones and Behavior* 44 (2003): 119–22; P. Gray, S. M. Kahlenberg, E. S. Barrett, S. F. Lipson & P. T. Ellison, "Marriage and Fatherhood are Associated with Lower Testosterone in Males," *Evolution and Human Behavior* 23 (3) (2002): 1–9; A. Mazur & J. Michalek, "Marriage, Divorce and Male Testosterone," *Social Forces* 77 (1) (1998): 315–30; A. Booth, & J. Dabbs, "Testosterone and Men's Marriages," *Social Forces* 72 (2) (1993): 463–77; S. M.

women are associated with the pursuit of sexual gratification and increased libido.[7] Among non-human primates higher testosterone, which is associated with plentiful food, is linked to increased sexuality and weakened pair bonds.

The link between testosterone and aggression is strong in both animals and humans.[8] For example, the most violent female and male offenders in prison have higher levels of testosterone than their less violent counterparts.[9] High-testosterone males are more likely to be delinquents, to use drugs, to abuse alcohol, to go AWOL from the military, and perform better in combat.[10] These findings are consistent with studies which link testosterone to sensation seeking and high-risk behavior.[11] High-testosterone people are also more gregarious and more likely to need and seek the company of others to be happy.[12]

---

van Anders & N. V. Watson, "Relationship Status and Testosterone in North American Heterosexual and Non-Heterosexual Men and Women: Cross-Sectional and Longitudinal Data," *Psychoneuroendocrinology* 31(6) (2006): 715–23.

[7] S. R. Davis & J. Tran, "Testosterone Influences Libido and Well Being in Women," *Trends in Endocrinology and Metabolism* 12 (1) (2001): 33–7.

[8] D. Olweus, A. Mattsson, D. Schalling & H. Low, "Testosterone, Aggression, Physical, and Personality Dimensions in Normal Adolescent Males," *Psychosomatic Medicine* 42 (2) (1980): 253–69; J. Ehrenkranz, E. Bliss & M. H. Sheard "Plasma Testosterone: Correlation with Aggressive Behavior and Social Dominance in Man," *Psychosomatic Medicine* 36 (6) (1974): 469–75; Y. Delville, M. Mansour & C. F. Ferris, "Testosterone Facilitates Aggression by Modulating Vasopressin Receptors in the Hypothalamus," *Physiology & Behavior* 60 (1) (1996): 25–9; J. S. Berg & K. E. Wynne-Edwards, "Changes in Testosterone, Cortisol and Estradiol Levels in Men Becoming Fathers," *Mayo Clinic Proceedings* 76 (6) (2001): 582–92.

[9] D. Olweus, A. Mattsson, D. Schalling & H. Low, "Testosterone, Aggression, Physical, and Personality Dimensions in Normal Adolescent Males," *Psychosomatic Medicine* 42 (2) (1980): 253–69; J. J. Dabbs & M. Hargrove, "Age, Testosterone, and Behavior among Female Prison Inmates," *Psychosomatic Medicine* 59 (5) (1997): 477–80; J. J. Dabbs, R. L. Frady, T. S. Carr & N. F. Besch, "Saliva Testosterone and Criminal Violence in Young Adult Prison Inmates," *Psychosomatic Medicine* 49 (2) (1987): 174–82.

[10] J. M. Dabbs, *Heroes, Rogues and Lovers: Testosterone and Behaviour* (New York: McGraw Hill, 2000); C. Gimbel & A. Booth "Who fought in Vietnam," *Social Forces* 74 (4) (1996): 1137–57.

[11] A. Booth, D. R. Johnson & D. A. Granger, "T and Men's Depression: Each and Social Behavior," *Journal of Health and Social Behavior* 40 (2) (1999): 130–40.

[12] Dabbs, *Heroes, Rogues and Lovers*.

High-status animals have higher testosterone levels than low-status animals, a relationship mirrored in human prison populations.[13] Outside of prison, however, humans with high occupational status tend to have *lower* testosterone. Large scale investigations of military veterans establish that testosterone levels are highest in the unemployed, next in blue collar workers, next in those working in sales and professions, and lowest of all among farmers, although a recent study of Australian males found the highest levels among the self-employed.[14] Ministers of religion also tend to have low testosterone levels. And, on average, people living in the country have lower levels than city dwellers. J. M. Dabbs, in his book Heroes, Rogues and Lovers: Testosterone and Behavior, summarizes a half-century of research:

> High levels of testosterone evolved when the human race was young and people needed the skills of youth. High testosterone helped them compete, but it also led them to take risks, fight, get injured, and die young and now it interferes with many modern activities. High testosterone individuals are energetic but impatient; they do poorly in school and end up with fewer years of education; they can dominate others in face-to-face meetings, but they have trouble handling the complexities of business. They lean toward harsh and competitive activities and away from subdued and thoughtful ones. High testosterone is a drawback when careful planning, reliable work habits and patience are needed, or when workers must attend to the needs of others. Except for a few of the top jobs in sports and acting, high testosterone, to my knowledge, does not contribute to financial success.[15]

The link between testosterone and aggression helps explain why animals in food-abundant environments are more aggressive and less fearful than those living in food-restricted environments. In particular, it helps to explain why baboons often mob or even attack predators, rather than fleeing as gibbons and monogamous langurs do. They do so at least partly because higher testosterone, as a result of more plentiful food, makes them more aggressive.

A more complex picture applies to parenting, where the effect of testosterone varies across species. In some species, such as tamarins,

---

[13] Ehrenkranz et al., "Plasma Testosterone," 469–75.

[14] J. M. Dabbs, "Testosterone and Occupational Achievement," *Social Forces* 70 (3) (1992): 813–24; F. J. Greene, L. Han, S. Martin, S. Hang & G. Wittert, "Testosterone is Associated with Self-Employment among Australian Men," *Economics & Human Biology* 13 (2014): 76–84.

[15] Dabbs, Heroes, Rogues and Lovers, 150–151.

testosterone is essential for paternal care,[16] while in gerbils levels drop when males begin caring for their young.[17] Humans are similar to gerbils in that the testosterone levels of men decline after their children are born.[18] Reduction in testosterone thus suggests an explanation for the better child care provided by males in food-limited environments.

Testosterone is not the only hormone influenced by CR. As noted in chapter one, vervet monkeys living in a food-limited environment have less leptin in their feces than those living in a food-abundant environment. Like testosterone, leptin is a hormone which influences the sexual behavior of rats. Its administration can reverse the disruption of ovulation in CR animals and restore fertility and sexual behavior.[19] Our studies show that leptin, like testosterone, is reduced by CR, but only for the 50% CR group.[20] Fig. 2.2 below summarizes findings from these studies.

---

[16] T. E. Ziegler & C. T. Snowdon, "Preparental Hormone Levels and Parenting Experience in Male Cottontop Tamarins, Saguinus Oedipus," *Hormones and Behavior* 38 (3) (2000): 159–167; C. B. Trainor & C. A. Marler, "Testosterone, Paternal Behavior, and Aggression in the Monogamous California Mouse (Peromyscus californicus)," *Hormones and Behavior* 40(1) (2001): 32–40; E. R. Brown, T. Murdoch, P. R. Murphy & W. H. Moger, "Hormonal Responses of Male Gerbils to Stimuli from their Mate and Pups," *Hormones and Behaviour* 29 (4) (1995): 474–91.

[17] M. M. Clark, D. V. Desousa, B. G. Galef & T. Bennett, "Parenting and Potency: Alternative Routes to Reproductive Success in Male Mongolian Gerbils," *Animal Behavior* 54 (3) (1997): 635–42; C. J. Reburn & K. E. Wynne-Edwards, "Hormonal Changes in Males of Naturally Biparental and Uniparental Mammals," *Hormones and Behavior* 35 (2) (1999): 163–76; J. S. Berg & K. E. Wynne-Edwards, "Changes in Testosterone, Cortisol and Estradiol Levels in Men Becoming Fathers," *Mayo Clinic Proceedings* 76 (6) (2001): 582–92.

[18] T. C. Burnham, J. F. Chapman, P. B. Gray, M. H. McIntyre, S. F. Lipson & P. T. Ellison, "Men in Committed, Romantic Relationships have Lower Testosterone," *Hormones and Behavior* 44 (2003): 119–22; A. S. Fleming, C. Corter, J. Stallings & M. Steiner, "Testosterone and Prolactin are Associated with Emotional Responses to Infant Cries in New Fathers." *Hormones and Behavior* 42 (4) (2002): 399–413; R. Boyd & P. J. Richerson, *Culture and the Evolutionary Process* (Chicago: University of Chicago Press, 1985).

[19] D. A. Schreihofer, J. A. Amico & J. L. Cameron, "Reversal of Fasting-Induced Suppression of Luteinizing Hormone (LH) Secretion in Male Rhesus Monkeys by Intragastric Nutrient Infusion: Evidence for Rapid Stimulation of LH by Nutritional Signals," *Endocrinology* 132 (5) (1993): 1890–7.

[20] J. Korczynska, E. Stelmanska & J. Swierczynski, "Differential Effect of Long-Term Food Restriction on Fatty Acid Synthase and Leptin Gene Expression in Rat

Fig. 2.2. Mean serum leptin levels of calorie-restricted adult rats.[21] CR reduces leptin for animals with 50% restriction.

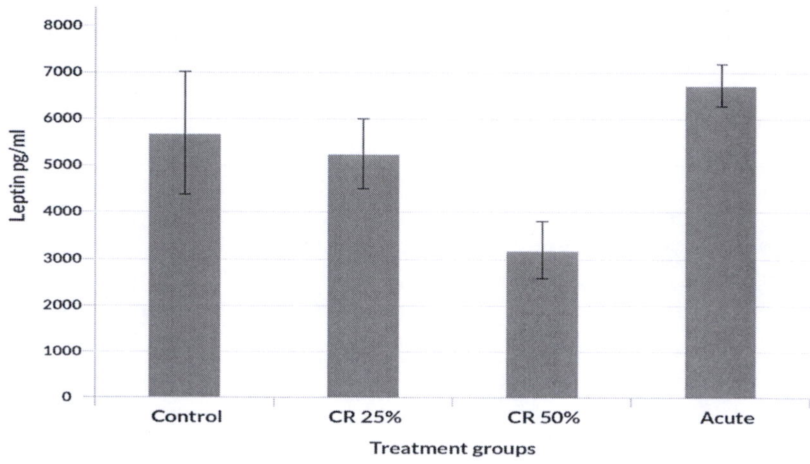

Among the three treatment groups only CR 50% led to a reduction in leptin levels. Neither CR 25% nor 50% restriction for three days had any significant effect.

Other hormones also respond to CR. For example, CR reduces luteinizing hormone (LH) and follicle-stimulating hormone (FSH), both of which trigger ovulation in females.[22] In males LH stimulates the production of testosterone and FSH promotes sperm production.

Calorie restriction and stress hormones

CR also affects stress hormones. Because hunger is stressful it might be predicted that it would increase stress hormones such as corticosterone (CORT) and adrenocorticotropic hormone (ACTH). But while mild CR

White Adipose Tissue," *Hormones and Metabolic Research* 35 (10) (2003): 593–7; D. S. Weigle, P. B. Duell, W. E. Connor, R. A. Steiner, M. R. Soules & J. L. Kuijper, "Effect of Fasting, Refeeding, and Dietary Fat Restriction on Plasma Leptin Levels," *The Journal of Clinical Endocrinology & Metabolism* 82 (2) (1997): 561–5; Govic et al., "Alterations in Male Sexual Behaviour," 140–6.

[21] Govic et al., "Alterations in Male Sexual Behaviour," 140–6.

[22] A. W. Root & R. D. Russ, "Short-Term Effects of Castration and Starvation upon Pituitary and Serum Levels of Luteinizing Hormone and Follicle Stimulating Hormone in Male Rats," *Acta Endocrinologica* 70 (1972): 665–75.

increases CORT in rats, it *reduces* or has no effect on ACTH.[23]

Stress is known to have damaging effects on health and well-being. Studies have shown repeatedly that it is a major contributor in the development of metabolic disorders. For example, in humans, cumulative psychological stress has been linked to coronary risk factors such as resting blood pressure and insulin resistance. Stress also severely impacts an organism's ability to fight infection and illness, which increases susceptibility to the common cold.[24]

But CORT also has a protective role, preventing these effects from going too far.[25] It facilitates recovery from stress by winding down stress

---

[23] Levay et al., "Calorie Restriction at Increasing Levels Leads to Augmented Concentrations of Corticosterone and Decreasing Concentrations of Testosterone in Rats," 366–73; Amorio et al., "Chronic food restriction and the circadian rhythms of pituitary-adrenal hormones, growth hormone and thyroid-stimulating hormone," 81–7; Seeley et al., "Behavioural, endocrine and hypothalamic responses to involuntary overfeeding," 819–23; Heiderstadt et al., "The Effect of Chronic Food and Water Restriction on Open-Field Behaviour and Serum Corticosterone Levels in Rats," 20–8; Chacón et al., "24-Hour Changes in ACTH, ACTH (adrenocorticotropic hormone) Corticosterone, Growth Hormone, and Leptin Levels in Young Male Rats Subjected to Calorie Restriction," 253–65.

Brady et al., "Altered expression of hypothalamic neuropeptide mRNAs in food-restricted and food-deprived rats," 441–7; Han et al., "Hyperadrenocorticism and food restriction-induced life extension in the rat: Evidence for divergent regulation of pituitary proopiomelanocortin RNA and adrenocorticotropic hormone biosynthesis," B288–94; Lindblom, "Differential Regulation of Nuclear Receptors, Neuropeptides and Peptide Hormones in the Hypothalamus and Pituitary of Food Restricted Rats," 37–46; Schneider, "Leptin and metabolic control of reproduction," 306–26; Flier, "Obesity and the hypothalamus: Novel peptides for new pathways," 437–40.

[24] R. Dantzer & K. W. Kelley, "Stress and Immunity: An Integrated View of Relationships Between the Brain and the Immune System," *Life Sciences* 44 (26) (1989): 1995–2008; S. Cohen, A. J. D. Tyrrell & A. P. Smith, "Psychological Stress and Susceptibility to the Common Cold," *The New England Journal of Medicine* 325 (9): 606–12.

[25] M. P. Mattson, S. L. Chan & W. Duan, "Modification of Brain Aging and Neurodegenerative Disorders by Genes, Diet and Behaviour," *Physiological Reviews* 82(3) (2002): 637–72; M. P. Mattson, W. Duan, R. Wan & Z. Guo, "Prophylactic Activation of Neuroprotective Stress Response Pathways by Dietary and Behavioural Manipulations," *NeuroRx* 1 (1) (2004): 111–16; A. Munck, M. P. Guyre & N. J. Holbrook, "Physiological Functions of Glucocorticoids in Stress and their Relation to Pharmacological Actions," *Endocrine Reviews* 5 (1) (1984): 25–44.

responses. For example, people who have made a full recovery from a stressful experience show higher levels of cortisol (the human equivalent of corticosterone) than those with post-traumatic stress disorder.[26]

Other stress-related hormones are not affected consistently. One study found an increase in adrenaline with 50% CR.[27] Our studies found no effect with 12.5% CR and 37.5% CR, but significant reductions with 25% CR and 50% CR.[28] In these studies differences between individual subjects were striking and may account for the varying results.

Calorie restriction and sexual behavior

CR can explain the reduced sexuality and delayed breeding of animals in food-limited environments, in that it reduces testosterone, leptin, luteinizing hormone (LH), and follicle-stimulating hormone (FSH). All of these affect sexual activity and sexual maturation.

Laboratory studies of rats have found that CR produces the same behaviors found in food-restricted primates. CR animals are less fertile, reach puberty later, and engage less frequently in sexual intercourse.[29]

---

[26] R. Yehuda, M. H. Teicher, R. L. Trestman, R. A. Levengood & L. J. Siever, "Cortisol Regulation in Posttraumatic Stress Disorder and Major Depression: A Chronobiological Analysis," *Biological Psychiatry* 40 (2) (1996): 79–88.

[27] T. Hilderman, K. McKnight, K. S. Dhalla & H. Rupp, "Effects of Long-Term Dietary Restriction on Cardiovascular Function and Plasma Catecholamines in the Rat," *Cardiovascular Drugs and Therapy* 10 (1) (1996): 247–50.

[28] Levay et al., "Calorie Restriction at Increasing Levels Leads to Augmented Concentrations of Corticosterone and Decreasing Concentrations of Testosterone in Rats," 366–73.

[29] R. J. Nelson, A. C. Marinovic, C. A. Moffatt, L. J. Kriegsfeld & S. Kim, "The Effects of Photoperiod and Food Intake on Reproductive Development in Male Deer Mice (Peromyscus maniculatus)," *Physiology & Behavior* 62 (5) (1997): 945–50; K. M. Vitousek, F. P. Manke, J. A. Gray & M. N. Vitousek, "Caloric Restriction for Longevity: II—The Systematic Neglect of Behavioural and Psychological Outcomes in Animal Research," *European Eating Disorders Review* 12 (6) (2004): 338–60; R. Weindruch & R. L. Walford, *The Retardation of Aging and Disease by Dietary Restriction* (Springfield, IL: Charles C. Thomas, 1988); J. F. Nelson, R. G. Gosden & L. S. Felicio, "Effect of Dietary Restriction on Estrous Cyclicity and Follicular Reserves in Aging C57BL/6J Mice," *Biology of Reproduction* 32 (3) (1985): 515–22; G. D. Hamilton & F. H. Bronson, "Food Restriction and Reproductive Development in Wild House Mice," *Biology of Reproduction* 32 (4) (1985): 773–8; C. J. Gill & E. F. Rissman, "Female Sexual

Although an obvious reason for these behaviors is reduced testosterone, our studies suggest that lower levels of leptin may also be important. For example, in one study a 25% CR diet reduced testosterone but not leptin levels and no decline in sexual activity was observed. However a 50% CR diet led to a decline in leptin as well as testosterone and subjects took longer to mount females, although their sexual activity was otherwise the same.[30]

The more general finding is that even moderate CR reduces sex drive and sexual desire. These effects have been observed in humans as well as animals.[31] Practitioners of CR diets commonly report a reduction in libido and a general loss of interest in sex.[32] Food reduction effects are far more striking when CR is severe. An example is found in the Minnesota Starvation Experiment, a study designed to gain insight into the physiological and psychological effects of semi-starvation by giving men 50% of their normal caloric intake.[33] The great majority of men who participated in the study reported a complete loss of sex drive soon after the initiation of diet.

Our studies also found that male CR rats were less attractive to females, in that females spent less time investigating CR males. These findings are mirrored by other studies which show that females of many species, such as hamsters and mice, are less interested in the smell of CR males than of

---

Behavior is Inhibited by Short- and Long-Term Food Restriction," *Physiology & Behavior* 61 (3) (1997): 387–94.

29 A. Govic, S. Kent, E. A. Levay, A. Hazi, J. Penman & A. G. Paolini, "Testosterone, Social and Sexual Behaviour of Perinatally and Lifelong Calorie Restricted Offspring," *Physiology & Behaviour* 94 (3) (2008): 516–22.

30 Govic et al., "Alterations in Male Sexual Behaviour, Attractiveness and Testosterone Levels Induced by an Adult-Onset Calorie Restriction Regimen," 140–6.

31 Davis & Tran, "Testosterone Influences Libido and Well Being in Women," 33–7.

32 Vitousek et al., "Caloric Restriction for Longevity: II—The Systematic Neglect of Behavioural and Psychological Outcomes in Animal Research," 338–60;
J. F. Morgan, J. H. Lacey & F. Reid, "Anorexia Nervosa: Changes in Sexuality During Weight Restoration," *Psychosomatic Medicine* 61 (4) (1999): 541–5; A. Keys, J. Brožek, A. Henschel, O. Mickelsen, & H. L. Taylor, *The Biology of Human Starvation* (2 volumes) (Minneapolis: University of Minnesota Press, 1950).

33 Keys et al., The Biology of Human Starvation.

those who are not food restricted.[34] Weight is unlikely to be the reason, since females paired with well-fed males showed no preference for heavier males.[35]

Similar findings have been reported for humans. Women who are ovulating, and thus in the most fertile period of their cycle, select the faces of men with higher levels of testosterone. But during the less fertile days of their cycle they prefer faces indicative of lower testosterone.[36]

Such findings suggest that women are able to detect men's testosterone levels from facial cues and make judgments regarding their mating quality. From a woman's perspective there is evolutionary sense in this. In non-fertile times she can pair with a lower testosterone male—one who is more likely to support her and her offspring (as we shall see)—while maintaining the option of being fertilized by a male with higher testosterone who is thus, at least in animal terms, fitter. High testosterone indicates that a male is likely to be better fed and socially dominant, suggesting superior fitness.

---

[34] H. Liang, J. Zhang & Z. Zhang, "Food Restriction in Pregnant Rat-Like Hamsters (Cricetulus triton) affects Endocrine, Immune Function and Odor Attractiveness of Male Offspring," *Physiology & Behavior* 82 (2–3) (2004): 453–8;

D. Meikle, J. H. Kruper & C. R. Browning, "Adult Male House Mice Born to Undernourished Mothers are Unattractive to Oestrous Females," *Animal Behavior* 50 (3) (1995): 753–8;

P. J. White, R. B. Fischer & G. F. Meunier, "The Ability of Females to Predict Male Status via Urinary Odors," *Hormones and Behavior* 18 (4) (1984): 491–4;

J. C. Wingfield & P. Marler, "Endocrine Basis of Communication in Reproduction and Aggression," in *The Physiology of Reproduction*, E. Knobil & J. Neil, 1647–77 (New York: Raven Press, 1988).

J. X. Zhang, B. Z. Zhang & Z. W. Wang, "Scent, social status, and reproductive condition in rat-like hamsters (Cricetulus triton)," Physiology & Behavior 74 (4–5): 415–20;

G. T. Taylor, J. Haller & D. Regan, "Female Rats Prefer an Area Vacated by a High Testosterone Male," *Physiology & Behavior* 28 (6) (1982): 953–8;

M. H. Ferkin & R. E. Johnston, "Roles of Gonadal Hormones in Control of Five Sexually Attractive Odors of Meadow Voles (Microtus pennsylvanicus)," *Hormones and Behavior* 27 (4) (1993): 523–38.

[35] Govic et al., "Testosterone, Social and Sexual Behaviour of Perinatally and Lifelong Calorie Restricted Offspring," 516–22.

[36] J. R. Roney & Z. L. Simmons, "Women's Estradiol Predicts Preference for Facial Cues of Men's Testosterone," *Hormones and Behavior* 53 (1) (2008): 14–9.

High testosterone levels also suggest that an animal has a strong immune system, in that they have been associated with depressed immune function. If so, a male with a high level who looks healthy must have very "good genes" because he has resisted infection, despite the handicap of high testosterone.[37] By analogy, if two men are equally fast at running a race but one is weighed down by a heavy pack, the man with the pack will be viewed as more fit. Like a heavy pack, testosterone is an immune system handicap.

The same explanation has been offered for anecdotal reports that "jocks" in high school have greater dating success than "nerds." Football players tend to have relatively high testosterone levels while, as noted, professionals and other white-collar workers have lower levels.[38] But high testosterone levels do not always improve mating success. In the previous chapter we noted that mice given extra sugar were less successful at breeding because they were less able to defend a territory. The same applies to gibbons, which are normally unable to attract a mate without having established an exclusive territory. An analogy would be to the high school jock who grows up to drive trucks for a living, compared with his nerd classmate who founds a multi-million dollar software company. Even with his dominant personality and high testosterone, the truck driver may have less long-term success in finding the most desirable mate.

Calorie restriction and health

Many studies have shown CR to be good for health. CR animals tend to have increased lifespans, greater resistance to disease, and fewer age-related diseases.[39] Fig. 2.3 below shows the results from a study designed

---

[37] I. Folstad & A. J. Karter, "Parasites, Bright Males, and the Immunocompetence Handicap," *The American Naturalist* 139 (3) (1992): 603–22; M. K. Angele, A. Ayala, W. G. Cioffi, K. I. Bland & I. H. Chaudry, "Testosterone: The Culprit for Producing Splenocyte Immune Depression after Trauma Hemorrhage," *American Journal of Physiology—Cell Physiology* 274 (6) (1998): C1530–6; A. Zahavi, "Mate Selection: Selection for a Handicap," *Journal of Theoretical Biology* 53 (1) (1975): 205–14.

[38] J. M. Dabbs, D. de La Rue & P. M. Williams, "Testosterone and Occupational Choice: Actors, Ministers, and Other Men," *Journal of Personality and Social Psychology* 59 (6) (1990): 1261–5.

[39] G. Fernandes, J. T. Venkatraman, A. Turturro, V. G. Attwood & R. W. Hart, "Effect of Food Restriction on Life Span and Immune Functions in Long-Lived Fischer-344 Brown Norway rats," *Journal of Clinical Immunology* 17 (1) (1997): 85–95; J. O. Holloszy, "Mortality Rate and Longevity of Food-Restricted

to determine if CR rats recover more quickly from a fever induced by injection with lipopolysaccharide (50 μg/kg). This experiment is a standard test of disease resistance that poses minimal danger to an animal's health.

The findings in Fig. 2.3 can be understood in terms of animals adapting to limited food availability by transferring resources from reproduction to body maintenance. If breeding is difficult in its current environment, an animal will expend more effort in maintaining itself so as to be alive and healthy when more food is available. This is part of a strategy that is likely to pass on the maximum number of genes to the next generation. In effect, when food is plentiful and females are available for mating, this means breeding as much as possible, even if it means a shortened lifespan. Otherwise, look after yourself and wait for better times.

Other studies have shown CR to improve learning, memory and motor performance among aged rats and mice, and to some extent in younger ones.[40] As indicated in the last chapter, better learning and memory may help animals to search for and remember the locations of food in a food-poor environment.

Exercising Male Rats: A Re-Evaluation," *Journal of Applied Physiology* 82 (2) (1997): 399–403; M. Hubert, P. Laroque, J. Gillet & K. P. Keenan, "The Effects of Diet, Ad Libitum Feeding, and Moderate and Severe Dietary Restriction on Body Weight, Survival, Clinical Pathology Parameters, and Cause of Death in Control Sprague-Dawley Rats," *Toxicological Sciences* 58 (1) (2000): 195–207; M. A. Pahlavani, "Caloric Restriction and Immunosenescence: A Current Perspective," *Frontiers in Bioscience* 5 (2000): D580-7; M. P. Mattson, W. Duan & Z. Guo, "Meal size and Frequency Affect Neuronal Plasticity and Vulnerability to Disease: Cellular and Molecular Mechanisms," *Journal of Neurochemistry* 84 (3) (2003): 417–31; D. K. Ingram, R. Weindruch, E. L. Spangler, J. R. Freeman & R. L. Walford, "Dietary Restriction Benefits Learning and Motor Performance in Aged Mice," *Journal of Gerontology & Geriatric Research* 42 (1) (1987): 78–81; J. Stewart, J. Mitchell & N. Kalant, "The Effects of Life-Long Food Restriction on Spatial Memory in Young and Aged Fischer 344 Rats Measured in the Eight-Arm Radial and the Morris Water Mazes," *Neurobiology of Aging* 10 (6) (1989): 660–75; N. Pitsikas, M. Carli & S. Fidecka, "Effect of Long-Term Hypocaloric Diet on Age-Related Changes in Motor and Cognitive Behavior in a Rat Population," *Neurobiology of Aging* 11 (4) (1990): 417–23; A. Wu, X. Sun & Y. Liu, "Effects of Caloric Restriction on Cognition and Behavior in Developing Mice," *Neuroscience Letters* 339 (2) (2003): 166–8.

[40] Ibid.; E. J. Masoro, "Overview of Caloric Restriction and Ageing," *Mechanisms of Ageing and Development* 126 (9) (2005): 913–22.

Fig. 2.3. Recovery from fever by control and calorie-restricted rats.[41] While the control group showed a fever for six days, the 25% CR group began to recover rapidly after three days, and the 50% CR group showed little evidence of fever at all. This suggests that calorie restriction increases resistance to disease.

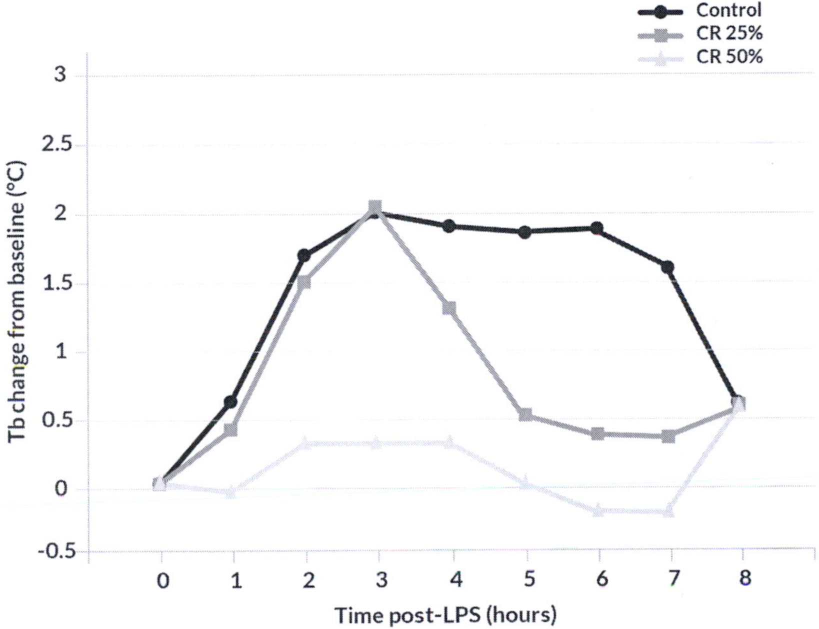

Calorie restriction, exploration and activity

CR animals are more exploratory. As a rule, rats prefer closed to open spaces as a way of avoiding predators, and take time to begin exploring open areas. The key difference is that CR rats are quicker than control rats to venture out. They spend more time in open areas as well as more time exploring novel environments.[42] This behavior is not driven by hunger as

---

[41] L. MacDonald, M. Radler, A. G. Paolini & S. Kent, "Calorie Restriction Attenuates LPS-Induced Sickness Behaviour and Shifts Hypothalamic Signalling Pathways to an Anti-Inflammatory Bias," *American Journal of Physiology. Regulatory Intagrative and Comparative Physiology* 301 (2011): 172–84.

[42] R. F. Genn, S. A. Tucci, A. Thomas, J. E. Edwards & S. E. File, "Age-Associated Sex Differences in Response to Food Deprivation in Two Animal Tests of Anxiety," *Neuroscience & Biobehavioural Reviews* 27 (1–2) (2003): 155–61. Heiderstadt et al., "The Effect of Chronic Food and Water Restriction on Open-

such, because exploration persists for up to ten days after unlimited food is restored.[43] Related studies find that 50% CR rats rear up on their hind legs—a sign of exploration—more often than controls.[44]

Other studies have found CR animals to be more mobile and spontaneously active than controls.[45] Again, this does not appear to be primarily a search for food as it persists even when there is no available food and the "activity" involves running on a wheel. These behaviors are quite distinct and even opposite to the effects of starvation, which typically induces lethargy.

Field Behaviour and Serum Corticosterone Levels in Rats," 20–8; K. Inoue, E. P. Zorrilla, A. Tabarin, G. R. Valdez, S. Iwasaki, N. Kiriike & G. F. Koob, "Reduction in Anxiety after Restricted Feeding in the Rat: Implication for Eating Disorders." Biological Psychiatry 55(11) (2004): 1075–81; E. A. Levay, A. Govic, J. Penman, A. G. Paolini & S. Kent, "Effects of Adult-Onset Calorie Restriction on Anxiety-Like Behaviour in Rats," *Physiology & Behavior* 92 (5) (2007): 889–96.

[43] Inoue et al., "Reduction in anxiety after restricted feeding in the rat: Implication for eating disorders," 1075–81.

[44] A. Govic, E. A. Levay, S. Kent & A. G. Paolini, "The Social Behaviour of Male Rats Administered an Adult-Onset Calorie Restriction Regimen," *Physiology & Behavior* 96 (4–5) (2009): 581–5.

[45] D. K. Ingram, J. Young & J. A. Mattison, "Calorie Restriction in Nonhuman Primates: Assessing Effects on Brain and Behavioural Aging," *Neuroscience* 145 (4) (2007): 1359–64; A. J. Dirks & C. Leeuwenburgh, "Caloric Restriction in Humans: Potential Pitfalls and Health Concerns," *Mechanisms of Ageing and Development* 127 (1) (2006): 1–7; D. Chen, A. D. Steele, S. Lindquist & L. Guarente, "Increase in activity during calorie restriction requires Sirt1, " *Science* 310 (5754) (2005): 1641; J. L. Weed, M. A. Lane, G. S. Roth, D. L. Speer & D. K. Ingram, "Activity Measures in Rhesus Monkeys on Long-Term Calorie Restriction," *Physiology & Behavior* 62 (1) (1997): 97–103; S. Fraňková & R. H. Barnes, "Influence of Malnutrition in Early Life on Exploratory Behavior of Rats," *The Journal of Nutrition* 96 (4) (1968): 477–484;
J. Hebebrand, C. Exner, C. Hebebrand, C. Holtkamp, R. C. Casper, H. Remschmidt, B. Herpertz-Dahlmann & M. Klingenspor, "Hyperactivity in Patients with Anorexia Nervosa and in Semistarved Rats: Evidence for a Pivotal Role of Hypoleptinemia," *Physiology & Behavior* 79 (1) (2003): 25–37; T. D. Moscrip, D. K. Ingram, M. A. Lane, G. S. Roth & J. L. Weed, "Locomotor Activity in Female Rhesus Monkeys Assessment of Age and Calorie Restriction Effects," *The Journals of Gerontology Series A: Biological Sciences and Medical Sciences* 55 (8) (2000): B373–80.

Calorie restriction and fear of predators

Willingness to explore and enter open areas is a standard test for low anxiety in rats, which suggests that CR animals are less anxious. However, studies have shown that 25% CR animals are *more* cautious when encountering cat urine. They are slower to enter areas marked with urine, more likely to flatten their backs and extend their hind paws (signs of caution and risk assessment), and more likely to freeze. They also groom themselves less, which for rats is another indication of caution.[46] All of which suggests that CR rats are more anxious than controls. How can these findings be reconciled?

The key is to recognize that *neither* behavior is indicative of anxiety, but both are adaptive responses to food shortage. When food is scarce and scattered, animals must spend more time and effort seeking it out—even before they feel hunger. This makes them more exploratory. But shortage of food indicates that predators are less of a danger, because if they were then the population would be controlled and food more plentiful. So rather than fight the predator or give warning, the best response is to flee. Both these behaviors are seen in food-restricted populations in the wild. As we saw in the last chapter, animals in such populations spend more time searching for food *and* are quicker to flee from predators—gibbons more than baboons, and monogamous langurs more than polygynous langurs.

Calorie restriction and reduced sociability

The same approach explains another aspect of CR behavior—that CR animals are less sociable. A group of rhesus monkeys which experienced a severe three-week food shortage became less sociable and spent less time grooming and playing than when well fed.[47] A study of men on a 50% CR diet for six months also found them to be significantly less social. And obese women on a severely restricted diet were less likely to indulge in social eating.[48]

---

[46] Unpublished paper.

[47] J. Loy, "Behavioural Responses of Free-Ranging Rhesus Monkeys to Food Shortage," *American Journal of Physical Anthropology* 33 (2) (1970): 263–72.

[48] Keys et al., *The Biology of Human Starvation*; G. D. Foster, T. A. Waddan, F. J. Peterson, K. A. Letizia, S. J. Bartlett & A. M. Conill, "A Controlled Comparison of Three Very-Low Calorie Diets; Effects on Weight, Body Composition, and Symptoms," *The American Journal of Clinical Nutrition* 55 (4) (1992): 811–17.

CR animals are less tolerant of members of their own species, attacking strangers more fiercely and persistently.[49] CR rats initiate interactions with unfamiliar rats sooner than controls. They also engage in more ongoing contact including "walking over" unfamiliar rats, which is a sign of dominance.[50] Traditionally, contact frequency is considered a sign of sociability, but from the above discussion it is clear that it can better be interpreted as wariness of strangers.

Like exploration and fear of predators, these behaviors are adaptive responses to food shortage. When food is scarce and scattered, feeding in smaller groups or as individuals is more efficient. Food shortage also indicates that predators are not as dangerous, so there is less need of larger groups for warning or defense. This means that other members of the species are more likely to be seen as competitors for food, best kept at a distance. These findings are consistent with the observed lower social tolerance of animals living in food-limited environments and their tendency to spread out and form smaller groups. The extreme form of such behavior is the territorial monogamy of the gibbon.

Calorie restriction and better maternal care

CR also improves maternal behavior. Relative to controls, 15% CR and 30% CR females spend more time close to their pups, suckle them more intensively and for longer, and engage in a greater number of maternal activities.[51] Further, females on a 25% CR diet while nursing maintain better nests and are quicker to retrieve pups when scattered.[52]

Levels of parental care vary enormously across species. Some, including most fish, do nothing to protect or nurture their young. Others, of which humans are perhaps the best example, provide protection, nourishment and

---

[49] Vitousek et al., "Caloric Restriction for Longevity," 338–60.

[50] Govic et al., "The Social Behaviour of Male Rats Administered an Adult-Onset Calorie Restriction Regimen," 581–5.

[51] M. K. McGuire, H. Pachon, W. R. Butler & K. M. Rasmussen, "Food Restriction, Gonadotropins, and Behavior in the Lactating Rat," *Physiology & Behavior* 58 (6) (1995): 1243–9; H. Pachon, M. K. McGuire & K. M. Rasmussen, "Nutritional Status and Behavior During Lactation," *Physiology & Behavior* 58 (2) (1995): 393–400.

[52] E. A. Levay, A. G. Paolini, A. Govic, A. Hazi, J. Penman & S. Kent, "Anxiety-Like Behaviour in Adult Rats Perinatally Exposed to Maternal Calorie Restriction," *Behavioural Brain Research* 191 (2) (2008): 164–72.

intensive training for years or decades. Species' levels of parental care correlate with lifespan, environmental stability, body and litter size, and other factors.[53] What CR seems to do is improve parental care so that the young have a better chance of survival in tough conditions, and also provide a competitive advantage in an environment where only the fittest individuals have a chance to breed.

Summary of the effects of calorie restriction on individuals

In the previous chapter we saw that animals in food-restricted environments show differences in behavior that adapt them to the needs of those environments. In this chapter we have seen that all of these behaviors can be explained as a direct response to mild food restriction. Calorie restricted animals become more active and exploratory, more fearful of predators, less sociable, and less tolerant of other members of the species. They reduce mating and breeding but take better care of their young. Their body maintenance systems cause them to live longer. They have reduced testosterone but higher levels of cortisol.

Effects of maternal calorie restriction on offspring

It is now time to consider how restricting the food of mothers affects their adult offspring. The first point to make is that severe food restrictions on mothers can harm their adult offspring. Rats whose mothers experienced 50% CR during gestation and nursing were more easily stressed and showed higher levels of CORT and ACTH as adults.[54] But the mild or short-term CR of mothers has very different effects, as indicated in Fig. 2.4.

---

[53] T. H. Clutton-Brock, *The Evolution of Parental Care* (Princeton: Princeton University Press, 1991).

[54] N. Sebaai, J. Lesage, C. Brenton & S. Deloof, "Perinatal Food Deprivation Induces Marked Alterations of the Hypothalamo–Pituitary–Adrenal Axis in 8-Month-Old Male Rats both under Basal Conditions and after a Dehydration Period," *Neuroendocrinology* 79 (4) (2004): 163–73; N. Sebaai, J. Lesage, D. Vieau, A. Alaoui, J. P. Dupouy & S. Deloof, "Altered Control of the Hypothalamo–Pituitary–Adrenal Axis in Adult Male Rats exposed Perinatally to Food Deprivation and/or Dehydration," *Neuroendocrinology* 76 (4) (2002): 243–53.

Fig. 2.4. ACTH levels in the adult offspring of CR mothers.55 ACTH levels are lower in adults whose mothers were exposed to CR before conception, gestation or lactation. CR restriction was at 25%, except for the pre-conception group whose mothers experienced 50% CR for three days only.

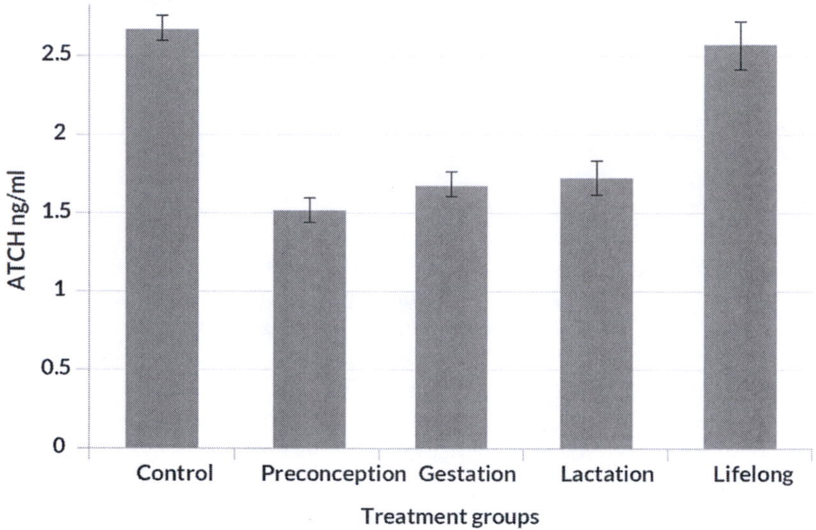

* Denotes a significant difference from the control group at $p<0.001$.

While severe food restriction of mothers is associated with higher levels of ACTH in the adult offspring, mild or short-term food restriction is associated with *lower* ACTH. Fig. 2.5 below finds a similar pattern for CORT.

Not only is this the opposite effect to severe food restriction in early life, it is also opposite to the effect of CR on adults. Adult CR increases CORT, while experience of mild CR in early life reduces it.

[55] E. A. Levay, A. G. Paolini, A. Govic, A. Hazi, J. Penman & S. Kent, "HPA and Sympathoadrenal Activity of Adult Rats Perinatally Exposed to Maternal Mild Calorie Restriction," *Behavioural Brain Research* 208 (1) (2010): 202–8.

Fig. 2.5. Serum corticosterone in in adult rats whose mothers were calorie restricted. CORT levels are lower in adults whose mothers were exposed to CR pre-conception, during gestation or during lactation. CR restriction was at 25%, except for the pre-conception group whose mothers experienced 50% CR for three days only. This effect is *opposite* to that of CR on adults, which increases CORT.

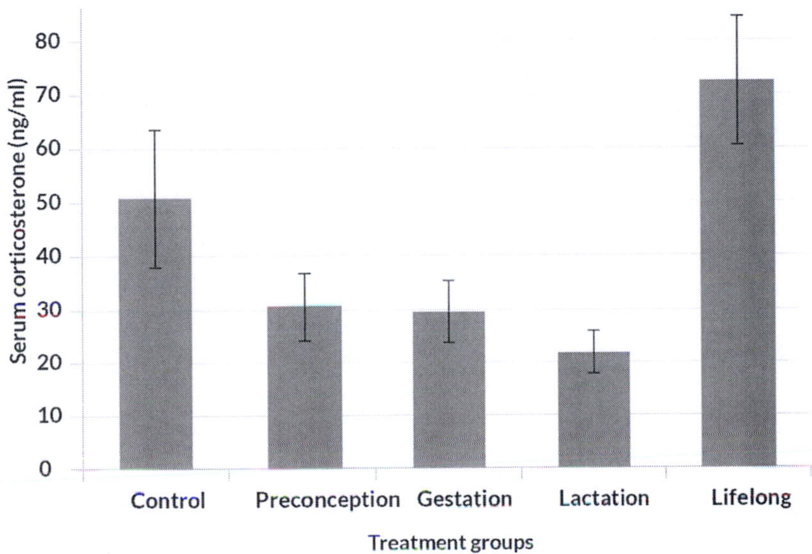

More surprising and significant findings emerge from studies measuring the effects of maternal CR on leptin and testosterone, both of which are reduced by adult CR. Figs. 2.6 and 2.7 show the results of these studies in which maternal CR selectively *increases* the levels of these hormones in adult offspring.

Fig. 2.6. Serum testosterone levels in rats calorie restricted in early life.[56] While CR reduces testosterone in adults, there are indications that maternal CR *increases* testosterone in the adult offspring.

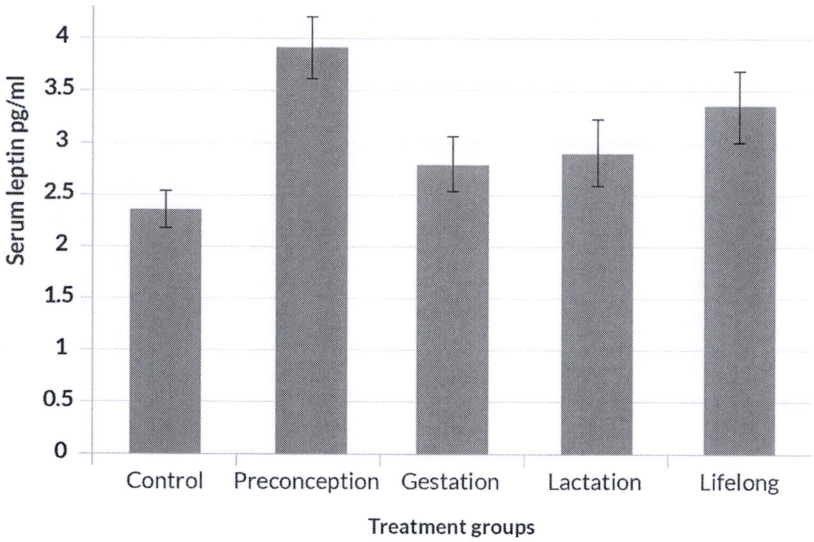

[56] Govic et al., "Testosterone, Social and Sexual Behaviour of Perinatally and Lifelong Calorie Restricted Offspring," 516–22.

Fig. 2.7. Serum leptin levels in adult rats whose mothers were calorie restricted.[57] Calorie restriction reduces leptin in adults, but three days of calorie restriction before a female conceives *increases* leptin levels in her adult offspring.

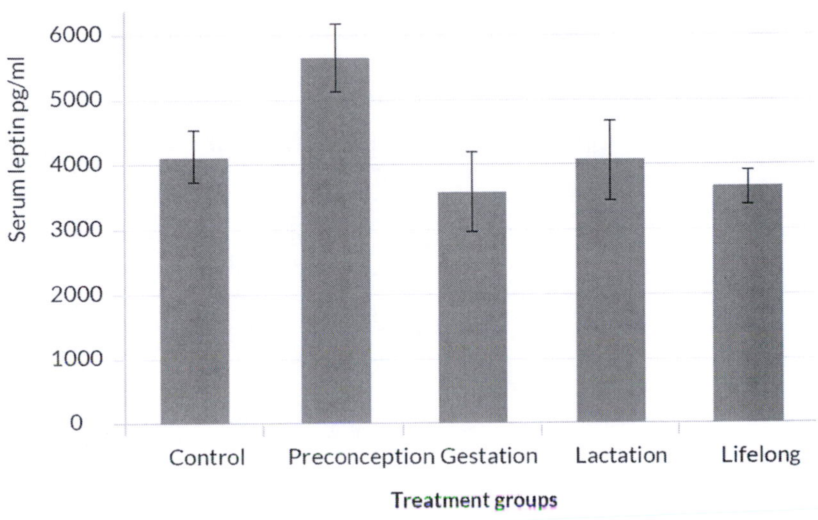

The findings from Figs. 2.4–2.7 are summarized in Table 2.1 below.

Table 2.1. Effects of calorie restriction in adults, and in the adult offspring of CR mothers, on testosterone, leptin, ACTH, CORT, and adrenaline.

	Testosterone	Leptin	ACTH	CORT	Adrenaline
CR in adults	Lower	Lower	Stable or lower	Higher	Variable
Effect on adults of severe maternal CR in infancy			Higher	Higher	Higher
Effect on adults of mild maternal CR in infancy	Higher	Higher	Lower	Lower	Lower

---

57 Govic et al., "Testosterone, Social and Sexual Behaviour of Perinatally and Lifelong Calorie Restricted Offspring," 516–22.

Better maternal care in the form of licking and/or grooming has been linked to lower ACTH and CORT in the adult offspring,[58] but this does not appear to be the reason for these findings. Although females experiencing CR while nursing were better mothers, those experiencing it before conception or during gestation were not, yet the same pattern appears in all three groups.

Effects of maternal calorie restriction on offspring behavior

The adult offspring of CR mothers clearly have different hormone patterns from controls. Their behavior also differs. Studies have found that 50% CR during gestation and lactation cause adult offspring to become more aggressive and dominant, as indicated by walking-over behavior in which animals crawl over one another. These animals are also more active in initiating interactions and responding to partners than controls[59]—behaviors which last well beyond the period of food restriction.

In our experiments, adult CR rats spent more time in the open arm of a maze than controls. This behavior is taken as indicating greater interest in exploration. Studies of 25% maternal CR found no difference in this test. However, the combined results from pre-conception, gestation, and lactation groups revealed fewer entries into the dangerous (for rats) central and middle zones of open fields during the first five minutes of testing.[60] The pre-conception group was lowest on entries and slowest to emerge into the open. This behavior profile is again opposite to that of adult CR.

---

[58] D. Liu, J. Diorio, B. Tannenbaum, C. Caldji, D. Francis, A. Freedman, A. Sharma, D. Pearson, P. M. Plotsky & M. J. Meaney, "Maternal Care, Hippocampal Glucocorticoid Receptors, and Hypothalamicpituitary-Adrenal Response to Stress," *Science* 277 (5332) (1997): 1659–62.

[59] Govic et al., "Alterations in Male Sexual Behaviour, Attractiveness and Testosterone Levels Induced by an Adult-Onset Calorie Restriction Regimen," 140–6; D. A. Levitsky & R. H. Barnes, "Nutritional and Environmental Interactions in the Behavioural Development of the Rat: Long-Term Effects," *Science* 176 (30) (1972): 68–71; , T. S. Whatson, J. L. Smart & J. Dobbing, "Social Interactions Among Adult Male Rats after Early Undernutrition," *British Journal of Nutrition* 32 (2) (1974): 413–19.

[60] Levay et al., "Anxiety-Like Behaviour in Adult Rats Perinatally Exposed to Maternal Calorie Restriction," 164–72.

Other studies revealed that adult CR rats are more active in investigating strange animals, a behavior that was also present in the adult offspring of the lifelong and gestation groups. The lifelong group also showed more dominance behavior such as pinning, walking over, and wrestling. (Pinning is one animal holding another down, with the latter's chest touching the floor.) This is consistent with the observed effects of adult CR.

However, the offspring of the pre-conception group spent *less* time wrestling than the offspring of the control group. Again, this is opposite to the effect of adult CR and opposite to the finding noted above, that the adult offspring of 50% CR mothers were more aggressive. The point to emphasize here is that the effects of mild maternal CR on offspring are in most ways the opposite to those of adult CR. Adult CR makes animals more willing to explore but their adult offspring are less willing to explore. This outcome depends on the offspring receiving plentiful food in later life, however. The offspring of the lifelong CR group, which also experienced food shortage as adults, behaved much like the adult CR group. They were more exploratory and more likely to investigate and dominate strangers by walking over and pinning them.

Effects of maternal calorie restriction on offspring sexual behavior

Recall that adult CR reduces interest in sex and sexual activity, which is consistent with the finding of reduced testosterone. However, mild maternal CR has the opposite effect here also. With 25% maternal CR, a clear increase is observed in sexual activity among the offspring of mothers experiencing calorie restriction in the pre-conception period. These subjects also showed the biggest rise in testosterone.[61] All maternal CR groups started copulation by mounting the females sooner than animals from the control group. Further, the lactation and pre-conception offspring reached ejaculation sooner and more often than control subjects. The only offspring group not to show increased sexual activity was the lifelong group, which was similar in its behavior to the control group. Because adult CR reduces sexual activity and maternal CR increases it, the effects cancel each other out.

---

[61] Govic et al., "Testosterone, Social and Sexual Behaviour of Perinatally and Lifelong Calorie Restricted Offspring," 516–22.

Effects of maternal calorie restriction on gene expression

We now turn to possible causes. How is it that the same stimulus, mild CR, can have opposite effects on both hormones and behavior when experienced in early and later life? This brings us to gene expression, or epigenetics.

Genes have their effects on the body by the production of proteins. Proteins can be in the form of hormones, such as testosterone or cortisol, or other substances such as enzymes. Various factors affect gene activity, meaning how much (if any) of a protein is produced (or 'expressed'). Recent research has shown that the activity of genes can be influenced by the environment, especially in early life. This is called epigenetics. An epigenetic effect occurs when a stimulus in the environment changes the activity of a gene and the activity is passed on in cells when they divide. One way this happens is through the attachment of methyl groups to sections of DNA, a process known as methylation. Heavily methylated sections of the genome tend to be less active in producing proteins.

In some situations epigenetic effects can be 'inherited' in the sense that they transfer directly from parent to child, although the effects fade over generations if the environmental stimulus fails to continue. For example, one study from Sweden found that the experience of famine or abundance during late childhood (roughly ages 9–12 in boys and 8–10 in girls) had a major impact on the lifespan of grandchildren. Boys whose paternal grandfathers experienced famine lived far longer than those whose grandfathers had access to ample food.[62] In other words, early life experience had an epigenetic effect that carried across two generations. In a similar study, epigenetic effects were observed among individuals conceived during the Dutch famine of 1944. Overall, they were more likely to suffer from obesity and those affected during the second trimester had a higher incidence of schizophrenia.[63]

Studies conducted by the author and his associates have been more limited in scope. Much of our work has been directed at identifying the effects of early experience on adult character. This research has also focused on the effects of mild food restriction rather than extreme deprivation states such

---

[62] L. O. Bygren, G. Kaati & S. Edvinsson, "Longevity Determined by Paternal Ancestors' Nutrition During their Slow Growth Period," *Acta Biotheoretica* 49 (1) (2001): 53–49.

[63] Vitousek et al., "Caloric Restriction for Longevity," 338–60.

as starvation. Currently we are measuring the activity level of a number of genes with known effects on hormones.[64] While this research is in a preliminary stage, early findings point to profound epigenetic effects on the offspring of calorie restricted rats.

For example, one gene strongly affected by maternal CR is the androgen receptor. Similar to a thermostat, it reacts to high levels of testosterone by shutting down testosterone production, thereby preventing testosterone levels from rising beyond a certain level. Maternal CR appears to down-regulate or dampen the activity of this gene in offspring, which should allow testosterone to rise higher than otherwise. This may explain why maternal CR is associated with higher testosterone and greater sexual activity in the offspring.

An interesting finding from this study is that the pre-conception, gestation and lactation groups all show a significant reduction in the activity of this gene compared to controls. The lifelong group is an exception in that it does not show a reduction, which raises the possibility that CR in the later life of offspring may partially reverse the effect of maternal CR on the androgen receptor.

Our results suggest that the epigenetic effects of maternal CR are complex yet can be significant, in some cases doubling and in other cases halving the activity level of specific genes. Changes of this magnitude can account for many of the observed changes in hormones and behavior.

The one critical point stemming from these studies is that early CR has an effect which is dramatically different and even opposite to the effects of CR in later life. In later chapters, especially chapter 5, we discuss evidence that CR-type influences on infants can have quite different and even opposite effects to those in older children (ages 6-12) and adolescents. Such differences provide a basis for understanding many aspects of human social behavior.

Summary and conclusions

This chapter has reviewed studies on the effects of mild food shortage on laboratory rats. One key effect of CR is reduction in testosterone, a hormone associated with aggression, sociability, and sexual activity. CR

---

[64] J. Penman, et al., Report Series on Studies on Rats Conducted at Latrobe University, 2012, (forthcoming).

also reduces leptin, a hormone associated with sexual activity. In neither case does this appear to be a stress reaction.

In terms of behavior, CR rats tend to be less sexually active. They explore more actively and are more willing to enter the exposed arm of a maze, behaviors which are usually taken as signs of low anxiety, yet they are more fearful of cat urine. They are less sociable and quicker to investigate and climb over strange animals (a sign of dominance).

On balance, female rats exposed to CR become better mothers. They spend more time their young, build better nests, and are quicker and more efficient at gathering pups back into the nest.

The effect of mild maternal CR on offspring is not only different from that of adult CR but in some ways opposite. In particular, offspring show less evidence of stress. (By contrast, severe maternal CR results in infants with higher levels of stress as adults). The offspring of CR mothers were also less exploratory but more sexually active than controls, behaviors that are the opposite of the effects of CR on adults. The offspring also showed a dramatic change in the activity level of certain genes, although there are indications that later CR may partially reverse these effects.

These findings indicate that the behavior of animals in food-restricted environments, as described in the last chapter, can be explained largely as a direct physiological response to mild hunger. The experience of mild food shortage brings about exactly the kind of changes that are adaptive in such environments. It causes animals to spread out, be more active in their search for food, flee from predators, limit and delay breeding, and provide better parental care.

In the next chapter we discuss how these biological systems can help explain much that is otherwise puzzling and inexplicable in human societies. For example, people in more complex societies tend to show family and social patterns characteristic of food shortage, even when food is plentiful. Occupationally successful people tend to have lower testosterone than then those who are low-skilled and unemployed, and testosterone levels may drop as food becomes more plentiful, findings that are opposite to observations in animals. It also helps explain why our species struggles so much with sexual behavior. Finally, we will see how changes in attitude and behavior associated with food shortage underpin civilization.

CHAPTER THREE

THE CIVILIZATION FACTOR

In chapter one we saw that certain forms of family and social behavior are characteristic of societies with large political units and complex economies. These behaviors include nuclear monogamous families, curbs on sexual behavior, late marriage and control of children. We proposed that, among many non-human primate species, similar behaviors can be explained as a direct response to food shortages which allow animals to adapt to food limited environments. When food is chronically short, social bonds break down so that troop size decreases. At the extreme, as with gibbons, a single mated pair with their immature young defends an exclusive territory. As part of the behavioral adjustment to scarce food, animals more frequently patrol their territory to repel possible intruders and protect their food supply. Such behavior is carried out at the expense of rest, socializing and sex. In the sense that they spend more time searching for food, animals in food-restricted environments can be seen as more industrious. They also make better parents, spending more time with their young and looking after them more carefully. And they are more likely to flee from predators.

In chapter two we reviewed studies of the biochemical effects of calorie restriction (CR) primarily on laboratory rats. We showed that restricting food to females before and after giving birth (maternal CR) has epigenetic effects on their offspring. Maternal CR affects the expression of certain genes which leads to changes in the physiology and behavior of offspring in later life. Maternal CR has a similar effect to adult CR in rendering animals more exploratory, but it can also have opposite effects such as increasing testosterone. It is important to note that these effects are quite distinct from those of stress, such as might result from starvation. For example, mild adult CR *reduces* levels of the stress hormone ACTH, while severe stress *increases* it. And, maternal CR also reduces ACTH levels in adult offspring.

Relating animal studies to humans

How do these findings relate to human behavior? In particular, why do people in civilized societies behave—in some ways—as if they are short of food? And how can they do so even when food is relatively plentiful?

A useful departing point for answering these questions is the well-established finding that CR reduces testosterone in both humans and animals. Moreover, the studies cited in chapter two show that humans with low testosterone behave differently than do those with higher testosterone. They engage in significantly less sexual activity, physical aggression and face-to-face confrontation. They have more stable marriages, stronger interest in their children, and are more likely to closely govern the behavior of their children. Further, they tend to control their own impulses, are less likely to use drugs, and are less outgoing and sociable.

Crucially, they are also more likely to be employed and to excel in occupations that require repetitive working routines and delayed gratification, such as farming or white collar occupations. Successful high testosterone individuals tend to sports stars or entertainers—professions that employ relatively few people.[1] Thus, in civilized societies, *low* testosterone is more likely to be associated with economic success.

The behavior of humans with low testosterone is in many ways similar to that of food restricted animals. Like low-testosterone humans, food restricted animals are less sociable, more interested in their offspring, more timid, and more likely to form exclusive pair bonds. The greater capacity of low testosterone people to work and to do so for future rewards also has analogs in animal behavior. These include low impulsivity and a greater willingness to explore and search for food even when not hungry. Comparable behavior is observed among hunter-gatherers and primitive horticulturalists where food is chronically scarce, in that monogamous marriage and paternal involvement in the provision of food are more common in populations living in these environments.[2] These findings

---

[1] J. M. Dabbs, *Heroes, Rogues and Lovers: Testosterone and Behaviour* (New York: McGraw Hill, 2000).

[2] P. Draper & H. Harpending, "A Sociobiological Perspective on the Development of Human Reproductive Strategies," in *Sociobiological Perspectives on Human Development*, edited by K. B. MacDonald, 340–372 (New York: Springer, 1988); D. C. Geary, "Evolution and Proximate Expression of Human Paternal Investment," *Psychological Bulletin* 126 (2000): 55–77; H. S. Kaplan & J. B. Lancaster, *An*

suggest close similarities between humans and other primates in the effects of CR.

However, as also mentioned in chapter one, CR *behavior* in humans is usually associated with advanced and complex societies in which food is not always scarce. To account for this apparent contradiction it is necessary to recognize some new factors, as findings from English history illustrate.

Food-restricted behavior and economic success

Turning to England of centuries past, the individuals most likely to restrict sexual activity and form nuclear monogamous families were members of the industrious middle classes who were unlikely to be short of food and who had above-average numbers of children. In his well-researched book A Farewell to Alms, Gregory Clark documents how for hundreds of years better-off families tended to have more surviving children than the poor. For example, a study of English wills in 1585–1638 establishes that men with an estate of more than £100 had 40% more children than those with smaller estates. Occupation accounted for only one fifth of the difference in wealth. In terms of surviving children, wealth mattered far more than education or literacy.[3]

The story does not end here. Differences in wealth occur for a reason. Contrary to popular opinion, wealth in the past was only partly the result of inheritance. Pre-modern England had a relatively high rate of social mobility.[4] There was a ready market in land and it was possible for an energetic and frugal laborer to buy property and move up the social ladder.[5] But the main form of social mobility (in contrast to much of the modern world) was downward, because wealthier families had more

---

Evolutionary and Ecological Analysis of Human Fertility, Mating Patterns, and Parental Investment, in "Offspring: Human Fertility Behavior in Biodemographic Perspective," edited by K. W. Wachter & R. A. Bulatao, 170–223 (Washington, DC: National Academies Press, 2003).

[3] G. Clark, *A Farewell to Alms: A Brief Economic History of the World* (Princeton, NJ: Princeton University Press, 2007), chapter six.

[4] Gregory Clark's extensive study of English surnames suggests that levels of social mobility have not changed significantly since the Middle Ages. G. Clark, *The Son Also Rises: Surnames and the History of Social Mobility* (Princeton: Princeton University Press, 2014).

[5] Clark, *A Farewell to Alms*, chapter eight.

surviving children. This meant that their property tended to be divided among a greater number of heirs so that each child received less. A study of seventeenth century wills in Suffolk records that at the time of their deaths, nearly half the sons of higher-class testators ended up with less wealth than their parents.[6]

The effect of downward mobility can compound over generations. For example, of 70 families owning small parcels of land in Halesowen in the years 1270–82, only 25 had any direct descendants owning land in 1348.[7] The rest of the descendants had lost their lands, emigrated, or their line had died out.

Better-off men were more likely to have better-off sons. Why this was so Clark does not explain, but a good possibility is that it had more to do with inheritance of character and work ethic than inheritance of wealth—points addressed in detail in later chapters. In one study of 147 father-and-son pairs, the sons with multiple brothers and sisters (and thus children with potentially less wealth to be inherited) were just as likely to end up as well-off as those with fewer siblings.[8] Given their system of primogeniture, aristocrats were able to maintain family wealth without a powerful work ethic, but this was only a small segment of the population.

Thus, for the great majority of people, the only way to maintain family wealth was through the CR-related behaviors of industry and frugality. These people were not wealthy by modern standards or even those of the time, but they were less likely than the poor to be short of food. And yet, in a seeming paradox, it was this group which showed the industry and frugality which we have associated with food shortage.

Introducing "C"

Here, a new terminology is needed. Because CR-like behavior is only partially related to food shortage in human societies, we need a term for the biological system that responds to environmental stimuli and leads to CR-like changes in temperament and behavior. In human societies this system is strongest in large states with complex economies, or what we commonly refer to as "civilizations." Because of this we will refer to it as

---

[6] Ibid., chapter six.

[7] Z. Razi, "Family, Land and the Village Community in Later Medieval England," *Past and Present* 93 (1981): 3–36.

[8] Clark, *A Farewell to Alms*, chapter six.

"C." Civilized societies, and animals chronically short of food, tend to be high C. Smaller-scale societies, and animals with plentiful food, tend to be low C.

The advantages of low C

In chapter two we discussed how behavior linked to food restriction in animals—now identified as low C behavior—can be beneficial to humans in some situations. For example, low C men have high testosterone. They tend to be dominant in face-to-face interactions, attractive to women, are often successful in sports and entertainment, and are bolder if less-disciplined soldiers.

Low C is well suited to life as a hunter-gatherer. In these societies men have little need to do routine work. Studies of 13 societies in which people live by hunting, gathering and shifting agriculture document that men work an average of 5.3 hours per day compared, for example, with 8.2 hours per day in Great Britain in 1800.[9] In such societies men have little need to do routine work and a great deal to gain by developing strong social bonds. The most successful men will be those who are aggressive, dominant and impulsive, good at hunting and war, able to form strong bonds with their fellows, and skilled at seducing fertile women. In these societies lower C men, like many dominant animals, may have shorter lives but more offspring.

The same points apply to the struggle for survival between small groups. Hunter-gatherers and small-scale agricultural societies are usually insecure as a result of constant feuds and raids. Under these conditions, bonds between group members and group defense are vital for survival and offer a distinct advantage over scattered families who are less able to cooperate with each other.

Nor is there any need to spread out in search of food. Most hunter-gatherers, and early-stage agriculturalists who practiced herding and horticulture, could feed themselves with a few hours of work per day, leaving ample time for socializing, leisure and sex. Constant small-scale warfare is one reason why these populations rarely increased in size to the point that their food supplies were compromised. In his book on Yanomamo villages in the Amazon basin, W. J. Smole explained that

---

[9] Ibid., chapter three.

people needed a much larger territory than their subsistence requirements so that they could hide from enemies.[10]

Viewed this way, our ancestral environment was more like that of a baboon troop with abundant food, rather than that of a gibbon family searching out scarce and scattered food in a tropical forest. In short, most hunter-gatherer societies bear a closer resemblance to cooperative baboon troops than to competing and intolerant gibbon families. The one key difference relates to sexual behavior, as discussed in chapter one. Primate species like baboons that form multi-male troops tend to be promiscuous, with dominant males getting better access to fertile females. Hunter-gatherers are more likely to indulge in promiscuity or polygyny than people in civilized societies, but they also develop strong pair bonds. Apart from this one area, however, humans are a naturally low-C species—sociable, cooperative and not especially hard-working.[11]

High C and farming

The low-C temperament was ideally suited to the needs of our hunter-gatherer ancestors, but this changed when people began to settle down and become farmers. The rise of farming is usually seen as stemming from the development of food crops and domesticated animals. To a large extent this explains why agriculture first arose in areas such as the Middle East that had wild plants and animals suited to domestication, an interpretation Jared Diamond proposed in his seminal work *Guns, Germs and Steel*.[12] But farmers also need to have the *temperament* for routine hard work, and to plan carefully for the future by, for example, selecting and putting aside quality seeds for next year's crops. This requires a different temperament

---

[10] W. J. Smole, *The Yanomamo Indians: A Cultural Geography* (Austin: University of Texas Press, 1976), 210.

[11] Ester Boserup, in *The Conditions of Agricultural Growth* (1965), and George Zipf, in *Human Behavior and the principle of least effort* (1965), have noted that people are not naturally inclined to produce surpluses, but instead follow a "law of least effort", preferring to complete their subsistence activities by expending a minimal amount of time and energy, enjoying leisurely activities rather than making the most productively of their time. According to Stephen Sanderson, "humans everywhere are psychobiologically constructed so as to follow a Law of Least Effort;" see *Social Transformations: A General Theory of Historical Development* (Oxford, Cambridge, Mass: Blackwell, 1995), 342.

[12] J. Diamond, *Guns, Germs and Steel: The Fates of Human Societies* (London: W.W.Norton, 1997).

from that of a hunter-gatherer—one with higher C. Thus, while low-C behavior is beneficial in hunter-gatherer societies, the advantage shifts to higher C with the rise of farming.

An example of the importance of temperament in performing routine work is found in the behavior of the forest-dwelling Mbuti pygmies of Central Africa. In recent times, Bantu-speaking farmers have colonized parts of the forest by converting land for agriculture. The Mbuti appreciated the benefits of farming but were far less keen on the hard physical work it demanded. As a result, they only rarely worked in the fields and were more likely to be employed as plantation guards.[13]

The Mbuti simply did not have the temperament for the monotonous tasks involved in farming. This same type of temperament can be seen in hunter-gatherers at the fringes of Western society. Jack McLaren's account of an attempt to run a coconut plantation in north Queensland a century ago indicates the problems that arose from the use of aboriginal labor. The workers were keen on trade goods but reluctant to work for more than a day at a time, and were inconsistent even on the days they did work. They tended to doze the afternoon away if permitted, and would down tools at any time to dig yams or chase a wallaby, needing constant supervision.[14]

Hunter-gatherers have proved far more adept at herding cattle, which has a great deal in common with hunting, than more monotonous tasks such as farming. For example, aboriginal Australians were accustomed to hunting emus by herding them into purpose-built corrals, and so took readily to stock work.[15] Pastoral work is clearly more congenial to a low-C temperament than farming.

The biological basis of work

But why should a temperament that adapts animals to food-restricted environments suit human beings to routine work such as that involved in farming? The answer can be found in studies showing that CR animals search for food even when they are not hungry, as suggested by the more active exploratory behavior of calorie restricted rats. In human terms, they

[13] C. M. Turnbull, *Wayward Servants: The Two Worlds of the African Pygmies* (Westport: Greenwood Press, 1976), 39–40.

[14] Henry Reynolds, *The Other Side of the Frontier: Aboriginal Resistance to the European invasion of Australia* (Ringwood; Harmondsworth: Penguin, 1982), 143.

[15] Ibid., 159.

are "working" for future, rather than immediate, benefit. This is exactly what a farmer must do—preparing the field now in the expectation of food harvested several months in the future. It is also what a student does, sacrificing present consumption for a better income in the future. Such activity may also be at the expense of socializing and sex, and in this sense a hard-working human also acts much like an animal in a food-restricted environment.

It is even possible that the human race was prepared for the rise of farming by a genetic change to slightly higher C. Between two-hundred and fifty thousand years ago, human skulls became rounder and brow ridges less prominent, both signs of lower testosterone.[16] The change could have been driven by the benefits of slightly higher C, such as in making possible greater care by fathers. This could help explain one of the greatest puzzles of the human past—why agriculture developed in several different regions within a few thousand years, using completely different plants and with no plausible contact between at least some of them. Points of origin include the Middle East, East Asia, New Guinea, the African Sahel and Meso-America. By modern standards all humans were low C before the rise of agriculture, but they may have been slightly higher C than in the distant past.

The biological basis of the market

Civilized peoples not only work harder at routine tasks, they are also more likely to relate to each other through the market. Relating this to biology is more difficult. No animal has the intellectual equipment to build and operate in a market economy, which involves spending effort to earn something that only has value in exchange for something else. But speculating for a moment, what kind of temperament would allow a rat or monkey to flourish in a market economy once it developed the intellectual capacity to build one? A possible answer can be found in the concept of impersonality. A territorial animal is oriented towards a piece of ground rather than to other members of the species outside the nuclear family. This is a relatively impersonal attitude, and it differs from that of hunter-gatherer bands and primitive agriculturalists. These are territorial in the sense that they drive intruders away from their tribal territory. But their

[16] *Science Daily*, "Society bloomed with gentler personalities, more feminine faces: Technology boom 50,000 years ago correlated with less testosterone," (August 1, 2014).

primary relationships are with individuals within the band, and these tend to be close and reciprocal. For example, a successful hunter shares the meat from his kill in the knowledge that others will do likewise with their kills. Or, as in Melanesia, a man may give away food during competitive feasts to gain status.[17] These primary economic relationships are personal between the people that participate.

Money and market economies, on the other hand, mean distributing goods in ways that are largely independent of personal relationships. They are *impersonal*. If someone has money they can usually buy the goods even if not related to the vendor. It is not necessary that the buyer even know the vendor prior to the transaction. This impersonal exchange is critical to the way that market economies work. A stall holder in a public market would not last long if they gave away produce to friends and family and refused to sell to strangers.

All known human societies engage in trade. Tribal peoples, for example, are quite capable of trading with people outside their band and tribe, and at times even with enemies. Margaret Mead gives many examples of such trade in New Guinea, including the Mundugumor buying mosquito nets and the Tchambuli selling baskets and buying captives.[18]

But people in less complex societies tend to focus on more personal forms of distribution, with both temperament and culture making it difficult to adjust to the market economy. For example, present-day Indigenous Australians often find it difficult to accumulate property because of the obligation to share their money with close kin. A study of urban Aboriginal households in Melbourne during the early 1960s reveals a culture of high reciprocity where money, meals, tasks and leisure time were shared not only between the extended family but also with boarders and acquaintances. Aboriginal notions of decent behavior required them to help anyone who asked, especially if young children were in need. This was also a response to help they had received in the past, and to ensure

[17] D. L. Oliver, *A Solomon Island Society* (Cambridge: Harvard University Press, 1955).
[18] M. Mead, *Sex and Temperament in Three Primitive Societies* (New York: Morrow Quill, 1963).

future support for themselves.[19]

Refusal to give or share known resources commonly resulted in public accusations and loss of face. For this reason, individuals not wanting to share assets would hide them. While reciprocal behavior may be commendable it is too personal to be effective in a market economy.[20] Market behavior requires a more impersonal temperament—one that strengthens bonds with the nuclear family and weakens ties with others, including close kin. At the extreme, it is the monogamous territoriality of a gibbon.

With this in mind one might speculate that gibbons with the intellect of a human could make effective traders because their high-C temperament means they lack strong preferences for one neighbor over another. They could offer to swap a particular type of nut in their own territory for a desired fruit in another. If only one of their neighbors has the desired fruit then that would be the one to trade with. And, supposing that they had the ability to travel more widely while still preserving a territory, they might then make such exchanges with a wide range of other gibbons

Personal relationships do matter in a market economy, but those that do are based on trust. A stallholder who sells bad fruit will not get my business tomorrow. Or in gibbon terms, if I gave you nuts and you fail to hand over fruit, I won't deal with you again. But it is a relationship created by an impersonal exchange which helps to facilitate future exchanges.

A baboon with human intelligence, on the other hand, would probably be a less effective trader than a gibbon. The relationships of a baboon are personal, primarily with other members of the troop. What matters is whether an animal is friend or foe, higher or lower in the troop hierarchy, or a potential mate. A more powerful animal could take food and a less powerful one may be forced to give it up. Friends might share food and thus build a coalition. An enemy might be challenged for it. A potential mate might be presented with it in return for sex. Baboon troops may have vaguely defined home ranges but are not territorial as such. There is no

[19] D. Barwick, "A Little More than Kin: Regional Affiliation and Group Identity among Aboriginal Migrants in Melbourne." Unpublished PhD thesis (Canberra: Australian National University Canberra, 1963).

[20] R. G. Schwab, "The Calculus of Reciprocity: Principles and Implications of Aboriginal Sharing," Centre for Aboriginal Economic Policy Research Discussion Paper 100 (1995): 8–10.

attachment to anything as impersonal as a piece of ground. Baboon society is broadly similar to the way that hunter-gatherer societies work, because both have low-C temperaments and are thus personal rather than impersonal in orientation.

Other explanations for the rise of agriculture

There have been suggestions in recent decades that the rise of such behavior with the development of agriculture has been at least partly driven by genes, especially given the evidence that humans have been evolving rapidly over the past 10,000 years in areas such as lactose tolerance and disease resistance.[21] While such a possibility cannot be totally dismissed, C-type behavior and attitudes could not be primarily genetic in origin. C can rise very fast in the general population, as happened in Europe between 1200 and 1850 (see chapter seven). Given that hard-working and thrifty people tended to have more surviving children, as indicated by Gregory Clark, it is barely possible that genes could have changed enough to account for the rise in C during that period. But the *fall* of C can be very much faster (chapters twelve, thirteen and seventeen) and there is no possible way that genetic change could account for this. Changes in C must therefore be primarily *epigenetic* rather than *genetic* in origin.

One of biohistory's fundamental hypotheses is that the political structure of a society reflects the temperament of its people, especially those who are high in status. Thus, it is the increase of the higher C temperament rather than any other economic or social factor that makes the development of more powerful and organized states possible. These states in turn have a clear advantage over weaker and more fragmented groups. An extreme example is the way high-C European peoples overran the lower C indigenous groups of North America after 1607 and of Australia after 1788. Other examples include the southward migration of northern Chinese farmers in the centuries bracketing the birth of Christ, and the expansion of agricultural Bantu speakers into southern Africa over the past thousand years.

---

[21] G. Cochrane & H. Harpending, *The 10,000 Year Explosion: How Civilization Accelerated Human Evolution* (New York: Basic Books, 2009); Nicholas Wade, *A Troublesome Inheritance: Genes, Race and Human History* (New York: Penguin, 2014).

Other benefits of high C

The advantages of higher C are not limited to making people better farmers or traders. Individuals with low testosterone levels are less aggressive and more law-abiding. As states become centralized and increasingly require more ordered behavior, violence and disregard for rules may bring punishment and even death, rather than extra breeding privileges. In effect the more advanced a society, the greater the advantages of higher C.

High C also plays a role in maintaining populations by increasing the care of children. Recall that CR mothers spend more time with their offspring, and low testosterone men are more devoted fathers. Better parental care may be a necessity because, as archaeological evidence shows, when people shift their lifestyle from hunting and gathering to farming their standard of living drops. People become shorter and malnutrition and disease increase.[22] When populations are dense, disease spreads more easily, especially when poor hygiene contaminates food and water supplies. In this harsher environment, children require the support and hard labor of both parents if they are to have the best chance in life. It is not surprising then that hard-working family men tend to have more surviving offspring than careless philanderers, a point suggested earlier in the discussion of English wills.

Later chapters present evidence that levels of C have tended to rise steadily over the past ten thousand years. States have increased in size, population density has increased, and trade networks have become larger. But what would cause C to rise so high? The early stages of agriculture did not require a dramatic change in temperament, though the Mbuti found even the early stages difficult to sustain. Further, lifestyles do not change overnight. The early stages of the adoption of agriculture were gradual, starting with the intensification of foraging and taking minute steps into farming and the domestication of animals. Foragers who adopted agriculture could hardly have been aware that they were embarking on a "revolution," but as population numbers increased and farming became a more serious and full time occupation, the potential value of a higher C temperament increased.[23] Not only does elevated C give individuals and

[22] Renfrew, C. and P. Bahn (2004). Archaeology: Theories, methods and practice. London, Thames & Hudson, p. 460.

[23] Esther Boserup makes the point that hunter-gatherers tend to accept a materially poorer lifestyle even when farming technology is available, because they dislike

families an advantage over their neighbors with lower C, it also benefits whole societies in their competition with other societies. As farming and other technologies develop, societies capable of raising C among their members gain an advantage over their rivals. Such societies are more productive and more likely to create an economic surplus which can be spent on military and other technologies. They also readily organize into larger and more effective political units.

Then there is the issue of violence. Populations increase, not only because people work harder and adopt better technology, such as improved food crops and more efficient farming methods, but also because there are fewer deaths from violence. For example, even before 1350, it is estimated that English males were about one-tenth as likely to die from violence (including wars) as the average hunter-gatherer. Death by violence was even less likely by the eighteenth century.[24]

This causal mechanism works both ways. Not only does higher C make possible larger states and more complex economies, but the rise of larger states (for example, when imposed by a more powerful neighbor) causes competitive pressures that increase the value of a high C temperament.

Physical technologies such as better plants and metal-working can be seen as acting in tandem with this change in temperament. They are not only more likely to be developed by well-organized, higher C peoples, but they increase the value of the high-C temperament. For example, a farmer with domesticated animals and quality seeds can use land far more productively than a slash-and-burn horticulturalist, allowing greater population densities. One important reason for the rapid advance of European settlers in North America and Australia is that they were far more numerous than the indigenous peoples in the area of settlement. By contrast, the Norse settlers in Newfoundland around 1,000 A.D. were too few to gain a foothold, despite superior weapons and a more advanced political organization.

---

the effort involved in horticulture. She believes population pressure was a key reason for the development of farming.

E. Boserup, V. D. Abernathy & N. Kaldor *The Conditions of Agricultural Growth: The Economics of Agrarian Change under Population Pressure* (London: Allen & Unwin, 1965); E. Boserup, *Population and Technological Change: A Study of Long-Term Trends* (Chicago: University of Chicago Press, 1981).

[24] Clark, *A Farewell to Alms*, chapter six.

Biohistory is compatible with technological explanations for Western success, such as expressed in Jared Diamond's *Guns, Germs and Steel.*[25] Diamond proposes that civilization in the Americas was held back by the lack of animals that could be easily domesticated, and the north-south configuration of the Americas meant that there was less opportunity for crops to spread east and west to areas of similar climate. Fewer crop varieties and lack of domesticated animals meant that the high C temperament was less of an advantage and so slower to develop. Similar explanations have been given for the slower development of sub-Saharan Africa including lack of suitable crop varities, the effect of tsetse flies on cattle, and the lack of navigable waterways which limited communication and trade.[26] But Africa at least had contact with the advanced cultural technologies of the Middle East. The Americas lacked such an advantage, and with less land area than Eurasia contained fewer competing societies to develop cultural technologies. Thus their native peoples were largely wiped out or absorbed, while Africans still dominate Africa.

How culture supports C

Allowing that higher C provides advantages to individuals and groups, how might it come about? The short answer is that cultures have evolved norms of behavior that persuade and/or pressure their members to act in ways that raise C. Biohistory labels these methods of elevating C as "C-promoters," and the key point to note about C-promoters is that they are usually associated with *religion*.

Fasting

One obvious C-promoter is fasting, which is a feature of many religious systems. Jews have been fasting for more than 3,000 years. Catholic and Orthodox Christians are expected to observe the Lenten fast during the weeks leading up to Easter. Other Christian groups have developed their own traditions, such as Mormons who are expected to fast for 24 hours once each month. These practices are not confined to Judaism and Christianity. Buddhist monks and nuns are expected to not eat after midday, and Hindus, Jains and Bahais prescribe fasting at different times of the year.[27]

---

[25] Diamond, *Guns, Germs and Steel*.

[26] T. Sowell, "The Geography of Africa" *The Thomas Sowell Reader* (New York, Perseus Books, 2011)

[27] E. Westermarck, "The Principles of Fasting," *Folklore* 18 (4) (1907): 391–422.

The most rigorous and best-known example is the Muslim A gradual rise in both C and V culminated in the Arab conquest and the rise of Islam, seen as an especially powerful and durable cultural system which promotes long-term success at the expense of economic progress. A clear implication is that the Muslim populations of the Middle East will spread and gradually assimilate most of the world into their own faith and culture, beginning with Europe. Patriarchy and purdah, not liberal democracy, will be the true "end of history." fast of Ramadan, during which Muslims are not to eat or drink during daylight hours for an entire month. They believe that fasting promotes patience, self-discipline and self-control.[28] The effect would not be as powerful as year-round fasting, but any form of food restriction should serve to increase C. Annual fasting also has another effect, which will be discussed in the next chapter.

Changing the type of food to reduce calories will have an effect similar to that of eating less. Certain foods may be forbidden for a specific period, such as meat during Lent. Or specified foods may be avoided altogether. Pork, a typically fatty meat high in calories, is an example. Avoiding it and consuming lower calorie food amounts to a mild form of CR. The same applies to complete or partial vegetarianism as practiced by many Hindus, Buddhists, and Christian groups such as Seventh Day Adventists.

Of all major religions, the Jains are the only ones who are rigorously vegetarian. As is typical of a people with high C, they are commercially successful. Although they make up less than 0.5% of the population of India, they pay 24% of the country's income tax.[29]

Fasting is not enough—the "effect feedback cycle"

So one C-promoter is fasting. But fasting alone is not enough to achieve and sustain very high C. Humans are a naturally low-C species, and high levels of C are required to support complex political and economic systems. The degree of fasting required to bring about very high C in humans would almost certainly undermine health. It is also clear, as indicated earlier, that high-C individuals are not necessarily short of food. This suggests that another type of C-promoter is required to account for the high level of C characteristic of complex societies.

---

[28] F. Azizi, "Medical Aspects of Islamic Fasting," *Medical Journal of the Islamic Republic of Iran* 10 (3) (1996): 241–246 www.muslim.org/islam/ramadan.htm (accessed September 4, 2014).
[29] *The Hindu*, Online Edition (August 20, 2007).

How, then, can C be raised to the level required? The answer to this question can be found in a peculiar quirk of our mammalian heritage which has enabled humans to "hijack" the biological systems which evolved to adjust behavior to food supplies, and use these to adjust temperament and behavior in ways that make civilization possible.

Its essence is this—any behavior that results from high C, *if reinforced by external pressures*, further raises C. Biohistory refers to this as the "effect feedback cycle." Its key requirement is that external pressures must change behavior in ways that go well beyond that dictated by temperament. In other words, people who act as if they have very high C will raise their level of C. For example—and these numbers are purely arbitrary—suppose that an individual has a C level of 4 but, because of social pressures, acts as if they have a C level of 6. This might be enough to raise their C to 5. More concretely, if someone's natural inclination is to have sex once a week, to raise their level of C they may need to restrict themselves to once a month.

One characteristic of all C-promoters is that they reduce testosterone. Mild food shortage reduces testosterone and is thus a C-promoter. But *any* change in behavior that reduces testosterone is a C-promoter. This is why groups such as Mormons avoid tea and coffee. Even though few calories are involved, caffeine increases testosterone.[30]

Restricting sex

The most powerful and effective C-promoter for humans is curbing sexual activity, which reduces testosterone levels and has none of the calorie-restriction drawbacks of fasting. The evidence for this is that sexual activity has been widely found to increase testosterone. Rabbits and pigs show a clear increase in the testosterone levels of males following

---

[30] I. Pollard, "Increases in Plasma Concentrations of Steroids in the Rat after the Administration of Caffeine: Comparison with Plasma Disposition of Caffeine," *Journal of Endocrinology* 199 (1988): 275–80; J. Svartberg, M. Mitdby, K. H. Bonaa, J. Sundsfjord, R. M. Joakimsen & R. Jorde, "The Associations of Age, Lifestyle Factors and Chronic Disease with Testosterone in Men: The Tromso Study," *European Journal of Endocrinology* 149 (2003): 145–52; C. D. Paton, T. Lowe & A. Irvine, "Caffeinated Chewing Gum Increases Repeated Sprint Performance and Augments Increases in Testosterone in Competitive Cyclists," *European Journal of Applied Physiology* 110 (6) (2010): 1243–50.

ejaculation,[31] and for bulls the mere presence of a receptive female has the same effect.[32] When rhesus monkeys are allowed access to receptive females their testosterone levels increase significantly.[33] The same applies to rats, which show a rise in testosterone levels with sexual activity, the presence of a receptive female, or even environmental clues associated with sex.[34]

These findings suggest that sexual activity may have an even bigger impact on testosterone than abundant food. Studies cited in chapter two noted that calorie restriction reduced testosterone levels by up to half. By comparison, sexually experienced male rats register 2.6 times their baseline testosterone levels after sexual activity.[35] Mere exposure to receptive females had a similarly dramatic effect. Taken as a whole, these studies suggest that sexual activity, or even exposure to receptive females,

---

[31] K. E. Borg, K. L. Esbenshade & B. H. Johnson, "Cortisol, Growth Hormone and Testosterone Concentrations during Mating Behavior in Bull and Boar," *Journal of Animal Science* 69 (8) (1991): 3230–40; A. Agmo, "Serum Luteinizing Hormone and Testosterone after Sexual Stimulation in Male Rabbits," *Acta Physiologica Scandinavica* 96 (1) (1976): 140–2.

[32] C. B. Katongole, F. Naftolin & R. V. Short, "Relationship Between Blood Levels of Luteinizing Hormone and Testosterone in Bulls, and the Effects of Sexual Stimulation," *Journal of Endocrinology* 50 (1971): 457–66.

[33] J. G. Herndon, A. A. Perachio, J. J. Turner & D. C. Collins, "Fluctuations in Testosterone Levels of Male Rhesus Monkeys during Copulatory Activity," *Physiology & Behavior* 26 (3) (1981): 525–8.

[34] F. Kamel & A. Frankel, "Hormone Release during Mating in the Male Rat: Time Course, Relation to Sexual Behavior, and Interaction with Handling Procedures," *Endocrinology* 103 (1978): 2172–9; F. Kamel, W. Wright, E. J. Mock & A. I. Frankel, "The Influence of Mating and Related Stimuli on Plasma Levels of Luteinizing Hormone, Follicle Stimulating Hormone, Prolactin, and Testosterone in Male Rat," *Endocrinology* 101 (2) (1977): 421–9;

J. M. Graham & C. Desjardins, "Classical Conditioning Induction of Luteinizing Hormone and Testosterone Secretion in Anticipation of Sexual Activity," *Science* 210 (4473) (1980): 1039–41;

H. Bonilla-Jaime, G. Vázquez-Palacios, M. Arteaga-Silva & S. Retana-Márquez, "Hormonal Responses to Different Sexually Related Conditions in Male Rats," *Hormones and Behavior* 49 (3) (2006): 376–82; J. V. Matuszczyk & K. Larsson, "Experience Modulates the Influence of Gonadal Hormones on Sexual Orientation of Male Rats," *Physiology & Behavior* 55(3) (1994): 527–31; A. I. Frankel & E. J. Mock, "Time Course of Hormonal Response to Sexual Behavior in Aging Male Rats," *Experimental Gerontology* 16 (1981): 363–9.

[35] Bonilla-Jaime et al., "Hormonal responses to different sexually related conditions in male rats," 376–82.

increase testosterone far more than food shortage reduces it.

Turning this argument around, if sexual activity increases testosterone then curbs on sexual activity should reduce it. A reasonable guess is that by combining fasting with strict limits on sexual activity, testosterone levels could be reduced by more than 80%.

Similar effects have been observed among humans. One study found that couples having sex on a particular evening had higher levels of testosterone (both men and women) than those who did not, even though initial testosterone levels were similar.[36] In another study, the testosterone levels of women who were temporarily physically separated from their partners peaked the day before they met their partners and after sexual activity. Levels were the lowest when they were separated from their partners for at least two weeks.[37] Salivary testosterone has also been found to increase in men attending sex clubs, with one study reporting a rise of 11% for observers and 72% for participants.[38]

Similar findings apply to luteinizing hormone (LH), which is increased by sexual activity. Both human and animal studies establish that sexual arousal, or the anticipation of sexual activity, increases the level of LH.[39] Sexually experienced rams also have been found to have higher levels of LH.[40]

Sexual activity immediately after puberty has an especially powerful and enduring effect. One study found that rats that were sexually active at 12 weeks (i.e. shortly after puberty which for rats is 6–8 weeks) had up to double the levels of testosterone of those without such experience, which

[36] Dabbs, *Rogues and Lovers*, 102.

[37] L. D. Hamilton & C. M. Meston "The Effects of Partner Togetherness on Salivary Testosterone in Women in Long Distance Relationships," *Hormones and Behavior* 57 (2) (2010): 198.

[38] Escasa, M. J., J. F. Casey and P. B. Gray, "Salivary Testosterone Levels in Men at a U.S. Sex Club," *Archives of Sexual Behavior* 40 (5) (2011): 921–6.

[39] Graham & Desjardins, "Classical Conditioning Induction of Luteinizing Hormone and Testosterone Secretion in Anticipation of Sexual Activity"; S. G. Stoléru, A. Enaji, A. Cournot & A. Spira, "Pulsatile Secretion and Testosterone Blood Levels are Influenced by Sexual Arousal in Human Males," *Psychoneuroendocrinology* 18 (3) (1993): 205–18.

[40] K. E. Borg, K. L. Esbenshade, B. H. Johnson, D. D. Lunstra & J. J. Ford, "Effects of Sexual Experience, Season, and Mating Stimuli on Endocrine Concentrations in the Adult Ram," *Hormones and Behavior* 26 (1) (1992): 87–109.

persisted for eight months even without further exposure to females.[41]

Applying this to humans, if early sexual experience is associated with elevated testosterone, then restricting sexual activity in the years following puberty should be an especially powerful C-promoter. Guarding the chastity of young women is the most obvious way to achieve this, and virtually all civilizations have prescribed this until recently. This does not eliminate the sexual activity of young men, of course, but the C of men is usually less important than that of women, in that men are not the primary caretakers of children.

Female chastity also reduces sexual opportunities for men, which increases their C. Whether this is an advantage or not depends on what society expects of men. For farming and especially commercial occupations, higher C offers an advantage because it promotes hard work and enterprise. This explains why religions whose members are heavily involved in commerce, such as Jains and Jews, tend to rigorously limit the sexual activity of their male members.[42]

But when men are warriors rather than workers, low C with its high testosterone has benefits. The more warlike the society and the less men are involved in jobs requiring hard work or business acumen, the greater the benefits of low C. Because women do not normally fight, there are no similar advantages to their having high testosterone. Thus it is adaptive for the most warlike societies to have double standards in which women's chastity is carefully controlled while men are free to indulge themselves with non-breeding women such as prostitutes, or even with one another. We will discuss this further in the next chapter.

Of course, male jealousy and fear of cuckoldry serve as powerful emotional incentives for controlling the sexual behavior of women. But jealousy alone does not explain why some societies are so concerned with cuckoldry, while others are far less so. If jealousy were the critical issue, then patriarchal societies would always control female sexuality. However,

---

41 D. Wu & A. C. Gore, "Sexual Experience Changes Sex Hormones but not Hypothalamic Steroid Hormone Receptor Expression in Young and Middle-Aged Male Rats," *Hormones and Behavior* 56 (3) (2009): 299–308.

42 A. C. Kinsey, W. B. Pomeroy & C. E. Martin, *Sexual Behavior in the Human Male* (Philadelphia: W.B. Saunders Company, 1948), 466–467; M. Zborowski & E. Herzog, *Life is with People* (New York: International University Press, 1952); Glasenapp, H. v., *Jainism: An Indian Religion of Salvation* (Delhi: Motilal Banarsidass, 1999), 228–31.

as we will see, they do not.

We noted earlier that in rats the period immediately following puberty is especially important in setting the level of C in later life. Sexual activity at this time doubles their testosterone in later life. The *Kinsey Report* finds a similar pattern for humans. Single males who reported substantially less voluntary sexual activity (that is, excluding nocturnal emissions) during their adolescence and twenties were far more likely to achieve a tertiary education, an indication of lower testosterone. Fig. 3.1 below documents this relationship for unmarried males from ages 15 to 30.

Fig. 3.1. Mean weekly total sexual outlet (minus nocturnal emissions) of unmarried males, by educational level.[43] Adolescents and young men with lower levels of sexual activity, including masturbation, heterosexual and homosexual intercourse, are more likely to achieve educational success.

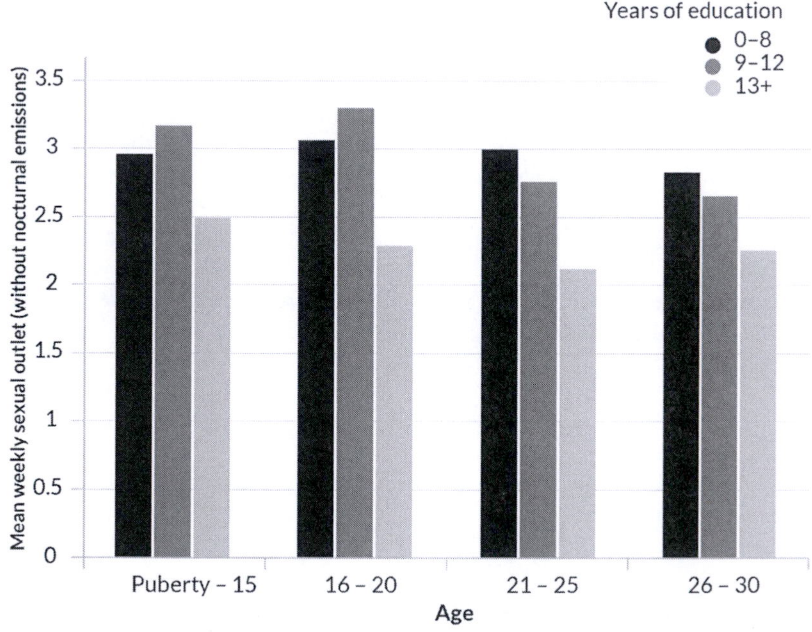

Fig. 3.1 relates the weekly frequency of sexual activity (including masturbation, heterosexual and homosexual intercourse) of unmarried males aged 15–30 to years of education. Note that the subjects who were

[43] Kinsey et al., *Sexual Behavior in the Human Male*, 342.

most educated had the lowest frequency of sexual activity in each of the four age cohorts.

One possible explanation for these findings is that males who go on to higher education have a lower sex drive, but the frequency of their nocturnal emissions suggests otherwise. As Fig. 3.2 indicates, males who completed a tertiary education had far more nocturnal emissions between ages 15 and 30.

Fig. 3.2. Mean weekly nocturnal emissions of single males, by educational level.[44] Years of education are linked even more strongly to nocturnal emissions, a likely measure of the degree of sexual frustration.

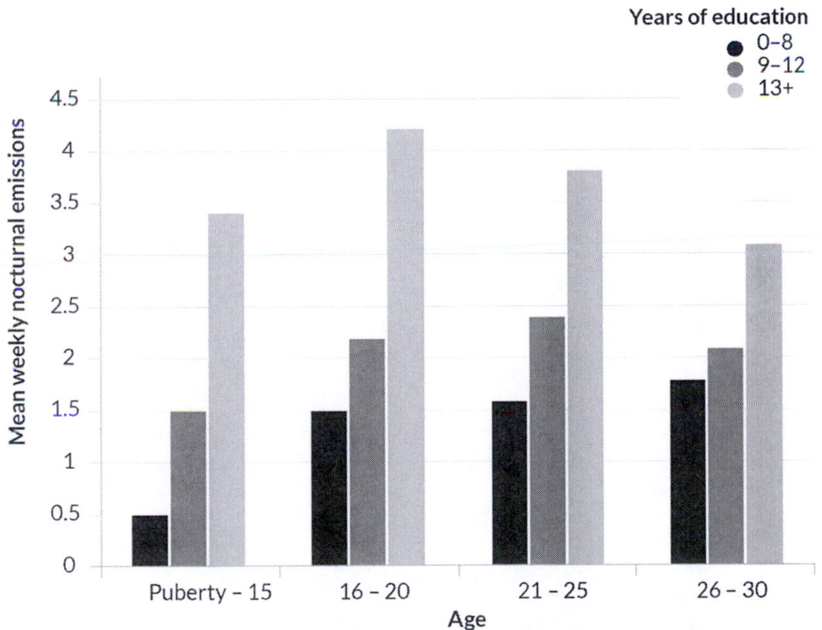

Nocturnal emissions serve as a measure of sexual frustration, meaning that sexual activity is below the level set by temperament. Fig. 3.2 suggests that it is sexual frustration, rather than a lower level of sexual activity as such, which best predicts educational achievement. It is also significant that the strongest links occur at the youngest ages. Men who went on to complete tertiary education had nearly three times the nocturnal emissions

---

[44] Ibid.

between the ages of 16 and 20 compared to those who completed only eight years of education. But between puberty and 15 they had nearly *seven times* more emissions.

If lack of a sexual outlet in early adolescence reduces testosterone in humans, as it does in rats, these findings make perfect sense, given the strong link between lower testosterone and occupational success.

Remarkably, the *Kinsey Report* indicates that tertiary educated men reported more emissions than those who achieved less education, even when they were married (Fig. 3.3).

Fig. 3.3 Mean weekly nocturnal emissions of married males by educational level.[45] Men who were sexually restrained completed more years of education than those who were not, even when married.

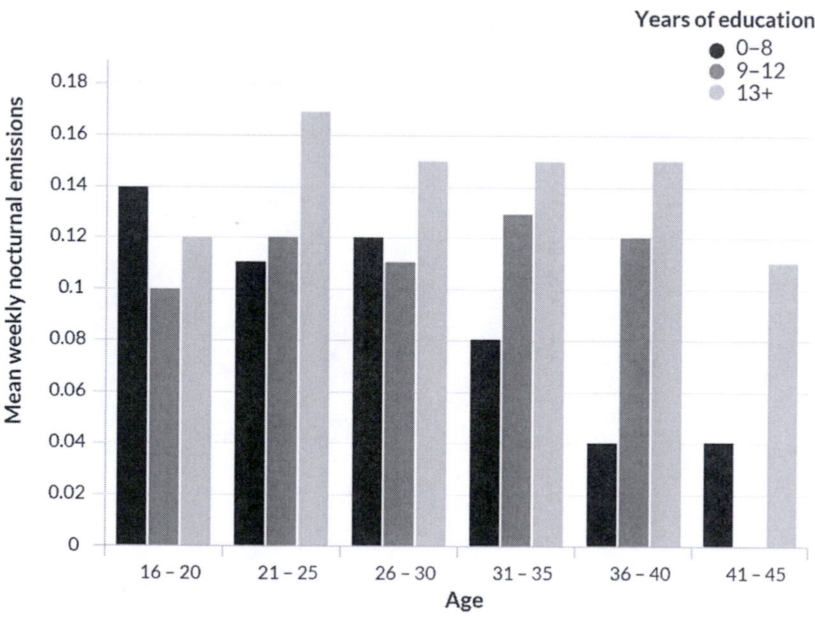

These findings imply that men who restrict their sexual activity below its natural level, thus raising their level of C, are more likely to achieve educational success. According to Kinsey, a similar pattern applies to

[45] Ibid.

occupation. Teenagers who ended up in higher-level occupations had lower total sexual activity and more nocturnal emissions than those who did not. It is noteworthy that these findings are related to the occupational status they achieved, not that of their parents. Teenagers from working-class backgrounds who achieved middle-class occupations had middle-class sexual patterns, while teenagers from middle-class backgrounds who ended up in working-class occupations had working-class sexual patterns.[46] In fact, the highest level of nocturnal emissions, indicating perhaps the greatest control of sexual behavior, was found in the sons of skilled laborers who became professionals. Restricting adolescent sex is thus a powerful C-promoter.[47]

A recent study has found similar results at the college level. Male and female college graduates were more likely to report never having had sex and were less likely to report having sex during the previous 12 months compared to age- and sex-matched non-college graduates. Their lifetime histories also contained fewer sexual partners.[48]

It is commonly thought that the reason successful parents have successful children is that the parents teach their children the value of work. Biohistory suggests another possibility—successful parents also influence their children's behavior in such a way that the children have fewer opportunities for sexual activity. For example, parents might require their children to stay at home in the evenings rather than go out to parties, take them to church, or discourage friends thought to be bad influences. Reduced sexual activity increases C, which renders teenagers temperamentally more likely to study hard and stay out of trouble. Parents' influence thus has physiological, psychological and cultural effects.

Because the effects of early sexual activity are so lasting, it is likely that there is an epigenetic effect—the level of testosterone (and presumably other hormones) in the period immediately after puberty has a relatively permanent effect on the epigenome. This is something yet to be investigated, but is a strong candidate for future research.

---

[46] Ibid., 420–5.

[47] These findings are especially striking given that many people, including Kinsey himself, used the reports as an argument for greater sexual freedom. A careful study of his statistics, however, suggests the opposite.

[48] New Strategist Publications (eds.), *American Sexual Behavior: Demographics of Sexual Activity, Fertility and Childbearing* (Ithaca, New York: New Strategist Publications, 2006), 57, 58, 171, 172.

The preceding studies offer insights as to why some children fail to maintain their parents' social and economic statuses. A teenager joining a "faster" social group will tend to adopt their attitudes and behaviors, including sexual behaviors, leading to a decline in C and thus an aversion to middle-class values associated with hard work and enterprise. Similarly, adolescents from deprived backgrounds who adopt stricter codes of behavior may raise their C and thus achieve occupational success.

Kinsey's findings also indicate that religious people of all denominations, and especially Orthodox Jews, have lower rates of sexual activity even when married.[49] Because religious teachings curb sexual activity it might be thought that religious people would have more nocturnal emissions, but this is not the case. This may be because religious men have higher C as a result of parental control in late childhood, which lowers testosterone and thus reduces their sex drive.

The lowering of C by early sexual activity can have other effects. Studies have found that people who engage in sexual activities early in life, and especially those with multiple partners, are far more likely to abuse drugs. This is a very plausible connection given that early sexual activity increases testosterone, and higher testosterone people are more likely to abuse drugs. Although, of course, drug use itself may also promote promiscuous sexual behavior, such as by motivating women to engage in prostitution.[50]

The focus of these studies has been on higher education, which is the primary route to occupational success in the modern world. But irrespective of educational attainment, curbing sexual activity should promote success in any endeavor that requires hard work, discipline, tolerance of routine and a willingness to forgo short-term gratification for the sake of long-term goals.

Turning back to the cross-cultural study first discussed in chapter one, we noted that civilized peoples are more likely to restrict sexual activity than

---

[49] Kinsey et al., *Sexual Behavior in the Human Male*, 466–7.

[50] S. F. Tapert, G. A. Aarons, G. R. Sedlar & S. A. Brown, "Adolescent Substance Use and Sexual Risk-Taking Behavior," *Journal of Adolescent Health* 28 (3) (2001): 181–9; R. E. Booth, J. K. Watters & D. D. Chitwood, "HIV Risk-Related Sex Behaviors among Injection Drug Users, Crack Smokers, and Injection Drug Users who Smoke Crack," *American Journal of Public Health* 83 (8) (1993): 1144–8; L. S. Zabin, J. B. Hardy, E. A. Smith & M. B. Hirsch, "Substance Use and its Relation to Sexual Activity Among Inner-City Adolescents," *Journal of Adolescent Health Care* 7 (5) (1988): 320–31.

those in smaller scale societies. One interpretation of this difference is that people with high C are less interested in sex, and there is some evidence for this. Societies where women are reported to *not* enjoy sex tend to have more advanced political and economic systems than those in which women report that they enjoy it (see Table 3.1 below).

Table 3.1 shows that both restriction of sexual activity and lack of interest in sex are strongly related to aspects of civilization, including larger political units, the market economy and hard work.

Table 3.1. Correlations in the cross-cultural survey dealing with women's dislike of sexual activity and restrictions on sexual behavior.[51] Both sex restrictions and lack of interest in sex are associated with civilization, indicating that reduced sexual activity is both a consequence and cause of high C.

	Women dislike sex	Premarital sex restricted	Adultery restricted	Divorce restricted
Size of political unit	.40	.53**	.51**	.35*
Hereditary status		.39**	.43**	.41
Market economy	.47	.45*	.54*	
Marriages based on status	.70*	.39*	.43**	,41
Status from wealth versus generosity		.46*	.47*	.30
Routine work		.42**	.48**	
Deities enforce morality	.51	.30*	.30*	
Modesty in dress	.45	.51**	.53**	
Shortage of food	.44			.25
Late age of puberty	.69			
significance	**.001	* .01	Others:	.05

This suggests both that restricting sexual activity increases C and that

[51] See www.biohistory.org.

higher C renders women less interested in sexual activity. The table also shows that women are less likely to be interested in sex when the society is short of food and when the age of puberty is later, both of which are biological aspects of high C.

The cross-cultural survey further suggests that restricting sexual activity is also associated with the desire for children, in the sense that having more children increases a woman's status (see Table 3.2 below).

Table 3.2. Societies which restrict sexual activity are more likely to want more children.[52]

	Premarital sex restricted	Adultery restricted	Divorce restricted
Children wanted	.32*	.25	.31
significance	* .01	Others: .05	

This is consistent with the biological picture of high C that includes a greater interest in children, necessary to rear successful offspring in a tough, food-restricted environment.

Control of children

Another C-promoter is the systematic control of a child's behavior, which is one way in which C is transmitted between generations. A child's C appears to be set in large part by parental control, with some support from the wider community. Control thus joins restrictions on sexual activity and epigenetic inheritance as prominent C-promoters.

The degree to which children are controlled largely reflects the parental level of C. High C parents exercise self-discipline and restraint in their own lives and find it natural to control and direct their children. However, social and religious pressures may encourage parents to be stricter than temperament alone would prescribe. For example, even parents who are not inclined to discipline their children may experience social censure if their children are rude, dishonest or lazy. Such pressures may help to increase C in the next generation.

There are societies in which C declines across generations, as appears to

[52] See www.biohistory.org

be the current case in the West, a matter examined in detail in later chapters. As it proceeds, parents become less capable of raising their children in a high-C manner, and C-promoting methods seem less "natural" to them. They are likely to accept the advice of experts, such as Dr Spock, who advocate less parental control, behavior which is congenial to their temperament. This line of analysis suggests that books advising parents about how to behave reveal less about how children *should be* treated than about how they *are* treated. The authors of such books reflect current values because books that do so are likely to sell more copies than those that reflect other values.

Of course, parents are not the only people who influence children. Other adult caregivers including relatives, nannies, neighbors and teachers, have an impact, as do other children. This point is underscored by studies of identical twins reared together where, for example, only 35% of the brothers of violent criminals or rapists engage in the same types of behavior.[53] Temperament therefore is set not by parents alone but by peers and the community.

Pre-natal influences

It is also very likely that C is transmitted between generations before birth, either through the environment of the womb or through epigenetic effects on sperm or ova prior to conception. The rat experiments described in the previous chapter show that CR has a significant effect during gestation or before conception, but do not determine whether this was a direct effect or a result of changes to the mother which influenced her behavior after birth. The extent of such influences will be determined by exposing fathers but not mothers to CR to eliminate environmental effects, an experiment currently being run.[54]

Rural living

We noted earlier that living in the countryside tends to promote C, whereas urban living tends to reduce it. This is consistent with Kinsey's figures which show that rural males have lower rates of sexual activity than city dwellers, that farmers have lower testosterone, and that country

[53] S. Mednick, "Crime in the Famly Tree," *Psychology Today* (March, 1985).

[54] Updates of experimental results will be posted on www.biohistory.org.

people tend to reach puberty later. [55] This helps to explain, for example, why hedonistic groups, from the decadent aristocrats of ancient Rome to flappers of the 1920s, are mainly found in cities. By implication, rural living can thus be seen as a C-promoter.

Work, Sabbath keeping

Earlier we mentioned the critical importance of the *effect feedback cycle.*Any behavior that is a result of high C, if reinforced by external pressures, will increase the level of C. High-C people tend to be less sexually active, so social pressures to restrict sexual activity tend to increase C. In other words, acting high C causes people to have higher C.

Another C-promoter is work. High-C people tend to be good at routine work, which means that financial, social or religious pressures to work hard tend to promote C. This is especially the case if the timing of work is determined by a code such as Sabbath keeping, which enjoins working hard for six days while taking the seventh day off. The more strictly the code is observed, the more likely that C will be promoted. Commercially successful groups, from Orthodox Jews to Mormons, not only place strong limits on sexual activity but are also strict Sabbatarians.[56] It is a strange irony, but seemingly an inherent part of human physiology, that the best way to get people to work hard is to introduce a rigid requirement that on certain days they must not work at all.

Ritual

Religious rituals such as the Catholic Mass are also C-promoters, particularly if performed in a precise and detailed manner. When people are required to act in a highly programmed way, such as repeating set prayers and chants, C is likely to increase. This is especially so for priests and monks from whom precise performance of rituals is expected. Monastic orders may require attendance at several services a day. Religious rituals are especially effective C-promoters because many of them are public, and social pressures can lead people to perform them far beyond levels congenial to temperament.

Any detailed code of behavior can have the same effect. Jewish ritual law, with its complex of rules that control many areas of life to a minute

[55] Kinsey et al., *Sexual Behavior in the Human Male*, 448.
[56] Schumpeter, "The Mormon Way of Business," *The Economist* (May 5, 2012).

degree, is an outstanding example. It includes strict rules about what foods may be eaten and how they are prepared, purification rituals such as those that follow menstruation, requirements for set prayers and ceremonies, circumcision of infants, and even principles for the way hair may be cut or beards trimmed. These rules serve as a powerful system of C-promoters which have created the highly successful Jewish temperament, and thus success in business and intellectual pursuits. The time and effort spent in studying the law and supporting full-time specialists can be regarded as a sound investment for a people focusing on business and the professions.

While Jewish ritual law is an extreme example, any code that controls or influences behavior should increase C. Honesty, obedience to the law, non-violence, courtesy, philanthropy and even altruism in a general sense are examples of these codes. The one common and essential point about C-promoters is that they require real effort. To maintain high C, people need to eat less than they would like, have less sex than they would prefer, and control their children more tightly than they are inclined to do. They not only need to work hard but to do so at appropriate times. And depending on their particular tradition, they need to participate in rituals and observances.

Social and religious pressures

Clearly, to behave in these ways requires powerful influences which act to thwart individual desires. Social disapproval is one such influence, as for example the way chastity has been maintained in many societies. A girl from a respectable family who strays attracts widespread condemnation. If she becomes pregnant she might be cast out by her family. If she marries it will most likely be to a poorer man, with the prospect of a harder and shorter life. Or she might become a prostitute, a dangerous and unhealthy occupation in most societies. She might even be killed. Whatever happens, her family suffers disgrace and humiliation. Given these potential consequences, an unmarried girl and her family have every incentive to keep her chaste during the all-important period following puberty, when the level of C is largely determined. A similar scenario applies today in much of the Muslim world.

While social censure is a powerful force influencing behavior, an equally potent one is religion. Recall that many of the C-promoters mentioned above are part of religious systems. Religion codifies many of these behaviors into rituals and laws that are highly resistant to change, especially when written down in sacred books. A clear example is the resistance of most Christian churches to accepting homosexuality, which the Bible condemns in no uncertain terms, despite growing acceptance in the general community.

For believers there is the added sanction of divine anger for aberrant behavior coupled with divine blessing as a reward for virtue. For example, consider the following quote from the Ten Commandments:

> I the Lord your God, am a jealous God, punishing the children for the sins of the fathers to the third and fourth generations of those who hate me, but showing love to a thousand generations of those who love me and keep my commandments.

The message is that those who obey God's laws will be more successful in their progeny than those who fail to obey. This principle works quite well at any level, since hard-working and devoted parents are likely to prosper and raise more children. And to the extent that their children accept the same values and work ethic, they too are likely to succeed.

The benefits of religion

Many scholars believe that religions confer competitive advantages on believers. For example, Judaism has been seen to enhance group cooperation and help its followers economically.[57] Group loyalty is largely maintained over generations by religious rituals and symbols that commemorate sacred values, defeats and victories.[58] The ordeal of community survival is continually refreshed in the memory of each generation.

To this may be added the proposition that religious codes can have a profound impact on temperament and behavior via their C-promoting effects. They promote industrious and disciplined behavior, the willingness

---

[57] K. B. MacDonald, *A People that Shall Dwell Alone: Judaism as a Group Evolutionary Strategy* (Westport Conn.: Praeger, 1994); Emile Durkheim writes of religion as the cement of society, uniting people around common goals and beliefs. See K. Thompson, *Emile Durkheim* (London: Tavistock Publications, 1982); Richard Dawkins, by contrast, sees religion as unnecessary and harmful, acting as a substitute for thought. See R. Dawkins, *The God Delusion* (London: Bantum Books, 2006).

[58] E. H. Spicer, "Persistent Cultural Systems," *Science* 174 (4011) (1971): 795–800.

to sacrifice present consumption for future benefit, better parenting and more cooperation at the community or national levels.

Religions also tend to be effective C-promoters because of the development of high-C religious specialists. One reason for this is that restrictions on sexual activity tend to be stronger for such people. Buddhist and Christian monks and Catholic priests are denied any legitimate sexual outlet. Others, such as Protestant ministers and Jewish rabbis, are not required to abstain sexually but they are held to a stronger standard than applies to the general community.

Sexual restraint can be required of religious specialists even by low C peoples. For example, the Yanomamo of the Amazon basin have few constraints on sexual behavior; but their shamans are required to fast rigorously and abstain from sexual intercourse for a year before initiation.[59] Religious specialists also carry the strongest obligations to perform rituals, and in general to behave in an exemplary manner. This in turn gives them relatively high statuses in most societies, which makes them a powerful pressure group for C-promoting behavior.

Until recently, most religious specialists in complex societies were supported by the authorities, thus adding political weight to divine displeasure and the forces of social conformity. Moreover, political leaders have had a shrewd understanding of the value of religion in buttressing their rule. Edward Gibbon captures this point in discussing the Roman Empire:

> The various modes of worship, which prevailed in the Roman world, were all considered by the people, as equally true; by the philosopher, as equally false; and by the magistrate, as equally useful.[60]

In this he echoes statements by such eminent Roman writers as Seneca and Lucretius. While religion has been the main vehicle for promoting C in most of the world, the philosophy of Confucianism has played much the same role in East Asia.

---

[59] Smole, *The Yanomamo Indians.*

[60] Edward Gibbon, *The Decline and Fall of the Roman Empire*, Vol. 1, chapter 2 http://ancienthistory.about.com/library/bl/bl_text_gibbon_1_2_1.htm (accessed September 4, 2014).

Competition between religions

Biohistory and the model of C offer a new way of understanding changes in human societies. From the earliest times, civilizations have invested immense resources in religion, ranging from Mayan temples to Sumerian ziggurats and Catholic cathedrals. From one perspective it might seem more sensible to direct productive surpluses into warfare or agriculture or trade rather than religion. Further, we might expect societies that "waste" resources on religion to be defeated by more efficient states that dispensed with priests and replaced them with soldiers.

But if religion is viewed as an essential driving force of social evolution, investment in it makes good sense. Societies that spend heavily on religious priorities and thereby increase the C of their members will be more successful than those that fail to make such investments. Viewed in this way, the advance of civilization is not just a matter of superior physical technologies but of more advanced and effective religions which constitute "cultural technologies." In the competitive struggle for survival, peoples with better cultural technologies have crushed or swamped their rivals, one effect of which is an overall rise in C across the past few millennia.

Stress as a C-promoter

The last C-promoter to consider is stress, such as that caused by overcrowding, danger or starvation. Studies have found that for rats, an increase in CORT in response to stress is associated with a decline in testosterone, and that the higher the CORT the more testosterone levels decline. Blocking the increase in CORT during stress partially nullifies this effect, in that testosterone levels are similar to those without increases in CORT.[61] Baboons respond in a similar manner. Wild baboons living freely in Kenya under conditions of social stress have been found to have high cortisol (the equivalent to CORT in primates) and reduced testosterone production in their testes.[62]

Direct infusion with cortisol or CORT mirrors these findings. When

---

[61] T. E. Orr & D. R. Mann, "Role of Glucocorticoids in the Stress-Induced Suppression of Testicular Steroidogenesis in Adult Male Rats," *Hormones and Behavior* 26 (3) (1992): 350–63.

[62] R. M. Sapolsky, "Stress-Induced Suppression of Testicular Function in the Wild Baboon: Role of Glucocorticoids," *Endocrinology* 116 (1985): 2273–8.

injected it reduces testosterone in men, rhesus monkeys, songbirds and lizards.[63] Stress also reduces levels of follicle-stimulating hormone and luteinizing hormone (both hormones also reduced by food shortage) and injection with CORT reduces the level of LH in hens and rats.[64] Further, stress acts as a C-promoter by reducing interest in sex.[65]

C-demoters

A "C-demoter" is the opposite of a C-promoter. C-demoters are anything

---

[63] D. C. Cumming, M. E. Quigley & S. S. C. Yen, "Acute Suppression of Circulating Testosterone Levels by Cortisol in Men," *Journal of Clinical Endocrinology and Metabolism* 57 (1983): 671–3; P. Doerr & K. M. Pirke, "Cortisol-Induced Suppression of Plasma Testosterone in Normal Adult Males," *Journal of Clinical Endocrinology and Metabolism* 43 (3) (1976): 622–9; C. P. Puri, V. Puri & T. C. A. Kumar, "Serum Levels of Testosterone, Cortisol, Prolactin and Bioactive Luteinizing Hormone in Adult Male Rhesus Monkeys following Cage-Restraint or Anaesthetizing with Ketamine Hydrochloride," *Acta Endocrinologica* 97 (1): 118–24; R. L. Norman, "Effects of Corticotropin-Releasing Hormone on Luteinizing Hormone, Testosterone, and Cortisol Secretion in Intact Male Rhesus Macaques," *Biology of Reproduction* 49 (1): 148–53; J. C. Wingfield & B. Silverin, "Effects of Corticosterone on Territorial Behavior of Free-Living Male Song Sparrows Melospiza Melodia," *Hormones and Behavior* 20 (4) (1986): 405–17; R. Knapp & M. C. Moore, "Male Morphs in Tree Lizards have Different Testosterone Responses to Elevated Levels of Corticosterone," *General and Comparative Endocrinology* 107(102) (1997): 273–9.

[64] K. P. Briski, "Stimulatory vs. Inhibitory Effects of Acute Stress on Plasma LH: Differential Effects of Pretreatment with Dexamethasone or the Steroid Receptor Antagonist, RU 486," *Pharmacology Biochemistry and Behavior* 55 (1) (1996): 19–26; F. H. Bronson, "Establishment of Social Rank Among Grouped Male Mice: Relative Effects on Circulating FSH, LH, and Corticosterone," *Physiology & Behavior* 10 (5) (1973): 947–51; R. J. Etches, J. B. Williams & J. Rzasa, "Effects of Corticosterone and Dietary Changes in the Hen On Ovarian Function, Plasma LH and Steroids and the Response to Exogenous LH-RH," *Journal of Reproduction & Fertility* 70 (1984): 121–30; F. Kamel & C. L. Kubanak, "Modulation of Gonadotropin Secretion by Corticosterone: Interaction with Gonadal Steroids and Mechanism of Action," *Endocrinology* 121 (2) (1987): 561–8.

[65] E. O. Laumann, A. Nicolosi, D. B. Glasser, A. Paik, C. Gingell, E. Moreira & T. Wang, "Sexual Problems among Women and Men aged 40–80 y: Prevalence and Correlates Identified in the Global Study of Sexual Attitudes and Behaviors," *International Journal of Impotence Research* 17 (2005): 39–57; J. Bancroft, E. Janssen, D. Strong, L. Carnes, Z. Vukadinovic & J. Scott Lang, "The Relation Between Mood and Sexuality in Heterosexual Men," *Archives of Sexual Behavior* 32 (3) (2003): 217–230.

that lowers C including sex, alcohol, recreational drugs, caffeine, idleness and high-calorie food. Each of these, and especially drugs and high-calorie food, have opposite effects to calorie restriction. In chapter one we saw the effects of one C-demoter when mice given extra sugar were less adept at defending their territories than those with a mouse-normal diet.

Most studies of high-calorie food concentrate on the physical effects, including increased risk of obesity, high blood pressure, diabetes and stroke. But there is also evidence of *psychological* effects that are opposite to C-promoters, in the sense that they reduce self-discipline and self-control. One study found that adolescents who drink more than five cans of soft drink a week are more likely to carry a weapon and attack peers, family members, and dates. Other studies have linked soft-drink consumption with violence and mood problems in adolescents, and with aggression, social withdrawal and attention problems in five-year-olds.[66]

C-demoters such as soft drinks and junk food boost mood in the short term, but in the long term they reduce C and raise anxiety. This often leads to a further "fix" which reduces C still further and creates more anxiety. Biohistory proposes that addictive drugs such as heroin and alcohol work this way. In humans, the C-demoting effects of drugs are so powerful that they can more than offset the C-promoting lack of nourishment experienced by many addicts. One reason drugs and alcohol may be so difficult to give up is that the anxiety stemming from the decline in C increases the desire for a short-term fix, a problem that may be more difficult to overcome than the physical addiction. Excessive sexual activity, of course, works in much the same way.

C-promoters have opposite effects in that they can bring about improvements in mood. As noted in a diet study discussed in chapter two, there was a long-term decline in anxiety associated with increased C.

---

[66] S. J. Solnick & David Hemenway, "The 'Twinkie Defense': the Relationship Between Carbonated Non-diet Soft Drinks and Violence Perpetration among Boston High School Students," *Injury Prevention* 040117 (2011); S. J. Solnick & David Hemenway, "Soft drinks, aggression and suicidal behavior in US high school students," *Int J Inj Contr Saf Promot. Epug* (July 8, 2013), http://healthland.time.com/2013/08/16/soda-contributes-to-behavior-problems-among-young-children/?iid=hl-main-lead (accessed September 4, 2014).

Summary and conclusions

The complex of hormonal, behavioral, temperamental and epigenetic changes associated with C-promoters has made the rise and maintenance of complex societies possible, particularly those focused on trade and business rather than war. It increases the capacity for work and delayed gratification. It promotes law-abiding behavior and reduces violence. It also increases interest in children and promotes their survival and success in competitive environments. It helps to maintain and increase populations.

To achieve and sustain high C, societies have adopted systems of belief and moral codes that have physiological and behavioral effects similar to those of calorie restriction. To the degree that people adhere to these systems they allow C to rise to a higher level than would otherwise be possible, even when food is relatively abundant. These influences on behavior—especially limits on sexual activity but also religious rituals and other behavioral codes—are key features of major religions. In this sense, religion can be seen as a central factor contributing to the rise of human civilization, at least as important is physical technologies such as metalworking and writing.

A critical point to understanding C-promoters is the workings of the *effect feedback cycle*,wherein any behavior that is a result of higher C can add to the level of an individual's C, provided that it is carried out *beyond* the requirements of an individual's temperament. By their nature, C-promoters take effort and discipline, and require powerful cultural and religious sanctions to maintain. Meanwhile, C-demoters such as recreational drugs and sex work in the opposite way, reducing C and tending to increase anxiety in the medium to long term.

CHAPTER FOUR

AGGRESSION

C is a physiological system that allows animals to adapt their behavior to the needs of a food-limited environment. In animals it is primarily a direct response to food shortage. The main reason for higher C among humans is that societies have developed religious and other cultural systems that change behavior in ways that mimic the effects of food restriction. These practices are termed C-promoters. Through the principle of the *effect feedback cycle* any behavior resulting from C, if reinforced by outside pressures, can increase C.

High C people tend to be industrious, skilled at agriculture and commerce, and more accepting of distant political authority. This means that individuals with higher C tend to prosper in complex societies which have social hierarchies and economies based on trade. In addition, societies with higher C members tend to overrun and absorb those with lower C. From this perspective, the rise of civilizations can be seen not just as the development of physical technologies such as agriculture and metal working, but of cultural systems that promote C—mainly religions.

C, however, is not the only example of a complex of behavioral traits found in advanced human societies. In this chapter we will consider a second set of traits relating to aggression, morale, and a high birth rate—this is referred to as "V" (for vigor).

This chapter extends the comparisons between gibbon and baboon behavior and C, and documents how baboon social behavior can serve as a model for V as well as for low C.

Baboons compared with domesticated animals

Variation in C-type behavior is not the only distinction to make between animal societies. Baboons have been used as a model for low C, but domesticated animals also have very low C. They are less timid than wild animals, which is what makes them "tame." They are socially tolerant, in that they are capable of living together in large groups. They make poor mothers, which is why a "broody hen" is an oddity in the farmyard. They are also highly sexed and indiscriminate in mating. Wild ducks, for example, have specific courtship rituals and flocking signals, and in the wild they will not normally mate with other varieties. These rituals often break down with domestication, which suggests that for some species they are acquired rather than innate behaviors.

Farmers tend to selectively breed for tame animals that can tolerate crowding and produce young easily, but some changes take place even when they are not desired. Breeders who wish to retain differences between varieties of domesticated birds often have to physically separate them so as to prevent inter-breeding.[1]

But in other ways, domesticated animals behave quite differently from baboons. Domesticated animals display a remarkable lack of aggression, which is what allows them to live in crowded conditions. Baboons, on the other hand, are known for their ferocity. As described in chapter one, a baboon troop can drive off and even kill the leopards which are their main natural predators. Early primate studies suggested that dominant males act as protectors to the troop, although more recently it has become clear that they only exhibit self-sacrifice when protecting close kin.[2]

Baboons also fight ferociously among themselves. The following description is from Robert Sapolsky's colorful account of baboon life, and tells of how a long-term alpha male, Saul, was deposed by a coalition of other males:

> Joshua and Menasseh, another big male, soon bound to be enemies, teamed up first. They spent a morning making coalitional appeasement gestures to each other, cementing a partnership, and finally worked up the nerve to challenge Saul, who promptly kicked their asses, slashed Menasseh's haunch, sent them both running. By most predictions, that should have settled that. Instead, the next day, Joshua and Menasseh formed a coalition with Levi … Saul dispatched the trio in seconds. And they came back the next day with the vile Nebuchanezzar in tow. Nebuchanezzar and Menasseh managed to hold their own for a few

---

[1] E. B. Hale, "Domestication and the Evolution of Behaviour," in *The Behavior of Domestic Animals*, edited by E. S. E. Hafez (London: Balliere, Tindall & Cox, 1962).

[2] R. M. Sapolsky, *A Primate's Memoir: Love, Death and Baboons in East Africa* (London: Vintage Books, 2002), 238.

> seconds fencing against Saul before he scattered them.
>
> The next day they were joined by Daniel and, as a measure of how much they just needed cannon fodder for this great enterprise, Benjamin. Six against one. I was betting on Saul. He emerged at the edge of the forest, and they surrounded him … It seemed like the assassination of Caesar …
>
> Saul made his decision, launched himself at Levi and Joshua. I'm sure he would have gotten away with it, scattered the six, but Menasseh got in a lucky shot from behind. He lunged at Saul's back as the latter leapt, managing to hit Saul's haunches. It knocked him off balance, and he missed Levi and Joshua, landing on his side. And everyone was on him in an instant.
>
> For three days afterward, he lay on the forest floor. Why he wasn't killed by hyenas then, I'll never know. He'd lost a quarter of his weight, his shoulder was dislocated and his upper arm broken, and his stress hormone levels were soaring. He recovered, though it was iffy for a while. [But] he was never in another fight, never mated again, disappeared to the bottom of the hierarchy. And he returned from whence he came, back into the wilderness.[3]

Note both the savagery and the coalition building in this description. Sapolsky goes on to note that the victorious coalition lasted for just one morning, followed by months of chaos as ranks flip-flopped.

Besides their aggression, a second difference between baboons and domesticated animals is that baboon troops have an impressive level of organization. The typical troop has a hierarchy of males with an alpha male at the head of the troop. The alpha has priority access to desirable females during their most fertile period, and thus tends to father most of the young. Once an alpha male has occupied his position long enough to regard the young as likely to be his progeny, he becomes highly protective of the troop.

When they travel, baboon troops move in a highly organized fashion. They set off in a coordinated manner in which males, especially the alpha male, are likely to initiate the direction of travel.[4] The most vulnerable troop

---

[3] Ibid., 204.

[4] S. Stueckle & D. Zinner, "To Follow or Not to Follow: Decision Making and Leadership During the Morning departure in Chacma Baboons," *Animal Behavior* 75 (6) (2008): 1995–2004; A. J. King, C. M. S. Douglas, E. Huchard, N. J. B. Isaac

members are protected in the center, and when the troop is faced with danger the higher status males race forward to face it.[5] Domesticated animals also tend to have a hierarchy or "pecking order," but are less cooperative.

Another difference between baboons and domesticated animals is that baboons seem to be intolerant of crowding. Early in the twentieth century a colony of mainly male hamadryas baboons was established at Whipsnade Zoo in England. The result was carnage. Many males and virtually all females and young were killed over a period of months and there were other differences characteristic of highly stressed societies, notably a status hierarchy that was both steep and unstable.[6] It is perhaps not surprising that animals as aggressive as baboons should be stressed when confined to a small space. While hamadryas baboons differ from savannah baboons in that they form one-male rather than multi-male troops, it is likely that savannah baboons would also be intolerant of crowding.

Another distinction between baboons and domesticated animals concerns relations between the sexes. Baboon males dominate females, largely because they are much bigger. Females may be threatened and attacked, although most of the time dominant males protect them.

Finally, we may consider attitudes towards offspring. As noted earlier, domesticated animals are typically low C and generally make poor parents. Baboons, on the other hand, are good parents—for a while. For instance, baboons are hugely interested in newborn babies, which have a distinctive black coloring, and make every effort to touch and handle them. In one population, adult females attempted to handle a newborn infant every 9 minutes, which had fallen to once every 30 minutes by the time the infant was one year old.[7] In human terms they are intensely devoted and protective mothers. But once weaned, which typically takes place when the mother goes into estrus, offspring are largely rejected by

---

& G. Cowlishaw, "Dominance and Affiliation Mediate Despotism in a Social Primate," *Current Biology* 18 (23) (2008): 1833–8.

[5] C. Sueur, "Group Decision-Making in Chacma Baboons: Leadership, Order and Communication during Movement," *BMC Ecology* 11 (2011): 26.

[6] S. Zuckerman, *Social Lives of Monkeys and Apes* (London: Kegan Paul, 1932).

[7] J. B. Silk, D. Rendall, D. L. Cheney & R. M. Seyfarth, "Natal Attraction in Adult Female Baboons (Papio cynocephalus ursinus) in the Moremi Reserve, Botswana," *Ethology* 109 (8) (2003): 627–44.

their mothers.[8] It is as if a two-year-old boy or girl, formerly a protected treasure, were left to run wild.

Baboon parenting is thus distinct from low-C parenting. Low-C mothers have less interest in their young at any age, which in humans also implies a lack of control or discipline. Baboon mothers combine the intensive care of infants with relative indifference to older offspring.

Baboon fathers also behave differently from low-C fathers in that they are more likely to protect their young. Once the alpha male loses his position, which is typically after less than a year, he continues to try and safeguard the offspring that are likely to be his. He defends them against both predators and immigrant males, who try to kill infants so as to bring females back into estrus and allow such males to sire more young. Infanticide presumably persists because it allows males to sire more of their own offspring.[9] Protection by the likely father guards against infanticide and it is often carried out at great personal risk, in that confronting an immigrant male or predator may cause injury or even death.

Introducing V

Thus, while baboons behave in ways that are characteristic of low C, there are other aspects of their behavior which cannot be explained in such terms. These include aggression, intolerance of crowding, hierarchical organization, male dominance, and intensive care of infants with rejection of juveniles. This is a constellation that we will also find in many human societies, such as the warlike pastoral tribes which have raided the settled lands throughout recorded history. The label given to this complex of behaviors and attitudes is “V,” indicating vigor and aggression.

What are the benefits of high V for baboons? The answer can be found in the baboon’s physical and social environment. Life on the savannah is dangerous. Baboons feed mostly on the ground and cannot outrun predators such as lions or leopards. Aggressive group defense greatly increases the chance of survival.

---

[8] L. T. Nash, "The Development of the Mother-Infant Relationship in Wild Baboons (Papio anubis)," *Animal Behavior* 26 (1978): 746–759.

[9] D. L. Cheney & R. M. Seyfarth, *Baboon Metaphysics: The Evolution of a Social Mind* (Chicago: University of Chicago Press, 2007), 55–58.

There is also starvation. Food on the savannah is normally plentiful, which is one reason why baboons have low C, but that is not always the case. Food can be scarce at certain times of the year, and even more so during extended periods of drought.[10] By making animals intolerant of crowding, high V encourages them to migrate, which is vital when food supplies are unstable. Animals which migrate readily are more likely to survive the next drought by finding a refuge area, while those staying in one place until all food is gone may starve. Intolerance of crowding impels migration even when food is locally plentiful, and high-V groups are more likely to survive the journey.

V-type behavior also helps baboons meet the challenges of migration: rival troops, predators, and limited knowledge of refuges such as suitable trees. Higher V makes baboons vigorous and aggressive, actively cooperating in foraging and defense, and organized to care for the young and weak.

This pattern of behavior is *not* required in the food-restricted environment of a tropical forest. Here, food resources are scattered but fairly constant so that population rises to the level of food supply. Predators cannot be a major cause of death, because heavy predation would reduce the population below the carrying capacity of the environment and food would be plentiful. This means there is less need for active defense such as mobbing, so animals with low V and high C such as gibbons flee from danger. Nor do animals in such environments normally need to migrate, so there is less need for group organization and defense.

Genes versus environment as a source of V

A difference in the level of V accounts for some differences in behavior between chimpanzees and bonobos. Chimpanzees are highly aggressive, and males have often been observed to attack and kill the males of neighboring troops. Chimpanzee males are strongly dominant over females. Bonobos are far less aggressive and have never been observed to kill each other. Also, and very unusually for primates, females tend to be dominant over males. All this suggests that chimpanzees have higher V

---

[10] S. C. Alberts & J. Altmann, "The Evolutionary Past and the Research Future: Environmental Variation and Life History Flexibility in a Primate Lineage," *Developments in Primatology: Progress and Prospects Part II* (2006): 277–303; D. K. Brockman & C. Van Schaik, *Seasonality in Primates: Studies of Living and Extinct Human and Non-Human Primates* (Cambridge: Cambridge University Press, 2005), 159.

than bonobos. Bonobos remain less aggressive than chimpanzees even in captivity with plentiful food, so that the difference is presumably genetic (as it is for gibbons, which maintain high C behavior in captivity). This pattern fits the connection between high V and famine, given that food seems to be more plentiful and nutritious in bonobo areas so there is less likelihood of famine.[11]

Baboons, like chimpanzees, appear to be genetically primed for high V. Not only are baboons aggressive by nature but their physical characteristics promote high-V behavior. The distinctive black coloring of infant baboons releases an innate protective response, and much larger males are likely to be dominant over females. Both of these factors, indulgence of infants and dominant males, will be seen as V-promoters. Animals with a genetic predisposition for high V are adjusted to life in dangerous and unpredictable environments.

Yet, just as levels of C can adapt to the environment in many species, so levels of V may vary quite widely. A study of a baboon troop in a forest fringe area, which had not suffered food shortage for a number of years, found the troop to be much less hierarchical than typical savannah troops. Males were less aggressive and more likely to flee from danger rather than protect the troop by rushing to face it.[12] In another case, baboons feeding at a garbage dump became far less aggressive, especially after a TB epidemic wiped out much of the troop (which presumably meant there was plentiful food).[13]

Japanese macaques provide another example of major differences in V-type behaviors such as hierarchy, aggression and intolerance of crowding. A study of five troops of macaques on Shodoshima Island found that they varied widely in the strength of their hierarchies, as measured by the exclusion of subordinate males from the center of the troop. The most hierarchical troops had a larger individual feeding area and defended a larger territory per monkey, an indication of aggression.[14] Findings from this study are shown in Fig. 4.1 below.

---

[11] Matt Kaplan, "Why Bonobos Make Love, Not War," *New Scientist* 2580 (December 2006); Vanessa Woods, "Humans Have a Lot to Learn From Bonobos, Scientists Say," *National Science Foundation* (May 12, 2010).

[12] T. Rowell, *The Social Behavior of Monkeys* (London: Penguin, 1972).

[13] Kaplan, "Why Bonobos Make Love, not War."

[14] M. Yamada, "Five Natural Troops of Japanese Monkeys on Shodoshima Island: II; A Comparison of Social Structure," *Primates* 12 (2) (1971): 125–150.

Fig. 4.1. Hierarchy strength, troop range per monkey and individual feeding area in five troops of Japanese macaques.[15] The more hierarchical troops controlled larger territories, an indication of high V. Compared with the least hierarchical troops, the most hierarchical had four times the range per monkey, and four times the individual feeding area.

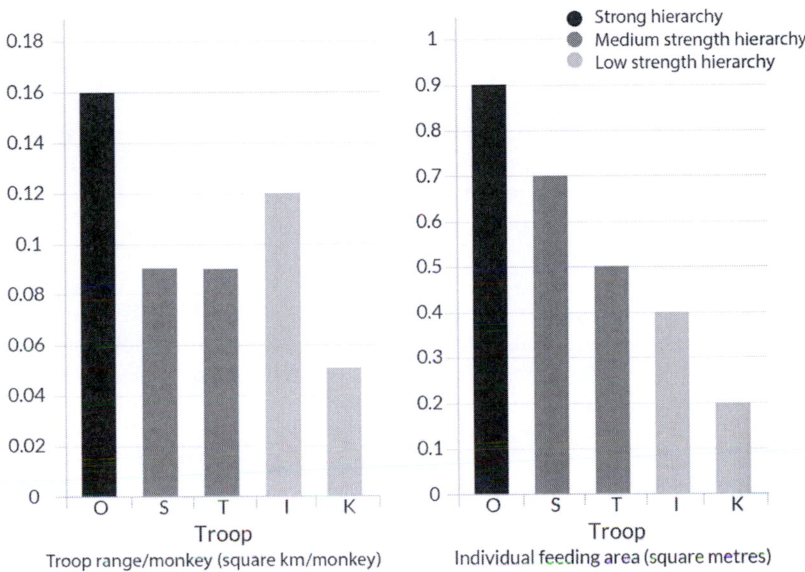

The one exception was I troop, which was less hierarchical but had a relatively large home range. The authors explained this by the fact that home ranges shrink when troops are provisioned, and the I troop was the only troop not to have been provisioned.

In addition, the more hierarchical troops were less tolerant of other troops in their home range. The biggest overlaps in home ranges were of the K troop (low hierarchy) with the I and T troops (low and medium hierarchy respectively), which points to a greater tolerance of crowding in these three groups. They required less personal space per troop member and were more tolerant of neighboring troops. They are, in our terms, lower V. This supports the idea that V behavior is variable and not fixed genetically.

[15] Ibid.

V increased by intermittent stress

Granting that V is partly set by genes, it is important to explain what causes it to increase. The trigger we propose is intermittent stress. Later, we will consider the physiological evidence for how this happens, but here we focus on features of the savannah environment that inflict stress on baboons.

First, there are stresses inherent in baboon society, such as aggressive conflicts within and between troops. Dominant animals threaten and attack subordinate ones, and even dominant males are stressed when established hierarchies break down and new ones are being established. This is attested to by high levels of cortisol measured in baboon troops at such times.

Even more fundamentally, the environment itself is highly dangerous. Baboons live on open grassland with the constant threat of attack by lions and leopards. The level of resulting anxiety is illustrated by this account of baboons crossing a dangerous water channel:

> Water crossings … are fraught with anxiety. Long before they enter the water, the baboons sit at the island's edge, nervously grunting and looking out towards the island they hope to reach. Any movement on the water's surface elicits a chorus of alarm calls and brief flight. Once they seem satisfied that the coast is clear, adults begin to cross. Reluctantly, the juveniles follow, some grunting nervously, others moaning or screaming, and others running to leap on their mothers' backs, anxious to get a ride … The whole spectacle is chaotic and amusing to the human observer but deeply distressing for the baboons, who are out of their element and vulnerable to any predator that lurks in the water or along the too-well-travelled path.[16]

Actual attacks are even more traumatic. For example, analysis of feces shows that female baboons experience a dramatic rise in glucocorticoid levels after a close relative is taken.[17]

---

[16] Cheney & Seyfarth, *Baboon Metaphysics*, 42.

[17] A. L. Engh, J. C. Beehner, T. J. Bergman, P. L. Whitten, R. R. Hoffmeier, R. M. Seymarth & D. L. Cheney, "Behavioral and Hormonal Responses to Predation in Female Chacma Baboons (Papio hamadryas ursinus)," *Proceedings of the Royal Society of London—Series B: Biological Sciences* 273 (1587) (2006): 707–12; Cheney & Seyfarth, *Baboon Metaphysics*, 57.

While further research is needed to test the idea that famine and predator threat increase V in non-human primates and rodents, there is suggestive evidence for this in human populations. A recent cross-cultural study found that societies with a history of unpredictable natural disasters are far more likely to engage in warfare—not especially at times when disasters strike but as a general pattern of behavior. Significantly, chronic food shortage (a C-promoter rather than V-promoter) was *not* found to correlate with warfare. The study also concluded that encouraging aggression in children was a consequence of war rather than the cause of it, because it declined when warfare ceased.[18]

Increased V as a response to predation helps animals to defend themselves against predators. As discussed earlier, this is partly because predators lower C indirectly by keeping populations down so that food remains plentiful. Lower C in turn causes animals to become bolder and group together, so they have a better chance of successfully warning troop members before an attack. But on the other hand, predation of troop members is highly stressful, and any reaction to short-term stress that causes animals to become more aggressive and well organized will also be very useful. In short, predation makes animals such as baboons better able to resist predators by potentially lowering C (if food becomes more plentiful) *and* increasing V.

Likewise, an increase in V as a response to famine helps animals to succeed in environments where famine is common. Famine, like predator attack, is highly stressful. Increased V as result of famine would encourage migration by increasing tolerance of crowding, and bring about aggressive and cohesive groups that would make migration successful. The advantage of the V system is that it works even when an individual has not directly experienced famine. The experience of parents or even grandparents can be transmitted through the upbringing of young baboons in the crucial years before puberty.

The transmission of V starts with baboon mothers, which are highly nurturing and protective of infants. Mothers may be anxious but they are not normally abusive or neglectful. In effect, infancy is a relatively stress-free period for baboons, apart from the anxiety transmitted by the mother. At weaning, however, a young baboon is abruptly rejected by the mother and enters a fraught and dangerous world. Apart from predation, he or she

---

[18] C. R. Ember & M. Ember, "Resource Unpredictability, Mistrust and War: a Cross-cultural Study," *The Journal of Conflict Resolution* 28 (2) (1992): 242–62.

is a low status animal in a fiercely aggressive and competitive society. It is this combination of an anxious but nurturing mother, followed by abrupt rejection and severe intermittent stresses after weaning, that transmits the high V temperament to the next generation. The reason for this will become clear when we study physiology.

Stresses continue in adult life. Although fights involving physical contact may be rare in a baboon troop with an ordered hierarchy, there is still conflict and competition. Subordinate males are often threatened and tend to have high resting levels of cortisol, which undergo only a relatively modest rise when they are challenged. This is not a healthy response.

Dominant males, on the other hand, only have chronic high cortisol levels when the dominance hierarchy is contested.[19] In other circumstances their cortisol levels are low, rising quickly under a direct challenge but then falling rapidly once the challenge is over. It is these dominant males who are the fiercest and most effective fighters. In other words, V is increased in adults by severe episodes of stress combined with relatively low stress at other times.

Thus, the V mechanism makes it possible for animals which experience famine or predation to migrate when population density rises, to organize tightly, and to react to threats with warlike ferocity. By transmitting V to the next generation through a set pattern of upbringing, it allows the next generation to act in a similar way even without any direct experience of famine or predation.

With time, abundant food causes V to weaken, just as it does C. Well-fed animals lose high-C characteristics such as good maternal care, pair bonds, territoriality, timidity, and selective mating. They also lose V characteristics such as intense care of infants, hierarchical organization, aggression and intolerance of crowding. As we will see, these are the same physiological and temperamental changes that happen to people when civilizations collapse.

Stress

To gain a deeper understanding of how famine and predation might bring about the characteristics of V, we turn to laboratory studies on the physiology and effects of stress.

---

[19] R. M. Sapolsky, "Stress in the Wild," *Scientific American* 262 (1) (1990): 106–113.

Stress can best be viewed as a response that evolved to help animals physically cope with danger. This is often referred to as the "fight or flight" mechanism. When faced with a challenge such as a hungry lion or a dangerous member of our own species, stress hormones including adrenaline, noradrenaline and cortisol are released into the bloodstream. These function to shut down bodily systems that are not related to fight or flight by diverting resources into muscles and other systems needed to cope with immediate danger.[20] Usually this reaction shuts down quickly when the danger is past, but in response to some conditions stress becomes chronic. Examples include overcrowding and constant harassment by more powerful rivals.[21]

Our concern here is the effect of stress in early life. As with calorie restriction, the effects on non-adults differ from those on adults. Further, the type and degree of stress matter a great deal. For example, a stressed mother who neglects or abuses her infant will have a very different impact on the infant from a mother whose anxiety causes her to be intensely warm and protective. Chronic and unavoidable stress has a very different, even opposite, impact from occasional stress that offers the individual some prospect of avoidance or escape. Studies of maternal neglect in rats and monkeys illustrate all of these points.

Maternal neglect

Arguably the most damaging form of stress is maternal neglect or abuse of infants. Male rats separated from their mothers during infancy develop depression-like behavior and are more aggressive to other males.[22] The same response profile has been found among primates. For example, in a series of experiments rhesus monkey infants were reared in controlled environments without mothers or contact with peers, with the intent to

---

[20] S. H. Dhabhar & B. S. McEwen, "Enhancing Versus Suppressive Effects of Stress Hormones on Skin Immune Function," *PNAS* 96 (3) (1999): 1059–1064.

[21] B. S. McEwen, "Central Effects of Stress Hormones in Health and Disease: Understanding the Protective and Damaging Effects of Stress and Stress Mediators," *European Journal of Pharmacology* 583 (2–3) (2008): 174–185; M. Kristenson, H. R. Eriksen, J. K. Sluiter, D. Starke & H. Ursin, "Psychobiological Mechanisms of Socioeconomic Differences in Health," *Social Science & Medicine* 58 (8) (2004): 1511–1522.

[22] A. H. Veenema, A. Blume, D. Niederle, B. Buwalda & I. D. Neumann, "Effects of Early Life Stress on Adult Male Aggression and Hypothalamic Vasopressin and Serotonin," *European Journal of Neuroscience* 24 (6) (2006): 1711–20.

raise animals there were hardy and healthy. The project failed. The animals reared alone were unable to relate to others, incapable of normal sexual behavior and liable to self-harm.[23] Related studies have shown that many rhesus females reared in isolation will not voluntarily mate. Not surprisingly, these females made extremely poor mothers.[24] Other studies have found that monkeys experiencing maternal abuse make poor mothers and have lasting emotional problems.[25]

Monkeys reared in social isolation exhibit bizarre and often ritualistic behavior. They are usually fearful but occasionally fearless. Hyper-aggressive or self-destructive behaviors may alternate with extreme passivity. The one common theme throughout these studies is that their social behavior makes it very difficult for them to cooperate with other members of their species. This is, of course, the opposite of the cohesive and organized high-V behavior.

The time of isolation or neglect is crucial, with the first six months being associated with more devastating effects than the second six months, even though both periods are before the normal age of weaning.[26]

The Mundugumor—a case study in infant neglect

No human society isolates infants to the degree just described for monkeys, but some give remarkably little attention and affection to infants. One such group is the Mundugumor, which was visited by the

---

[23] H. F. Harlow & M. K. Harlow, "Maternal Behavior of Rhesus Monkeys Deprived of Mothering and Peer Associations in Infancy," *Proceedings of the American Philosophical Society* 110 (1): (1966): 58–66.

[24] S. J. Suomi, "Maternal Behavior by Socially Incompetent Monkeys: Neglect and Abuse of Offspring," *Journal of Pediatric Psychology* 3(1) (1978): 28–34; Harlow & Harlow, "Maternal Behavior of Rhesus Monkeys Deprived of Mothering and Peer Associations in Infancy."

[25] A. H. Veenema, "Early Life Stress, the Development of Aggression and Neuroendocrine and Neurobiological Correlates: What Can we Learn from Animal Models?" *Frontiers in Neuroendocrinology* 30 (2009): 497–518.

[26] G. D. Mitchell, E. J. Raymond, G. C. Ruppenthal & H. F. Harlow "Long-Term Effects of Total Social Isolation upon Behavior of Rhesus Monkeys," *Psychological Reports* 18 (2) (1966): 567–80; H. F. Harlow, R. O. Dodsworth & M. K. Harlow, "Total Social Isolation in Monkeys," *Proceedings of the National Academy of Sciences of the United States of America* 54 (1) (1965): 90–7; V. Reinhardt, "Artificial Weaning of Old World Monkeys: Benefits and Costs," *Journal of Applied Animal Welfare Science* 5 (2) (2002): 151–6.

anthropologist Margaret Mead during the 1930s.[27] Infants were held in a stiff basket, often hung from a peg, and nursed quickly and roughly, which usually resulted in minimal contact with their mother. After infancy the neglect continued. Children were punished severely and inconsistently and treated with little affection. An unexpected death of a child from drowning was likely to be seen less as a personal tragedy than as an annoying mishap causing unwanted trouble.

The Mundugumor were riven by hostility and mistrust between parents and children, husbands and wives, and brothers and neighbors. What social coordination existed was the result of a few amiable men and women who were generally despised. Fear was a constant theme, ranging from the fear of other people to that of drowning. As we would expect from such a stress-filled childhood, the Mundugumor were ferociously aggressive and feared by their neighbors. In short, the Mundugumor showed all the features associated with early isolation in rhesus monkeys, including fearfulness and extreme aggression, poor social adjustment and poor parenting behavior.

While this is one case study, we may tentatively state that there is an important lesson to drawn from the Mundugumor. The story of humanity is frequently a story of struggle for survival or supremacy with more powerful societies conquering, expelling or exterminating their neighbors. Being aggressive can be an advantage in this respect and the Mundugumor were no exception. They lived in an exceptionally fertile area from which all other groups had been expelled, and they enjoyed plentiful food for which there was minimal need for work.

---

[27] Nancy McDowell, *The Mundugumor: From the Field Notes of Margaret Mead and Reo Fortune*, Smithsonian Series in Ethnographic Inquiry (Washington, DC: Smithsonian Institution Press, 1991); M. Mead, *Sex and Temperament in Three Primitive Societies* (New York: Harper Perennial, 2001). Margaret Mead's studies have numerous problems. Her Samoan work has been strongly criticized for its misunderstanding of Samoan culture, based on unreliable informants. See Derek Freeman, *Margaret Mead and Samoa: the Making and Unmaking of an Anthropological Myth* (Canberra: Australian National University Press, 1983). Her work on the Tchambuli has also been criticized as exaggerating both their pacifism and level of female influence. Biohistory also completely rejects Mead's concept of the "blank slate"— the idea that human nature is almost completely malleable. However, her observations of childrearing and character, approached with caution, have considerable value.

But their aggression came at a high price. The Mundugumor lacked social cohesion and were poor parents—exactly the same behaviors found in monkeys deprived of maternal care. One consequence was that their population had been dropping markedly in recent years despite plentiful food. By contrast, baboons are effective mothers and their groups are cohesive and well-organized. So while both societies suffer extremes of stress, Mundugumor are low V and baboons tend to be high V.

Cortisol and aggression

In chapter two we suggested that food restriction can have different effects on infants and adults, most notably with respect to testosterone. Restricting the food of adults reduces testosterone, while restricting the food of mothers has no effect or may even increase the testosterone of offspring.

We have seen that neglect or abuse in infancy can have damaging effects in later life, including hyper-aggression. A number of studies have found that experience of stress in later childhood and adolescence also makes children and animals more aggressive.[28] On the other hand, chronically

---

[28] Veenema et al., "Effects of Early Life Stress on Adult Male Aggression and Hypothalamic Vasopressin and Serotonin"; Veenema et al., "Early Life Stress, the Development of Aggression and Neuroendocrine and Neurobiological Correlates: What Can we Learn from Animal Models?"; S. Brummelte, J. L. Pawluski & L. A. M. Galea, "High Post-Partum Levels of Corticosterone given to Dams influence Postnatal Hippocampal Cell Proliferation and Behavior of Offspring: A Model of Post-Partum Stress and Possible Depression," *Hormones and Behavior* 50 (3) (2006): 370–82; J. Kaufmann, B. Birmaher, J. Perel, R. E. Dahl, S. Stull, D. Brent, L. Trubnick, M. Al-Shabbout & N. D. Ryan, "Serotonergic Functioning in Depressed Abused Children: Clinical and Familial Correlates," *Biological Psychiatry* 44 (10) (1998): 971–81; C. S. Widom, "Mother-Child Interactional Style in Abuse, Neglect, and Control Groups: Naturalistic Observations in the Home," *Psychological Bulletin* 106 (1) (1989): 3–28; M. J. Essex, M. H. Klein, E. Cho & N. H. Kalin, "Maternal Stress Beginning in Infancy May Sensitize Children to Later Stress Exposure: Effects on Cortisol and Behavior," *Biological Psychiatry* 52 (8) (2002): 776–84; T. D. Barry, S. T. Dunlap, S. J. Cotton, J. E. Lochman & K. C. Wells, "The Influence of Maternal Stress and Distress on Disruptive Behavior Problems in Boys," *Journal of the American Academy of Child and Adolescent Psychiatry* 44 (3) (2005): 265–73; E. J. Mash & C. Johnston, "Sibling Interactions of Hyperactive and Normal Children and their Relationship to Reports of Maternal Stress and Self-Esteem," *Journal of Clinical Child & Adolescent Psychology* 12 (1) (1983): 91–99.

stressed adults with higher levels of stress hormones tend to be less aggressive.

Studies on rats are consistent with this view. Rats with a dysfunctional glucocorticoid system, and thus low levels of circulating CORT, are more aggressive to intruding rats than controls. The study also found a greater level of activity in areas of the brain that control fear and stress, suggesting that they feared the intruder. Once these rats were treated with an injection of glucocorticoids they were no more aggressive than controls.[29] In studies of mice and hamsters, stressed animals with higher baseline CORT were found to be less aggressive.[30]

The effect of CORT in inhibiting aggression seems to apply mainly when it is chronically high. Trout given a single cortisol treatment are more aggressive towards intruders than before the injection, but became less aggressive following three days of treatments.[31] There is other evidence showing that low doses of CORT increase aggression, whereas high doses reduce it.[32]

The same pattern has been found in humans. Low cortisol levels are associated with aggression in school-age boys.[33] This also applies to adolescent boys for verbal and physical aggression, cruelty to people and

---

[29] J. Halász, Z. Liposits, M. R. Kruk & J. Haller, "Neural Background of Glucocorticoid Dysfunction-Induced Abnormal Aggression in Rats: Involvement of Fear- and Stress-Related Structures," *European Journal of Neuroscience* 15 (3) (2002): 561–9.

[30] J. A. Politch & A. I. Leshner, "Relationship between Plasma Corticosterone Levels and Levels of Aggressiveness in Mice," *Physiology & Behavior* 19 (6) (1977): 775–780; J. C. Wommack, A. Salinas, R. H. Melloni & Y. Delville, "Behavioral and Neuroendocrine Adaptations to Repeated Stress During Puberty in Male Golden Hamsters," *Journal of Neuroendocrinology* 16 (9) (2004): 767–75.

[31] Ø. Øverli, S. Kotzian & S. Winberg, "Effects of Cortisol on Aggression and Locomotor Activity in Rainbow Trout," *Hormones and Behavior* 42 (1) (2002): 53–61.

[32] D. K. Candland & A. I. Leshner, "A Model of Agonistic Behavior: Endocrine and Autonomic Correlates," *Limbic and Autonomic Nervous Systems Research*, Edited by L. V. DiCara, 137–163 (New York: Plenum Press, 1974).

[33] K. McBurnett, B. B. Lahey, P. J. Rathouz & R. Loeber, "Low Salivary Cortisol and Persistent Aggression in Boys Referred for Disruptive Behavior," *Archives of General Psychiatry* 57 (1) (2000): 38–43.

pets, destructive behavior, lying, truancy, vandalism and stealing.[34]

Testosterone is associated with aggression, but it is clear that cortisol also plays its part. For example, one study found that prisoners who had committed violent crimes had higher levels of testosterone, but the link was strongest among prisoners with low cortisol levels. The findings suggest that cortisol may directly moderate the relationship between testosterone and aggressive behavior.[35] Similarly, a study of boys in a delinquency program showed that those exhibiting high testosterone were more aggressive than those with low testosterone levels, but only if their cortisol was low.[36]

People and animals that respond effectively to challenge tend to have low resting cortisol levels which ramp up quickly in response to crisis, and then decline when the crisis is past.

A study of air traffic controllers found that the best-adjusted controllers had generally low levels of cortisol coupled with a rapid and effective response to stress (such as an unusually heavy workload). Other controllers showed high chronic cortisol levels and a less active response to stress. Members of the latter group were less happy with their work and showed more psychological problems than the low cortisol controllers.[37]

Thus it is that while past experience of stress *increases* aggression, chronic ongoing stress (indicated by higher cortisol) *reduces* aggression. This is why dominant male baboons are so ferocious—they have experienced severe stresses in the past but are currently high status, and thus less likely to be chronically stressed. They were also indulged and protected as

---

[34] I. van Bokhoven, S. H. M. Van Goozen, H. van Engeland, B. Schaal, L. Arseneault, J. R. Séguin, D. S. Nagin, F. Vitaro & R. E. Tremblay, "Salivary Cortisol and Aggression in a Population-Based Longitudinal Study of Adolescent Males," *Journal of Neural Transmission* 112 (2005): 1083–96.

[35] J. M. Dabbs, G. J. Jurkovic & R. L. Frady, "Salivary Testosterone and Cortisol among Late Adolescent Male Offenders," *Journal of Abnormal Child Psychology* 19 (4) (1991): 469–78.

[36] A. Popma, R. Vermeiren, C. A. M. L. Geluk, T. Rinne, W. van den Brink, D. L. Knol, L. M. C. Jansen, H. van Engeland & T. A. H. Doreleijers, "Cortisol Moderates the Relationship Between Testosterone and Aggression in Delinquent Male Adolescents," *Biological Psychiatry* 61 (3) (2007): 405–11.

[37] R. M. Rose, C. D. Jenkins, M. Hurst, B. E. Kreger, J. Barrett & R. P. Hall, "Endocrine Activity in Air Traffic Controllers at work III. Relationship to Physical and Psychiatric Morbidity," *Psychoneuroendocrinology* 7 (2–3) (1982): 125–34.

infants by highly anxious mothers. Thus they are able to cooperate and their aggression is appropriately channeled. Their cortisol levels are normally low, but rise rapidly when challenges arise, whether from a leopard or a competing male.

Intermittent stress: the toughening effect

So far we have a somewhat paradoxical picture. People and animals are more aggressive following early experience of stress, which is associated with elevated CORT. Aggression is also increased by a single infusion of cortisol, so long as it is not too high. But aggression is associated with *lower* levels of cortisol. How can the same hormone have seemingly opposite effects? An answer can be found in what may be termed the "toughening effect," a process by which short-term exposure to stress trains the body to cope better with future stresses.[38]

We saw earlier that maternal neglect or separation has a lasting impact on behavior. Separating infant rats from their mothers for more than three hours makes them more reactive to stress, more fearful, and less willing to explore novel environments or foods. But handling rats for brief periods up to fifteen minutes has the opposite effect. Their adrenal glands become larger but they are less reactive to stress as adults, less fearful, and more adventurous. One possible explanation is that short separations make the mother more attentive to pups in the form of licking and grooming while longer separation makes her less attentive.[39]

But the toughening effect is a more likely reason, as indicated by studies of the direct administration of CORT, which simulates many of the effects of stress. Studies have found that administering high levels of CORT to mothers after birth reduces maternal care and produces offspring which are more fearful as adults and have an impaired adrenocortical response to stress. However, moderate CORT in drinking water has beneficial effects. It improves maternal care, protects against brain damage, and improves learning and memory.[40] And it has also been found to protect the offspring

---

[38] J. Coates, *The Hour Between Dog and Wolf: Risk-Taking, Gut Feelings and the Biology of Boom and Bust* (London: Harper Collins, 2012).

[39] M. J. Meaney, "Maternal Care, Gene Expression, and the Transmission of Individual Differences in Stress Reactivity Across Generations," *Annual Review of Neuroscience* 24 (2001): 1161–92.

[40] Brummelte et al., "High Post-Partum Levels of Corticosterone given to Dams influence Postnatal Hippocampal Cell Proliferation and Behavior of Offspring: A

against Ischemic brain damage, a form of stroke.[41] Our own studies have even found a slight increase in the sperm count of the male offspring. Further, CORT is thought to help protect infants against infection.[42] Thus while severe stress, especially combined with maternal neglect of infants, is harmful, moderate stress without neglect or abuse can be beneficial.

This effect can also be seen after infancy. When hamsters are stressed during puberty by exposure to a dominant male, they became more aggressive as adults than animals not experiencing such stress, while the same exposure as adults is associated with lower levels of aggression.[43] Note that the mere exposure to a higher status animal has this effect, even when there is no actual attack or even threat. In the next chapter we will see evidence of V-promoting effects when parents systematically control children, even in the absence of punishment. The effects are less strong in the absence of punishment, but they still exist. Thus, parental control is a V-promoter as well as a C-promoter but only when applied in *late* childhood. Parental control in infancy works to undermine the indulgence required for the maximum level of V.

Apart from their effects on cortisol, repeated short-term stresses may increase the capacity of amine-producing cells that are responsible for

---

Model of Post-Partum Stress and Possible Depression"; A. Catalani, G. S. Alemà, C. Cinque, A. R. Zuena & P. Casolini, "Maternal Corticosterone Effects on Hypothalamus–Pituitary–Adrenal Axis Regulation and Behavior of the Offspring In Rodents," *Neuroscience & Biobehavioral Reviews* 35 (7) (2011): 1502–17; S. Brummelte & L. A. M. Galea, "Chronic Corticosterone during Pregnancy and Postpartum affects Maternal Care, Cell Proliferation and Depressive-Like Behavior in the Dam," *Hormones and Behavior* 58 (5) (2010): 769–79; P. Casolini, M. R. Domenici, C. Cinque, G. S. Alemà, V. Chiodi, M. Galluzo, M. Musumeci, J. Mairesse, A. R. Zuena, P. Matteucci, G. Marano, S. Maccari, F. Nicoletti & A. Catalani, "Maternal Exposure to Low Levels of Corticosterone during Lactation Protects the Adult Offspring against Ischemic Brain Damage," *The Journal of Neuroscience* 27 (26) (2007): 7041–6.

[41] Casolini et al., "Maternal Exposure to Low Levels of Corticosterone during Lactation Protects the Adult Offspring against Ischemic Brain Damage," 7041–7046.

[42] J. Newman, "How breast milk protects newborns," http://www.bobafamily.com/pdf/Breastfeeding/HowBreastmilkProtectsNewborns.pdf (accessed September 4, 2014).

[43] Wommack et al., "Behavioral and Neuroendocrine Adaptations to Repeated Stress During Puberty in Male Golden Hamsters."

noradrenaline, adrenaline and dopamine.[44] When released from the adrenal gland, adrenaline and noradrenaline act very quickly in focusing attention and initiating the release of glucose into the blood, thus readying the body for challenge. Unlike cortisol, however, they do not reduce the body's repair and maintenance systems, such as the immune system.[45] They are absorbed very quickly, so their effects are short-lived. Nor do they cause emotional distress.

In contrast, chronic stress depletes amines and allows them little time to recover.[46] Depleted amines are associated with many psychiatric disorders.[47] Reduced dopamine is also found among people who are lacking in motivation and unable to experience enjoyment, even such familiar pleasures as favorite foods. Depleted noradrenaline availability can lead to a lack of arousal and enthusiasm, and depressed people commonly combine low levels of noradrenaline and dopamine with chronically high cortisol.[48]

Short-term stresses give the amines time to recover, and the experience makes the amine-producing cells more productive. The net effect is that people are made capable of coping with stresses via an amine response

---

[44] A. Adell, C. Garcia-Marquez, A. Armario & E. Gelpi, "Chronic Stress Increases Serotonin and Noradrenaline in Rat Brain and Sensitizes their Responses to a Further Acute Stress," *Journal Of Neurochemistry* 50 (6) (1988): 1678–1681.

[45] S. C. Segerstrom & G. E. Miller, "Psychological Stress and the Human Immune System: A Meta-Analytic Study of 30 Years of Inquiry," *Psychological Bulletin* 130 (4) (2004): 601–630.

[46] M. J. Weiss, P. A. Goodman, B. G. Losito, S. Corrigan, J. M. Charry, W. H. Bailey, "Behavioural Depression produced by an Uncontrollable Stressor: Relationship to Norepinephrine, Dopamine, and Serotonin Levels in Various Regions of Rat Brain." *Brain Research Reviews* 3 (2) (1981): 167–205; Sunanda, B. S. Rao & T. R. Raju, "Restraint Stress-Induced Alterations in the Levels of Biogenic Amines, Amino Acids, and AChE Activity in the Hippocampus." *Neurochemical Research* 25 (12) (2000): 1547–1552.

[47] B. S. McEwen, "Central Effects of Stress Hormones in Health and Disease: Understanding the Protective and Damaging Effects of Stress and Stress Mediators," *European Journal of Pharmacology* 583 (2–3) (2008): 174–185.

[48] M. Tichomirowa, M. Keck, H. J. Schneider, M. Paez-Pereda, U. Renner, F. Holsboer & G. Stalla, "Endocrine Disturbances in Depression," *Journal Of Endocrinological Investigation* 28 (1) (2005): 89–99; K. A. Roth, I. M. Mefford & J. D. Barchas, "Epinephrine, Norepinephrine, Dopamine and Serotonin: Differential Effects of Acute and Chronic Stress on Regional Brain Amines," *Brain Research* 239 (2) (1982): 417–424.

rather than the more damaging cortisol response.[49]

When an amine response to a challenge is insufficient and a cortisol response is optimal to ready the body for action, repeated short-term stresses also seem to render the cortisol response more effective.[50] The body is conditioned so that cortisol ramps up when needed and declines rapidly when the immediate challenge has passed. This toughened response to challenge avoids the damaging effects of long-term stress. Almost any kind of stressor appears to have this toughening effect on rats, including handling by humans, vigorous exercise, mild shock or cold.[51] The important point is that each episode must be relatively short-term, allowing ample time to recover.

The effect of a stressor depends not only on its duration but on how much the stressed individual feels in control of it. This is illustrated by a series of experiments in which two monkeys were yoked together in the laboratory. At random intervals both were given an electric shock. One, the "coping" monkey, could turn off the shock for both monkeys. The other, the "passive," monkey, could not. Though both animals experienced the same pain, the passive monkey had no way to control or affect the outcome.[52]

Electric shocks are painful. Thus, it is not surprising that both monkeys were stressed in the experiment. Response to stress, however, is complex and involves different hormones and parts of the brain. In this experiment, each monkey reacted differently. The coping monkey showed more of what is known as a Cannon response, the passive monkey more a Selyean response. Table 4.1 below summarizes these response differences.

---

[49] Adell et al., "Chronic Stress Increases Serotonin and Noradrenaline in Rat Brain and Sensitizes their Responses to a Further Acute Stress," 1678–1681; S. Jordan, G. Kramer, P. Zukas & F. Petty, "Previous Stress Increases in Vivo Biogenic Amine Response to Swim Stress," *Neurochemical Research* 19 (12) (1994): 1521–1525.

[50] P. H. Wirtz, U. Ehlert, M. U. Kottwitz, R. La Marca & N. K. Semmer, "Occupational Role Stress is Associated With Higher Cortisol Reactivity to Acute Stress," *Journal of Occupational Health Psychology* 18 (2) (2013): 121–131.

[51] J. Irwin, A. Pardeep, R. M. Zacharko & H. Anisman, "Central Norepinephrine and Plasma Corticosterone Following Acute and Chronic Stressors: Influence of Social Isolation and Handling." *Pharmacology Biochemistry and Behavior* 24 (4) (1986): 1151–1154.

[52] J. P. Henry & P. M. Stephens, *Stress, Health and the Social Environment* (New York: Springer-Verlag, 1977), 118–41.

Table 4.1. Cannon and Selyean responses to stress.[53] Stresses over which the individual feels some sense of control have very different effects on hormones, brain activity and behavior from those which are perceived as uncontrollable. Cannon responses have the "toughening effect," which allows the body to respond more effectively to future challenges. In terms of Biohistory, they increase V. Selyean responses are more likely to lead to chronic stress.

	Cannon Response	Selyean Response
Neuro-endocrine system	Sympathetic-adrenal-medullary	Pituitary-adrenal-cortical
Hormones	Adrenaline, noradrenaline, testosterone	ACTH, corticosterone, testosterone falls
Area of brain	Amygdala	Hippocampus septum
Evoked more by	Threat to control; can fight back, or threat is of known and fixed duration	Loss of control; unable to respond effectively or predict outcome
Behavior	Arousal, defense, displays aggression and anger	Depression, withdrawal, low mobility, lower sex and maternal drives
Personality of people responding in this way	Ambitious, driving, vigorous, strive for dominance	Subordinate animals, people who lack self-acceptance and have a poor sense of their worth
Some related pathologies	More likely to have heart disease	Tumors and peptic ulcers, viral and bacterial infections

The features shown in Fig. 4.1 above explain why punishment or neglect in infancy, when the individual is too young to understand why it is being administered and is helpless to avoid it, can lead to lifelong anxiety. The infant's response is Selyean, and so future response to challenge is likely to be Selyean. By contrast, a brief period of separation or the high CORT of an anxious but nurturing mother are likely to produce a Cannon response, and future response to challenge is more likely to be of the Cannon type.

---

[53] Ibid.

The situation changes when the child is older. Chronic and unavoidable stress is still likely to produce a Selyean response, but occasional punishment may be met by a Cannon response. Older children, unlike infants, know why they are being punished. It may be that they have done something wrong, or simply failed to stay out of reach of an angry parent. They also understand that the punishment will end, just like the yoked monkey which can turn off the electric current. Thus, punishment in infancy tends to undermine V, while intermittent punishment in later childhood may increase it.

This toughening effect is likely to be stronger in larger families, which are consistently linked to more authoritarian parents, physical punishment and child neglect.[54] If nothing else, having another baby in short order is likely to reduce the protection and indulgence of older children, most of whom will be well past infancy, so that the effect is to increase rather than reduce V.

This pattern is well understood by people in some cultures. In an Egyptian village discussed in greater detail in chapter six, the youngest and only children were seen as lacking the experience of competition and conflict in childhood that produces aggressive and competitive adults. As a result, parents positively encouraged sibling rivalry and what we would term "bullying" as necessary experiences to produce the aggressive and competitive temperament required for success in their culture.[55]

Maternal anxiety, high status males

There is one final V-promoter to consider, which will play a hugely important role in human societies. Anxious mothers, provided they are not neglectful, seem to produce adult offspring with a toughened response to challenge. As discussed in chapter two, food restriction increases CORT in mothers but *lowers* it in their adult offspring. This has a great effect, as shown in Fig. 4.2. The adult offspring of food-restricted mothers have less than half the CORT levels of controls, implying that exposure to CORT in infancy reduces the chronic level of CORT in adults.

---

[54] S. J. Zuravin, "Unplanned Childbearing and Family Size: Their Relationship to Child Neglect and Abuse," *Family Planning Perspectives* 23 (4) (1990): 155–161

[55] H. Ammar, *Growing up in an Egyptian Village* (New York: Octagon, 1966), 108–9, 129.

Fig. 4.2. Serum corticosterone levels in rats that were calorie restricted in early life.[56] Rats whose mothers were calorie restricted while nursing them have lower levels of the stress hormone CORT as adults. Since CR increases CORT, this suggests that exposure to higher CORT in infancy causes rats to have lower CORT levels as adults.

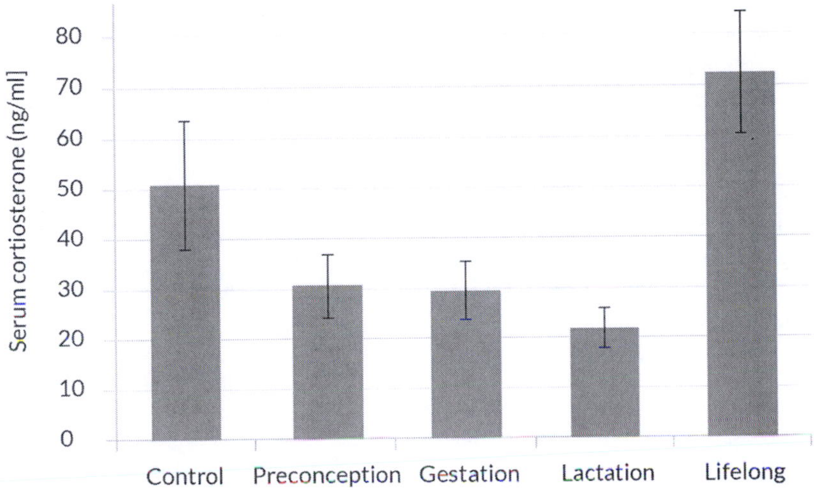

As we have seen, if an anxious mother is also abusive and neglectful then the infant experiences a Selyean-like stress response which renders it less able to cooperate as an adult. This is low V. But an anxious mother who is doting and affectionate creates a Cannon-like response in her infant so that it becomes more confident, cooperative, assertive and aggressive when the need arises. This is high V.

Another important factor behind V is high adult status, because cortisol inhibits aggression. This means that male V is maximized when males are strongly dominant over females. A male infant is reared by a subordinate and thus anxious mother, and grows up to be a dominant and thus less anxious male. This helps to explain the extremely high V of baboons, since males tend to be strongly dominant over females (aided by their much larger size). Putting this in human terms, V is increased by patriarchy.

---

[56] E. A. Levay, A. G. Paolini, A. Govic, A. Hazi, J. Penman & S. Kent, "HPA and Sympathoadrenal Activity of Adults Perinatally Exposed to Maternal Mild Calorie Restriction," *Behavioral Brain Research* 208 (2010): 202–8.

To summarize—four influences are required to maximize the aggressive but cooperative high V temperament. One is that the mother be anxious, and transmit this anxiety to her infants. The second is that the mother also be indulgent and protective of her infants so that their stress response in later life is Cannon-like rather than Selyean-like. The third is that the juvenile experiences frequent but intermittent stresses after infancy, even if only by exposure to powerful and higher status individuals. The fourth is relatively high status as an adult.

Epigenetic Changes

It is likely that all these influences have epigenetic effects, and this even applies to adults. Rats given access to a running wheel were found to react much better to stress than controls, and showed clear differences in the epigenetic mechanism of methylation, which influences the expression of certain genes expressed in the brain.[57]

Support for this can be found in our study of the epigenetic effects of maternal food restriction.[58] In chapter two we saw significant effects on genes relating to the control of testosterone, but there was also an impact on genes relating to CORT. This is significant because food restriction causes a moderate increase in CORT and thus could act as a V- promoter.

One such change was found in the glucocorticoid and mineralocorticoid receptors, which react to high levels of CORT by dampening down CORT production and preventing its rise beyond a certain level. In our maternal CR animals the activity of these receptors was reduced, which should allow the levels of CORT to rise higher than otherwise. Similarly, the POMC gene has the effect of signaling the release of ACTH, so activation of this gene should increase ACTH. POMC was significantly more active in the pre-conception group and should therefore be associated with higher levels of ACTH. Since these animals had *lower* levels of CORT and ACTH than controls, the likely conclusion is that changes to these genes

---

[57] A. Collins, L. E. Hill, Y. Chandramohan, D. Whitcomb, S. K. Droste & M. H. M. Reul, "Exercise Improves Cognitive Responses to Psychological Stress through Enhancement of Epigenetic Mechanisms and Gene Expression in the Dentate Gyrus," (2009), http://www.plosone.org/article/info%3Adoi%2F10.1371%2Fjournal.pone.0004330 (accessed Septmber 4, 2014).

[58] J. Penman et al., Report series on studies on rats conducted at Latrobe University 2012, (forthcoming).

help make possible the more active and efficient stress response characteristic of higher V individuals. This contrasts with that of animals or people who are chronically stressed and show only a muted stress response to danger.

Genetic "set points" for V

We saw in chapter two that each species has a genetically-based "set point" for C arising from the environment for which it has been adapted. For example, gibbons are adapted to a tropical forest environment where food supplies are normally stable and have a high set point for C. Baboons are adapted to the savannah environment where food is normally plentiful and have a low set point for C.

The same applies to V. As discussed above, baboons are naturally high in V, a state reflecting the threat of predators and famine to which they are also adapted. Along with the behavioral tendency to high V they also have distinct physical features that support V. These include the larger size of males, which helps make them dominant over females, and a distinctive black coloring of infants that promotes the intensive care of the very young. By contrast, gibbons lack the physical characteristics to support V. Males and females are similar in size and infants have no obvious differences in coloration.

Like baboons, humans evolved for life on the dangerous savannah, so it seems reasonable to suspect that we too would have high V. However, based on our knowledge of egalitarian hunter-gatherer communities, humans are likely to be moderately low in terms of V.[59] Compared with a typical baboon troop, most hunter-gatherer societies treat children with mildness and affection. Also, the size difference between men and women is more like that of gibbons than baboons, and human infants lack a distinctive coloration. On balance, it is likely that humans are genetically primed for moderate to low V.

---

[59] This is not to suggest that hunter-gatherers are uniformly peaceable. Acts of violence may be more common in such societies than in civilized communities.
A. Gat, *War in Human Civilization* (Oxford: Oxford University Press, 2006); S. Pinker, *The Better Angels of our Nature* (New York: Penguin, 2011). However, even in the most violent and unequal of societies, such as the Yanomamo described in chapter six, no single male dominates in the way that a dominant baboon male does, controlling access to all females.

But as with C, the set point for V can change via natural selection in response to any long-term change in the environment. In species such as gibbons, which are narrowly adapted to their environment and seem to vary little in social behavior, the level of V seems to be relatively fixed. But V in other species, such as humans and baboons, can be changed by environmental conditions. In later chapters we will return to these differences when we examine human societies in which members are as fierce and hierarchical as any baboon troop, and other societies that are as mild and tolerant of crowding as domesticated animals. Just as high V in baboons is associated with the fierce aggression and tight organization needed to fight off leopards and other baboon troops, high V gives a human society the characteristics that make it successful in war. Fig. 4.3 illustrates the differences between species in the level and variability of V.

Fig. 4.3. Hypothetical set points and range of variation for V. Gibbons are predisposed to low V and baboons to high V. The behavior of hunter-gatherer groups suggests that humans are predisposed towards moderately low V. But compared with other primates, human societies show a startling range of variation in levels of V.

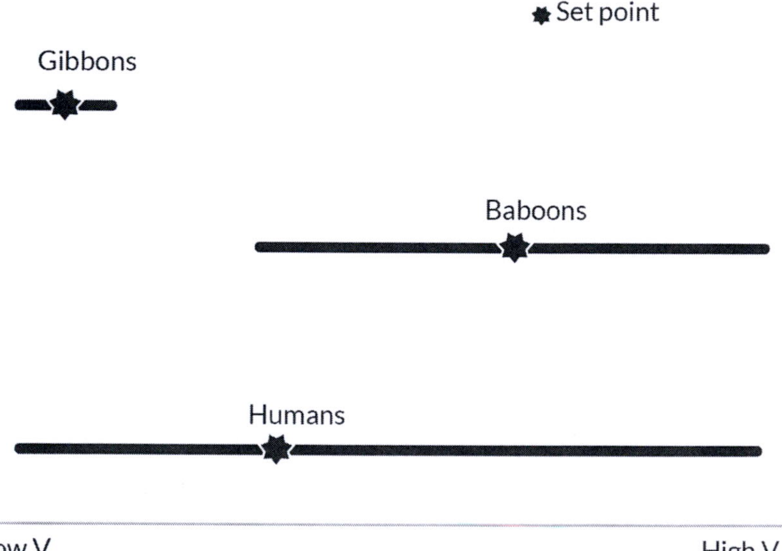

Understanding V

In its most fundamental sense, V helps animals to thrive in a dangerous and unpredictable environment. It is a cluster of behaviors and attitudes that includes aggression, high morale and the psychological ability to effectively respond to challenges without long-term stress. It is the result of offspring being reared by anxious mothers, provided that mothers are not neglectful or abusive. It is increased by intermittent and relatively controllable stresses in childhood after the age of weaning. For a child a controllable stress is one that can be relieved by a specific behavior such as behaving as parents require, or that is of known and fixed duration. For adults, high status lowers resting cortisol levels and thus promotes V.

High V individuals have low resting levels of cortisol and other stress hormones, but with the potential to mount a rapid and effective stress response in the face of challenge. They have the capacity to be aggressive, but in appropriate circumstances and in cooperation with other members of the species. They are less likely to be socially maladjusted. They react to threat with a Cannon-like rather than a Selyean-like response.

V is subject to the same *effect feedback cycle* that applies to C. Any behavior resulting from V will cause V to increase. For instance, the aggression of high-V males can lead to confrontation and conflict, which in turn can lead to periodic stresses that increase V.[60]

V is reduced by chronic and uncontrollable stresses, and most strikingly by maternal neglect or separation in infancy. The biochemical effects of such early experience include chronically high cortisol and ineffective responses to stress in later life. Low-V individuals tend to be passive, lacking in confidence, and have poor social adjustment.

V in human societies

In chapter three it was suggested that gibbons are genetically primed for high C and humans for low C, but human C is far more variable. Raising C beyond this level is highly advantageous for some societies, though it requires that people act in ways at odds with temperament, such as by eating less food or having less sexual activity than they would prefer.

---

[60] See chapter three for a more extensive discussion of this concept.

The same applies to V. Humans are both genetically primed for low V *and* highly variable in their level of V. And V can be highly beneficial if it helps a society gain success in war. Stress alone is not enough to achieve this state, as indicated by the Mundugumor headhunters discussed earlier. Although feared by their neighbors they were poorly organized and would only fight when the odds were very much in their favor. Success in war requires tight organization and a higher order of courage.

A high level of V can help tribes and nations in their struggle for survival. High-V peoples are warlike, confident, well organized at the local level, and frequently conquer or drive out neighboring groups. This point is captured by the medieval Muslim philosopher Ibn Khaldun, who describes the recurrent influx of patriarchal and warlike peoples into the settled lands with dense farming populations and cities, where they took the local women for their own and imposed elements of their culture and sometimes language.[61] Perhaps the best example of a high-V culture in recent times would be the peoples of Afghanistan who have fought two superpowers to a standstill during the past few decades, although many Afghans fought for both the Soviets and the Americans.

Throughout history, the advantages of high V have been equally as great for complex, large-scale civilizations. In later chapters we will propose that, along with violence and destruction, barbarian invaders brought civilization a long-term benefit by imposing V-promoting customs on the peoples they conquered. Such practices not only maintain birth rate but allow local groups in complex civilizations to organize and defend themselves. In turbulent times, local defense is often the key to survival. For this reason, we must consider the ways in which V is raised in human societies.

Higher V from harsh and unstable environments

Harsh environments with unstable food supplies tend to increase V. Warlike tribes ranging from the Elamites of the twenty-fourth century BC to the Arabs of the eighth century AD repeatedly overran the settled lands of the Middle East. These tribes came from harsh and unforgiving environments, whether mountains or deserts. They were pastoralists, living on marginal land not suited to farming because of poor water supplies or

[61] Ibn Khaldūn, *The Muqaddimah : An Introduction to History*, translated from the Arabic by Franz Rosenthal, 3 vols. (New York: Pantheon Books, 1958).

rugged terrains.[62] The same pattern of aggressive pastoralists overrunning farming peoples can be seen in other areas, including Africa and the Asian steppes.[63]

Once they became established in wealthy and fertile areas they lost much of their martial vigor and became victims of more warlike peoples in turn. Ibn Khaldun also noted this, believing that barbarian tribes, because of their relative cohesion, youth and vigor, conquered decadent civilizations that had become soft through easy living.[64]

Raising V through culture

A more significant source of V in human societies can be found in culture. Just as cultures raise C by such methods as restraints on sexual activity and religious rituals, so they have developed ways to raise the level of V.

One such method is fasting, which is a V-promoter as well as a C-promoter, especially when it is severe but infrequent, as with the annual Lenten fast of Catholic and Orthodox Christianity. The most rigorous and effective V-promoter is the Ramadan fast of Islam which forbids eating or drinking during daylight hours for an entire month. Such fasts have the effect of V-promoting famines, as compared to milder but more frequent fasts with have the C-promoting effects of chronic mild food shortage.

V is also increased by commandments to honor parents, and by behavior associated with principles such as "spare the rod and spoil the child." Both of these reinforce hierarchy, and punishments inflict the kind of short-term stresses that toughen the stress response for later life.

The society of ancient Sparta can be seen as a system designed to maximize the level of V. The Spartans were, in their time, the best warriors in Greece. Their whole way of life was a preparation for war, and it incorporated exactly the kind of short-term stresses described earlier.

---

[62] Gat, *War in Human Civilization*, 189–221.

[63] T. Barfield, *The Nomadic Alternative* (New Jersey: Prentice Hall, 1993); K., Fukui, D. Turton, I. Karp, "Warfare among East African Herders," *American Ethologist* 9 (1): 203–204;
E. M. Thomas, *Warrior Herdsmen* (New York, London: W. W. Norton, 1981); E. Hildinger, *Warriors of the Steppe: A Military History of Central Asia, 500 B.C. to 1700 A.D.* (Cambridge, Mass.: Da Capo Press, 2001).

[64] Khaldūn, *The Muqaddimah.*

According to Plutarch, boys were trained by constant, vigorous exercise, running and swimming as well as wrestling and fighting. They were provided with only a single tunic and cloak in the coldest weather. They were left short of food and often had to steal or starve. They were trained to endure pain, such as in a public ceremony where boys ran a gauntlet of flogging to steal cheeses from an altar. Taken together, these factors together caused Spartan warriors to have higher V than their rivals, most of whom were part-time soldiers, giving them a key advantage in war.[65]

English public schools of the nineteenth century provide another example of high-V training. They had a similar (if less rigorous) system to the Spartans characterized by vigorous sport, cold showers, physical punishment, and institutionalized bullying through the fagging system. The Duke of Wellington is said to have observed that "the battle of Waterloo was won on the playing fields of Eton," indicating a connection between a Spartan upbringing and British Imperial success:

> Almost unconsciously the public school boy absorbed a complete code of behavior which would enable him to do the "right thing" in any situation. It involved obedience to superiors, the acceptance of a position in the hierarchy, team spirit, and loyalty. It produced the gifted amateur, trained for nothing but ready for anything, who had a relaxed air of command, a sense of duty and a feeling of the obligation of the superior to his inferiors. It also involved the traditional British phlegm, reserve, understatement, unflappability, the stiff upper lip, a result of the inculcation of modesty in victory and defeat, the all-male society in which emotion was sissy, the encouragement of restraint in the exercise of power.[66]

For the English, the mid-nineteenth century was a time of maximum C, reflected in the emphasis on reserve, discipline and restraint. But their focus on hierarchy, team spirit and toughness is the epitome of V. This was also specifically training for the English upper classes, whose higher status as adults would reduce chronic stress and bring V to a maximum.

Military Drill

Increasing the level of V is almost certainly the main value of military drill. While new recruits need to be able to use weapons and work as a team, it is not immediately obvious why marching in step and presenting

---

[65] Plutarch, *On Sparta* (London: Penguin, 2005), 21–22.

[66] J. Richards, *Visions of Yesterday* (London: Routledge and Kegan Paul, 1973), 40.

arms for several weeks might create better soldiers. The key is that such activities force the soldiers to behave in a way typical of high V, which means acting cooperatively within a steep and controlling hierarchy. Adopting high-V behavior tends to raise V, an example of the *effect feedback cycle.*

Unlike in the eighteenth century, marching in step is not the way modern armies fight, but experience has shown it to be an effective way of creating effective soldiers. In Goodbye to All That, Robert Graves describes his time as an instructor and front-line officer during the First World War, and his observations that proficiency at parade-ground drill correlated with combat effectiveness:

> We [the officers at the Harfleur "Bull Ring"] all agreed on the value of arms-drill as a factor in morale … I used to get big bunches of Canadians to drill: four or five hundred at a time. Spokesman stepped forward once and asked what sense there was in sloping and ordering arms, and fixing and unfixing bayonets. They said they had come across to fight, and not to guard Buckingham Palace. I told them that in every division of the four in which I had served … there were three different kinds of troops. Those that had guts but were no good at drill; those that were good at drill but had no guts; and those that had guts and were good at drill. These last, for some reason … fought by far the best when it came to a show … I told them that when they were better at fighting than the Guards, they could perhaps afford to neglect their arms-drill.[67]

Severe and intermittent stresses, such as a period of famine or attack by a predator attack, or simply by exposure to powerful authority, also increase V in adults. Traditional military training is designed to simulate just these kinds of stresses. An extreme example is the behavior of the Imperial Japanese army before 1945 where soldiers were physically beaten by their superiors. In Western armies abuse has traditionally taken the form of insults and dressing down by an NCO, but the effects are likely to be similar. Any such experience, especially combined with a program of rigorous exercise, will elevate stress hormones such as cortisol.[68]

Harsh environmental conditions provide another form of stress. The most rigorous military training is provided to elite units, exemplified in a

---

[67] R. Graves, *Goodbye to All That* (London: Penguin, 1958), 156.

[68] J. A. Smith, "Evaluation of Cortisol and DHEA as Biomarkers for Stress." Paper 626, (2008), http://scholarship.shu.edu/dissertations/1045/ (accessed September 3, 2014).

description of the US Navy SEAL training called "Hell Week":

> Trainees are constantly in motion; constantly cold, hungry and wet. Mud is everywhere—it covers uniforms, hands and faces. Sand burns eyes and chafes raw skin. Medical personnel stand by for emergencies and then monitor the exhausted trainees. Sleep is fleeting—a mere three to four hours granted near the conclusion of the week. The trainees consume up to 7,000 calories a day and still lose weight …
>
> Throughout Hell Week … instructors continually remind candidates that they can "Drop-On-Request" (DOR) any time they feel they can't go on by simply ringing a shiny brass bell that hangs prominently within the camp for all to see.[69]

An instructor at the San Diego facility had no doubt that the key factor in the training of SEALS is psychological rather than physical:

> The belief that [this training] is about physical strength is a common misconception. Actually, it's 90 percent mental and 10 percent physical … [Students] just decide that they are too cold, too sandy, too sore or too wet to go on. It's their minds that give up on them, not their bodies.[70]

If the purpose of SEAL-type training was purely to build fitness, this could be achieved far more easily and reliably by an ongoing regimen similar to that used in professional sports training. But its aim is also to build a toughness of mind and an aggressive confidence, which is exactly what these elite units require. They require high V.

The rigorous training also serves another purpose—to select people who already have the necessary toughness. Seventy percent of candidates fail stage one of SEAL Basic Training, which is striking given that the candidates specifically volunteered for what is known to be a tough and arduous course. In other words, candidates would normally have high V before volunteering, and those with the highest V would be the most likely to survive because their stress response has been honed by past experience

---

[69] About Careers, "SEAL Training Hell Week," http://usmilitary.about.com/od/navytrng/a/sealhellweek.htm (accessed September 4, 2014).

[70] Ibid.

to effectively meet the challenge.[71]

A case study—cult purchase in the New Guinea highlands

V can be increased in adults but, as with C, the most important influencing factors are the experiences of early life. The initiation ceremonies of preliterate peoples, which boys needed to go through to be considered men, seem designed to elevate V and at the same time inculcate cultural values. The peoples themselves are fully aware of the benefits of such ceremonies. In the Enga culture of the New Guinea highlands, cults were purchased by local leaders with the direct aim of improving the behavior of young men.[72]

Clan leaders chose initiation rites believed to produce longevity, strength and the capacity for hard work and responsibility. Also valued was the ability to raise many pigs, speak in public, and father many children. Clansmen met to compare the economic and political performance of other clans, and then approached the most impressive. Criteria included the bachelors' physiques, the clan's success in producing and exchanging goods and in defending its land, and whether it had produced prominent young men. Payment —in pigs, salt, axes and oil—was arranged for rites, spells, sacred objects, training for local experts, and supervision of the performance. The cost of the sacred objects alone could be as much as a young man's bride price, which allowed him to get married.

Initiation rituals imposed hardships designed to separate boys from their mothers—emotionally as well as physically. Candidates were confined in ritual shelters and forbidden to urinate or defecate for days. Their diet was restricted. During the ritual they were marshaled and subjected to frightening sights and sounds. Some boys fled in terror. Those that persisted would be made to sit in darkness for extended periods awaiting further trials. When the rituals were completed, the initiates were required to observe an extended period of strict food taboos, lifted one at a time over the following months and years. These practices would impose severe intermittent stress (V) and intense control of behavior (C).

---

[71] An account of the value of military drill, and other co-ordinated forms of movement, can be found in William McNeill, *Keeping Together in Time* (Cambridge MA.: Harvard University Press, 1997).

[72] P. Wiessner & A. Tumu, *Historical Vines: Enga Networks of Exchange, Ritual, and Warfare in Papua New Guinea* (Washington DC: Smithsonian Institute Press, 1998).

Bachelor cults were designed to prepare young men for adult responsibilities and leadership. Bachelors were judged in need of reform if they showed inappropriate sexual behavior or were generally ineffective. There were also ceremonies designed to deal with sickness and other calamities.

The toughest and most demanding cults, such as the Kaima ancestor cult, originated in the tougher environment of the western highlands. These were also the most successful, spreading widely through the Enga cultural area. The toughest environments produce the men with highest V who are most likely to develop high-V cults and religions. This is something we will examine later when studying the rise of Islam.

Patriarchy

Another function of the successful cults was to strengthen patriarchy, accomplished by focusing on male ancestors and providing men with a crucial ceremonial role. And of all V-promoting customs in human societies, patriarchy is probably the most important.

The cross-cultural survey discussed in the previous chapter shows this pattern (see also Table 4.2 below). Two measures of patriarchy were used: formal dominance, which reflects patriarchal customs, and actual dominance in the home. Societies in which men dominate women are more aggressive, both in terms of internal conflicts and in competition with neighboring groups, than those in which women have higher status.

Table 4.2. Aggression, male dominance, and punishment.[73] Patriarchal societies, and those which punish older children, tend to be more warlike. Both patriarchy and punishment are V-promoters but patriarchy appears to be the most effective.

	Formal male dominance	Home dominance	Punishment in late childhood
Conflict within community	.38**	.47**	.28*
Feuds with other communities		.46**	
significance	** .001	* .01	

[73] See www.biohistory.org.

These findings indicate that aggression is increased not by the customs that underlie patriarchy (formal male dominance), but by the actual control and domination of women (home dominance). A woman's anxiety is likely to be influenced by her relationship with her husband rather than what society says it should be. Home or actual dominance also correlates moderately with feuding, while formal dominance and punishment of older children do not. A likely explanation is that more civilized societies tend to suppress feuds.

The link between patriarchy and V applies to every level and type of society. When Margaret Mead visited the ferocious Mundugumor of New Guinea, she also spent time with a tribe called the Tchambuli who lived nearby. She describes them as a society where women were psychologically dominant over men.

According to her account, Tchambuli customs required each boy to kill a captive for his initiation, which indicates that they had once been warlike, but in recent times had lost all taste for violence. Their initiations could only be completed by purchasing captives such as orphans from neighboring tribes. As a result they became subject to increased raiding, and shortly before European contact they were driven from their fertile lakeside home by more aggressive peoples, returning only when warfare was prohibited by the colonial administration.[74] Later studies suggest this is an exaggeration. The Tchambuli did war on neighboring tribes on occasion, many men beat their wives, and neither sex was clearly dominant. But it seems clear that they were both less aggressive and less patriarchal than their neighbors.[75]

Anglo-Saxon England is another example of a society where declining patriarchy was accompanied by reduced capacity for warfare. English society was formally patriarchal, but it was a nation where women had unusually high statuses by the standards of mainland Europe at that time. A woman was not to be married against her will and she retained her property in marriage. Women could take oaths and were able to act as grantors, grantees and witnesses of charters. They could own and bequeath land, and surviving wills show no differences in the treatment of sons and

---

[74] Mead, *Sex and Temperament in Three Primitive Societies*, 232.

[75] D. E. Brown, *Human Universals* (New York: McGraw Hill, 1991); Education Portal, "Tchambuli Tribe: Culture, Gender Roles & Lesson," http://education-portal.com/academy/lesson/tchambuli-tribe-culture-gender-roles-lesson.html#lesson (accessed September 4, 2014).

daughters.[76]

But higher status for women was associated with growing military weakness, especially by contrast with high-V Norsemen from the bitterly cold and famine-prone lands of Scandinavia. King Aethelred, who reigned between 978 and 1016, paid massive sums to buy off Danish invasions. He was briefly deposed by the Danish king Sweyn in 1013, and Sweyn's son Cnut seized control after his death.

Eventually, in 1066, England fell to the Normans, Norsemen who had settled in northern France and whose aptitude for war made them rulers of territories ranging from southern Italy and Sicily to England. They were more patriarchal than the English, as illustrated by changes in English law after the conquest. A husband now had absolute right to control his wife's dowry and could give it away or sell it. Sons were favored over daughters in inheritance and widows were no longer the guardians of their own children. And their laws reflected the presumption that women were totally obedient and submissive to their husbands and unable to resist their will.[77] In other words, biohistory suggests that the high status of Anglo-Saxon women reduced English V and thus their capacity for war, which made conquest by a higher-V people possible.

Patriarchy is linked not only to conflict and warfare but to a number of other variables which we have previously linked to V (see Table 4.3 below). This indicates that patriarchal societies are not only more aggressive but show stronger local organization and hierarchy. When patriarchy is stronger local leaders have higher status, kin groups are more unified, and adult sons tend to obey fathers. This is similar to the pattern of behavior seen among baboons, which combine aggression with a cohesive, hierarchical social organization. It is this social pattern that adapts animals to a dangerous, changeable environment, and makes humans better at waging war. Patriarchal peoples, like baboons, tend to form aggressive, well-organized and hierarchical local groups. Such groups contrast strikingly with the Mundugumor, who are aggressive but not well organized.

---

[76] C. Fell, *Women in Anglo-Saxon England* (Oxford: Blackwell, 1984).

[77] F. G. Buckstaff, "Married Women's Property in Anglo-Saxon and Anglo-Norman Law and the Origin of the Common-Law Dower," *Annals of the American Academy of Political and Social Science* 4 (1893): 33–64.

Table 4.3 Variables linked to patriarchy and sexual restrictions in the cross-cultural survey.[78] Stronger male dominance is a sign of V and is linked to restrictions on sexual activity, to strongly organized local groups (local leaders high status, unity of kin groups, adult sons obey fathers), to children wanted (a measure of how much children add to a woman's status), and to punishment in late childhood—another V-promoter. Restrictions on sexual behavior are also related to patriarchy and other variables associated with V (local leaders high status, adult sons obey father, children wanted).

	Formal male dominance	Dominance at home	Premarital sex restricted	Adultery restricted
Premarital sex restricted	.29*	.28*	xx	.58**
Adultery restricted	.39**	.21	.58**	xx
Local leaders high status	.29	.31*	.32*	.30
Unity of kin groups	.37	.36*		
Adult sons obey father	.47**	.44**	.39*	.29
Children wanted	.36*	.37*	.25	.31
Punishment 0-12	.30*	.26	.36*	.30
significance	** .01	* .01	Others:.05	

Effective warfare requires not only aggression but coordination, and the willingness to risk or even sacrifice one's life in the service of one's group. It is the dominant baboon who puts the females and young behind him and turns to face the leopard; the soldier who covers the man on his left with his shield and has equal confidence that the man on his right will do likewise. Such behavior is the essence of high V.

---

[78] See www.biohistory.org.

Table 4.3 indicates that patriarchal societies also tend to restrict sexual behavior, usually explained by the need for men to have greater confidence in the paternity of their children. But this does not explain why patriarchy in the home shows a strong correlation to premarital chastity (.28) than to restrictions on adultery (.21). Nor does it explain why in some Muslim cultures a girl's chastity is such a concern to her family that they may even take her life if she offends. A more plausible explanation is that limiting sexual activity is also a V-promoter, as confirmed by significant correlations of sexual restraint with local leaders having high status and adult men obeying fathers. Women under stress are more likely to neglect and abuse infants, so raising a woman's C—by restricting her sexual activity or controlling her in other ways—has the potential to counteract the effects of stress by making her more maternal.

Patriarchal societies also tend to have a high birth rate, as suggested by the "children wanted" variable which measures how much the status of women depends on having more children. Once again there are similarities to baboon society, where intense care of young infants and rejection of infants after the first few months reduce the interval between births. This is an adaptation to a dangerous and changeable environment where fast breeding is vital to group survival.

This link between V and fast population growth explains why people from the deserts and steppes migrate repeatedly into more settled areas. From one perspective, this is contrary to common sense. Populations should grow where food is plentiful on the fertile plains and river valleys, and remain fixed or decline in mountains or deserts where food is scarce. Yet the opposite is more often the case, and V theory provides an explanation.

Finally, Table 4.3 indicates that patriarchal societies are more likely to punish children in late childhood. Both patriarchy and punishment of older children are V-promoters and, from the principle of the *effect feedback cycle*, are also consequences of high V. Thus, patriarchy should make people more likely to punish older children, and punishment of older children should make societies more patriarchal.

While punishment of older children is a V-promoter, the theory predicts that punishment of infants should *reduce* V. As indicated by monkey studies, neglect and abuse of infants leads to a Selyean-type response which undermines V. Table 4.4 shows that the effect of punishing infants is in some ways opposite to that of punishing older children.

Table 4.4. Variables that link to punishment of infants and patriarchy.[79] Punishment of infants is negatively correlated with unity of kin groups, adult sons obeying fathers, and children wanted—all aspects of V that correlate positively with patriarchy.

	Punishment 0–2		Dominance at home
Unity of kin groups	-.30		.36*
Adult sons obey father	-.26		.44**
Children wanted	-.44		.37*
significance	**.001	*.01	Others: .05

These findings are also consistent with an observation first developed in chapter two—the response of animals to a stimulus is highly dependent on the age at which the stimulus is experienced. For example, early food restriction increases or has no effect on testosterone, while later food restriction reduces it.

Harsh treatment of juveniles

There is another factor likely to increase V. We have noted that baboon mothers, though solicitous and protective of very young infants, tend to start rejecting them at a relatively young age. This is part of the "fast breeding" profile of high V—mothers are freed to become pregnant again as soon as possible. But given that early separation from mothers is a result of high V it may also be the cause of it due to the *effect feedback cycle*. One possible reason is that ongoing contact with a highly anxious mother after weaning acts to set lifetime anxiety levels higher, even though such experience before weaning does not. Also, maternal supervision may protect a child from rough play and bullying, which serve to increase V.

This interpretation implies that people with younger siblings should have higher V as long as the gap between offspring is no more than 2–3 years, because the birth of a new baby normally reduces the amount of time a mother spends with a child. This means that having more offspring should increase the V of the offspring, just as V increases the birthrate.

---

[79] See www.biohistory.org

Research by Frank Sulloway provides some support to this view. He found first-born sons to be more conservative and identify more with parents.[80] Children receiving a drop in attention at the birth of a younger sibling should have higher V and thus be more conservative.

The *effect feedback cycle* predicts that an increase in V from any cause, including occasional severe stresses, should increase male status relative to females. In other words, living in harsh environments such as deserts or mountains or the frozen north should, other things being equal, promote patriarchy. And because V promotes confidence and high morale, having high status will also promote V. This is the pattern characteristic of baboons—the most aggressive (high-V) animals are the high status males. Table 4.5 below summarizes these findings.

Table 4.5. Characteristics of V and the factors that increase it.

V increased by	V characteristics
Mother anxious but maternal to infants	Males dominant over females
Women's sex restricted	
Early weaning; another sibling	Higher birthrate
Occasional stresses: cold, hunger, predators, threat, exposure to powerful authority	Aggressive, intolerant of crowding
	Copes well with challenge
Military basic training, e.g. drill	Cooperates in hierarchical group
Higher status as adult	Confident, high morale

The entries in Table 4.5 are consistent with the view that V is a system that facilitates adaptation to unstable environments. People or animals react to occasional severe stresses, such as famine, by changing their behaviors to adapt to such conditions. They breed faster, become more aggressive and confident, and cooperate within hierarchical groups. These same behavioral changes cause them to treat their young in such a way as to elevate their V. More confident and dominant males can be expected to make females more anxious, for the same reason that lower status animals

[80] F. J. Sulloway, *Born to Rebel: Birth Order, Family Dynamics, and Creative Lives* (New York: Pantheon Books, 1996).

and people in any environment tend to be more anxious than those with higher status. Shorter birth intervals mean juveniles are released from maternal protection and exposed to intermittent stresses in the environment.

Among baboons, stresses are provided by powerful males. Although dominant males are not generally anxious they are inclined to threaten or even attack lower status animals. It is these outbursts that allow baboon males to fight off leopards and strive with each other for dominance. A similar outburst directed against a juvenile is a massively stressful experience, but one soon ended by flight. Even without any specific threat, their very power acts as a V-promoter for juveniles.

In human terms, the equivalent is a beating or lecture administered by a father or schoolmaster, or any form of systematic control. Other intermittent stresses result from incidents of conflict with similarly aggressive peers, or occasional bullying by older children. In all these cases the stress is severe but of limited duration which leads to the "toughening" effect described earlier. The overall effect is to produce the high-V character that is optimal for survival in an unstable environment.

The function of such behavior is that the generation that has experienced famine or severe predator stress transmits its experience to the younger generation, who may not have had the same experience. The younger generation thus has much of the aggression and group cohesion necessary to cope with an unstable and dangerous environment. However, their V will not be as strong as the older generation, and without further reinforcement will fade over time.

Summary and conclusions

The factor we have labeled V, which can be identified in the behavior of animals such as savannah baboons, is also found is humans and is one of the key factors in biohistory. V is created by an attentive mother who is made anxious by low social status, famine or social trauma. It is reinforced by abrupt separation from the mother at weaning (around age two in humans), such as can result from the birth of another infant. It is further strengthened by periodic stresses after weaning and to a lesser extent in adulthood, or any exposure to powerful authority. Thus, parental control in late childhood is not only a C-promoter but also a V-promoter, though only mildly so in the absence of punishment. Finally, V comes to a peak when adults have relatively high status and thus lower cortisol.

Animals and people with high levels of V are aggressive, confident and function well within small group hierarchies. Males tend to be dominant over females. Males with high V are more effective soldiers and tend to have more children than males with low V. Societies with high V are more likely to make war and conquer their opponents. Like C, V is strongly linked to the restriction of female sexual activity.

V is a mechanism that has evolved to adjust social behavior to resource-rich but unstable and dangerous conditions. When predation is common and there are periodic famines, animals are afflicted by intermittent stresses or by exposure to high status animals, and thus become better at breeding, fighting and migrating. Some human societies have developed religious and other cultural systems that reinforce V, such as patriarchy, annual fasts and close-order drills. This gives them a competitive advantage over peoples with lower in V.

In the next chapter we look more closely at the formation of C and V during childhood.

CHAPTER FIVE

INFANCY AND CHILDHOOD

In previous chapters we have seen that environmental influences, especially those early in life, cause epigenetic and hormonal changes which have broad effects on temperament and behavior. In this chapter we review key laboratory and field findings from earlier chapters and consider how they inform our understanding of human behavior and societies.

C is a trait that adapts populations to chronically limited food supplies. Mammals with very high C become less tolerant of each other, an extreme form of which is found among monogamous pairs of gibbons that defend exclusive territories. Humans tend to form pair bonds at all levels of C, but high-C humans have a strong bias towards monogamous nuclear families. They are temperamentally inclined towards hard, repetitive work. High-C people also tend to do better in trade and the professions, activities which require a willingness to sacrifice in the present for future benefit, and more impersonal social relationships. They also appear to be more law abiding and better at forming larger-scale political units. C-promoters include control of children at all ages, mild hunger, and sexual restraint.

V is a system that adjusts populations to dangerous environments with erratic food supplies. High-V animals and humans tend to form cohesive, hierarchical groups led by powerful and dominant males. They are aggressive, confident and fast-breeding. V-promoters include patriarchy, control and punishment of *older* children, any form of threat or stress to older children or adults, and sexual restraint.

While both C and V are triggered by environmental influences, they have cross-generational effects. High-C parents tend to act in a way that gives their offspring high C, while high-V parents transmit V to their offspring. C and V may also be transmitted in the womb by hormones and other biochemical influences, and by direct epigenetic inheritance.

Until now we have considered C-promoters and V-promoters as simply raising the level of C and V respectively. But, in fact, C-promoters and V-promoters in infancy have very different effects from those in later childhood. In this chapter we will be introducing two new characteristics: infant C as a form of C and child V as a form of V. We will also be considering the effects of childrearing patterns on the overall level of stress.

Child V—authority and stress in late childhood

V-promoters include an anxious but attentive mother, relatively abrupt separation from the mother after weaning, experiences of authority and stress in later childhood, and occasional stresses in later life. While all of these contribute to V, the experiences of late childhood have a distinct effect which will be termed "child V." The main effect of this is to make people more accepting of authority in general. Cross-cultural studies have found that parents in politically complex societies are stricter and more controlling of children.[1] They train children to be obedient, to act dutifully and responsibly in performing tasks, to restrain themselves from showing emotion easily, and to dress modestly.

To gain a more detailed picture, the cross cultural survey discussed in chapter three and detailed at www.biohistory.org, focused on societies where it was possible to estimate the level of control and punishment as measured separately for the periods before and after weaning. Control was defined as the systematic direction of a child's behavior, and punishment was defined as the infliction of pain or stress. Control and punishment tend to be correlated, but it is quite possible for control to be exerted without punishment, such as by gentle direction or the appeal to a child's love. It is also possible (and regrettably not uncommon) for punishment to be exercised without control, such as a parent hitting a child out of anger rather than for any specific act.

The levels of punishment and control measured were for infancy and the peak level in childhood (ages 0–12), but in practical terms the peak levels are equal to those found in late childhood, since, with one exception (the Manus, to be dealt with below), that is where they are found.

Our own findings confirm those of other studies in showing that control and punishment of children is strongly related to political complexity. But they also made it clear that the crucial period is *late* childhood (see Table 5.1 below).

---

[1] H. Barry, I. L. Child & M. K. Bacon, "Relation of Child Training to Subsistence Economy," *American Anthropologist* 61 (1) (1959): 51–63.

Table 5.1. Variables linked to control, punishment and patriarchy. Punishment and control of older children is significantly linked to wider political authority. Patriarchy is only related to the power of local leaders and has a much stronger link with aggression.

	Control		Punishment		Men dominant at home
	0–2	0–12	0–2	0–12	
Large political unit	.25	.56**		.25	
Arbitrary government		.44**		.25	
Government feared		.25	.25	.23	
Leader's high status		.38**		.36**	.31
Deference		.28		.32	
Marriages based on status		.35			
Hereditary status	.28	.50**			
Conflict				.28*	.47**
Feuds	-.28				.46**
significance	**.001	*.01			Others: .05

Another V-promoter, patriarchy, is included for the sake of comparison. It is strongly linked to aggression and to the power of local leaders—the kind of aggressive behavior and tight organization which is characteristic of baboon troops, but *not* to the acceptance of wider political authority. Societies which punish children in late childhood are also somewhat more aggressive, but far more likely to accept wider political authority.

The effects of stress

Though both punishment and control in late childhood work to increase child V, they have other and distinct effects. Parental control, which is a C-promoter as well as a V-promoter, is unrelated to aggression. In fact, societies which control children in infancy are actually *less* likely to feud. All of this is consistent with our picture of C as inhibiting aggression and promoting more disciplined behavior. The one form of political behavior most characteristic of parental control is respect for hereditary rank, which is based on loyalty rather than fear.

Punishment, on the other hand, is most strongly related to acceptance of arbitrary authority, to deference, and the tendency to fear leaders. This is the behavior observed in an extreme form in the Whipsnade baboons, as discussed in chapter four, when crowding stress gave rise to a brutal leader who attacked and tyrannized the other animals.[2] This suggests that punishment not only increases child V more than control, but also raises the level of stress.

Conventional thinking is that powerful and brutal leaders create stress. By contrast, biohistory proposed that powerful and brutal leaders rise in stressed societies because people are more ready to obey them than in low-stress societies. For example, consider a political regime where two leaders—which we will call "B" (for brutal) and "C" (for charismatic)—are vying for power. B imprisons and murders his political opponents while C tries to win them over with favors and persuasion. In a low-stress society where people are motivated by loyalty or reward they find B repulsive and C attractive, so C gains supporters. In a stressed society where people are motivated more strongly by fear, they may like C but are more likely to obey B. Thus B gains supporters.

Dictators may not need to win elections but they do need supporters, and having a style of leadership that attracts support is the key to political power. Thus it is that stressed societies create brutal leaders, not vice versa. Failure to understand this point caused the utter fiasco of the recent wars in Iraq and Afghanistan, which tried to create democracy where the psychological underpinnings did not exist. The Russians solved their problems in Chechnya far more effectively by installing a brutal dictator of their own choosing.

Punishment in late childhood is compatible with powerful and authoritarian states because it increases child V and thus acceptance of authority. Punishment in infancy is even more strongly related to stress, as was the case with the Mundugumor and as suggested by the strong correlation between punishment in infancy and "government feared." But it does not correlate with size of political unit or any aspect of C or V because it undermines V and thus social cohesion. The next chapter will show how high levels of stress as a result of late childhood punishment

[2] S. Zuckerman, *Social Lives of Monkeys and Apes* (London: Kegan Paul, 1932).

play out in an analysis of political attitudes in an Egyptian village.

Child V and tradition

There is one further aspect of child V to be noted at this stage, and for this we must turn again to animal studies. A group of male rats was stressed before puberty and then conditioned to associate a sound with a shock. As adolescents, but not as adults, they showed an increase in conditioned fear, and as adults this conditioning was harder to extinguish than in rats which had not been stressed.[3] In a related study, rats stressed with a foot shock during their third week were also slower to abandon a conditioned fear in later life, something not found when rats experienced foot shock during their second week of life.[4] This suggests that stress in the juvenile period or at puberty makes conditioning easier to impose but harder to erase when animals become adults.

In human terms, what this means is that people experiencing authority in late childhood should be more conservative in their values, but only when combined with indulgence in infancy. The Mundugumor, whose childrearing methods stressed both infants and older children, showed elements of high-V behavior in their ferocity but were anything but conservative. A constant complaint among members of this society was that the old traditions and values were breaking down. This connection between child V and tradition is a finding that will take on major importance in later chapters.

Infant C

It is now time to consider C. For many purposes, C can be considered as a single entity, since the level of C in adults strongly correlates with the way infants are treated. But the effects of C-promoters on infants are different from their effects on older children and adults.

---

[3] M. Toledo-Rodriguez & C. Sandi, "Stress before Puberty Exerts a Sex- and Age-related Impact on Auditory and Contextual Fear Conditioning in the Rat." *Neural Plasticity* (2007) (article ID 71203:1–12); G. E. Hodes & T. J. Shors, "Distinctive Stress Effects on Learning During Puberty," *Hormones and Behavior* 48 (2) (2005): 163–71.

[4] M. Matsumoto, H. Togashi, K. Konno, H. Koseki, R. Hirata, T. Izumi, T. Yamaguchi & M. Yoshioka, "Early Postnatal Stress Alters the Extinction of Context-Dependent Conditioned Fear in Adult Rats," *Pharmacology Biochemistry and Behavior* 89 (3) (2008): 247–52.

The first type of C to be considered is termed "infant C," which stems from C-promoters in infancy. In the rat experiments discussed in chapter two, this was the group whose mothers experienced calorie restriction (CR) while nursing. Some aspects of infant C may be partially reversed by CR in later life. For example the androgen receptor, which was less active in rats only experiencing CR in infancy, was more active in the group experiencing CR throughout life. But most effects of C-promoters in infancy are irreversible, which suggest that infant C in humans is largely fixed by age four or five.

An individual's infant C tends to reflect the C of their parents, since an adult's C normally determines the extent to which they discipline a child, but there are exceptions. In this chapter we will present societies in which parental control is exercised only in late childhood, giving them C without infant C. This will be seen to have a profound impact on temperament and thus social structure.

C

The other form of C, the result of C-promoters from earliest infancy into adult life, is simply called "C." In terms of our experiments, C is the trait associated with food shortage in juvenile and adult rats. It can be increased by any form of control by authority figures or by personal conviction, mild food shortages, requirements to work or study, and any code which disciplines and regulates behavior. After puberty, restrictions on sexual activity constitute an especially powerful C-promoter. At this point the crucial point to note is that C-promoters in childhood appear to have a more permanent and pronounced effect than those following puberty, and C-promoters immediately following puberty have more impact than those in later life.[5] It is also possible to have C substantially lower than infant C, such as when a child is controlled in infancy but grows up in more prosperous and "liberal" times (as in our own age). However, the highest levels of C *always* involve training in infancy, which is why it is referred to simply as "C" rather than (for example) late childhood/adult C.

There are also likely to be differences between the effects of C-promoters in childhood and those after puberty. Children brought up strictly tend to be disciplined and hard-working but they are not necessarily religious or

[5] Recalling the experiment cited in chapter two when sexual experience immediately after puberty had a major and lasting impact on testosterone levels.

likely to have large families. These two aspects, religious commitment and an intense interest in children, would seem to be especially associated with what may be termed 'adult C'. Such differences will be better established by future research, including both animal and human studies, but for the purposes of this book the effects of C-promoters in late childhood and adult life will be considered together.

As indicated in chapter two, one key difference between C and infant is that C in general is associated with lower testosterone while infant C is not. There are even findings to suggest that infant C may *increase* testosterone, although this is yet to be confirmed.

C versus infant C

When comparing the effects of early versus late control, the cross-cultural survey shows some strikingly different patterns (see Table 5.2 below). Correlations for punishment and unaffectionate fathers are included for comparison.

We have already noted that control of older children, along with punishment, is correlated with virtually all measures of political complexity. This is presumably a child-V effect. Early control (infant C) is not a V-promoter and so has little effect on political complexity but, along with late control, it is related to all the family and economic behaviors linked to C in chapter two. These include restrictions on sexual behavior, monogamous nuclear families, later age of marriage, status from wealth rather than generosity, routine work, the market economy, and religions that stress morality.

However, matters were different in the past. Consider the situation where success and even survival require courage to the point of recklessness, such as hand-to-hand combat with a ferocious enemy. The brutal chaos of ancient and medieval battles can scarcely be imagined today, but it is in these situations that testosterone-fuelled aggression would provide a huge advantage, especially for a professional warrior. For many societies throughout history, war at its most brutal level was a way of life. The ruling class in the ancient world was almost always established by warfare. Success in war meant a life of luxury and power, multiple women, and numerous offspring with a good chance of surviving childhood. Failure in war meant insignificance, an ugly death or the terminal horror of slavery. It is in these societies, where aggression rather than business is the key to male success, that the double standard is most prominent.

Table 5.2. Variables linked to childrearing practices in the cross-cultural survey.[6] Punishment and control were rated for their peak level in childhood, but for all but one society (the Manus) this peak was in late childhood.

	Punish 0–2	Punish 0–12	Control 0–2	Control 0–12	Father not affectionate
Size of political unit		.25	.25	.56**	.44**
Government arbitrary		.25		.44**	.45**
Fear of government	.25	.23		.25	.32*
Leader's high status		.36*		.38**	.22
Deference		.32		.28	.35
Status-based marriages				.35	.43**
Hereditary status			.28	.50**	.32
Premarital sex restricted		.30*	.22	.38**	.36**
Adultery restricted		.36*		.47**	.36*
Divorce restricted	.36*	.30*		.26*	
Frigidity	.45	.51			
Monogamy			.28	.36**	
Late age of menarche			.63		
Late age of marriage			.33*	.54**	.26
Nuclear family		.27	.33*	.33*	
Status from wealth		.31*	.33	.56**	.38
Market economy		.33**	.48**	.59**	.46**
Routine work		.27	.35*	.58**	.31*
Achievement orientation				.28*	
Deities enforce morality		.34*	.20	.51**	.46**
Modesty in dress		.30	.24	.68**	.33*
significance	**.001	*.01		Others:	.05

By far the most effective way to achieve martial vigor is through high V, especially since this promotes group discipline as well as aggression. As detailed in the last chapter, subordination of women achieves this by

[6] See www.biohistory.org

making them more anxious, and boys reared by anxious mothers who grow up to be high-status adult males have the highest level of V. But rigid control of women's chastity, with looser standards for men, also serves to maximize men's testosterone.

In such societies, as for example those of modern Afghanistan, a breath of scandal for females can bring disgrace or even death so their sex lives are rigidly curtailed. Another way to maintain women's C is to put girls and women under heavy pressure to act in specified ways. Women may be required to defer to husbands, brothers and even sons in public. They may be required to dress in a careful and modest fashion, serve others before taking food for themselves, keep busy with routine tasks such as spinning and weaving (even where there is no economic necessity for such work), and tolerate philandering husbands while making few demands in the bedroom. In the extreme form, as in some Muslim societies and ancient Athens, women have been (at least ideally) confined entirely to the home.[7] All of this would give them extremely high C, while maintaining the V and child V of their sons.

Men in such societies, on the other hand, are free to avail themselves of prostitutes whose lower levels of C are relatively unimportant as they are unlikely to produce children. Men may even have some latitude to indulge in homosexual activity, as in ancient Greece or the samurai class of medieval Japan.[8] Men of the warrior caste could also be forbidden to engage in business, which tends to promote C and thus reduce testosterone.[9] This was the case for medieval knights and for Japanese samurai, who were discouraged from engaging in business and farming. Another problem with trade for warriors is that it commonly leads to wealth and luxury which undermine V. Martial peoples such as the Spartans were well aware of the connection between hardship and virtue, as Plutarch wrote about Lycurgus:

> The third and most masterly stroke of this great lawgiver, by which he struck a yet more effective blow against luxury and the desire for riches, was the ordinance that all should eat in common. They were to partake of the same bread and meat, of specified kinds. This prevented them from living at home, lying on costly couches at splendid tables, delivering

---

[7] P. Cartledge, (ed.), *The Cambridge Illustrated History of Ancient Greece* (Cambridge: Cambridge University Press, 2002), 130ff.

[8] R. C. Kirkpatrick, "The Evolution of Human Homosexual Behavior," *Current Anthropology* 41 (3) (2000): 385–413.

[9] Cartledge, *The Cambridge Illustrated History of Ancient Greece*, 213.

> themselves into the hands of their tradesmen and cooks, to be fattened like greedy brutes. For such indulgence and excess would ruin not only their minds but their bodies and make them feeble, needing long sleep and warm baths and unable to work, just like invalids.[10]

There are, however, limits to the advantages of lower C—even for men. The most effective soldiers tend to be ferociously aggressive but also well disciplined. Roman legionaries were no braver than the Gallic warriors they defeated, but managed to combine courage with a disciplined organization which made every soldier part of a well-oiled machine. Much the same may be said for many of history's most famous warriors, such as the Japanese samurai and Spartans, who were also renowned for their discipline. And discipline as such is a product of both forms of C.

There is then a trade-off between the discipline associated with high C and the testosterone resulting from lower C. The trade-off is least pronounced for societies in which there is a warrior caste which is not expected to work. The Spartan elite had helots to do the manual labor which allowed them to focus on becoming the best warriors in Greece. Likewise, medieval knights lived by the labor of their serfs, and Japanese samurai were a class apart. In most civilizations, and especially the warlike ones, soldiers have been an elite and specialized class. Thus, an individual warrior might do well by combining the discipline of infant C with the higher testosterone from somewhat lower C. But even for professional soldiers, some level of C is likely to provide an advantage.

The Benefits of higher C for Business and the Professions

For most occupations, higher C is a distinct advantage. It is associated with hard work, self-discipline, and the willingness to sacrifice present consumption for future benefit. The cross-cultural survey shows that control in later childhood correlates at a high level with work and the market economy. Another indication is that limiting the sexual behavior of men, as opposed to just women, is strongly linked to occupational success.

As detailed in chapter three, the *Kinsey Report* found that men with a higher frequency of nocturnal emissions during adolescence tend to be more successful in terms of education and occupation. Groups such as Orthodox Jews and Jains which specialize in commercial occupations tend

---

[10] Plutarch, *Lives of the Noble Grecians and Romans* (New York: Random House, 1992).

to be especially rigorous in controlling men's sexual activity, and also to follow strict codes of discipline which act as C-promoters.

The Puritans of the sixteenth and seventeenth centuries, who lived by a rigidly defined code of moral values, are a good example. They were strict Sabbatarians and successfully lobbied to ban Sunday recreations. Before services, ministers visited their parishioners to assess that they were spirituality prepared to partake in communion.[11] "Religious exercises" or bible readings were considered a defining source of joy and regularly conducted in the home both after Sunday service and in the evenings. Families gathered in each other's homes for readings and travelled enthusiastically to meet with other parishes for spiritual cultivation.[12]

The Puritans were notorious for the strict manner in which they upheld their moral order. For example, people charged with sexual misconduct, drunkenness, gambling during the service or using incorrect liturgy were sentenced to public penance or excommunication.[13] Following the migration to America, the Puritans' penalties escalated, as one observer in Boston wrote:

> For drunkenness Robert Cole was ordered to wear a red D on his chest for a year. Three persons were whipped and banished for adultery. In this case the church elders argued that the legal death penalty for this crime was sufficiently known and should be executed, but the civil magistrates gave the offenders the benefit of doubt.[14]

Or consider the Mormons. Today, young Mormon men carrying out their required missionary service accept a degree of discipline and austerity that feeds and underpins their enthusiasm. For two years they are forbidden to date, to listen to secular music, to visit the beach and attend most forms of entertainment. Their daily activities are rigorously controlled, including the amount of time spent on prayer, scripture study and evangelism. They must dress in a carefully prescribed manner. They are assigned a missionary companion from whom they are not to part and who must always be

---

[11] D. Cressy, *Birth, Marriage & Death, Ritual, Religion, and the Lifecycle in Tudor and Stuart England* (Oxford: Oxford University Press, 1997), 299.
[12] D. Rosman, *From Catholic to Protestant, Religion and the People in Tudor England* (London: UCL Press Ltd., 1996), 61–62.
[13] K. Wrightson, *English Society, 1580–1680* (London: Hutchinson & Co., 1982), 209–215.
[14] J. Adair, *Founding Fathers, the Puritans in England and America* (London: J.M. Dent & Sons Ltd., 1982), 165.

addressed as “elder” followed by a surname. At the direction of the mission president, companions and locations are changed periodically.[15] Their discipline and routine would be expected to raise C to an exceptional degree, perhaps only matched by the more rigorous monastic orders.

Even after completing their missions, adult members of the Mormon church accept a remarkable degree of discipline including a monthly fast, performing assigned services such as regularly visiting other members of the ward, and wearing specified undergarments. Worthiness is assessed by regular interviews with local church leaders.

The outstanding success of Mormons in modern America can be understood in these terms. Mormons are barely 2% of the American population but lead JetBlue, American Express, Marriott, Novell, Deloitte and Eastman Kodak. There are more than a dozen Mormons in Congress including Senate Majority Leader Harry Reid. Mitt Romney and John Huntsman were prime contenders for the White House. Other notable Mormons include news commentator Glen Beck, the creators of the *Twilight* Series, management guru Stephen Covey, and the Osmonds.[16] This is all the more striking given the deep suspicion of the Mormons felt by Evangelicals, many of whom would not even consider them to be true Christians.

High C and religion

All of these examples are of religious groups, and there is an obvious link between high C and religious sentiment. Ministers of religion tend to have high C, as indicated by their low testosterone levels.[17] Moreover, religious people in general tend to have lower levels of sexual activity even within marriage, as indicated by the *Kinsey Report* findings discussed in chapter three. The relationship works both ways. Religious principles tend to reduce sexual activity and thus increase C, while people with higher C are more likely to be religious. The link between asceticism and religious fervor is a commonplace of history, ranging from the desert ascetics of the late Roman Empire to medieval monks and seventeenth-century Puritans.

---

[15] Missionary Handbook, LDS Church (2006).

[16] S. Mansfield, *The Mormonizing of America: How the Mormon Religion became a Dominant Force in Politics, Entertainment and Pop Culture* (Brentwood, Tennessee: Worthy Publishing, 2012), Introduction.

[17] J. M. Dabbs, "Testosterone and Occupational Achievement," *Social Forces* 70 (3) (1992): 813–24.

This connection also explains a curious point about religious denominations in the modern West. There is a strong trend throughout Western societies towards more liberal views on moral issues such as homosexuality and sexual behavior in general. One might then expect that churches adopting such views would grow the fastest because they fit with what most people want to hear. However, in reality the fastest growing churches are those with the more rigorous and traditional views such as Mormons, Pentecostals, Southern Baptists and Jehovah's Witnesses.[18] In other words, these are the churches which hold on most tightly to the C-promoting customs of traditional religious belief. By doing so they attract people with higher C, and by raising the C of their congregations create a greater sense of religious fervor.

This inclination towards religiosity and religious discipline is part of the system that allows human societies to create high C. Doing so is not easy, however. It requires that people act in ways our hunter-gatherer ancestors would consider profoundly unnatural. This is why ideologies that promote "natural" behaviors such as free sexual activity are so popular and have become dominant in declining-C societies. What is more pleasant than to be told that doing exactly what you feel like doing is not only permitted but morally superior? The downside is that "natural" human behavior creates a temperament that is very well suited to living in a low-technology environment without formal education, in groups of fifty or so, hunting wild animals and gathering plants. This is not the formula for success in a world of competing and clashing civilizations.

The fact that high-C religious groups tend to have more children is another factor that helps to maintain a society's C, despite declining C in the majority population. For example, according to the 1990 Jewish Population Survey, the birth rate of Orthodox Jews in Israel is 3.3%, which is more than double the rate for Conservative and Reform Jews (1.4% and 1.3%, respectively). And it is widely believed that the ultra-Orthodox Haredim are much more fertile than the Orthodox. There is even

---

[18] D. M. Kelley, *Why Conservative Churches are Growing* (Macon: Mercer University Press, 1986);
A. Anderson, *An introduction to Pentecostalism: Global charismatic Christianity* (Cambridge: Cambridge University Press, 2004); R. T. Cragun & R. Lawson, "The Secular Transition: The Worldwide Growth of Mormons, Jehovah's Witnesses, and Seventh-day Adventists," *Sociology of Religion* 71 (3) (2010): 349–73.

evidence of fierce competition among Haredim regarding family size.[19] The explosive growth of the Haredi community has put considerable pressure on the policy exempting their young men from military service if they are engaged in religious study. The growth of such groups is limited only by the appeal of more liberal lifestyles, which draw away a portion of the young with each generation.

Creativity

A high level of C probably also plays a significant role in creativity. Malcolm Gladwell's book *Outliers* proposes that outstanding success in any field requires at least 10,000 hours of dedicated practice, using examples from Bill Gates to the Beatles to Ice Hockey players.[20] C helps people to achieve this kind of focus, and may also help to make minds more flexible.

Parental control in late childhood is a C-promoter but it also increases child V, which has been linked to more traditional thinking. Intense control with minimal punishment increases child V only modestly and has a much larger effect on C, so is quite compatible with independent thought and creativity.

But the highest level of C requires powerful C-promoters in adult life as well, which is why limiting sexual outlet might be expected to enhance creativity. It is striking how often men of genius appear to have had very limited sex lives, sometimes because their preferences were outlawed by society. For example, Leonardo da Vinci never married and was at one time charged with sodomy.[21] Michelangelo's paintings and love poetry suggest he was homosexual by orientation.[22] Isaac Newton never married and was commonly supposed to have died a virgin.[23] Stephen Hawking was a brilliant but lazy undergraduate who only began to focus after he developed motor neurone disease, a condition known to limit sexual

[19] C. I. Waxman, "The Haredization of American Orthodox Jewry," *Jerusalem Letters* 376, (1998).

[20] M. Gladwell, *Outliers* (New York, London: Little, Brown and Company, 2008).

[21] "How do we Know Leonardo was Gay?" www.bnl.gov/bera/activities/globe/leonardo_da_vinci.htm (accessed September 9, 2014).

[22] A. Hughes, *Michelangelo* (London: Phaidon, 1997), 326.

[23] A. Storr, "Isaac Newton," *British Medical Journal (Clinical Research Edition)* 291 (6511) (1985): 1779–1784.

activity.[24] Although his condition did not prevent Hawking from marrying and siring three children.

An understanding of how this might work on a psychological level can be gained from Raymond Cattell's "16 PF," a questionnaire that measures personality on 16 different traits.[25] Cattell identifies two traits that are commonly seen as aspects of extroversion. One is dominance, which involves forcefulness and strength of character. The other is surgency, which is an outgoing "life of the party" temperament. Although these traits correlate, meaning that dominant people are more likely to be surgent, Cattell found that creative people tend to be dominant but with a *low* level of surgency. Strength of character combined with introverted focus helps make exceptional achievement possible. And restricting sexual activity, especially in adolescence, should reduce surgency.

The problem that arises is that people who reject traditional thinking, and thus have the potential for flexible thought, are also very likely to reject traditional curbs on sexual behavior, especially since these are so at odds with temperament. Thus, a person with a flexible mind set is *less* likely than most to maintain high C. It is only when flexibility of thought is combined with severe limits on sexual outlet (including masturbation), and linked with high intelligence and opportunity, that true genius arises. This helps to explain why it is so rare.

Homosexual orientation as a C-promoter

It was mentioned earlier that a homosexual orientation might contribute to genius by limiting sexual activity in a society which disapproves of it.

There is no consensus on the causes of homosexuality, though most biologists attribute it to genetic factors or the intra-uterine environment, or

---

[24] M. White & J. Gribbin, *Stephen Hawking: A Life in Science* (2nd ed.) (Washington DC: National Academies Press, 2002); T. Bridget, "The Impact of Assistive Equipment on Intimacy and Sexual Expression," *The British Journal of Occupational Therapy* 74 (9) (2011): 435–442

[25] R. B. Cattell, *Personality and Motivation Structure and Measurement* (New York: World Book, 1957); R. B. Cattell, *The Scientific Analysis of Personality* (London: Penguin, 1965); R. B. Cattell, *Personality and Mood by Questionnaire* (San Francisco: Jossey-Bass San Francisco, 1973).

both.[26] Whatever the cause, people with an exclusively homosexual orientation are less likely to have children, so the question arises as to why the potential for such an orientation is not bred out of the population, so to speak, given that around 2% of the population identify themselves as being predominantly homosexual.[27]

One possibility is based on the observation that the female relatives of homosexual men have more children, suggesting that homosexual genes may make women more interested in men, or otherwise more fertile. Such genes could thus be maintained in the population by their benefit to women, compensating for their disadvantage to men.[28]

But another possibility is that bisexuality could act as a C or V-promoter in societies where homosexual relations are discouraged, by making heterosexual relations less satisfying and thus less frequent. Given that individuals with higher C and V have an immense advantage in the competitive struggle for survival, the benefits of some "homosexual" genes for bisexual men could far outweigh the disadvantage of having "too many" and thus having no children. After all, it takes only very occasional relations to sire a child every year or two. Cultural norms which outlaw homosexuality would greatly increase this effect, though at great cost to the exclusively homosexual minority.

High infant C maintained through nannies

Even the most extreme double standard cannot always prevent C declining in elites, especially when they are wealthy and urban. Women may be sexually controlled but they have better food and less need to work hard so their C tends to fall. However, the infant C of their offspring may be maintained if there is a class of women with very high C who can act as surrogate parents. This is what happens when the wealthy families of

---

[26] B. L. Frankowski, "Sexual Orientation and Adolescents," *Pediatrics* 113 (6): 1827–32;
Submission to the Church of England's Listening Exercise on Human Sexuality, Royal College of Psychiatrists,
http://www.rcpsych.ac.uk/workinpsychiatry/specialinterestgroups/gaylesbian/submissiontothecofe.aspx (accessed September 4, 2014); "Homosexuality is in the Genes, say Scientists," *The Times*, February 14, 2014.

[27] G. J. Gates, "How Many People are Lesbian, Gay, Bisexual, and Transgender?" Williams Institute, University of California School of Law, (April 2011).

[28] R. Kunzig, "Finding the Switch," *Psychology Today* (May 1, 2008).

Victorian Britain hired nannies to look after their children.

A study of nineteenth-century nannies suggests that children in their charge were under almost unbelievably close control.[29] A great number of regulations were enforced dealing with the most minute areas of life. Nannies began toilet training soon after birth and were often obsessed with constraining sexual behavior:

> Pot training began very early, in the first month or so. Thereafter it was continued, in a sense, right up to the moment the child left Nanny's care … Training took several forms. At first—innumerable placings on the pot, dozens a day. Once the baby or child had learnt, the most common method of persuading a child to go was to leave him or her there … till something "happened." Eleanor Acland describes in her autobiography how "It was nursery law that we might not quit the water closet till we were fetched." Once one of them was forgotten from tea until bedtime and sat there for five hours.[30]

That without supervision a very young child would sit for five hours shows the rigor of such training. The result was to maintain C at a much higher level than would be possible if these boys had been raised by their parents, thereby promoting self-discipline. The anxiety transmitted by these lower class caretakers would also increase V in adult life.

The same pattern of using nannies to maintain C and V in elites was not limited to England. In ancient Greece, Spartan nannies were highly esteemed for their tough, no-nonsense treatment of toddlers.[31]

It is now time to consider the link between childrearing and social behavior in a number of contemporary societies.

Case Study 1—Yanomamo

Until about 10,000 years ago, all of our ancestors made their living by hunting, gathering and fishing, and had a low-C, low-V temperament ideally suited to such patterns of subsistence. Loyalties were personal, inclining them to form the small co-operative bands that such a lifestyle

---

[29] J. Gathorne-Hardy, *The Unnatural History of the Nanny* (New York: Dial Press, 1973), 263.
[30] Ibid., 263.
[31] T. Holland, *Persian Fire: the First World Empire and the Battle for the West* (New York, Anchor, 2007), 82.

requires. Children were treated indulgently and sexual relations relatively free. This is the temperament we revert to when cultural and economic restraints are removed—it is human beings behaving "naturally." Examples can still be found in such societies as the Bushmen of the Kalahari and forest pygmies of central Africa.

The Yanomamo of the Amazon rain forest were not far from this low-C extreme. They were part-time horticulturalists who gained most of their calories from gardening, but this took the men only a few hours a day and they spent more time hunting. They were polygamous where possible and traditionally went naked.

They had no controls on sexual behavior apart from those imposed by male possessiveness. Affairs were commonplace, a frequent topic of conversation and a common cause of violence. Wives might be taken by force, loaned out or even given away.[32] In short, there was no sense that extra-marital sex was "wrong."

In all this they were a typical low-C society, with the lack of child training we would expect from such a people. As one observer commented:

> Children are inseparable from their mothers for the first few years of life. Boys are sometimes carried, in shoulder slings or tump straps, by their mothers and older sisters until they are several years old. Then, suddenly, the boys become mobile and independent, attaching themselves to play packs. Young girls stay close to their mothers much longer, learning the female role of the Yanomamo woman.
>
> Yanomamo are indulgent with children ... Children are punished infrequently. However, a severe beating is sometimes given suddenly by an angry parent. Spanking, or other formalized punishments, are not used. The Yanomamo are emotional rather than calculating in raising their children.[33]

Yanomamo child rearing practices also formed part of their spiritual belief system. For example, it was believed that children's souls are not firmly tied to their physical forms and could wander out of their bodies when they cried. As such, they were continuously coddled to protect them from

---

[32] N. M. Chagnon, *The Yanomamo* (Fort Worth: Harcourt Brace College Publishers, 1997).

[33] W. J. Smole, *The Yanomamo Indians: A Cultural Geography* (Austin: University of Texas Press, 1976), 73–74.

being in any distress.[34]

Boys in particular enjoyed a relatively control-free childhood that could stretch into their late teenage years, compared to girls who started work much earlier in the form of assisting their mothers with chores.[35]

But although low C, these people were distinctly high in V. This may have been given some boost by occasional harsh punishment of children, as indicated above, but far more by the subordinate position of women. Men were seen as superior to women, and husbands were strongly dominant over wives. A wife who was slow to prepare her husband's dinner might be scolded or beaten, with the hand or a piece of firewood. More severe punishments for women could include being cut with a machete, shot with an arrow in the leg or buttock, burned with a glowing stick, or even killed (the latter usually for infidelity or the suspicion of it). At times Yanomamo men beat their wives for no apparent reason, other than to "keep them on their toes." Occasionally, women might attempt to escape to a neighboring village, usually because of a particularly brutal husband. This, however, was a risky business since they might be severely beaten, mutilated or possibly even killed if their own village retrieved them.[36]

Starting as soon as they were physically able, girls were required to work much harder than their brothers to help their mothers with the daily collection of firewood and water. These onerous tasks required considerable effort. Life was generally tougher than for males, and females also had almost no say in who they should marry, with decisions being made primarily by their older kin.[37] The general brutality and inequality they experienced, quite distinct from the more egalitarian and thus low-V experiences of most hunter-gatherer women, had a serious psychological impact:

> By the time most women are thirty years old … they seem to have developed a rather unpleasant attitude towards life in general and toward men in particular. To the outsider, the older women seem to chronically speak in what sounds like a "whine," frequently punctuated with

---

[34] H. Inalcik, S. Faroqhi, B. McGowan, D. Quataert & S. Pamuk (eds.), *An Economic and Social History of the Ottoman Empire, 1300–1914* (Cambridge: Cambridge University Press, 1996), 783–784.

[35] B. Masters, *Christians and Jews in the Ottoman Arab World: The Roots of Sectarianism* (Cambridge: Cambridge University Press, 2001), 8, 54.

[36] Chagnon, *The Yanomamo*, 126.

[37] Ibid.

contemptuous statements and complaints.[38]

Although females had little if any political influence at the village level, older women did receive some level of respect. They were immune from raiders and so able to go from one village to another, acting as messengers or on errands (such as to recover dead bodies). Apart from this, women were considered utterly inferior to men in all aspects of Yanomamo society.[39]

Boys were reared very differently. From an early age they learned "masculine" behavior by watching and imitating their fathers. Boys as young as four were encouraged to be fierce, and were rarely punished for beating girls in the village with toy clubs or other implements. Fathers frequently goaded infant boys into hitting them, and a household would rejoice at his "ferocity" with cheers if the infant struck him. Infant boys soon learned that quick and violent action was an appropriate response to any form of frustration or displeasure.[40]

All these forms of behavior, from subordination of women to the encouragement of violence towards them, acted to keep women in a high state of anxiety. We have seen that anxious but indulgent mothers, combined with a high status for adult males, maximizes the level of V in their high status adult sons. So it should be no surprise that the Yanomamo society showed the typical high-V characteristic of ferocious aggression.

This could take a number of forms. A relatively moderate expression was mock warfare, where members of opposing groups struck each other on the chest. Competitors often coughed up blood for days after, and fatalities were not uncommon. Beyond this, club fights broke out at the slightest provocation and could quickly engulf an entire village.

There was also outright war. Raiding between villages was constant, usually justified by some past grudge. Even if there had been no actual offense to justify revenge, men might attack on the belief that the other village had used sorcery against them. Fear and suspicion of outsiders was endemic within Yanomamo society, and not without cause. It has been estimated that 25% of all deaths among adult males were due to violence.

---

[38] Ibid.
[39] Ibid.
[40] Ibid.

The following conversation illustrates the various causes of warfare. A Yanomamo man is trying to understand the causes of the Second World War.

> "Did any of your kinsmen get killed by the enemy?"
>
> "No."
>
> "You probably raided because of women theft, didn't you?"
>
> "No."
>
> At this point he was puzzled. He chatted for a moment with the others, seeming to doubt my answer.
>
> "Was it because of witchcraft?" he then asked.
>
> "No," I replied again.
>
> "Ah! Someone stole cultivated food from the other!" he exclaimed, citing confidently the only other incident deemed serious enough to provoke man to wage war.[41]

Success in warfare gave great benefits. Men who had killed produced, on average, three times as many children as those who had not, due to their success in seizing and keeping multiple wives.[42] In order to succeed, a Yanomamo man needed to be extremely ferocious—in other words, he needed high V. And that is exactly what his upbringing, and the brutal treatment by his father of his mother, were designed to give him.

On the other hand, there was little consistent control or punishment in late childhood—the type of treatment that would produce child V. Thus, the Yanomamo had no reverence for authority and no political organization beyond the village. Groups simply split off as a result of disputes when the population rose past a certain level. Villages formed alliances with one another, but no lasting political union was seen between them, even when connected by marriage, kinship or descent from a common ancestor.

---

[41] Ibid.
[42] Ibid.

Case study 2—Japanese

The village of Niiike is a rice growing community in south-western Honshu, and was studied in the early 1950s. By every measure, Japanese at this time had very high C. Beginning in the late nineteenth century, Japan had industrialized with impressive speed, far outstripping any other non-European country. This village has been part of a market economy for several centuries, growing rice and cotton as commercial crops well before the age of industrialization.[43]

The people of Niiike were not only hard working but also had a positive attitude to work. They were inclined to see work as a positive opportunity to achieve success, rather than a burden. They took a great pride in their work so that even mundane jobs such as the erection of a temporary fence were done with beautiful craftsmanship. This also is an indication of high C.

Their family patterns were typical of high C—nuclear families with tight control of sexual activity. All marriages were monogamous, and brothers who married moved out of the family household to form new nuclear families, though one son (usually the eldest) continued to live with and look after his parents. There was little premarital sex, at least for women, and the stigma of sexual misbehavior badly damaged a girl's marriage chances. Marriages were arranged at a relatively late age, and marital sex seemed unrewarding for women.[44]

> The sex act itself usually is a brief, businesslike affair with a minimum of foreplay. The husband, after waiting in the quilts at night for the rest of the household to settle into slumber, grasps his wife and satisfies himself as quietly and inconspicuously as possible.[45]

The best time of a woman's life was seen as the period before she married, though it was imbued with a sense of melancholy because of the hard time to come.[46]

---

[43] R. K. Beardsley, J. W. Hall & R. E. Ward, *Village Japan* (Chicago: The University of Chicago Press, 1959), 54, 65–68. In chapter seven, "The Rise of the West," we show evidence of increased control of children in the Tokugawa period (seventeenth to early nineteenth centuries) and after.

[44] Ibid., 312–318.

[45] Ibid., 332.

[46] Ibid., 313.

The people were generally reserved and inhibited and lived according to a deeply entrenched code of behavior:

> Control is an attitude—and simultaneously a positive value—expressed in a variety of conventionally approved ways, including restraint, discipline, endurance and self-sacrifice. The sense of quiet tranquility in Niiike comes in good part from the restraint put on all expression of emotion. Exuberance is repressed; so is every form of violence … Noisy argument or quarrelling is as rare as riotous laughter, and physical violence among adults is shocking even to think about … Self-control carries implicitly the assumption that one gains power over the world by disciplining one's self. This assumption can be carried into action through self-sacrifice, which thus can be seen as an opportunity rather than a duty.[47]

These Japanese also had a strong sense of loyalty to their nation—a group of people of similar ethnicity, language and culture.[48] We will come to see this as one of the defining characteristics of high infant C. It is also a product of child V, which by promoting acceptance of authority favors large political units.

From the cross-cultural survey we saw that control of children at all ages is associated with high C, though control of children in later childhood is also linked to child V. Control of children in Niiike started in infancy and continued throughout childhood:[49]

> Until well into the toddling age, a child "does not understand" attempts to train him, so Niiike people do not hold it against a youngster if he fails to follow instructions nor do they blame him for disobeying. Actually, of course, many principles and habits are instilled in the baby before the age of "understanding," and the real consequence of the Niiike point of view … is a minimum of early punishment or negative discipline.
>
> Training of children beyond the toddler stage is a conscious goal of Niiike parents and grandparents. They have certain well-defined goals to which they direct their efforts … The child must learn early and well, however, to obey and conform to certain inflexible rules. Obedience is easier, perhaps, in that these are rules for a way of life followed by everyone he knows, not rules made especially for children. Many express the pattern of hierarchy. Speech is one example.

---

[47] Ibid., 66–67.
[48] Liah Greenfeld, in *The Spirit of Capitalism, Nationalism and Economic Growth* (2001), argues that the rise of nationalism and the frugality of Japanese people can be traced to the Tokugawa era.
[49] Beardsley, *Village Japan*, 294.

Girls were even more tightly controlled than boys. A little girl faced more numerous and restrictive compulsions than does her brother. She must be careful to use polite speech and answer politely when spoken to. She must sleep in the approved posture, lying on her back without spread-eagling. On the other hand, children were rarely punished or even scolded. A mother angered at her child might pinch him painfully, or a father might cuff a child on the head when exasperated beyond endurance.

This is a childrearing pattern that would create high C, including infant C, since control started very early and continues throughout childhood. Children were simply expected to behave, rather than to obey out of fear. At the same time, the degree of control in infancy was clearly less than in case studies to be given later, notably Americans in the 1950s.

A case study of a remote rural village in the 1930s showed an even less rigorous attitude to childrearing, with younger children commonly defying their parents. But even here, toilet training was typically completed in the first year of life, and polite manners taught from infancy by constant repetition and instruction.[50]

Thus, by every measure so far linked to C, this twentieth-century Japanese village society rated high. They were market oriented and had an unusually positive attitude to work. They formed nuclear, monogamous households. Sex was tightly controlled and seems to have provided little pleasure for women. They married late. They lived according to a tough moral code of discipline and self-sacrifice.

They also showed indications of high V, including child V. This was a highly patriarchal society and also, as we have seen, one in which children were rigorously controlled in late childhood. All of this can be seen in the ritual of family meals:

> Traditionally the seating order at family meals is as rigidly prescribed as the protocol at government banquets. The household head, in the place of honour (yokoza) on the "upper" side opposite the kitchen area, usually is flanked by his male heirs in order of birth, while females of the household are arranged opposite, along the "lower" side … Women and girls eat with the men and boys, but they are served last, are skimped when food is short,

---

[50] R. J. Smith & E. L. Wiswell, *The Women of Suye Mura* (Chicago: University of Chicago Press, 1982), 212–220; J. F. Embree, *Suye Mura: a Japanese Village* (Chicago: University of Chicago Press, 1939), 185.

> and do not take the initiative in mealtime conversations.[51]

There was another custom which would serve to increase V. The social ideal was that couples live with the man's parents, a pattern common in the long-established civilizations of India, China and the Middle East. The effect of this was to greatly increase the anxiety of a young mother, since a woman's mother-in-law was traditionally harsh, difficult and fault-finding. Along with patriarchy this is a powerful V-promoter, and an important part of the V-forming tradition. Mothers-in-law in Niiike increased the stress on a young wife so much that nearly half of all first marriages failed.

Given such powerful V-promoters we might expect these villagers to be highly aggressive, but there was in fact little evidence of conflict within the community. Japanese at this time were certainly capable of aggression, as indicated by their recent history in which they had briefly conquered much of East Asia. This conundrum can best be explained in terms of an inhibition of aggression by powerful C-based controls on behavior. Once such controls were removed, as they were in Nanjing and other cities of China during the previous war, the Japanese could be violent and brutal.

The Japanese also had higher child V than the Yanomamo, a result of parental control in late childhood. Given the lack of punishment it was only at a moderate level, but still with a significant impact. This was a hierarchical society, with a reverence for authority at every level up to the Emperor. But it was authority based on respect rather than fear, reflecting the lack of childhood punishment and thus of stress. Japan at this time was a democracy, and had been even before the War.

Moderate child V was also reflected in a certain level of reverence for tradition. These people were attached to their ancestral values and traditional ways of life. And yet, they were quite open to change. Reflecting the attitudes of a typical villager:

> He has new horizons through education, assurance of occupational freedom, and a heightened sense of participation in the national destiny. He is accepting the new and looking to the future while living with

[51] Beardsley, *Village Japan*, 161. Compared with other Japanese, Okinawans are much less rigorous about training in early childhood. But Okinawa was only annexed in 1879, and until 1945 Okinawans were considered an ethnic minority rather than Japanese as such. T. Maretzki & H. Maretzki, "Taira, An Okinawan Village," *Six Cultures, Studies of Child Rearing* (New York: Wiley, 1963), 481.

equanimity amid much that is old.[52]

The Japanese at this time were a remarkably flexible and open-minded people. In the previous century they had carefully adopted exactly the elements of Western ideas that they wished, turning their nation from an agrarian backwater into an industrial state. In the decades that followed they would become one of the wealthiest countries in the world. These villagers, though in a rural area far from any major city, were far from limited in their thinking. For example, they totally depended on modern science and technology for all their health needs.[53]

But if any level of child V promotes traditional values, how could these Japanese be so open to change? For example, the Yanomamo and many other such societies neither control nor punish older children and thus should have low child V. And yet they are typically far less open to change than the Japanese. The proposed answer is that infant C makes people more willing to break with the past and consider new ideas. In the case studies to follow we will see that societies which control infants tend to be very positive about change, while those that control and especially punish only older children are far more traditional.

Another point to make is that the Japanese at this time had a remarkable talent for machinery. Japan was the first non-Western country to industrialize, and did so with startling speed in the eighty years before this study was done. This is another characteristic we will be linking to infant C and is consistent with its "impersonal" aspect.

Case study 3—Egyptians

The Yanomamo and Japanese were both high V but low and moderate in terms of child V. An example of a society with high V and high child V can be found in a study done some sixty years ago in a traditional Egyptian village.[54]

The Egyptian villagers in this study had several indications of high C. They were hardworking farmers, owned land, used money and were involved to some extent in the market economy. In times past, Arabs have been noted traders in areas such as East Africa. As Muslims, they were

---

52 Beardsley, *Village Japan*, 57.
53 Ibid., 61.
54 H. Ammar, *Growing up in an Egyptian Village* (New York: Octagon, 1966).

devotees of a highly prescriptive moralistic religion.

On the other hand, the villagers were less involved in the market than the Japanese, or than Europeans have been for many hundreds of years. Also, work was seen as a burden, to be given up gratefully as soon as one's sons were old enough to take it on, something which sons often resented.[55] This is distinct from the Japanese attitude that work was valuable in itself. There were other indications that C was below Japanese levels. These Egyptians accepted polygyny, and brothers sometimes owned and worked land in common. All of this indicates only moderate C, because the highest levels of C require training from infancy.

When it comes to V, the picture is very clear. These Egyptian villagers, like Yanomamo and Japanese, were intensely patriarchal. Women were largely confined to the home, and were required to be submissive, obedient and respectful. A wife was expected to offer her husband "the best food when he knocks, pour out the water when he washes his hands after the meal, and give him priority in every respect."[56]

As discussed earlier, patriarchy increases V by subordinating women. This increases their anxiety, which is transmitted to their infant sons. This effect was amplified, as in Japan, by having young wives live under the tyranny of a critical and fault-finding mother-in-law.

There was also the most rigorous control of women's sexual activity. It is commonly supposed that this is a natural result of patriarchy, which allows men to ensure the paternity of their offspring by controlling their wives' sexual activity. But, as we saw with the Yanomamo, the most brutal dominion of men over women is quite consistent with frequent affairs. Patriarchy can even promote extra-marital sex in some situations, such as by allowing powerful men to make use of other men's wives. In Egypt, this was not the case at all.

So, rather than being simply a natural result of patriarchy, limiting sexual activity (especially that of women) has a quite different function—that of maintaining high C and V in a civilization. Both these traits require highly unnatural behaviors given that our genetic "set level" is quite low. And as chapter nine will show, there is a strong tendency for C and especially V to break down as population density increases. The greater the density,

---

[55] Ibid., 28.
[56] Ibid., 150.

especially when combined with any level of food security, the greater the pressure on these traits.

In some societies, control of women's behavior is reinforced by the practice of female circumcision, which acts to deny women sexual pleasure. This and other practices of extreme patriarchy are not only brutal but come at a great cost to society, in terms of women's health and their ability to contribute economically. For example, women are less able to work outside the home and must be accompanied by male relatives on any excursions, which also gives the men less time for productive activity. That these customs have survived and become stronger suggests that they have a benefit large enough to outweigh these costs.

It must be emphasized that this "benefit" is not measured in terms of human health, wealth or happiness, but in terms of biological survival. People with high V tend to be aggressive, co-operative in small groups, stubbornly conservative, and to have more children. Such groups tend to expand at the expense of those with lower V, which means that their customs become more widespread.

Egyptian methods of childrearing were typical of a high V people, starting with extreme indulgence in infancy. Infants were neither punished nor controlled, indicating low infant C: "Children are treated permissively and with very few controls until they start to respond orally to the external world."[57] The only constraint on infants was to stop them biting their mother's breast while nursing. However, in later childhood parents became harsher and more controlling, creating high levels not only of child V but of stress:

> While in the first five or six years the child's respect for its parents and elders is inculcated mainly through physical gestures and intimacy, the respect later on is expected in terms of a decrease in word intimacy and physical proximity … The general pattern of behavior between the child and its elders is thus largely one of dominance, submission couched in the term "filial piety."[58]
>
> Connected with producing fear in the children is the violent and bad-tempered manner in which adults administer punishment to them. Punishment … may be in the form of fulminations or curses, or it may be corporal … Corporal punishment is not uncommon either by beating,

[57] Ibid., 140.
[58] Ibid., 130.

> striking, whipping or slapping … In administering punishment there is no consistency or regularity; for the same offense the child might be beaten harshly, or his offense allowed to pass unnoticed.[59]

Another source of child V and stress was other children. Children suffered an abrupt loss of attention once another baby was born, stressful in itself. Further stress was contributed by the roughness and aggression of children's play, and by the fact that parents were more likely to ridicule than help a child asking for protection. In fact, sibling rivalry was encouraged as a way of making a child, especially a boy, strong and aggressive. A case was cited of a 7-year-old boy who was considered weak as a result of being an only child. He was sent to live with a large family until he became competitive and "lively," after which he was brought back home.[60]

This example illustrates how a high birth rate is not only the result of high V but also increases V by pushing the child out into the world at the birth of another sibling. And village families are large because of the people's intense desire for children:

> The extent of the desire of women for children can be gauged by the multifarious devices to which they resort to induce pregnancy. A woman who does not bear children is called "akir"—one who literally "kills" her offspring—and the woman who has stopped bearing is referred to as "mushahira"—being bound or literally affected (in some way) by the moon.[61]

The high value placed on children is both a cause and effect of high V due to the "effect feedback cycle," as is typical of all V characteristics. High-V people are temperamentally inclined to want more children, and social values favoring frequent births serve to reinforce V. Such values are an important part of the V-forming tradition, and of considerable value in themselves in promoting population growth. In the competitive struggle for survival, peoples who breed fast must eventually overrun those who do not, given relatively free migration.

These stresses must have been avoidable and escapable to some extent, in order to produce the "toughened" stress response described in chapter four. But they would also generate not only high child V but a considerable

---

[59] Ibid., 137–138.
[60] Ibid., 108–109, 129.
[61] Ibid., 88.

amount of ongoing stress in the society. Both these aspects were evident in their political attitudes. The people thought of authority as necessarily involving an assertion of power and dominance, and could not respect those who did not display these attributes. Writing of the eighteenth century it was observed that, "if the peasants were administered by a compassionate multazin they despised him and his agents, delayed payment of taxes, called him by feminine names … They still consider both Government and Government officials as agencies of imposition and control, and hence to be feared."[62]

To the author of the study, who had grown up in this village, this attitude towards authority had an obvious effect on government:

> Various political struggles have occurred throughout the history of Egypt, but the common people, tied down to their land, submitted to any social order enforced upon them. They rarely took part in any resistance and left the battles to the military and ruling classes. To them, a military defeat meant the ousting of one ruling power by another.[63]

Acceptance of powerful and arbitrary authority is characteristic of societies which punish children severely in late childhood, combining child V with a high level of stress. It is also characteristic of eras in which children are punished severely, as we will see when discussing Tudor England in chapter seven. The political setup of the state reflects the temperament of the people, and such societies must struggle to maintain democratic forms of government, no matter the prestige and international pressures behind it. Though Egypt was "freed from tyranny" by the revolution of 2011, the victorious Muslim Brotherhood showed signs of authoritarian ambitions and was soon overthrown by the army.

Another key aspect of V is the solidarity of local groups, and this was certainly a powerful factor in the village. Solidarity started with the extended family where adult sons were expected to obey and respect their fathers, but extended to the clan, the section and the village. And as consistent with the characteristics of stress, each level was held together by respect and deference to those higher up. It was an organization based on hierarchy:

> The second important norm that governs the whole social structure is the weight and respect given to, and the authority wielded by, the person who

[62] Ibid., 80–81.
[63] Ibid., 70–71.

> plays the role of the senior, normally chosen on age basis as well as on capacity to speak and argue well, besides other factors such as economic status and social prestige.[64]

There is another aspect of Egyptian political attitudes to highlight at this stage. Unlike the fiercely nationalistic Japanese, the Egyptians in this village had a relatively weak sense of national identity, something that has been evident for more than two thousand years. Egypt has been ruled by foreigners (Persians, Greeks, Romans, Arabs and Turks) since the time of the Persian king Cambyses in the sixth century B.C. The villagers referred to themselves and other Muslims by the term "nation of Muslims" and felt that ethnically they were Arabs first and foremost.[65]

This cannot be explained by the punishment of children. We will see in the next chapter that sixteenth century Europeans also punished their children severely and yet were strongly nationalistic. From the French driving out the English in the late fifteenth century to the Dutch and English resisting the Spaniards in the sixteenth, Europeans at this time were highly stressed yet fiercely resisted foreign rule.

Thus, the relative indifference of Egyptian villagers to the source of the ruling authority cannot be explained by child V or by stress. Biohistory proposes that nationalism, which is essentially a strong preference for rulers of one's own ethnic group, is another characteristic of high infant C. Compared to Europeans, the low infant C of Egyptians has made them much less resistant to foreign rule. Even the form of nationalism that allowed Nasser to seize control of the Suez Canal, which occurred at about this time, was pan-Arab rather than narrowly Egyptian. This also fits with the impersonal orientation associated with infant C.

Apart from accepting authority, another aspect of child V is that it promotes resistance to change. From the Japanese example it was also suggested that infant C is associated with more flexible thinking. These Egyptian villagers indulged infants *and* strongly disciplined older children, so on both accounts they should be strongly traditional. And this is exactly what we find—a deep resistance to change. Religious belief and behavior were pervasive, governing all aspects of life, and social patterns and customs had changed little, despite decades of governments with modernizing ideas:

---

[64] Ibid., 47.
[65] Ibid., 72.

> Although one finds that the villagers have adopted certain aspects of material culture from the town, such as the use of manufactured goods, new types of vegetables, and some medicines, such adoption is a surface substitution. Their ideas about land ownership, the importance of children, family and clan solidarity, categories of respect relationships, sex dichotomy, supernatural sanctions, and the importance of ritual and ceremony in both social and religious life, remain almost unaltered.[66]

Finally, V is associated with aggression, though the form this takes depends on other factors. All three societies considered so far were patriarchal and thus aggressive, though Japanese aggression within the society was inhibited by very high C. The Egyptians had lower C than the Japanese, so were less likely to be inhibited for this reason. But the stress and tension associated with the severe punishment of children was another factor driving aggression. Thus, conflict and competition were pervasive features of village life, even if not to the extent that they stopped the community acting in a cohesive and well-organized manner.

Thus we see in a single village all the characteristics so far linked to V, child V and stress. These include a childrearing pattern combining indulgence and protection of infants with the early weaning and stressing of older children, an intense desire for children, patriarchy and patterns of residence that stressed women, solidarity of local groups based on hierarchy, acceptance of powerful authority, resistance to change, and aggression.

The study was written more than sixty years ago, but since then the Muslim Middle East has been strikingly resistant to "modernization." This applies not only to the Arab world but also to Iran, Afghanistan and Pakistan. As birth rates in many countries have fallen below replacement levels, Muslim countries have in general experienced much less of a decline. And while religion continues to retreat in industrial nations, Islamic fundamentalism is as strong as ever. In recent decades it has even gained increased political power in countries ranging from Iran to Turkey. Resistance to secularism and assimilation applies even to the growing numbers who migrate to Western countries.

These countries also show unusually high levels of conflict, unless held down by fear. Political regimes tend to be brutal and authoritarian, and the alternative being not so much stable democracy as chaos and civil war. This was seen in Lebanon in the civil war of 1975—91, and in Iraq once

[66] Ibid., 79.

the US invasion removed the harsh hand of Saddam Hussein. Afghanistan is a further example, though here local warlords and Taliban groups have been rather more the norm than stern central control. The potential violence in Arab countries is also indicated by civil wars that broke out in Libya and Syria following the Arab Spring of 2010–11, quite unlike the transition to democracy in Eastern Europe after Soviet domination was withdrawn. All this is characteristic of high V, including high child V, and a very high level of stress.

Case study 4—Chinese

The countries of the Middle East are poor and fractious by Western standards, but this is a highly successful type of society. The inhabitants of these lands are fertile, aggressive and fervently attached to their cultures and religions. And all these factors can be accounted for by a family pattern that involves patriarchy, strict control of women's sexuality, indulgence of infants and the harsh treatment of older children.

The power of this pattern is made clear by the fact that it is, with minor variations, the family pattern of much of mankind. In particular, it is the standard for traditional families throughout India, Pakistan and China, as well as the Middle East. Thus, our next example is a society from almost the other end of the world—an ethnic Chinese village in Taiwan in the 1960s—which provides a very similar picture of childrearing.

Infants were not punished, controlled or trained in any way. If they cried they were immediately offered the breast. Toilet training aroused little concern or interest.[67] Perhaps one difference is that, compared to Egypt, there seems to have been a little more discipline after weaning, which indicates some slight level of infant C.

> Taiwanese parents assume that children cannot really "understand" until they are around six years old. They claim that until then they do not try to teach them anything and expect little in the way of obedience. At most they hope that the preschool child will not injure herself or cause her parents too much trouble. Obviously even such a low level of expectation requires some socialization on the parents' part, and actually a good deal of training goes on before a child enters school. It is not unusual for a four-year-old girl to be put in charge of her two-year-old-brother, though the

[67] M. Wolf, *Women and the Family in Rural Taiwan* (Stanford: Stanford University Press, 1972), 58–60.

> mother will insist that both stay within her hearing range. Parents may think they do not "expect" obedience of preschool children, but a mother will severely scold or even beat a four-year-old girl who does something that endangers her small brother.[68]

However, as in Egypt (but unlike Japan), there was an abrupt increase in discipline at the age of five or six:[69]

> Whether or not the age at which parents expect their children to suddenly become obedient, responsible, and helpful, and the age at which they first attend school are anything other than coincidence is moot. In the past this was the age at which girls gave up their freedom of movement by having their feet bound … Little boys whose disobedience was a source of amusement or at most brought a laughing swat suddenly find themselves hit with a ruler for not sitting down when told to. Fathers who used to be affectionate become distant, with a tendency to lecture … When their sons reach the age of "reason," fathers must withdraw to become dignified disciplinarians.

Also as in Egypt, punishment of older children tended to be harsh:

> A beating administered by a Taiwanese parent is often severe, leaving the child bruised and in some cases bleeding. Parents prefer to use a bamboo rod to discipline children, but they will use their hand or fist if there is no bamboo available, and if they are really angry, they will pick up whatever is at hand. Crueller forms of physical punishment are also used by a few parents, such as making the offending child kneed on the ridged surface of an abacus or tying the child in a dark corner.

Mothers commonly punished children in a violent fit of anger, so uncontrollable at times that family members and even outsiders might intervene to prevent the child from becoming seriously injured.

Traditional Chinese society, like that of Egypt, was, at least in outward form, intensely patriarchal:

> A preschool Taiwanese girl learns her first subtle lessons about the second-class status of her sex. She has heard from the time she could understand that she was a "worthless girl" … By age five most little girls have learned to step aside automatically for boys, at least when their parents are watching.

---

[68] Ibid., 65.

[69] Quotes on childrearing, see ibid., 67–70.

Men not uncommonly beat their wives, and such lower status should cause women to have a higher level of anxiety transmitted to their infant sons, resulting in higher V. This was reinforced by residence with a critical and fault-finding mother-in-law, as seen in Egypt and Japan. In this culture, young married women in their 20s had the highest suicide rate of any group.

On the other hand, the actual relationship between husband and wife seems to have been somewhat less unequal in China than in Egypt. Women were not confined to the home, and within the home they seem to have had a relatively stronger position. For example, though a woman could not directly control a husband who drank and gambled, she could do much to put pressure on him, such as influencing their children to have no respect for him, and using the community of women to publicize his abuses. This was sometimes effective in changing behavior. And when husbands behaved well the couple were described as becoming friends, with husbands consulting wives on important decisions, such as those regarding the education of their sons. It is also important to note that the position of women had improved in the last two generations, such as in being freer to move around.

Thus Chinese childrearing and family patterns were like those of Egyptians in the emphasis on patriarchy, total indulgence of infants, and an abrupt increase in control and punishment at the age of five or six. The similarity is especially striking if it is considered that these were cultures thousands of kilometers apart, with totally different religious systems and relatively little contact until recently. It is a clear case of convergent evolution, coming about because it creates a hardworking people with a high birth rate and strong traditions, who readily accept powerful authority. Families and societies like these tend to do well in the competitive struggle for survival.

On the other hand, there are important differences. Based on these studies, Chinese were not quite as patriarchal as Egyptians which suggests lower V, something confirmed by the lower level of aggression in this community. While Egyptian parents encouraged their sons to be aggressive, Taiwanese parents strongly discouraged it.

The other difference is that there seemed to be more control of children before the age of five or six, suggesting higher infant C. Also consistent with this is that the Chinese, despite accepting Manchu domination for several centuries, have been less tolerant of foreign rulers than Egyptians.

There were indications of higher C as well, including the enormously strong work ethic of Chinese people, a somewhat lesser practice of polygyny, and perhaps again the intolerance of aggression. These differences will be important when we come to examine Chinese economic success in recent decades.

Case study 5—Manus

Of the cases given so far the Yanomamo tend not to exert significant control over their children at any age, giving them low C, including low infant C. Japanese tend to control children in early and late childhood, so they have both infant C and moderate child V. Egyptians and Chinese tend to control and especially punish older but not younger children, giving them child V with low infant C.

The only variation remaining is control in infancy but not later childhood, which would produce infant C without child V. This is so rare that only one society has been identified as having such a pattern—the Manus people of the Admiralty Islands (strictly speaking, the Manus Titan people, distinct from the other inhabitants of Manus Island). Understanding this unique people will help us not only to illustrate our current picture of infant C but also to extend it.

At the time they were studied by Margaret Mead in 1928, the Manus were living in stilt houses above a lagoon, a dangerous environment for young children which required parents to show constant vigilance and careful training to keep them from falling into the water. These comprised a form of control that acts as an effective C-promoter for infants. In this sense, the Manus treated their infants like the calorie restricted rats described in chapter two, giving them a lot of attention and retrieving them back into the nest. For example:[70]

> When [the child] is about a year old, he has learned to grasp his mother firmly about the throat, so that he can ride in safety … The decisive, angry gesture with which he is reseated on his mother's neck whenever his grip tended to slacken has taught him to be alert and sure-handed … For the first few months after he has begun to accompany his mother about the village the baby rides quietly on her neck or sits in the bow of the canoe while his mother punts in the stem some ten feet away. The child sits

[70] M. Mead, *Growing Up in New Guinea* (New York: Perennial Classics, 2001), 22–28, 37–39.

> quietly, schooled by the hazards to which he has been earlier exposed.

Children must also learn to walk as soon as possible:

> Side by side with the parent's watchfulness and care goes the demand that the child himself should make as much effort, acquire as much physical dexterity as possible. Every gain a child makes is noted, and the child is inexorably held to his past record … So every new proficiency is encouraged and insisted upon.

The same intensive training also applied to property:

> Before they can walk they are rebuked and chastised for touching anything which does not belong to them. It was sometimes very tiresome to listen to the monotonous reiteration of some mother to her baby as it toddled about among our new and strange possessions: "That isn't yours. Put it down. That belongs to Piyap [Mead]. That belongs to Piyap. That belongs to Piyap. Put it down." But we reaped the reward of this eternal vigilance: all our possessions … were safe from the two and three-year-olds who would have been untamed vandals in a forest of loot in most societies.

Training in modesty, another powerful C-promoter, began at a very early age:

> Children must learn privacy in excretion almost by the time they can walk; must get by heart the conventional attitudes of shame and embarrassment. This is communicated to them not by sternness and occasional chastisement, but through the emotions of their parents. The parents' horror, physical shrinking and repugnance is communicated to the careless child. The adult attitude is so strong that it is as easy to impregnate the child with it as it is to communicate panic.

Given the rigor of this early training, the complete lack of obedience training in later years is striking:

> The parents who were so firm in teaching the children their first steps have become wax in the young rebels' hands when it comes to any matter of social discipline. They eat when they like, play when they like, sleep when they see fit. They use no respect language to their parents and indeed are allowed more license in the use of obscenity than are their elders. The veriest urchin can shout defiance and contempt at the oldest man in the village … They do no work … The community demands nothing from them except respect for property and the avoidance due to shame.

Since the lessons of modesty and property were well learned at a very early age, there was very little of any kind of training after early childhood. Boys in particular were notably free in almost every sense until married.

Biohistory indicates that such an upbringing would give the Manus people very high infant C but very low child V and low stress. In other words, they would be strongly market oriented but poor at accepting authority. And this is exactly what we find.

At the time they were studied, the Manus were highly successful traders, obsessively hard working and ambitious and collecting a great deal of wealth. But they had no political organization beyond the village level, and local power was based only on wealth and the abrasive, forceful personality brought by wealth and success:

> They have learned neither real control nor respect for others … They have learned only that riches are power and that it is purgatory not to be able to curse whom one pleases.[71]

It has been suggested that the "impersonal" attitudes associated with territoriality might explain why people with high infant C are oriented to the market. This is because buying and selling depend on an impersonal attribute—the value of the goods—rather than the relationship with the buyer or seller. In the section on Japan it was also suggested that the Japanese talent for machinery might also be explained by the "impersonal" aspect of infant C.

It is thus of particular interest to note that the Manus also had an extraordinary passion for machinery. This was especially striking in a people who had absolutely no prior experience with such things. They were particularly fascinated by American machinery, to which they gained access during the Pacific War:

> And the Manus watched, fascinated. They seem to have got into every installation, and I never knew when I would encounter either a superior toleration for my quite good field glasses because they didn't have a search light attached, or, as I squatted on the floor trying to stop a case of arterial bleeding, an account of the magnificent equipment of the operating room of an American military hospital. Manus were down in the engine room and up on the bridge; a people who enjoyed machinery as much as they did presented constant entertainment to the American troops.[72]

---

[71] Ibid., 149.
[72] M. Mead, *New Lives for Old* (New York: Perennial Classics, 2001), 169.

There is only one society on record with the same intensive training of infants as seen among the Manus—Northern Europe in the late eighteenth and nineteenth centuries. This was, of course, the time for the when the tinkering of countless inventors and engineers launched the world into the machine age. In the next chapter we will trace the rise of this childrearing pattern from Medieval times, proposing that it was the principle cause of the Industrial Revolution.

Thus, it is of considerable interest to observe that the Manus shared another feature of nineteenth-century Europeans—a moralistic religion with a heavy focus on honesty and property rights. This had no relationship to Christianity or any other "advanced" religion, but was a form of ancestor cult.

> With this emphasis upon work, upon the accumulation of more and more property, the cementing of firmer trade alliances, the building of bigger canoes and bigger houses, goes a congruent attitude towards morality. As they admire industry, so do they esteem probity in business dealings. Their hatred of debt, their uneasiness beneath undischarged economic obligations is painful … Finally, their religion is genuinely ethical; it is a spiritualistic cult of the recently dead ancestors who supervise jealously their descendants' economic and sexual lives, blessing those who abstain from sin and who labour to grow wealthy, visiting sickness and misfortune on violators of the sexual code and on those who neglect to invest the family capital wisely.[73]

This code can be seen as another set of impersonal attitudes, in that it applies regardless of personal relationships. Just as reliance on the market is an impersonal way of distributing goods, and machinery an impersonal skill, so integrity and repayment of debt are impersonal principles which have the added bonus (in a developed civilization) of being very useful in a market economy.

This is in contrast to moral principles with a more personal base, such as giving to the destitute, which is likely to depend heavily on the appearance and age of the recipient. Manus people, especially in contrast to others in their culture area, were far more likely to distribute property according to impersonal ideas such as market exchange or repayment of debt, and thus showed markedly impersonal attitudes. In this also they had much in common with northern Europeans, as will be discussed in the next chapter.

---

[73] Mead, *Growing Up in New Guinea*, 11.

Manus religion also had a strong component of sexual morality, which is always associated with high levels of C. This was a wealthy society by pre-industrial standards, so C could not have been maintained by hunger. It is of particular interest to note that women were not only constrained from having sex but did not like it:

> Married women are said to derive only pain from intercourse until after they have borne a child … They confide little in each other. Each conceals her humiliating miserable experience as did the Puritan women of the Victorian era. Every woman, however, successfully conveys to her daughters her own affective reaction to the wearisome abomination which is sex.[74]

Young men in the traditional society had access to sex through female captives from a neighboring tribe or even another Manus village, treated so badly that they usually died in captivity or soon after release. This clearly would not have given the men the same sexual outlet as they would get from willing girls in a more permissive society, and of course had no impact on the women who would raise the next generation. A brutal reality of the double standard is that it focuses male sexual activity on a class of women, whether paid or forced, who for reasons of higher mortality, disease or some form of birth control, are less likely to become mothers. This minimizes the sexual activity of other women, increasing their C and thus the infant C of the next generation.

It is striking that Manus C was so high, even though there was almost no parental control in late childhood. This again illustrates the importance of infant control, combined with the strict control of sexual behavior, in bringing C to a maximum.

Thus we have a society with high infant C and C, and with very low child V. Child V is the product of punishment or control in late childhood and is associated with acceptance of authority and strength of tradition. The Manus did not punish *or* control their older children, so for them child V was at an absolute minimum.

On purely theoretical grounds, then, this society should have two characteristics. First, the people should have very little respect for authority and an extremely weak political organization. In fact, there was no political organization beyond the village level, and even within the village little formal organization. There was no formal chief or leader, but

---

[74] Mead, *Growing Up in New Guinea*, 118–119.

simply a pattern where successful and forceful men had a great deal of influence. Neither were the Manus much in awe of their new colonial masters, at one stage organizing a widespread strike of Work Boys.

Second, their sense of tradition should be peculiarly weak. This is especially so if we accept the notion, proposed in the Japanese section, that infant C is also linked to acceptance of change. It might further be noted that northern Europeans in the nineteenth century, the only other society with equivalent levels of infant C, were also peculiarly open to new ideas.

Thus it should be no surprise that Manus were also exceptional in their openness to change. This was shown not only by their success as traders, which requires a great deal of flexibility and creative thinking, but their extraordinary willingness to adopt—almost overnight—an ideology that reversed almost every tenet of their culture. Margaret Mead describes their obsession with change and new ideas:

> The great avidity with which they seized on the new inventions which came with European contact, was partly rooted in their driving discontent with things as they were.[75]

Within a few decades they had completely abandoned their religion, economic system and ideas of social relationships in favor of a radical system called the "New Way."

What makes this case study so useful is that there was a generation of Manus who passed late childhood and adolescence during the Japanese occupation, and so had a much more stressful time. This is exactly the experience that should, according to our theory, increase child V and thus make them more accepting of authority and more rigid about their beliefs. Mead makes this connection quite explicit:

> They had had, in fact, the same early childhood as the older men, but had lacked the kind of late childhood and adolescence in which the older men's habits of companionship and friendliness, and their capacity to feel free, had been born. They were stiff and difficult fathers, with less tenderness and indulgence to their children than the older or much younger men ... In the brittle, dogmatic orthodoxy of these young men, there is limited refusal to accept anything new.[76]

---

[75] Ibid., 154.
[76] Ibid., 358–359.

This "orthodoxy" was in relation to the New Way, a form of authoritarian cargo cult which most Manus adopted after the Pacific War. There are descriptions from this time of people listening for hours to the cult leader, Paliau, without "batting an eyelid." But this rigid orthodoxy applied only to that birth cohort affected by the conflict.[77] Most Manus abandoned the movement during the 1960s and it also became a lot less authoritarian, which is exactly what would be expected from a new generation arising with lower child V. The older generation with higher child V was still around, of course, but less influential. It is a pattern we will observe in many areas—that young people in their twenties have an impact on society out of all proportion to their numbers. This will have particular significance when we come to examine the origins of wars and economic recessions in chapters nine and ten.

The combination of high infant C and low child V explains why societies like the Manus are so rare—they are extraordinarily change-prone. The Manus were successful as a people, wealthy and expanding in population before European contact. But their extreme flexibility and desire for change meant that the merest nudge was enough to obliterate their traditional belief system, culture and society, which formed the basis of their unusual temperament.

Manus society had an extreme level of features we can now link to infant C. They tightly controlled infants and limited sexual activity. They were hard working, successful traders. They had a moralistic religion relating to impersonal values such as honesty and avoidance of debt. And they also had a passion for machinery and an openness to change.

The origins and decay of the Manus character

There is one more lesson from the Manus case—the question of how such a society might have arisen. Manus traditions stated that their religion and culture were of quite recent origin:

> The ideal of every man in the community is the golden age, which every generation believes to be just a generation behind him, when the spirits took no interest in mortal amours and whenever one met a woman alone,

[77] K. Kais, "The Paliau Movement," Buai Digital Project (April 1998) http://www.pngbuai.com/100philosophy/paliau-movement/ (accessed September 4, 2014).

one could take her by the hair.[78]

Such a recent origin is quite probable, in that Manus numbers were growing rapidly in the period before Europeans arrived. Oral traditions suggest that the Manus people originated in a single settlement on Kasta Reef, which seems to have become submerged as part of a volcanic catastrophe around 1850. The refugees settled on Peri and a number of other island and coastal settlements, something also confirmed by archaeological evidence.[79] By the time Margaret Mead arrived, only eight decades later, there were several thriving settlements. By 1980 there were 3,654 Manus Titan speakers on and around Manus Island, making them the largest of the thirty linguistic groups on the Island and almost 10% of the total population.[80] This does not include the substantial number of Manus Titan people living elsewhere in New Guinea, indicating an exceptionally rapid population growth.

The origin of this society's peculiar traits could have been something as simple as a change of residence. People who began living in stilt houses in a lagoon would need to discipline and train their infants to keep them safe, leading to a rise in infant C. As a result, some people might become more interested in trade, and better at it. There would also be an increase in guilt and an inclination to act according to certain values, interpreted as a change in the demands of ancestral spirits.

The types of values adopted could be quite variable in the beginning. But men who saw the spirits as limiting sexual behavior would become better traders, helped, of course, by the fact that women were becoming less interested in sex. Their success would be seen as evidence that they were right, so others began to adopt the same values. This would, in turn, increase the level of C and thus control of infants, making the feelings of guilt even stronger in the next generation. The same could be said of men who believed the spirits required them to be honest. They would become more successful, and therefore their interpretation of the spirits more dominant. We thus have two independent factors—trading success and early childhood experience—reinforcing each other.

---

78 Mead, *Growing Up in New Guinea*, 122.

79 B. Minoi, "Manus from the Legends to Year 2000: A History of the People of Manus," (UPGN Press, September 2000), 35, 46.

80 P. Demerath, "Negotiating Individualist and Collectivist Futures: Emerging Subjectivities and Social Forms in Papua New Guinean High Schools," *Anthropology & Education Quarterly* 34 (2) (2003): 136–145.

Excellent education and motivation mean that in the past many Manus found employment as skilled workers outside the province. When the anthropologist Lola Romanucci-Ross visited Peri in the early 1960s they were still showing evidence of enterprise and success, being widely employed in more "backward" areas as "doctor boys" (medical assistants), teachers, students and workers.[81]

Manus Titan people still feel themselves superior to, and more dynamic than, other peoples of the Admiralty Islands. But once they moved to land and abandoned their traditional childrearing practices, as happened after the Pacific War, we would expect their peculiar temperament to disappear within a generation or two. And there is some evidence of this, especially for the young.

In recent times the gap between the achievement level of Manus people and others has declined, leading to a surplus of educated people. Thus, fewer people are leaving Peri to find work and there is a re-emphasis on traditional subsistence activities. Perhaps partly as a result, interest in education has declined, and performance of year 10 students in the Manus Province has gone from well above to well below national levels. It must be admitted that Manus Titan people are only a small percentage of the people of Manus Province, and the decline may be partly a result of keeping more students in school. But at the same time there has been a reduction in ethnic differences among the young, who are also widely regarded by parents and teachers as more insolent and disobedient than in the past.[82]

Case study 6—Americans

The final society to be considered is a New England town in the early 1950s. This community study was done by cultural anthropologists, using the same kinds of approach as anthropologists normally use in studying non-Western societies.[83] America, like China and Egypt, belongs to the

---

[81] L. Romanucci-Ross, *Mead's other Manus* (South Hadley, Mass.: Bergin and Garvey, 1985), 155.

[82] Demerath, "Negotiating Individualist and Collectivist Futures: Emerging Subjectivities and Social Forms in Papua New Guinean High Schools."

[83] J. L. Fisher & A. Fischer, "The New Englanders of Orchard Town, U.S.A.," *Six Cultures: Studies in Child Rearing*, edited by B. B. Whiting, 939–1009 (New York: John Wiley, 1963). The underlying concept of the studies in *Six Cultures* is

family of long-civilized nations. But in one respect these Americans were totally unlike Chinese and Egyptians and more like the Manus villagers described in the last section. This was in their tough treatment of infants.

Crying, for example, might be deliberately ignored as a way of teaching a baby not to indulge in unpleasant behavior. Also of particular significance is that personal contact with a baby was not marked by close bodily contact, as in many societies.[84] Babies slept separately from their mothers in a crib, which was often in a separate room.

Like Manus mothers, these American parents began child training at a very early age. Toilet training typically began in the first year of life. Common techniques included shaming, putting the child on the pot at regular times, praising for performance, and promising rewards for not wetting the bed. Out of a sample of 24 children, 8 were fully trained by 18 months and 19 by 2½. Babies were encouraged and trained to walk quickly. They were punished for biting, and might also be punished for picking up another child's toys or other objects from the floor.

This all took place before the age of two, a time when Chinese and Egyptian parents considered training impossible.

> In Orchard Town, the newborn infant is thought of as a "potential." The central concept of the child as a potential involves beliefs about the inheritance of characteristics, beliefs about the influence of parental training on the child, beliefs about the influence of the social environment and education, and the beliefs about stages and norms.[85]

For these parents, the time between the ages of two and five was considered to be the most important stage of character formation. Physical punishment, normally spanking, was used more commonly at this age than at any other. Children were commonly put to bed earlier than they would like. Other forms of training included modesty, table manners, eating a balanced diet, and doing simple chores. They were taught to recognize property rights and not to take other children's toys, and learned to respect adults and not show aggression against their parents. The transition from the control of infants to the discipline of young children was rarely smooth, but the toddler usually adapted quickly to the new set of

---

that the ecology and economy of the region determines childrearing methods, which in turn influence personality (Introduction, 4–5).

[84] Ibid., 947.

[85] Ibid., 921.

expectations:

> Initially children respond to the stronger discipline with temper tantrums. With these there is a shift in the parents' feelings toward the child from indulgence to some hostility towards his anti-social acts. Negatives become more common in the parents' speech with the child: "No! No!"[86]

In Egypt and China, control became markedly stricter after the age of 5. Among the Manus, the level of control declined in later childhood. In this American town, as in Japan, there was no obvious change. School-age children experienced greater discipline in the classroom than at home, and were expected to perform more chores by the age of seven or eight. On the other hand, and in stark contrast to China and Egypt, physical punishment was *less* prevalent after the age of 5.

Parents were relatively equal in status, wives did not live under the control of overbearing mothers-in-law, and there was little punishment in later childhood. All this means that these people should be relatively low V, high infant C (more than the Japanese but less than Manus), and moderate child V.

America in the 1950s was very much the kind of society we would expect from these childrearing and family patterns. High infant C implies hard work, mechanical aptitude and strong impersonal loyalties, as well as (from the Manus example) a sense of commercial honesty and a capacity for guilt. High infant C, combined with lack of punishment in late childhood, should make these people open to new ideas. Economically, such a society should be innovative and prosperous, good with machines and with low levels of corruption.

In political terms, moderate child V would make the people positive about their traditions and accepting of wider political loyalties, far more than the Manus. Infant C would also enhance national loyalties. Combined with the lack of deference to arbitrary authority implied by low levels of punishment, we would expect such people to form a stable, democratic government.

One important feature to note about this society is that it was in a period of rapid change, stable as the 1950s might appear from our time. Most parents would have grown up in the Great Depression of the 1930s, and their childhood had been a great deal tougher in terms of discipline and

---

[86] Ibid., 949.

life experience. Such a background would produce significantly higher C and child V, accounting for the conservative and consensual spirit of the age.

The children of the study would grow up to be the baby boomer generation, far less traditional in outlook and with many participating in the sexual revolution of the 1960s and 1970s. This would, of course, drop their level of C below that of their parents.

Table 5.3 below summarizes the childrearing patterns of these five societies, and Table 5.4 the characteristics that can be linked to each of the five traits.

Table 5.3. Infant C and C, child V and V in six cultures. 1 represents the minimum of known societies and 6 the maximum.

	Control 0-5 (infant C)	Control and punish 6-12 (child V)	Punishment (stress)	Control/sex restraint (C)	Patriarchy (V)
Yanomamo	1	1	2	1	6
Japanese	3	3	1	6	6
Egyptians	1	6	6	4	6
Chinese	2	6	5	5	5
Manus	5	1	1	6	2
Americans	4	3	1	5	2

Table 5.4. Characteristics of infant C, child V, stress, C, and V

Infant C	Flexible, market oriented, hardworking, machine skills, intense preference for rulers similar in language, culture, religion and ethnicity
Child V	Traditional, accepts authority, larger states
Stress	Authority based on fear, large status differences
C	Hardworking, disciplined, delayed gratification,
V	Aggressive, high birth-rate, small group cohesion, strong local leadership

Society Reflects Temperament

The commonplace idea of the modern world is that people in different societies have more or less the same temperament. Thus, it is thought that Egyptians and Japanese and Chinese and Americans have the same underlying emotional and psychological make-up, and it is only differences in education and circumstance that make them different. If a nation has been taken over by a dictatorship, it is due to economic or political contingencies, combined with the opportunism of the dictator. This gives rise to the widespread belief that autocratic governments are aberrant, and that if they can be overthrown or democratic ideas introduced, the eventual end result will be a peaceful and stable democracy. Such views are so prevalent that they have shaped the foreign policy of many Western governments, with disastrous consequences for nations such as Iraq and Afghanistan.

The same is generally said of differences in wealth and industrialization between different nations. Given some initial aid and access to advanced education, it is believed any nation can become wealthy as well as stable and democratic. Similar explanations are given for differences between ethnic groups within nations. The relative poverty of one group compared to another is believed to reflect the prejudice of the majority or an inherited disadvantage, such as the legacy of caste or slavery. Given this belief, any differences can and must be overcome by anti-discrimination laws, financial aid and affirmative action.

Contrary to these widely accepted ideas, biohistory proposes that governments are not forced upon societies, nor are they embodiments of values and ideas which can be changed according to circumstance and opportunity. Rather, a government reflects the underlying temperament of the population, even in cases where that population loathes and fears it. This temperament is very hard to change and strongly predisposes the people and their government toward certain values and ideas. When people defer to powerful authority, governments tend to be authoritarian. When loyalty to individuals takes precedence over loyalty to institutions and laws, influential individuals can disregard those same institutions and laws.

The same applies to attitudes that affect economic achievement, such as capacity for work, innovative thinking and the preference for immediate versus delayed gratification. People who are epigenetically primed for success are far more likely to be successful.

This temperament reflects the effects of early life experience, especially different methods of childrearing. But it also reflects experience in the womb, and even very likely direct epigenetic inheritance from previous generations. For this reason, only limited changes to temperament are possible in later life.

In recent decades certain East Asian nations such as South Korea, Taiwan and Singapore have become stable and prosperous democracies, supporting the idea that all nations can follow the same path. Francis Fukuyama's The End of History[87] is the best-known example of this type of thinking. But these examples are misleading. East Asian nations have exceptionally high C based on millennia of Confucian teaching, so their people are strongly inclined to be disciplined and hardworking. A recent decline in child V, from the combined influence of western ideas and growing affluence, has also helped to make them more flexible and less authoritarian. We have seen that, even in the 1950s, Taiwanese family patterns were more conducive to economic growth than Egyptian ones. The sexes were less unequal, punishments in late childhood less severe, and child training seems to have started earlier.

Peoples such as the Yanomamo or the inhabitants of New Guinea, apart from odd exceptions such as the Manus, are low in every form of C and thus find wealth and democratic stability much harder to achieve. And their already low C is further undermined by the modest affluence resulting from Western technology, aid and institutions, including the political and judicial systems that halt warfare and allow food shortages to be relieved. Aid also undermines C by granting rewards without disciplined work, and therefore in some instances may increase poverty and corruption.[88]

But the key problem is that all such efforts focus on creating Western-type economic development, but *without the temperament that was responsible for it*. Raising C requires people to act in a manner quite unnatural to the hunter-gatherer mindset, given that the set point of C for our species is low. Historically, Christian missionaries traveling to these areas of the world worked to encourage the kind of beliefs and behaviors that would

---

[87] Francis Fukuyama, *The End of History and the Last Man* (New York: Free Press, 1992).

[88] W. Easterly, *The White Man's Burden: Why the West's Efforts to Aid the Rest have Done so Much Ill and so Little Good* (Oxford: Oxford University Press, 2006).

raise C. But now that the dominant culture in the West is one of falling C, the long-term impact of its influence on the economic development of these societies will be largely negative.

The notion that temperament governs culture and the capacity for industrialization is all the more relevant when considering the Arab nations of the Middle East, which apart from oil wealth remain poor. This is striking considering that this is the oldest area of civilization, with thousands of years of social complexity behind it. Centuries ago this area was more advanced than Europe in knowledge and technology. In recent times its peoples have had easy access to Western ideas, and strong governments with the potential to implement them, yet even when earnest attempts have been made to do so, they have failed, despite the clearly demonstrated benefits.

The most striking illustration of this can be found in the oil-rich states. If wealth alone were able to bring about general commercial success, then Saudi Arabia would have achieved it. Almost unlimited funds have been made available for education and to bring in outside experts to teach the locals. Yet, despite strict quotas regulating the number of Saudis in private sector jobs, 9 out of 10 work in the public sector and youth unemployment has reached 30%. Retail and construction jobs, for which Saudis lack either the aptitude or the interest, are dominated by expatriates.[89]

There are three characteristics of the Egyptian village which tell us why this might be so. The first is a lack of infant C, indicated by the complete lack of parental control in infancy found in Middle Eastern societies. The Japanese, Manus and American case studies all controlled infants and thus had higher infant C, which is linked to market orientation and openness to new ideas. Lack of infant control should make Egyptians, and by extension other Arabs, more conservative and less oriented to the market.

The second characteristic is that the level of C, based on the systematic control of children at all ages, seems to be less than that found in the West (at least until recently), Japan, or even China.

Finally there is the factor of brutal punishment and other stresses in late childhood, which increase both child V and stress, making people not only more rigid in their thinking but accepting of harsh, autocratic governments. It also causes them to descend into anarchy when such

[89] "Saudi Could Face Higher Unemployment—IMF," *Gulf Business*, July 25, 2013.

authority is lacking.

The Arab Middle East, including of course Egypt, is the most perfect example of a society with powerful traditions that maintain child V and stress, together with low infant C. Islam, with its severe annual fast and patriarchal traditions, is the archetypal high V religion. And the same pattern can be found in non-Arab Muslim states such as Iran, Afghanistan, Turkey and Pakistan.

But it is only Western ethnocentric thinking, with its focus on wealth-creation, which considers such societies as "backward." These societies are powerful, successful and, above all, durable. They form dense agricultural populations with high birth rates and resilient traditions. Even wealth, which act to undermine both C and V, has limited ability to change these cultures. Thus, Saudi Arabia and the Gulf states have maintained their patriarchal traditions and fervent devotion to Islam despite several decades of oil wealth.

Such societies tend to be poor, for the reasons given above. But in terms of the success of the society, rather than the well-being of individuals, poverty is not necessarily a problem. High V peoples are aggressive and well organized and tend to have many children, even if in normal conditions most of them die from hunger or disease. As long as enough are born to replace them with a few to spare, the population will grow. And rigid patterns of thought are also not a problem in a society where most people live by subsistence farming. They discourage rash innovation and allow people to hold onto the values and traditions that maintain high levels of V. Thus it is that, at this level, a group with higher V but lower C would tend to win out over another with lower V but higher C.

The struggle for survival is a grim business and takes little account of human health, wealth or even happiness. What matters in the end is the number of surviving children.

Higher C minorities in the Muslim world

Given that higher V people have a competitive advantage, and that Islam is the ultimate high V religion, it might be asked why Christians and Jews have survived and even flourished (at least until recently) in the Muslim world. This is all the more striking given the frequent hostility of the majority population, and various financial and political incentives to convert.

The answer is that Christianity, and even more Judaism, have effects on temperament that tend to favor C at the expense of V, and that this is an advantage in occupations such as trade and the professions that favor flexible thinking. We saw in chapter three that Jewish law is a cultural technology that has been effective at raising C to exceptional levels. It does this by the minute control of everyday behavior, as well as tight control of sexuality for both men and women. We would expect physiological tests to show that of these three groups in Muslim lands, Muslims will have the highest V and lowest C, Jews the highest C and lowest V, and Christians will be intermediate on both.

One indication of this is that Christians and Jews, though a shrinking minority of the general population, formed most of the merchant class of the Ottoman Empire.[90] Considering that V is associated with a higher birth rate and C with more attention to children and thus a lower death rate, it is significant that nineteenth-century European observers universally noted that Christians had a lower birth rate but also a lower deathrate than their Muslim neighbors (although firm demographic information is lacking).[91]

Much the same can be said of the position of Jews in Eastern Europe. As a higher C and lower V minority they were outstandingly successful in commerce and (where permitted) the professions. Demographic information from the nineteenth century shows that Jewish fertility tended to decline faster than for most other groups, but this was more than made up for by lower levels of infant and child mortality.[92] This allowed a rapid increase in the Jewish population despite persecution, intermarriage and emigration.

Despite the economic benefits of Christianity and Judaism, the majority Muslim populations of the Middle East have proved a firm basis for large and powerful empires, not least because lack of national loyalties allows

---

[90] Masters, *Christians and Jews in the Ottoman Arab World.*

[91] Inalcik et al., *An Economic and Social History of the Ottoman Empire.*

[92] E. Silber, "Some Demographic Characteristics of the Jewish Population in Russia at the End of the Nineteenth Century," *Jewish Social Studies* 42 (3–4) (1980): 269–280; S. W. Baron, "The Jewish Question in the Nineteenth Century," *The Journal of Modern History* 10 (1) (1938): 51–65; S. DellaPergola, "Patterns of American Jewish Fertility," *Demography* 17 (3) (1980): 261–273; L. A. Sawchuk, D. A. Herring & L. R. Waks, "Evidence of a Jewish advantage: A study of infant mortality in Gibraltar, 1870–1959," *American Anthropologist* 87 (3) (1985): 616–25. C. O'Grada, "Dublin Jewish Demography a Century Ago," *The Economic and Social Review* 37 (2) (2006): 123–47.

them to grow much larger than European nation states. On several occasions, from the eighth century AD to more recently in the sixteenth century when Turkish armies approached the gates of Vienna, Muslim powers have come close to overrunning Europe.

However, in recent centuries the tables have turned in favor of the West. This success has been based on a form of temperament including high C and especially high infant C, represented in this chapter by the American and Japanese case studies. It is also the prevailing pattern of Europe. Over the past five-hundred years Western culture has risen to dominate much of the world—first militarily and more recently through institutions, ideas and culture.

In the next chapter we will examine this culture—how it arose and why it has become so successful. This analysis will also help us to understand, eventually, why it is now in decline.

Summary and conclusions

Variant forms of V and C have been identified which stem from the way children are treated in early life. Child V is the result of experiencing powerful authority in late childhood, approximately age 6–12. It makes people more accepting of authority and more traditional in outlet. Imposed in the form of punishment it raises the level of stress, resulting in societies with extreme status differences and where leaders tend to be feared.

Infant C is the result of control and other C-promoters in infancy. It is associated with economic success, flexibility of though, impersonal loyalties such as to laws and republican institutions, and respect for hereditary rank. The more general form of C is set from late childhood onwards and is associated with success in business and the professions. For people who are open in their thinking, C can provide the focus that makes high-level creativity possible.

We have also seen that allowing C in men to drop below the level of infant C, such as by permitting them greater sexual license could benefit a warrior society by increasing male testosterone. High infant C can also be achieved by placing upper-class children in the care of lower-class and very high-C nannies. However, by far the strongest contribution to warlike aggression is the level of V, driven especially by contact with anxious but attentive caretakers in infancy.

This chapter has looked at case studies of societies with different levels of these traits, focusing on how people treat children and on relations between the sexes. Note that adult temperament is a product not only of parental behavior but the behavior of other members of the community, plus biochemical and direct epigenetic effects. This is why the connection between childhood experience and adult character seems to be a lot more consistent at the community or cultural levels than for individual families.

CHAPTER SIX

THE RISE OF THE WEST

Having studied the social, economic and political effects of different levels of C and V in a number of societies, the next step is to follow historical changes in C and V over many centuries to see if biohistory can throw light on the rise and fall of civilizations.

The rise of Europe

In the year 1450, Christian Europe showed few signs of future greatness. Populations had collapsed over the previous century. Russia had only recently broken free from centuries of rule by the Mongol empire. Islam was on the rise. Muslims had taken back Palestine from the Crusader states and were still entrenched in southern Spain. Byzantium was about to fall, ending the last remnant of Christian power in the East. Within a century, the victorious Ottoman Turks would begin making inroads into central Europe. At the same time in the Far East, a powerful and united China under the Ming dynasty was without rival or peer.

Five hundred years later, Europe controlled the world. Vast new continents had been discovered, conquered and colonized, their native cultures and languages largely wiped out. India was under British rule, the Ottoman Empire was overshadowed and would shortly be carved up. Western populations had risen dramatically and were still on the rise. China, long humiliated and bullied by Western powers in trade, diplomacy and warfare, was in turmoil.

Underpinning all this was a new kind of economy based on rapid technological development, global trade and large-scale manufacturing, bringing material wealth which was beyond anything previously known and increasing decade by decade. These economic factors gave the West a military and diplomatic power that dwarfed the rest of the world.

One striking point about these developments is the apparent reluctance or inability of non-Western powers to adopt Western technology. Even

Turkey, close neighbor to Europe and with its capital on the European continent, was left behind. The only significant exception was Japan, which in the late nineteenth century made a conscious decision to catch up with the West—an aim achieved with miraculous speed through the adoption of European economics and technology.

Over the past century and a half, writers and theorists have offered many explanations for the remarkable rise of Western civilization. The key factors identified by these thinkers include the unique nature of European geography, access to natural resources, sudden scientific advance, the appearance of new political institutions, wealth generated by colonies, and so on.[1]

Biohistory, by contrast, explains the rise of the West in terms of a change in *temperament*. A powerful expression of this argument can be found in Gregory Clark's book *A Farewell to Alms*, which will be quoted extensively in this chapter.[2] Clark leaves open the reason for this change in temperament, although suggesting that it might be genetic in origin. This is also the view taken by Nicholas Wade in a recent book.[3]

But this temperamental change can be much better explained in epigenetic terms. We have seen from cross-cultural and other evidence that politically and economically complex societies tend to show the distinct patterns of behavior and attitude which are linked together as C: later age of puberty,

---

[1] These theories are analyzed and documented in R. Duchesne, *The Uniqueness of Western Civilization* (Leiden, Boston: Brill, 2012). Engaging with all these theories, commonly identified under such labels as "Marxist," "Weberian," "Smithian growth," "Malthusian" or "World-Systems analysis," is beyond the scope of this chapter. Biohistory starts from very different premises and the focus of this chapter will be on outlining how biohistory can offer an alternative line of future research. The approach closest to biohistory would be Nobert Elias's *The Civilizing Process*, originally published in 1939. This was the first effort to write of the civilizational process that took place in Western Europe from the Middle Ages onward in terms of changes in the personality structure of Western people, including stricter control of impulses, instinctive drives and emotions, cultivation of feelings of shame and embarrassment regarding our animal nature. Elias, however, did not offer an adequate or scientific biological account of the causes of this temperamental changes.

[2] G. Clark, *A Farewell to Alms: A Brief Economic History of the World* (Princeton, NJ: Princeton University Press, 2007).

[3] N. Wade, *A Troublesome Inheritance: Genes, Race and Human History* (New York: Penguin, 2014).

delayed marriage, reduced sexual activity, greater control of children, nuclear families, inhibition and reserve. Loss of them in the past, such as in ancient Rome (see chapter eleven), seems to have been followed by economic and political decline. The link between restraints on sexual activity and more complex societies was noted as long ago as the 1930s by the anthropologist Joseph Unwin, who demonstrated this connection in a study of 86 societies.[4]

Conservatives of all ages have made the same connection, if less systematically, seeing the decline of traditional values as causing the decline of civilization. But they have not been able to explain why this should be so, nor why the greatest flourishing of a civilization occurs *after* these family patterns have begun to break down. For example, the greatest age of Latin literature and the greatest increase in Roman power can be found in the first century BC, when old-fashioned Roman morality had been in decline for more than a century. And Western civilization still dominates the world at the dawn of the twenty-first century after decades of decline in traditional Christian values. Indeed, many people today believe that sexual freedom and liberal childrearing are essential parts of being a truly modern society.

This chapter traces the changes in C and V in European societies over the last thousand years. We will see that aspects of C relating to personal behavior, such as control of children and restraints on sexual behavior among ordinary people, increase in lockstep with political and economic development—a process that has been noted by other writers.[5] We will also see that as the familial and personal aspects of C reached unprecedented heights in the mid-nineteenth century, so economic development took an unprecedented step forward with the Industrial Revolution.

We will see that most indications of C show a steady increase to the nineteenth century. The exceptions are political cohesion and population growth, which showed a general increase but went backwards in certain eras, notably in the fifteenth century. These exceptions will be dealt with in chapter eight.

---

[4] J. D. Unwin, Sex and Culture (Oxford: Oxford University Press, 1934).
[5] N. Elias, *The Civilizing Process*, Vol 1: The History of Manners (Oxford: Blackwell, 1969).

The following analysis will deal mainly with England (and later Britain), though most of the lessons would apply equally well to other European countries.

Delayed puberty

Our first indication of rising C is an objective one—the age of puberty. We have already seen in chapter two that calorie restriction delays the age of sexual maturation.

Much has been made of the declining age of puberty over the past century and a half, but in fact early puberty was once the norm. Until the sixteenth century the age of first menstruation was similar to today—medieval doctors placed it around age 12 or 13.[6] We can infer that the same applies also to the lower classes, even though they were presumably less well fed, since the fourteenth-century poll tax was imposed on males of the age of 14 and over, as indicated by the presence of public hair.[7] Historical evidence gives a similar picture—Margaret Beaufort gave birth to the future Henry VII in 1457 when she was 13.

By contrast, the age of menarche in nineteenth-century Europe, including England, was significantly later, as high as 17 in some countries.[8] It has been suggested that this was a result of the Little Ice Age, which made the climate of Northern Europe colder between 1300 and 1850.[9] But that does not explain why puberty was so much earlier in the fourteenth century, which saw widespread famines. And, as discussed in chapter three, indications from adult height and workers' wages suggest that ordinary people probably ate as well in 1800 as they had in 1200.[10] Also, food

---

[6] P. Laslett, *Family Life and Illicit Love in Earlier Generations* (Cambridge: Cambridge University Press, 1977), 214–232; J. B. Post, "Ages at Menarche and Menopause: Some Mediaeval Authorities," Popul Stud (Camb) 25 (1) (1971):83–7; W. D. Amundsen & C. J. Diers, "The Age of Menarche in Medieval Europe," *Human Biology* (1973): 45.

[7] L. M. Matheson, "The Peasants' Revolt through Five Centuries of Rumor and Reporting: Richard Fox, John Stow, and Their Successors," Studies in Philology 95 (2) (1998): 121–151

[8] P. Laslett, "Age of Menarche in Europe since the Eighteenth Century," in The Family in History, edited by T. K. Robb & R. F. Rotberg, 30, 40 (New York: Harper & Row, 1973).

[9] B. Fagan, *How Climate made History, 1300–1850* (New York: Basic Books, 2000).

[10] Clark, A Farewell to Alms, 815, 936, 961, 999, 1030.

shortage certainly cannot explain why middle-class women in the nineteenth century reached puberty later than peasants in the fourteenth century.

A far better explanation for the rise in the age of puberty is rising C, especially since (as we will see) this is consistent with so many other social, political and economic changes.

Later marriage

Another indication of rising C is a delay in the age of marriage until well past the age of puberty. This is characteristic of food-restricted species such as gibbons, among whom mating may be delayed for a number of years until a suitable territory can be claimed.

Literary evidence suggests that child betrothal and marriage in early adolescence were common in the Middle Ages, even for commoners, something that is not seen in the eighteenth century.[11] The age of marriage in the upper classes shows a clear increase in the period leading up to the early nineteenth century (see Fig. 6.1 below).

This pattern was not confined to the elite. Evidence from parish registers suggests that by the eighteenth and nineteenth centuries, middle-class and working-class women mostly married in their mid-twenties (see Fig. 6.2 below), which is very late by cross-cultural standards and by those of the Middle Ages. (Interestingly, the peak age of marriage was seen in the early eighteenth century rather than the mid-nineteenth, a variation from the pattern which is related to short-term population cycles and will be discussed in chapter eight.)

---

[11] J. Hagnal, “European Marriage Patterns in Perspective,” in *Population in History*, edited by D. V. Glass & D. E. C. Eversley (London: Edward Arnold, 1965), 120; E. A. Wrigley, "The Growth of Population in Eighteenth Century England: A Conundrum Solved," *Past and Present* 98 (1983): 131.

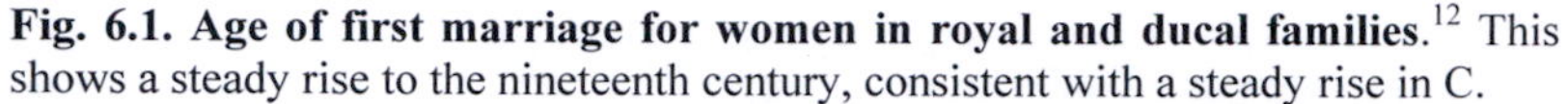

Fig. 6.1. Age of first marriage for women in royal and ducal families.[12] This shows a steady rise to the nineteenth century, consistent with a steady rise in C.

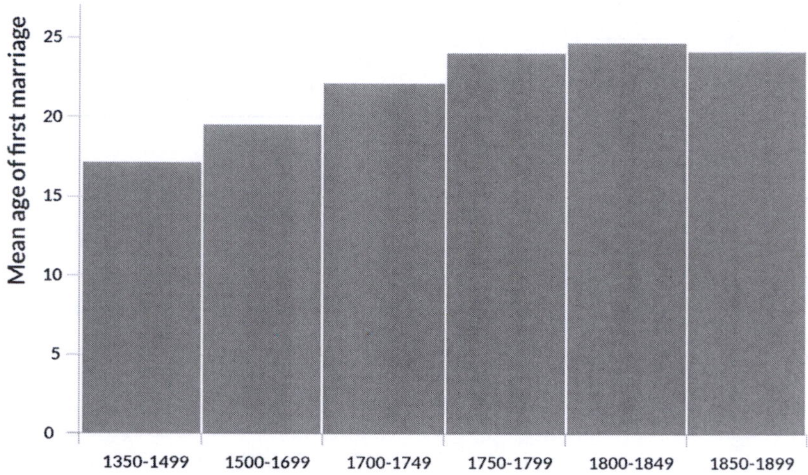

Controls on sexual activity

We saw in chapter two that calorie restriction reduces the sexual activity of rats. Also, cross-cultural studies suggest that high-C societies restrict sexual behavior. Thus, both lack of interest in sex and social controls of sexual activity can be used as indicators of C.

However, this relationship is complicated by the fact that attitudes to sex may also be influenced by stress, which tends to reduce sexual desire (see chapter three). When an ascetic religion such as Christianity is dominant and prestigious, abstinence may become attractive to an anxious minority, even in a relatively low-C society. Therefore, when assessing levels of sexual activity, it is important to distinguish between attitudes towards sex as practiced by the religious elite, and sexual habits prevailing in the general population. When the prevailing habits are restrictive towards sex we have clear evidence of high C. When prevailing habits are more lax but a small minority of people are extreme and even fanatical, we should suspect the presence of extreme anxiety.

---

[12] T. H. Hollingsworth, "A Demographic Study of the British Ducal Families," in *Population in History*, edited by D. V. Glass & D. E. C. Eversley (London: Edward Arnold, 1965).

Fig. 6.2. Age of first marriage for women from parish registers in England and Wales.[13] The age of marriage for commoners also rose with time, but the peak came somewhat earlier for reasons to be discussed in chapter eight.

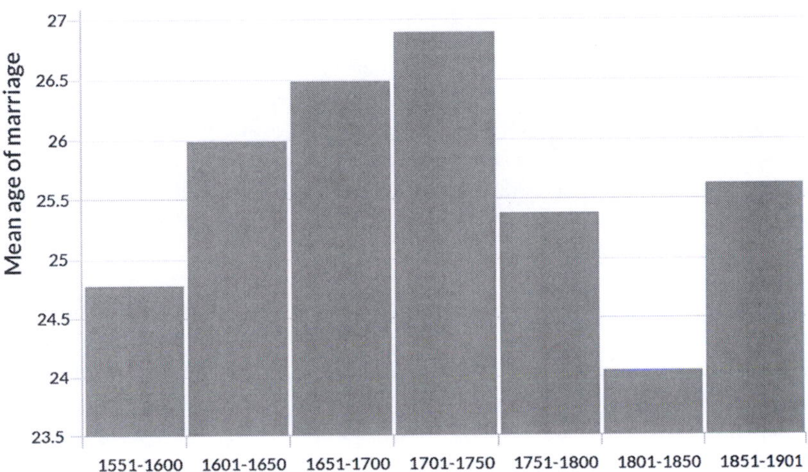

Since its inception, the Christian church was rigorous in denouncing sex before marriage for both males and females. Masturbation, adultery and all manner of sexual misbehavior were forbidden. Early theologians were torn between the need to condemn pleasure and the role sex had in reproduction, a problem that was partially resolved when Augustine declared sex within marriage legitimate only so long as it was performed without lust and with the mind in command.[14]

Early Christianity was noted for its extreme forms of asceticism. Monasticism, which involved renouncing all sexual and financial activity, spread to Europe from Egypt and Syria in the third and fourth centuries

---

[13] R. Woods, *The Demography of Victorian England and Wales* (Cambridge: Cambridge University Press, 2000), 108; E. A. Wrigley, "Variation in Mean Age of First Marriage Among 10 English Parishes, 1551–1837," in *The Population History of England 1541–1871: A Reconstruction*, edited by R. S. Schofield (London: Edward Arnold, 1984); Data from 1850–1901 from E. A. Wrigley & R. S. Schofield, *The Population History of England 1541–1871* (London: Edward Arnold, 1981), 437, and averaged out between six figures ranging from 1851–1901 to fit the format of the previous data. Data comprise England only, minus Monmouth.

[14] G. Hawkes, Sex and Pleasure in Western Culture (Cambridge: Malden, Polity Press, 2004), 63.

AD.[15] Other religious groups, such as the Stylites, whose members could spend years living atop a pillar as a form of religious devotion, also appeared around the fifth century.[16]

The severity of early Christian attitudes can also be seen in the church calendar, which forbade sex on Sundays, teaching that the resulting pollution would make it impossible to take communion. Sex was also forbidden on Thursdays and Fridays, which were to be devoted to pre-communion fasting and self-denial respectively. The early Christian calendar had not one but three Lenten periods of up to sixty days in which sex was not permitted. In addition, sex was forbidden on numerous feast days of the church calendar and on the days before communion on each of these.[17]

Despite the high level of devotion among some early Christians, historians have concluded that Christianity had less of an impact on sexual behavior than its leaders hoped for or even its critics complained of.[18] Until the ninth century, Christianity was mostly confined to larger towns and monastic centers. Since a bishop's real authority rarely extended far outside his city, Christianity held far less sway over the countryside where most of the population resided. Although monasteries, abbots and rural bishops would provide some influence, in many places peasants would rarely see a priest more than once a year.

It was only over time that Christianity began to work its way into the fabric of rural society. In the ninth and tenth centuries rural parishes were set up to bring the countryside under church administration. By the twelfth century a fairly dense network of parish churches covered much of Western Europe, cementing the authority of the church and its moral teachings in place.[19]

But even when Christianity penetrated these remote areas, pagan customs lingered for centuries. For example, in the sixth century Pope Gregory I advised missionaries to move gently so that Anglo-Saxon converts would

---

[15] D. M. Hadley, Masculinity in Medieval Europe (New York: Longman, 1999), 109.
[16] C. W. Hollister & J. M. Bennett, Medieval Europe, a Short History (New York: McCraw-Hill, 2002), 75.
[17] Hawkes, Sex and Pleasure in Western Culture, 67.
[18] M. E. Wiesner-Hanks, *Christianity and Sexuality in the Early Modern World, Regulating Desire, Reforming Practice* (New York: Routledge, 2000), 49, 52.
[19] D. Power, The Central Middle Ages, Europe 950–1320 (Oxford: Oxford University Press, 2006), 133.

feel comfortable in their new faith, gradually adding Christian content and converting old pagan festivals and customs into new Christian ones. The conversion of the pagan winter solstice festival into Christmas is one such fusion that occurred in those centuries.[20]

Popular attitudes towards sex remained far more relaxed than church teachings might indicate. During late antiquity and throughout the Middle Ages women were generally considered to be more lustful than men, a view shared by both medical practitioners and theologians. This was also reflected in folk stories. For every story of knaves seducing innocent maidens, there was one of women engaging in sexual acts simply because they enjoyed them. Women were seen as sexually voracious, and the theme of woman as amoral temptress was a common one. In addition, female pleasure was considered an important part of reproduction. The female orgasm was not only seen as desirable and morally legitimate, but widely believed to be necessary for conception.[21]

Expectations of sexual restraint in the Middle Ages seem to have been far more relaxed than in later times, especially before the twelfth century. Sexual activity before marriage could harm a woman's reputations, especially for wealthy merchants and members of the aristocracy, but did not necessarily ruin her marriage opportunities. Despite clear religious sanctions there was considerable ambiguity about premarital sex. "Simple fornication" between unmarried men and women was fairly common and generally accepted. People often married after having children or after the bride became pregnant. Overall, there was a considerable gap between the form of Christianity practiced by the church and the pious elite on the one hand, and that of the laity whose desires "could only be tamed through marriage" on the other.[22]

There was some tightening of standards in the late Middle Ages, with evidence of fines being levied for premarital sex. Though even these may have had less to do with punishing casual sex than making sure that it ended in marriage, or simply to prevent unmarried people from living

---

[20] Hollister & Bennett, Medieval Europe, 79–81, 168.

[21] V. L. Bullough & J. A. Brundage, Handbook of Medieval Sexuality (New York and London: Garland Publishing Inc., 1996), 86, 87; S. Garton, Histories of Sexuality (New York: Routledge, 2004).

[22] M. R. Karras, Sexuality in Medieval Europe, Doing Unto Others (New York: Routledge, 1985), 38, 39.

together.[23] There was far more leniency towards couples intending to marry, compared with those who had no such intention.[24] Even homosexuality might have been at least occasionally tolerated in the early medieval period.[25] All of this is consistent with relatively low C before about 1500.

Between the eleventh and fourteenth centuries lay people became noticeably more religious. Ordinary people, particularly those in towns who could read, began to take Christianity more seriously, as indicated by the increasing intolerance of other faiths, a growing number of heresies and the launching of the first crusade in 1096.

The eleventh and twelfth centuries also saw church reforms on a scale that would not be equaled until the time of the Reformation. For all its successes, the Catholic Church had been more or less dominated by secular powers since the time of Augustine. Secular rulers like Charlemagne did not even bother to consult Rome before carrying out monastic or theological reform, and regarded the Pope as merely someone who was supposed to set a good example.[26] Lay investiture—the practice of local rulers installing clergy—was common, and for several centuries the papacy was effectively a prized office that was passed between several powerful aristocratic families.

The flawed nature of this arrangement reached a crisis point in 1032 when a famously immoral young aristocrat named Theophylactus became Pope Benedict IX. An active homosexual with few qualifications, he held orgies and feasts in the Lateran palace and was accused of multiple rapes and murders, only to later become bored and sell his office to another and then attempt to reclaim it. By 1046, three different men claimed to be the Pope.

Following widespread public outrage, Pope Leo IX (1049-1054) carried out a program of reform. He freed the papacy from Roman politics and increased its power to the point that it had effective control of both ecclesiastical offices and church doctrine. One of his first reforms was to reinforce the notion that its clergy were somehow different, purer than and

---

[23] Ibid., 97.

[24] P. M. Rieder, *On the Purification of Women, Churching in Northern France, 1100–1500*. (Basingstoke: Palgrave Macmillan, 2006), 68.

[25] Garton, Histories of Sexuality, 76.

[26] C. Backman, The worlds of Medieval Europe (New York: Oxford University Press, 2003), 208–209.

superior to the rest of the population, as part of an overall attempt to redeem the image of the church in the eyes of the laity.[27]

Even though priests were officially required to be celibate, most did marry at this time. Their popular image before the eleventh century was of licentious men who often engaged in a great deal of womanizing. To combat this, reforms began focusing on the sexual attitudes of the clergy, especially the enforcement of celibacy. Opposition was strong, as many priests and bishops were members of the aristocracy who needed to marry for practical reasons. Despite this opposition, by 1125 no priest could formally marry within the Catholic Church, although in many cases they would simply demote their wives to concubines or visit brothels to satisfy their needs.[28]

As the secular clergy increasingly took on sexual and moral purity as part of their identity, the practice began to spread to the laity as well, although this applied mainly to men, since women were seen as too lustful to be chaste. Writers in the twelfth century also began to focus increasingly on the issue of women committing adultery, which was becoming an increasingly serious offence. One reason for this was that in the eleventh and twelfth centuries, the church had taken control of the marriage law and made divorce illegal. Before that time it was possible to repudiate an adulterous wife and remarry, but with the new laws people had to be far more vigilant of their spouse's sexual behavior. The twelfth century also saw further persecution of homosexuals and enforcement of sexual restrictions.[29]

In the thirteenth century, marriage was increasingly defined as a sacrament or visible evidence of God's grace, and was the expected norm for all Christians. Sexual transgressions became linked to occult practice, and a newly formalized Inquisition became vigilant against misbehavior. Also, for the first time, sermons were given in the local tongue and thus became important vehicles for relaying the church's ideas about sexual discipline.[30] However, despite the severity of the church's regulations, there was still a considerable divide between church teaching and actual

---

27 Hollister & Bennett, Medieval Europe, 205–209; H. Berman, *Law and Revolution: the Formation of the Western Legal Tradition* (Cambridge Mass.: Harvard University Press, 1984).

28 Karras, Sexuality in Medieval Europe, 20–22, 43, 44.

29 Ibid., 89; Garton, Histories of Sexuality, 77.

30 Wiesner-Hanks, *Christianity and Sexuality in the Early Modern World*, 39, 40.

behavior. For all its rigorous campaigning and threats of damnation, the medieval church never succeeded in obtaining a universal acceptance of its sexual regulations.[31]

But by the fourteenth century, fears of damnation, demons and the devil were growing more acute. Non-conforming practices such as homosexuality and pagan rituals now tended to be denounced as satanic. Accusations of sodomy were among the devices used by the Church to crush the Knights Templar, though their wealth and power may have been a more crucial motivation.[32] By late in the century, homosexuals in some Italian city states could be burnt at the stake.[33]

Public stigma against premarital sex was also gaining momentum, especially in Southern Europe. By the end of the fourteenth century, Venetian courts were treating the loss of a woman's virginity and even male fornication as an offense against God, rather than simply (at least for women) an affront to family honor. Despite this, prosecutions for premarital sex were still quite rare.[34]

Attitudes in England remained somewhat more relaxed, as illustrated by several episodes in Chaucer's Canterbury Tales. In The Reeve's Tale, two students take revenge on a dishonest miller by sleeping with his wife and daughter. The women are seen as impassive, and the act less a violation of sexual taboos than a claim for payment on the miller's property.[35]

This changed dramatically in the second half of the sixteenth century. The Protestant Reformation ushered in a new era of sexual austerity in northwestern Europe. Martin Luther and John Calvin preached that the individual, not the established church, was responsible for salvation, and there was a forceful re-emphasis on the evils of ungodly sexual pleasure outside marriage. Puritan pamphleteers like Philip Stubbes believed that Catholic courts and confessionals treated sexual misdemeanors too lightly. They recommended that adulterers, fornicators and those guilty of incest drink poison or be branded in public to distinguish them from the virtuous.

---

[31] Karras, Sexuality in Medieval Europe, 38, 39.

[32] Wiesner-Hanks, *Christianity and Sexuality in the Early Modern World*, 41.

[33] Ibid., 47.

[34] P. N. Stearns, *Sexuality in World History* (Oxon, Canada: Routledge, 2009), 48.

[35] Karras, Sexuality in Medieval Europe, 99.

This had a significant impact on public morality. By the end of the century, advice to young men on refraining from sexual activity spread beyond specifically religious tracts and began to enter popular culture. The correspondence of eminent men to their sons was increasingly focused on warning them about the dangers of vice. Reflecting the moral tone of the time, in 1619 Patrick Scot, author and occasional tutor to Prince Charles, wrote to the younger generation on the costs of "whoring":

> Lechery is no other thing than a furious Passion, shortening the life, hurting the Understanding, darkening the Memorie, taking away the Heart, spoyling Beauty, weakning the joints, ingendring Sciatica, Gouts, Giddinesse in the Head, Leprosie, and Pox.[36]

Immodesty of dress and conduct were increasingly frowned upon and branded as "unmanly." Losing self-control in particular, whether in lust, anger or drunkenness, was seen as losing humanity and becoming like a beast.[37]

In the seventeenth century there was an increased focus on marital sex and growing condemnation of adultery. Cases of premarital sex were taken to the "Bawdy Courts," set up solely to deal with sexual matters, and at their peak between 1450 and 1640. What is interesting about these trials is that they mostly involved commoners, marking the emergence of a new trend in European culture—a common sexual morality that applied across all social classes.[38]

In England in 1650 the Commonwealth Adultery Act laid down the death penalty for female adultery and incest, and three months' imprisonment for fornication. Brothel keepers were whipped, branded and jailed for three years, and executed if they reoffended. Although repealed after ten years on the Stuart Restoration, such laws indicate the degree of fanaticism in at least some sections of society.

The seventeenth century also saw growing distinctions between the public and private, making it easier for secular laws to control public intimacy. Attitudes towards self-censorship increased and violations of these

---

[36] P. Scot, *Omnibus & Singulus. Affording Matter Profitable for all Men, Alluding to a Fathers Advice or Last Will to his Sonne* (London, dedicated to King James and Prince Charles, 1619).
[37] A. Shepard, Meanings of Manhood in Early Modern England (Oxford: Oxford University Press, 2003), 21.
[38] Hawkes, Sex and Pleasure in Western Culture, 98, 99.

boundaries were increasingly accompanied by feelings of shame and embarrassment.[39]

Policing of sexual matters also moved steadily from church to state. Medical professionals and civil administrators set down laws and wrote medical textbooks on sex, whilst the authority of priests declined, a trend which would continue (except among minority Catholic communities) until the present day.

The late seventeenth and early eighteenth centuries mark something of a break in this trend. In 1694 the English parliament decided not to renew the Licensing Act, which had been set up to restrict the creation and sale of indecent material. A great influx of illicit erotic and pornographic literature subsequently came streaming in from Europe, coinciding with what seems to have been a briefly more liberal trend in sexual values.[40] Sensual pleasures came to be regarded as a good thing for both men and women and sexual activity outside marriage was widely tolerated. Sexually liberated women were frequently admired or praised, whilst the subversion of gender roles was seen as a legitimate source of amusement or leisure. (This can be attributed, at least in part, to a decline in religious feeling resulting from the secondary population cycle mentioned earlier, and to be discussed in chapter eight.)

But the degree of change, and the extent to which it applied to all sections of society, was limited. Homosexuality and masturbation were increasingly stigmatized in popular culture, and parish records suggest that the rate of premarital pregnancy was low. In fact, brides were less likely to go pregnant to the altar in the late seventeenth and early eighteenth centuries than in earlier or later periods. The level of restraint in this regard was not matched until the mid-twentieth century, when relatively conservative values coincided with widespread contraception.[41]

But even this degree of liberalization was reversing by the close of the eighteenth century, when there was a backlash against the sexual and

---

[39] Ibid.

[40] R. B. Shoemaker, Gender in English Society, 1650–1850, The Emergence of Separate Spheres? (New York: Addison Wesley Longman Inc., 1998), 59, 60.

[41] L. Stone, *The Family, Sex and Marriage in England: 1500–1800* (London: Weidenfeld & Nicholson, 1977), 610; D. S. Smith & H. S. Hindus, "Premarital Pregnancy in America: 1640–1971: An Overview and Interpretation," The Journal of Interdisciplinary History 5 (4) (1975): 169–70.

political freedom of women.[42] By the mid-nineteenth-century sexual modesty in middle-class culture was so severe that it was considered impolite for a woman to mention or even acknowledge the use of undergarments for fear it may lead to discussion of anatomical details. As one Victorian lady expressed it: "[Undergarments] are not things, my dear, that we speak of; indeed, we try not even to think about them."[43]

Even more significant, in terms of tracking the biological basis of this change in temperament, is that women seem to have become less interested in sex. Marriage manuals and most doctors in the Victorian era assumed that respectable women felt no sexual desire at all, and it is difficult to see how this opinion could have been widely accepted without some support from personal experience. In 1857 Dr William Acton, a popular authority on diseases of the urinary and generative organs wrote:

> Having taken pains to obtain and compare abundant evidence on this subject I should say that the majority of women (happily for them) are not very much troubled with sexual feelings of any kind. What men are habitually, women are only exceptionally … there can be no doubt that sexual feeling in the female is in the majority of cases in abeyance … and even if roused (which in many instances it can never be) is very moderate compared with that of the male … As a general rule, a modest woman seldom desires any sexual gratification for herself. She submits to her husband, but only to please him; and but for the desire of maternity, would far rather be relieved from his attentions.[44]

The doctor also wrote a caution about excessive sexual activity, reflecting the popular beliefs of the day:

> I maintain that debauchery weakens the intellect and debases the mental powers, and I reassert my opinion that if a man observes strict continence in thought as well as deed, and is gifted with ordinary intelligence, he is more likely to distinguish himself in liberal pursuits than one who lives incontinently, whether in the way of fornication or by committing marital excesses. The strictest continence, therefore, in the unmarried, and very moderate sexual indulgence in the married state, best befit any one

---

[42] Hawkes, Sex and Pleasure in Western Culture,112, 113.

[43] C. W. Cunnington, *English Women's Clothing in the Nineteenth Century: A Comprehensive Guide* (New York: Dover Publications, 1990), 20.

[44] William Acton, *The Functions and Disorders of the Reproductive Organs in Childhood, Youth, Adult Age, and Advanced Life: Considered in Their Physiological, Social, and Moral Relations*, 3rd Edition (London: Churchill, 1862), 101.

engaged in serious studies.[45]

Modesty in women, far from being imposed by external authority, came to be seen as an inner quality stemming from a lack of sexual desire. The medical "discovery" in the 1840s that conception did not require an orgasm simply reflected the assumptions about female sexuality that had been prevalent for two generations. From as early as the late eighteenth century, the image of a passionless and moral mother came to be seen as the epitome of femininity.

Cultural norms dictated that men were only meant to have sex with their wives if the goal was to reproduce, and to demand marital sex for any other reason was considered ungentlemanly. If a man wanted to have recreational sex, he was expected to hire a prostitute (which was considered immoral, but not unnatural) and not "treat his wife like one."[46] The other source of gratification was an extramarital relationship. There was an increase in the number of middle-class men keeping mistresses, a custom previously practiced mostly by the upper classes. Although still considered to be outside the bounds of acceptable morality, this sort of behavior became so common that it often barely attracted attention, and was considered preferable to homosexuality or masturbation.[47]

Victorians had especially extreme views on masturbation. No longer considered merely a bad habit, as in earlier times, it was now seen as a dangerous disease. Although there was serious medical doubt about the notion, it was widely believed that the male body was endowed with a fixed quantity of sperm, and that too much indulgence would cause either impotence or a draining of energy from the body or mind, as well as leading to consumption, curvature of the spine, insanity and even death. This view of male anatomy formed the basis of vigorous polemics against masturbation, nocturnal emissions and even overly frequent sex within marriage. Men who did not moderate their behavior were riddled with guilt and anxiety, and although sexual desire in men was not completely condemned, the ways in which it could be fulfilled were tightly

---

[45] Ibid., 196

[46] S. L. Steinbach, *Understanding the Victorians. Politics, Culture and Society in Nineteenth-Century Britain* (New York: Routledge, 2012), 197.

[47] W. D. Rubinstein, Britain's Century, A Political and Social History 1815–1905 (London: Edward Arnold, 1998), 331; Steinbach, *Understanding the Victorians*, 197.

constrained.[48]

Parents of respectable families were constantly on vigil for signs that their children had been misbehaving, and it was believed that masturbators could be recognized by their "stunted frames, their underdeveloped muscles, sunken eyes, pasty complexions, acne, damp hands and skin."[49] Parents of the time were advised to keep a close watch on their children for signs of this behavior, violating their privacy whenever required. If such behavior was suspected, they should employ a rigorous regime of severe treatments, including sponge baths and exercise to the point of exhaustion. Popular children's literature also reflected this condemnation of masturbation:

> May every school boy who reads this page be warned by the warning of their wasted hands from the burning marle of passion where they found nothing but shame and ruin, polluted affections and an early grave.[50]

The focus on sexual morality in stories designed for schoolchildren was a distinctive feature of the late eighteenth and nineteenth centuries, not seen in the sixteenth or seventeenth.[51] The result of such attitudes, in flat contradiction to their intention, was an explosion in the number of prostitutes. In 1839 it was estimated there were as many as 80,000 prostitutes in London.[52] But given the expense, moral disapproval and the danger of disease, as well as the horror of masturbation, it is likely that most respectable men (as well, of course, as the large majority of women) had very low levels of sexual activity.

It is important to mention, however, that this sexually restricted behavior did not reflect all of society, especially those who stood outside the prevailing church-going culture of the day. In the music halls visited by clerks and tradesmen, courtships and sexual liaisons occurred on the

---

[48] J. Tosh, A Man's Place: Masculinity and the Middle-class Home in Victorian England (New Haven and London: Yale University Press, 1999), 44, 45, 46.

[49] S. Marcus, *The Other Victorians: A Study of Sexuality and Pornography in Mid-Nineteenth Century England* (New Brunswick: Transaction Publishers, 2009), 19.

[50] A. N. Wilson, The Victorians (London: Hutchinson, 2002), 288.

[51] Ibid., 289.

[52] The Great Social Evil: "The Harlot's House" and Prostitutes in Victorian London, Rhianna Shaw '11, English 60, Brown University (2010), http://www.victorianweb.org/authors/wilde /shaw.html (accessed September 12, 2014).

premise that men and women had equally strong libidos.[53] This difference in sexual standards is also reflected by the fact that whilst middle class women almost never became pregnant outside of marriage, illegitimacy rates among the working class were somewhat higher than a century earlier, reflecting a greater tendency for the working classes to engage in premarital sex.[54] Prostitutes were also an exception, their sexual appetites being seen as evidence of their corruption.

The apparent lack of desire in middle class women reflects an unprecedented level of C in the mid-nineteenth century, and the austere values of middle-class Victorian society (prostitutes and mistresses aside) indicate not only this high-C temperament but the values that had acted to create it. This is entirely consistent with a peak of C in the mid-nineteenth century.

Control of young children

As we saw in chapter two, calorie restricted rats not only spend more time with their infants but are quicker and more efficient at gathering them back into the nest. This is a pattern of behavior suited to environments where food is limited. To be successful, parents must put greater effort into rearing relatively fewer young, helping them to not only survive but have a chance of breeding in the face of fierce competition. The cross-cultural survey, as indicated in chapter five, suggests that the equivalent human behavior is control and training of very young children, something we saw vividly in Manus and American parents, and to a lesser extent among the Japanese.

Training of young children seems to have been quite rigorous even in medieval times. From a book of advice on childrearing, written in 1475:

> Little children [from age three onwards] must not speak until spoken to, they must not chatter or stare about, they must stand until they are told to sit, they must not look sulky, they mustn't pick their noses or scratch their ears, pick their nails or teeth, they mustn't drink with their mouth full, mustn't lean against post or door, mustn't put their elbows on the table, nor wink or roll their eyes.[55]

---

[53] Tosh, A Man's Place, 44, 45, 46.
[54] Steinbach, *Understanding the Victorians*, 201.
[55] J. Gathorne-Hardy, The Unnatural History of the Nanny (New York: Dial Press, 1973), 45.

The crucial point to note here is the age. Control of children does not seem to have started before their third birthday, which would produce only moderate infant C. But this is still significantly earlier than was the case for Egyptian and Chinese parents, who did not begin serious discipline until ages five or six.

A study of parenting manuals suggests that the training of children was becoming increasingly rigorous in the sixteenth and early seventeenth centuries. Seager's School of Virtue, first published in 1557, warned children against gambling, swearing, drinking, lying, anger, malice and oversleeping, advised on how to behave in church and on what psalms should be learned by heart. Work came to be seen as a major moral good in itself, and any form of pleasure was increasingly viewed with suspicion.[56]

Discipline also seems to have started at a somewhat younger age. In early seventeenth-century Europe, from the limited sources available, severe discipline was imposed from the age of weaning at around two years. A contemporary diary shows that the future Louis XIII was first whipped within a month of his second birthday (in 1603). From that age on, children were rigorously trained. They could be punished for too much crying, refusal to eat, lying and even refusal to show proper forms of affection for parents. Discipline continued after the age of five or six but it certainly did not increase. In fact:

> Adults began to feel at home with the six- or seven-year-old child who was physically more robust and self-sufficient, and intellectually more approachable. In other words, he was "cured" of the malady of infancy which had made him so incomprehensible and frightening to his elders.[57]

Louis' father, Henry IV, began to have a stronger and more positive relationship with his son from age five. If anything, he became more lenient. This is a pattern very different from that of China or especially Egypt today, where there is relatively little control before the age of five or six but a dramatic increase at this age. Infant C is produced by parental control before the age of five, but the Manus examples suggests that the greatest effect is from the earliest control. Thus, this earlier start to training suggests a rise in infant C.

---

[56] Ibid., 46.
[57] D. Hunt, Parents and Children in History (New York: Basic Books, 1970), 89, 133–138.

But infant C was not yet at its peak because there was, up to this time, no training in the first two years of life. Until the late seventeenth century, infants had been fed on demand, weaned late, and seen as incapable of reason.[58] Although bedwetting at a later age was severely punished, there is no evidence of early toilet training or disapproval of play with genitals.[59] Thus, although infants might be punished, there is little evidence of control.[60]

But this began to change from the late-seventeenth century, and even more in eighteenth and nineteenth centuries. There was a growing emphasis on toilet and sex training of infants, and a move away from demand feeding. This was especially noteworthy at a time when parents were showing more affection and less punishment—a matter we will take up shortly.[61]

---

[58] L. Stone, *The Crisis of the Aristocracy* (Oxford: Oxford University Press, 1965), 591; C. Hill, *Society and Puritanism in Pre-Revolutionary England* (London: Becker & Warburg, 1964); Gathorne-Hardy, The Unnatural History of the Nanny, 46–8. Stone, *The Family, Sex and Marriage in England*, 175.

[59] Stone, *The Family, Sex and Marriage in England*, 102, 159–61; J. E. Illick, *Childrearing in Seventeenth Century England and America. The History of Childhood*, edited by L. DeMause (New York: The Psychohistory Press, 1974).

[60] Gathorne-Hardy, The Unnatural History of the Nanny, 48; Hunt, Parents and Children in History, 133; L. DeMause, "The Evolution of Childhood," in *The History of Childhood*, edited by L. DeMause, 40–1 (New York: The Psychohistory Press, 1974).

[61] DeMause, "The Evolution of Childhood," 39; Illick, *Childrearing in Seventeenth Century England and America*, 309, 318–9; Ariès and Stone both describe a growing level of affection shown by parents towards children in this period. Other authorities (Pollock) have failed to find such a pattern, while still others (Roberts) take a middle ground In terms of V and C, affection as such appears to have little impact. The key behaviors affecting them are control/education and punishment. By this criteria, there is evidence (Roberts) that parents, particularly from the seventeenth century onward, were committed to a new form of moral education (control) rather than physical punishment. The family among Puritans, for example, came to be seen as an institution dedicated to the salvation of souls through a proper education in the rules of good behavior. According to Ozment, with the Reformation, fathers became primarily responsible for the moral as well as the economic well-being of their families. While appropriate punishments were employed to enforce discipline, children were seen to be capable of character development through education . Stone, *The Family, Sex and Marriage in England* 1500–1800. London: Weidenfeld; P. Ariès, *Centuries of Childhood* (Harmondsworth: Penguin Books, 1960); L. H. Pollock, *Forgotten Children: Parent–Child Relations from 1500 to 1900* (Cambridge: Cambridge University

Some parents operated in a manner reminiscent of a military campaign. By a certain perverse logic, Mme Acarie in Seventeenth Century France consistently forced her children to eat dishes which they disliked, and denied them those they appeared to enjoy. John Locke cited with approval the example of a "prudent and kind mother" who whipped her infant daughter eight times in one morning to master her stubbornness. Had she stopped at the seventh, he commented, "she had spoiled the child for ever." The American Baptist minister Francis Wayland described in some detail his battles with his fifteen-month-old son during the 1830s, on one occasion depriving the boy of food for three days to win an argument over a crust of bread. He reported that: "Since this event several slight revivals of his former temper have occurred, but they have all been easily subdued," and that his son had become "mild and obedient." The stakes were high for the more pious parents—nothing less than eternal salvation for their sons and daughters.[62]

The same changes could be seen in Colonial America in the early eighteenth century, with parents spending more time and effort controlling their children at a very young age. One expression of this was that babies as young as seven or eight months were encouraged to stand up and walk rather than crawl, a practice reminiscent of Manus training. There was considerable interest in writers such as Frances Hutcheson, who believed that children had an innate moral sense which must be nurtured in infancy and childhood.[63]

Perhaps the most famous statement about the treatment of children in this period relates to Susannah Wesley (mother of the evangelists John and Charles Wesley) who had nineteen children in all:

> When turned a year old (and some before), they were taught to fear the rod, and to cry softly; by which means they escaped abundance of correction they might otherwise have had; and that most odious noise of the crying of children was rarely heard in the house; but the family usually

---

Press, 1983); B. Roberts, *Through the Keyhole: Dutch Child-Rearing Practices in the Seventeenth and Eighteenth Century* (Hilversum: Verloren, 1998).

[62] C. Heywood, *A History of Childhood: Children and Childhood in the West from Medieval to Modern Times* (Cambridge: Polity Press, 2001), 99.

[63] R. Middleton & A. Lombard, *Colonial America, a History to 1763*, 4th Edition (Wiley-Blackwell: United Kingdom, 2011), 267, 271

lived in as much quietness as if there had not been a child among them.[64]

General trends in childrearing between 1660 and 1820 are difficult to establish because of the differences between social groups. Some upper-class parents became more lenient after the Restoration in 1660, but prosperous farmers showed the opposite tendency.[65] Similarly, in the eighteenth century many wealthier families became more permissive while religious people increased the control of their children.[66]

But from the late eighteenth to the nineteenth century, most writers find an increase in parental pressure and supervision.[67] A study of nineteenth-century nannies suggests that children of the upper classes and at least one segment of the lower classes (from which nannies were drawn) were under very tight control. A great number of regulations were enforced over even the most minute areas of life. Nannies tended to pursue an obsessive drive against behavior such as touching of genitals. Rigorous toilet training began in the first few months of life and required very young children to sit for hours on the potty if unable to perform.

The concept of childhood also changed in this period, coming to be seen as a precious state which children ought to enjoy to some extent. The idea that each child had its own distinct character which should be valued and nurtured only appeared in the late eighteenth and early nineteenth centuries.[68] This did not lessen the degree of control, but merely placed more emphasis on training, as opposed to the punishment more common in earlier times.[69]

---

[64] Susanna Wesley, letter to her son, July 24, 1732, in Charles Wallace Jr (ed.), *Susanna Wesley: The Complete Writings* (New York: Oxford University Press, 1997), 369.

[65] DeMause, “The Evolution of Childhood,” 39; Illick, *Childrearing in Seventeenth Century England and America*, 318–9.

[66] Gathorne-Hardy, The Unnatural History of the Nanny, 46–8; Stone, *The Family, Sex and Marriage in England*, 405.

[67] P. Robertson, “Home as a Nest: Middle-Class Childhood in Nineteenth Century Europe,” in The History of Childhood, edited by L. DeMause, 422 (New York: The Psychohistory Press, 1974); M. Jaeger, Before Victoria (London: Chatton and Windus, 1956), 105; Stone, *The Family, Sex and Marriage in England*, 415, 669.

[68] Tosh, A Man’s Place, 39.

[69] G. K. Clark, Portrait of an Age, Victorian England (Oxford: Oxford University Press, 1977), 22.

Table manners could be taught by strapping a child's elbows to the table. The most obsessive discipline might be applied to such transgressions as taking off shoes without undoing the laces. Nannies might insist that bread could be eaten with butter or jam, but never both, or that yesterday's crusts must be eaten before today's bread was made available. Children were taught neatness and politeness to an extreme degree, that authority must be respected and obeyed, and that any form of self-indulgence was wrong.

It could be said that nannies went beyond, even insanely beyond, the discipline necessary to enforce these social lessons. But such behavior is not to be understood in rational terms. It is the expression of a deep-seated, emotionally driven need to control and direct, which is the essence of high C. It must also be stated that the behavior of nannies reflected not just the treatment of upper class children but the respectable working classes from which nannies were drawn. Parental advice books also reflect similar values. All this is clear evidence of rising C, and especially infant C, from medieval times to the nineteenth century.

Nuclear families

The small family unit is a type of social organization linked to calorie restriction and to food restricted environments. Not only are calorie restricted animals and humans less sociable, but their social groups typically consist of a mated pair and their immature offspring. The cross-cultural survey linked nuclear families and monogamy with other indications of C such as restricting sexual activity, and political and economic complexity. No human population is quite as extreme as gibbon societies, in which couples defend exclusive territories, but relative strength of the nuclear family is a clear indication of high C. It is thus of particular interest to observe a strengthening of the nuclear family over several centuries leading up to the nineteenth.

There is widespread agreement among historians that the nuclear family was growing in importance from the late Middle Ages onwards.[70] Even in the fourteenth century, people were building extra rooms in their houses for privacy. From the fifteenth to the nineteenth centuries there was an

[70] Although monogamy was always more emphasized in the West than in Eastern cultures, as for example in the differential treatment of illegitimate children. MacDonald, K. B. (1995). "The Establishment and Maintenance of Socially Imposed Monogamy in Western Europe." *Politics and the Life Sciences*, 14, 3–23. This is clearly related to Christianity, but also perhaps reflects higher infant C.

increased emphasis on privacy and intimacy between husband and wife.[71] The emotional emphasis on the family was also on the increase. The Victorian middle-class home has been described as a triumph of the sentimental and emotional over the purely economic aspects of family life.[72]

Not only were family ties becoming stronger, but other ties were weakening. There was a decline in kin and patronage ties, and a growing separation of servants and masters.[73] It became less and less common for middle class homes to have rooms devoted solely to business, with the exception of some professional men who required it. In addition, cellars for commercial storage, in-house workshops and offices all declined as the family was increasingly separated from the economy. The nineteenth century family home was more secluded and private than in the previous century, when callers might drop in casually for breakfast. Guests usually came by invitation only, with the front door shut and visitors required to state their business to a servant.[74]

Victorians increasingly separated their families not only from work but from the local community. An elaborate system of calling cards emphasized the degree of formality which regulated contact with outsiders. Although a consistent strand of domesticity was found in both aristocratic and bourgeois families in the eighteenth century, it was not until the 1830s and 1840s that the ideal of the home became a cultural norm. This ideal was expressed by countless writers during the period, as James Anthony Froude wrote in his novel, *The Nemesis of Faith* (1849):

> When we come home, we lay aside our mask and drop our tools, and are no longer lawyers, sailors, soldiers, statesmen, clergymen, but only men.

---

[71] P. Aries, G. Duby, R. Chartier & A. Godlhammer, *History of Private Life*, Volume III: Passions of the Renaissance (Cambridge. Mass.: Harvard University Press, 1988); P. Aries, G. Duby, R. Chartier & A. Godlhammer, *History of Private Life*, Volume IV: From the Fires of Revolution to the Great War (Cambridge Mass.: Harvard University Press, 1988).

[72] Tosh, A Man's Place, 13.

[73] F. R. H. DuBoulay, An Age of Ambition (London: Nelson, 1970), 116; H. M. Smith, Pre-reformation England (London: MacMillan, 1963), 235; L. B. Wright, *Middle Class Culture in Elizabethan England* (London: Methuen, 1935), 203; Stone, *The Family, Sex and Marriage in England* ; M. Girouard, *Life in the English Country House* (New Haven and London: Yale University Press, 1978), 219, 285, 298.

[74] Tosh, A Man's Place, 21.

> We fall again into our most human relations … We cease to struggle in the race of the world, and give our hearts leave and leisure to love.[75]

Outsiders also noted the emphasis on domesticity as being one of the defining features of English culture life during the era. French writer Hippolyte Taine summed up the prevailing view of England in the 1850s:

> Every Englishman has, in the matter of marriage, a romantic spot in his heart. He imagines a "home," with the woman of his choice, the pair of them alone with their children. That is our own little universe, closed from the outside world.[76]

Seclusion and refuge from the outside world, ideally with a rural ambience, reflected a quieter and simpler country life that defined the ideal middle-class Victorian household. It was also during this period that the celebration of Christmas, which had previously been more of a community festival, became domesticated, evolving into the family-centered occasion we know it as today. The idea of the Christmas tree, imported from Germany in the early nineteenth century, reflected this change.

Even the spirituality of the age came to reflect the importance of domestic life, with the concept of heaven increasingly thought of as an extension of home. Heaven was seen as a place where the recently departed would be reunited with those who had passed away before them and live in eternal domestic bliss. By the mid-nineteenth century the theme of the reunion of earthly families in the afterlife was the dominant popular image of heaven.[77] All these factors, including the increasing emphasis on privacy and seclusion of the monogamous family, are indicators of extremely high C during the nineteenth century,

Inhibition and reserve

We have seen that calorie restriction makes animals and people less gregarious and more socially inhibited. It was also suggested in the last chapter that C is associated with reserve. Thus it is of interest to see the impulsiveness and quick temper of men in the sixteenth and seventeenth centuries giving way to reserve and inhibition in the nineteenth.

Although the traditional caricature of the English is the stoic and reserved

---

[75] Quoted in Ibid., 33.
[76] Quoted in Ibid., 29.
[77] Ibid., 21, 23, 24, 29, 30, 32, 33, 38, 39.

gentlemen, this was not always the case. When visiting London at the turn of the sixteenth century, the Dutch scholar Erasmus described the English as being open and forward with their emotions. He noted their tendency to kiss one another: "In a word, wherever you turn, the world is full of kisses."[78] Elizabethan audiences could be moved to tears by a theatrical performance.[79] When Italians visited England in the Elizabethan period they remarked on how emotional people were and how little self-control they had. Similarly, when the French visited they remarked that "sang-froid" and "insouciance," meaning coolness and nonchalance respectively, were qualities that the English did not possess. Until as late as the eighteenth century it was not considered insulting to describe someone as "sentimental" and people of taste and refinement were praised for open displays of emotion.[80]

However, a change took place in the mid- to late-eighteenth century. Books on etiquette began to circulate among the rising middle classes, expressing the values of manners and refined conduct as a modern adaptation of ancient courtly values. This new culture of restraint and refinement included rules and regulations on everything from how to sit and walk correctly, to behavior at the dinner table, and personal codes of etiquette. Chief among these rules and modes of behavior was the concept of self-control where the regulation of the body, the mind, appetites and emotions was considered the first step in the civilizing process. In contrast to earlier periods, passionate displays of feeling in social conduct, for good or ill, were now frowned upon.[81]

---

[78] S. W. Crompton, Desiderius Erasmus (Philadelphia: Chelsea House, 2005), 22.

[79] L. C. Pronko, “Kabuki and the Elizabethan Theatre,” *Educational Theatre Journal* 19 (1) (1967): 9–16

[80] *BBC News Magazine*, “The End of the Stiff Upper Lip,” http://www.bbc.co.uk/news/magazine-19728214 (accessed September 5, 2014); B. Escolme, *Emotional Excess on the Shakespearean Stage: Passion's Slaves*, (London: Bloomsbury Arden Shakespeare, 2013); M. Deacon, “Ian Hislop's Stiff Upper Lip: an Emotional History of Britain, BBC Two, review,” *The Telegraph* (October 3, 2012), http://www.telegraph.co.uk/culture /tvandradio/9582671/Ian-Hislops-Stiff-Upper-Lip-an-Emotional-History-of-Britain-BBC-Two-review.html (accessed September 5, 2014).

[81] L. Young, *Middle-Class Culture in the Nineteenth Century, America, Australia and Britain* (Basingstoke: Palgrave Macmillan, 2003), 83, 95; H. Hitchings, *Sorry!: The English and their Manners* (London: John Murray, 2013); N. Elias, *The Civilizing Process,* Vol. 1, The History of Manners (Oxford: Blackwell, 1969).

The nature of the wedding ceremony also went through a dramatic change between the early modern period and the nineteenth century. The late medieval wedding was a community affair, both boisterous and interactive. Couples faced teasing, ritual joking and physical tests such as the jumping of the petting stone. This rite involved the placing of a stone or bench in the couple's path, usually at the church steps or gate. It was thought that the bride's form in performing the jump was a good indicator of what sort of wife she would be—a good mistress or an "ill-tempered shrew."[82] Clergy had no control of the festivities and as late as the eighteenth century might join in the fun.[83]

By the nineteenth century, weddings had become sober occasions where the immediate family took on a more significant role. The jests and physical rites of previous centuries were now deemed undignified and were replaced by an emphasis on modesty and reserve. Respectable women were no longer able to kiss, dance and touch.[84]

Reserve now became the middle class norm. Englishmen in that period were noted for being inhibited and cool—the proverbial "stiff upper lip"—with a great emphasis placed on the value of solitude.[85] By the reign of Queen Victoria, a typical Englishmen was typified as being "Strong and silent, earnest, matter of fact, sparing in his speech, scornful of theory, unemotional, manly and athletic."[86]

It was not, however, until the mid-nineteenth century, with the Crimean War and the veneration of the quiet virtue of the individual soldier, that stoicism came to be seen as a true national value. From the eighteenth century onward, excessive displays of emotion came to be regarded as bad

---

Elias traces an increase in self-restraint as a result of socialization from birth onwards.

[82] J. R. Gillis, "From Ritual to Romance: Toward an Alternative History of Love," in *Emotion and Social Change*, edited by C. Z. Sterns & P. N. Sterns, 100 (New York: Holmes and Meier, 1988).

[83] Ibid.

[84] Ibid., 108.

[85] J. A. Williamson, The Tudor Age (London: Longmans, Green & Co, 1953), 81–83; Stone, *The Family, Sex and Marriage in England*, 156; Ariès, Centuries of Childhood, 267–8.

[86] H. Tingsten, *Victoria and the Victorians* (London: George Allen & Unwin, 1972), 27.

form in public schools, with the main focus now on patriotism and duty.[87]

Poets of the Victorian era venerated the ideals of stoicism and self-control, as expressed in Rudyard Kipling's "If," from 1895:

> If you can keep your head when all about you
> Are losing theirs and blaming it on you;
> If you can meet with Triumph and Disaster
> and treat those two impostors just the same:

There had been a huge change in temperament over the previous five hundred years.

Impersonal loyalties and high infant C

Cross-cultural evidence suggests that larger political units are a product more of child V than C. As discussed in the previous chapters, the large Empires of India, China and the Middle East are the product of high child V with only moderate C. Consistent with this is that increasing C in Europe did not, on the whole, result in larger states but in states that were more centralized. This is because rising C, and especially rising infant C, increases impersonal attitudes and loyalties.

Until the sixteenth century, despite the decline of feudalism, a European man's strongest loyalties were very often to his local lord. This kind of loyalty is less personal than loyalty to a friend, neighbor or even a tribal chieftain, but far more personal than loyalty to a king or parliament. A local lord was someone most people would have seen and perhaps even met, and who their immediate retainers knew quite well. Even a peasant was expected to turn to his lord for protection and justice.

The feudal system, with its repressive power structures and potential for turmoil, has at times been regarded as essentially negative. We live in a time of centralized states upon which we rely to keep order. But that is not how most people in the Middle Ages viewed it. A local baron could be seen as a protector against royal tyranny, a king being a distant and unfamiliar figure most people would never have seen. So if a feudal baron took up arms in some struggle involving the monarchy, local people might well follow him. The more inherently personal the locals were in the way

---

[87] P. A. W. Collins, *From Manly Tear to the Stiff Upper Lip: The Victorians and Pathos* (Wellington: Victoria University Press, 1974), 17.

they gave their loyalties, the stronger the position of the local lord.

An example of such a man is Richard Neville, Earl of Warwick, who lived in the mid-fifteenth century. Originally a supporter of Henry VI, he became the chief supporter of the house of York and helped to put Edward IV on the throne. Finding his influence curbed by the queen's family, he switched sides again and helped bring back Henry VI, before being defeated and killed in a final battle which restored Edward to power. In all these changes his loyal followers seem to have simply gone along, fighting for and against whichever claimant their lord told them to. Yet, in the following century people became far less likely to support their lord in rebellion, giving the Tudor monarchs unprecedented power.

History tends to focus on great men, because that is what we know about. Ordinary soldiers and peasant levies did not keep diaries at that time, or at least not any that have survived, and yet it is their decisions that determined what the great men could get away with. Past events actually make better sense once we understand that the prevailing temperament of the population can change, sometimes quite quickly.[88] Even economic and institutional changes can best be understood as reflecting these changes in temperament.

For example, outbreaks of civil war are often seen as resulting from personal ambition, tyrannical rule or economic deprivation. But given the right political circumstances, ambitious and power hungry men will be sorely tempted to take up arms against the state if they think they can succeed. What matters is whether people choose to follow them. Strong personal loyalties increase the likelihood that they will, a concept which can account for the feudal chaos that periodically erupted in every European country until about the sixteenth century, when loyalties became impersonal enough for some countries, including England, to form stable nation states.

England provides many examples of how personal loyalties could undermine national loyalties, but the strongest evidence can be found in France. There was a time in the eleventh century when the authority of the king of France broke down to the extent that he could stand on the battlements of his castle in the Île de France and look out on the strongholds of his rebellious barons. In other words, his authority did not

---

[88] The restoration of royal authority was only partially a result of rising C. It had more to do with another cyclical pattern, to be discussed in chapter eight.

reach even as far as he could see. When loyalties are intensely personal, people follow and obey the leader they know. That leader then has no incentive to obey the king, since neither his own vassals nor those of anyone else accept the king's authority.

Historians have often defined such eras in terms of the personality of the kings. A weak king, or one who comes to the throne as a child, is less likely to be obeyed. The important point to make about eleventh century France is that this same condition applied throughout the country. The great nobles of Burgundy, Aquitaine and elsewhere faced *exactly* the same problems of rebellious vassals as did the king. If great leaders can restore unity from chaos and rebellion, then surely one out of France's thousands of contending nobles would have had that unique combination of character and ability?

Order was restored in the twelfth century—not so much in the country but within the great fiefs. Thus, as the king of France became supreme in his domain, so did the rulers of Burgundy, Aquitaine, Anjou and Normandy. (The situation was complicated by the fact that the greatest of these barons, the hereditary ruler of Aquitaine, Anjou and Normandy, was also the king of England). Only from the reign of Philip Augustus (1179–1223) and increasingly in the thirteenth century, did the king of France establish effective control over the country.

Once again, these changes make better sense if we accept that there was a change in the temperament of the population. As the people of France became more impersonal in their loyalties, so authority shifted to more distant individuals, who thus gained power. Nobles in the thirteenth century had less power relative to the king than they had in the eleventh century, and by the sixteenth century their descendants had even less.

In England and France national cohesion took a step back in the late Middle Ages, again for reasons to do with the secondary population cycles that will be discussed in chapter eight. But from the sixteenth century onwards, Europeans increasingly gave their strongest sense of loyalty to monarchs. This was still a personal tie, since the monarch was an individual, but a more distant and less well-known individual. This change in loyalties and attitudes was reflected in the strength of the Tudor monarchy, which was paralleled by the rise of powerful rulers in France and elsewhere. Once again, we see evidence of similar trends in different areas which argue against them being simply the reflection of more competent rulers.

Alongside these changes came growth in other, even more impersonal, loyalties. English parliaments were first set up in the thirteenth century and gradually gained influence. This happened even in times of political unrest, such as the fifteenth and seventeenth centuries.

The rule of law, which also depends on impersonal attitudes, shows much the same pattern. Legal solutions to conflict were only made customary by Henry II in the mid-twelfth century. Even the generally turbulent fifteenth century saw progress, such as in the setting up of formal training for lawyers in the Inns of Court, and the firm establishment of trial by jury. There may also have been an improvement in the standard of professional judges.[89]

By the nineteenth and early twentieth centuries these loyalties had become strong enough to support relatively stable, republican forms of government in Britain and many other European countries. Even absolutist rulers such as Frederick the Great and Napoleon Bonaparte developed elaborate and relatively impartial law codes administered by professional bureaucrats and judges. And nowhere do we see the feudal anarchy of earlier centuries.

Of course, all of these different loyalties mixed and melded within nations, regions and even individuals. The English Civil War of the 1640s reflected a combination of loyalties at all three levels. In many parts of the country, men would still turn out in support of their local leaders. Loyalty to the king was also, of course, widespread. And then there was support of Parliament—or, for many Parliamentarians, support for an ideal of parliamentary government that had never existed, as yet. As the most impersonal form of loyalty, this was strongest among the highest C section of the population—the Puritans, with their disciplined and austere self-control. It was also stronger in the more economically developed south and east, the capacity for economic development being another indicator of high C.

The most extreme form of impersonal attitudes is one that combines acceptance of law and representative government with adherence to a political party, in the sense of a platform of ideas and principles rather than the platform for a leader. British political parties were beginning to form in the late seventeenth and eighteenth centuries but were far from their modern forms. Factions tended to surround powerful and usually aristocratic leaders, and many members of parliament were not aligned

---

[89] H. S. Bennett, The Pastons and their England: Studies in an Age of Transition (Cambridge: Cambridge University Press, 1922), 3–7.

with any party. When William Pitt became prime minister in 1784 he was faced with a hostile majority in Parliament but managed to persuade a number of members to change sides.

Over the next century the modern party system became fully established, with virtually all MPs committed to one party or another. Today in Western countries, though rather less so in recent years, most people continue to vote for the same party despite changes in leadership and their local representative. This means that, for the average citizen, loyalty to an individual is taken almost completely out of the picture.

This is the reason that countries like Papua New Guinea and Zimbabwe, despite adopting the forms of Western democracies, have had great trouble forming stable democratic governments. Loyalties in such places are almost entirely personal, with a strong element of clan or tribal affiliation. The pre-contact political system in Papua New Guinea was based largely on face to face kin groups. Thus, when given the vote people tend to vote for tribal loyalties and for individuals rather than the parties they represent, so there is little to stop elected politicians from changing allegiance. In some cases, democracy is also undermined by corruption, indicating both the strength of personal ties and the weakness of impersonal codes of behavior.

The decline of personal political loyalties from the eleventh century to the nineteenth century was neither smooth nor continuous. Both the fifteenth and eighteenth centuries saw relative rises in aristocratic influence, leading in the former case to the feudal chaos of the 1450s. This is yet another example of the secondary population cycle to be discussed in chapter eight.

Overall, though, the political changes in Europe between 1100 and 1850 can be explained as resulting from a steady change from personal to more impersonal loyalties, consistent with a rise in C. Bearing in mind that loyalties may conflict in different segments of the population and even within the same individual, a progression can be laid out as shown in Table 6.1.

Table 6.1. Different levels of C reflected in political loyalties.

C	Loyalty to	Example
lowest	Face to face	Hunter-gather band
	Local leader	Eleventh-century France (feudal anarchy)
	More distant leader	Twelfth-century France (powerful Dukes)
	King	Sixteenth-century France or England
	Representative Body	Eighteenth-century England
highest	Political Party	Nineteenth- and twentieth-century England

Nation states

There is another factor that affected the European political scene in these centuries. Nationalism is a relatively impersonal loyalty but it also implies loyalty to leaders who are similar in ethnicity, language and culture.[90] In chapter one it was suggested that high-C animals and people would be more likely to choose mates who are similar to themselves, and preference for similar rulers may have similar psychological roots.

Nationalism was relatively weak in the Middle Ages. The English king Henry II controlled much of France in the late twelfth century, gained by inheritance and marriage, with little sign of popular resistance. This feat was repeated at various stages during the Hundred Years War of the fourteenth and fifteenth centuries, and only at the end of this time did French nationalism—as personified by Joan of Arc—become a serious political factor.

But once national loyalties were established in the sixteenth century, the borders of states changed remarkably little. Spain, Portugal, France, Switzerland and many German states stayed largely unchanged. Conquest of an entire country became more difficult because, with people developing loyalties to their nation, they fought harder to defend it. The English lost their last foothold in France in 1558. The Dutch fought a bitter and successful war of independence against Spain, the superpower of the day. Spain inherited Portugal but could not hold it. Louis XIV and

[90] A. Smith, *The Ethnic Origins of Nationalism* (Oxford: Blackwell, 1988).

Napoleon fought grueling wars of conquest with little net long-term change to national boundaries. Even small political units such as the northern Italian city states and a number of minor German principalities maintained their identity and independence until the late-nineteenth century.[91]

As infant C increased, rising nationalism actually reduced the size of some empires. It freed the Dutch from Spanish rule, and nineteenth-century Italians from the German-speaking Habsburgs. This is in contrast to peoples in the Middle East and India with lower infant C who have been less reluctant to accept a foreign ruler. In these cases rising C, and especially rising child V, tend to increase the size of empires. This is a phenomenon to be discussed in chapter fifteen.

As with every other index, nationalism indicates a rise in C to the nineteenth century.

Citizen soldiers

Another factor following the same pattern is that when C is high, ordinary men tend to make effective soldiers. In the Middle Ages the armored knight was the heart of military organization. By the nineteenth and early twentieth centuries, European nation states were fielding mass armies consisting of tens of thousands (eventually millions) of men who fought with extraordinary courage and determination.

It is common to attribute this change to industrial technology, especially the rise of firearms, but that is not a sufficient explanation. Ordinary men, well trained and disciplined, could defeat mounted knights well before the rise of gunpowder, as proven by Scots pikemen at Bannockburn in 1314 and Welsh archers at Agincourt in 1415. Likewise, the Swiss successfully fought off the Habsburgs to confirm their independence in the fourteenth century, and later became famed as mercenaries.

But it was only in the nineteenth century that mass citizen armies became the norm, a development that began in Revolutionary France in the 1790s. Napoleon's defeat in 1815 was followed by a century of relative peace when large armies were not required, though English soldiers of humble

---

[91] A general overview of the rise of nation-states in Europe can be found in C. Tilly (ed.), *The Formation of National States in Western Europe* (Princeton: Princeton University Press, 1975).

backgrounds fought with remarkable effect during the Crimean War. But when war broke out in 1914, for reasons to be discussed in chapter nine, the citizen soldiers fought with a ferocity and sense of discipline equal to that of any professional army.

Market economy

Economic distribution through the impersonal market economy, rather than through personal relationships, is a key aspect of high C and especially infant C. Early English society was based mainly on subsistence farming, with peasants offering services rather than rent to their feudal lord. But during the centuries following the High Middle Ages, the market economy gradually came to play a more and more important role in English life. Trade developed, medieval service in kind was translated into money rents, and land became a commodity to be bought and sold.

The growth of the market was not equally fast in all sections of the country, and it is notable that the more market-oriented areas were also those with the more impersonal political loyalties. We have already noted that during the English Civil War the more economically advanced areas such as the south-east (higher C economic attitudes) tended to support Parliament (higher C political attitudes), while areas that were less economically advanced (lower C economic attitudes) tended to support the king (lower C political attitudes). But by the nineteenth century the market economy was all but universal. People whose ancestors had been peasant smallholders were either working for wages or in business for themselves.

As the market economy became stronger, there was also a decline in non-market regulatory systems. In the Middle Ages, prices were largely determined by the boroughs and guilds. Loosening began in the fourteenth and fifteenth centuries when the prosperous wool industry came to be located in the countryside near sources of water power, and thus escaped guild control. Guilds declined still further in the sixteenth and seventeenth centuries, and government attempts to re-impose such controls were largely abolished between 1760 and 1850.[92]

---

[92] M. W. Flinn, An economic and social history of Britain: 1066–1939 (London: McMillan, 1961), 36, 83; A. Burnie, An Economic History of the British Isles (London: Methuen, 1961), 178–81; S. R. Epstein & M. Prak, *Guilds, Innovation and the European Economy: 1400–1800* (Cambridge: Cambridge University Press, 2010).

The same can be said of building regulations, which define what sort of buildings can be erected and where. The guilds effectively determined such matters in the Middle Ages, but by the nineteenth century there were few effective controls. In effect, the impersonal forces of the market determined who could build what buildings and where, rather than group agreements (guilds) or government control.

The rise of the market can also be traced in attitudes to usury, which was forbidden to Christians in the Middle Ages. Banking first came out of the shadows in fourteenth-century Florence, but in England it was not until 1545 that interest rates of up to 10% were made legal. Even this was repealed in 1552, until it was finally legalized in 1572.[93] From then on, banking became more and more central to economic activity.

All of these measures show a rise in free-market thinking to the nineteenth century, consistent with a rise in C.

Attitudes towards work

We have already seen that routine work is another aspect of high-C behavior, which can be related to the requirement that animals in food-restricted environments search incessantly for food.

The work ethic in the Middle Ages was less strong than it subsequently became. There was a general habit of idleness on Sundays, plus up to a hundred saints' days on which work was forbidden. In medieval England holiday leisure is estimated to have taken up around one third of the year. In France no work was to be done on 52 Sundays, 90 "rest days" and 38 additional holidays. Southern European states were even more relaxed. Travelers in Spain noted that holidays totaled about five months of the year.[94]

The first opposition to such "time wasting" was voiced by the Lollards of the fourteenth and fifteenth centuries, and saints' day holidays were greatly reduced in the sixteenth century. At this time there were accounts of employees being forced to work seven days a week, countered by

---

[93] Burnie, An Economic History of the British Isles, 111–12.

[94] E. Rodgers, *Discussion of Holidays in the Later Middle Ages* (New York: Columbia University Press, 1940), 10–11; C. R. Cheney, "Rules for the Observance of Feast-days in Medieval England," *Bulletin of the Institute of Historical Research* 34 (90) (1961): 117–47.

increasingly strict observance of the Sabbath.[95]

The rise of factories in the late eighteenth and nineteenth centuries represented an increase in both the regularity and total of hours worked, especially since there were no longer off-seasons. The widespread habit of making Monday an informal holiday also declined at this time.[96] By 1840 the hours of an average worker had risen to between 3,105 and 3,588 per year, only declining substantially in the twentieth century. By 1987 manufacturing workers in the UK worked for 1,856 hours a year.[97] Based on hours worked and especially in their regularity, we see a clear increase in the period up to the nineteenth century, another indication of rising C.

Falling time preference (thrift, investing for the future)

Another aspect of C, as discussed in chapter three, is an important key to economic and professional success. This is the habit of thrift, and the willingness to sacrifice present consumption for future benefit. Examples of the latter include investing in trade skills and other forms of education at the expense of current income.

The preference for immediate consumption over long-term benefit is known as "time preference."[98] Hunter-gatherer peoples tend to have a very high time preference. For example, a study of Mikea forager-farmers in Madagascar observed that farming was vastly more productive than foraging in terms of return on effort, and yet the people planted only half the land needed to feed themselves and so spent most of their time foraging.[99] Farming of course requires a much lower level of time

---

[95] W. B. Whitaker, Sunday in Tudor and Stuart Times (London: Houghton, 1933), 13, 16, 20.
Hill.

[96] D. A. Reid, "The Decline of St Monday: 1766-1876," Past and Present 71 (1976): 76–101;
E. P. Thompson, The Making of the English Working Class (London: Pelican, 1968), 321–322, 337–338, 345, 351.

[97] Based on 69-hour week: W. S. Woytinsky, Hours of labor. Encyclopedia of the Social Sciences, vol. III, (New York: Macmillan, 1935). A low estimate assumes a 45-week year, a high one assumes a 52-week year. Calculated from Bureau of Labor Statistics data, Office of Productivity and Technology.

[98] Clark, A Farewell to Alms, chapter nine.

[99] B. Tucker, "The Behavioral Ecology and Economics of Variation, Risk and Diversification among Mikea Forager-Farmers of Madagascar." PhD dissertation, Department of Anthropology, University of North Carolina, Chapel Hill. (2001).

preference to be viable, since the work of planting is done months before any benefit is achieved.

Between 1200 and 1800 the Englishman's rate of time preference fell to a remarkably low level, which was accompanied by a rise in thrift. One indication of this is the rise in the price of land, relative to the rents that could be taken from it. In the twelfth century land had a rate of return typically at 10% or greater. By 1800 it had fallen to 4–5%. This also applied to rent charges, which were effectively loans secured against the land. A similar decline in rates of return can be seen in Genoa, the Netherlands, Germany and Flanders.[100]

The same pattern can be seen in the value of skilled labor. Since gaining a skill is an investment in time and a loss of immediate income, people with a lower time preference will be more willing to do this. That means that if time preference falls there will be more skilled workers relative to unskilled, so their wage advantage will fall. And this is what happened in England. In the thirteenth century a building craftsman earned nearly double the wage of the laborer supporting him. In later centuries the difference was far less.[101]

Literacy and numeracy

Another very concrete measure of low-time preference is investment in education. It is thus significant that levels of literacy and numeracy rose steadily in England from the Middle Ages to 1800 without any form of government support and at a time when most people worked as manual laborers. In the Middle Ages a man could claim "benefit of clergy" and be tried in the (more lenient) church courts simply by demonstrating his ability to read a passage from the bible. Literacy was so rare outside the clergy that this was considered a valid test. An indication of growing literacy is the number of people who signed their marriage certificates, rising from around 40% of men in the sixteenth century to almost 80% by 1800. By this time most women were also signing their names, compared with virtually none two-hundred years earlier.[102]

This change was especially striking given that (as indicated above) the wage premium of skilled labor was actually lower than it had been. In

---

[100] Clark, A Farewell to Alms, chapter nine.
[101] Ibid.
[102] Ibid.

other words, literacy was actually less valuable than in earlier times, and yet people were increasingly willing to spend the time and money to achieve it.

Religion of practical morality versus ritual and charity

We have seen that a key aspect of C is a code of behavior that causes people to act in a way different to their "natural" hunter-gatherer temperament. Such codes not only increase C, but their strength is a useful indicator of C. The more people are pressured or persuaded to act in a way that is artificial or constrained by custom, the stronger C will become. We saw examples of this in the last chapter with the honesty and debt avoidance of the Manus and the elaborate courtesy of the Japanese.

But some codes are more indicative of high C than others. When C is low to moderate, giving tends to be based on sympathy without any overriding moral principle, such as being concerned with *why* the individual is in need. The extreme example of this is the high level of sharing that goes on in hunter-gatherer societies. Though a moral precept of the highest order, and one emphasized by all the major religions, giving on the basis of personal sympathy is a relatively personal form of moral action. It depends on the situation of the recipient, and commonly on the degree of emotion they arouse.

Religious rituals are another code of behavior, and one that has probably been vital in raising C in times past. The extent to which they reflect or support C depends on how difficult they are and how pervasive. At the low end, attending mass once a week for an hour is a fairly mild requirement. Attending it every day would be tougher. And at the extreme high end come codes such as the Jewish ritual law which govern and guide behavior on a daily or even hourly basis.

Strict codes of moral behavior act in a similar way, since they affect every aspect of life. We have seen that morality in religion is strongly linked to other aspects of C in the cross-cultural survey, and the Manus people had a strongly moralistic religion. Based on their example we can infer that honesty is the moral value most characteristic of very high C and especially of infant C. It is relatively impersonal, in that it applies to all people regardless of relationship, and (like Jewish ritual law) is likely to affect behavior on a daily or even hourly basis. It also perhaps fits our picture of high-C people as less socially oriented, so likely to be less skilled at lying.

For well over a thousand years the English people were adherents of Christianity, a religion with profound effects on behavior. But in tracing religious beliefs and practices over this time, there is a clear change in emphasis on different aspects of behavior.

Popular religion in the Middle Ages, though not without ethical content, was largely concerned with ritual. Life was seen as a battle between God and the Devil, with the mass and other sacraments as vital weapons. Saint worship and the relics of saints played an important role.[103] The bible was in Latin and most people could not read anyway, even in their mother tongue. Church services were also in Latin and sermons rare.[104] All of which suggests that moral teaching was not as prominent as in later centuries.

The strongest moral content was on the business of charity, both in terms of almsgiving and the endowment of monasteries, which played an important role in helping the needy. During the thirteenth century communities were seen as open-handed in their alms-giving to the monastery hospitals that provided relief for the poor. From the twenfth century onwards, provision to the poor was seen as inseparable from pious expression and was viewed as "one of the seven works of mercy in the penitential process of making satisfaction for sin." For the clergy, care for the poor was mandatory as well as virtuous. Monasteries were expected to follow the example of St Benedict to receive the poor stranger and visitor alike.

The bounds of this medieval charity have perhaps been exaggerated in retrospect. In reality, provision for the poor would only amount to a fraction of the income of many monasteries, often as little as five per cent. Even so, in absolute terms this could be a substantial amount and so was viewed as a key means of relief for the poor at the time.[105] Traditions, like providing a meal for a poor man for thirty days on the anniversary of the death of a monk, were grounded in religious custom. In one monastery this involved tens of thousands of meals, later limited to fifty a day to avoid

---

[103] F. Heer, The Medieval World: Europe from 1100–1350 (London: Weidenfeld & Nicholson, 1962), 35–40.

[104] J. R. H. Moorman, Church Life in England in the Thirteenth Century (Cambridge: Cambridge University Press, 1955), 5–6, 71, 77–78.

[105] A. D. Brown, Popular Piety in Late Medieval England (Oxford: Clarendon Press, 1995), 181, 194, 196.

financial strain.[106]

Some change of emphasis can be seen from the thirteenth century when wandering friars began calling on people to repent their sins, and priests began preaching to meet this competition.[107] The popular devotional movement of the late Middle Ages, with its emphasis on living a good life within the world, rather than by retiring into a monastery, also suggests an increasing emphasis on morality.

The thirteenth century also saw the beginnings of a stress on giving only to the *respectable* poor. In 1229, for example, Bishop Bingham stressed that the inmates of St Nicholas's were to be "Christ's Poor," meaning they must have sound moral character. By the late-fourteenth century attitudes towards charity had changed markedly, perhaps partly exasperated by labor shortages brought about by the Black Death. During this time the poor had come to be viewed with such a degree of suspicion that monastery hospitals were increasingly left to decay or transformed into chantries that were less focused on poverty relief.[108]

In the later Middle Ages these early monastery hospitals were largely replaced by alms houses. Rules governing admission became increasingly specific. Charitable bequests left by testators were also becoming more specific, often specifying that the money was to go only to the local poor as opposed to vagrant beggars. An example of these new requirements can be seen in 1472, when Margaret Hungerford's alms house gave preference to the poor tenants of the Hungerford estate. It also required that recipients not be "lecherous, adulterous or a tavern-goer."[109]

The sixteenth-century Reformation put an increased emphasis on behavior. There was an emphasis on sermons, liturgies and scriptures in English, fewer sacraments, abolition of relics, and a lower status for priests. Salvation became less a matter of buying indulgences and masses, of intercession by priests and the sacraments of the church, and more a matter of faith and the undeserved grace of God. Significantly, the one moral

---

[106] L. J. R. Milis, Angelic Monks and Earthly Men (Woodridge: The Boydell Press, 1992), 53.

[107] J. R. H. Moorman, *A History of the Franciscan Order from its Origins to the Year 1517* (Oxford: Clarendon Press, 1968), 17.

[108] A, D. Brown, *Popular Piety in Late Medieval England* (Clarendon Press: Oxford, 1995), 184–5.

[109] Ibid., 184, 185.

teaching *less* emphasized by the Protestant church was charity.[110]

In theory, salvation by faith might seem to make good works irrelevant. In practice, as shown by the Puritans and especially Calvinists, the result was a considerable stress on morality, and a tendency to formulate all worldly problems as moral problems. Pilgrim's Progress is a supreme example of this seventeenth-century attitude. Through the parable of a journey Bunyan showed life as a constant struggle against sin, so difficult that most people were unable to follow it and were lost.

The Quakers took this thinking to an extreme. They had an elaborate moral code with a strong emphasis on honesty, but little interest in theology. To them the church was no longer necessary to salvation, and even pagans might possess the spirit of God.[111] It is a reflection of the times that Quakers were convicted almost entirely for civil offenses—calling clergy names, refusing to pay tithes and rates, not doffing hats, keeping businesses open on holidays, etc.[112] This is in sharp contrast to the reasons given for persecuting heretics in the Middle Ages, which were largely theological. Growing religious toleration in England was at least partly a result of this declining concern with theology and ritual and the increasing emphasis on morality and behavior.

This trend culminated in the nineteenth century. The strict moral standards of Victorian times were the classic ones of a high C society—honesty, a high level of sexual restraint and public order. By now these attitudes were at least partly independent of religious belief and would remain strong even as religion began to wane towards the close of the nineteenth century. Many working-class people were indifferent to religion and many middle-class people attended church largely for social reasons.[113] Moral earnestness, rather than religious piety, became the distinguishing characteristic of the time.[114] In other words, the emphasis on ritual and sacraments had been reduced, and that on morality increased.

---

[110] A. G. Dickens, The English Reformation (London: Batesford, 1964), 14–15, 69–71, 319.

[111] H. Van Etten, George Fox and the Quakers (New York: Harper Torchbooks, 1959), 18, 25, 68, 190.

[112] C. E. Whiting, Studies in English Puritanism: 1660–1688 (London: Frank Cass, 1968), 212.

[113] K. S. Inglis, Churches and the working class in Victorian England (London: Routledge & Kegan Paul, 1963), 322, 329.

[114] Tingsten, *Victoria and the Victorians*, 32.

People in the nineteenth century were acutely aware of human suffering. This was the age when first the slave trade and then slavery itself were abolished, and when Factory Acts began providing protection to workers, especially children. There was also a great sense of generosity, with many middle class households donating as much to charity as they paid for food or rent. Londoners in 1885 are estimated to have donated more money to charity than the national budget of Sweden.[115]

The great difference to medieval charity, and also that of modern times, is that giving was centered on the idea of moral rehabilitation. It was meant to foster the values of respectability, domesticity and often evangelical Christianity. Most middle-class people saw poverty primarily as the fault of the individual, not of circumstance. For example, they might be lazy or drunken. So while working to alleviate some of the worst symptoms, they expected the poor to adopt their values and live thriftily.[116] To receive help, the poor had to agree to attend church, live austerely, give up alcohol, change their sexual habits and improve their housekeeping. Philanthropists went to great lengths to differentiate between the "deserving" and "undeserving" poor, and staff at homes and institutions often spent hours interviewing candidates to judge their worthiness to receive assistance. A far greater emphasis was placed on the concept of self-help, character and practical morality than the virtue of giving handouts.[117]

Nowhere was this more apparent than in the operation of the workhouse, a grim institution that made any form of work preferable to living off the state. The first act towards this new system occurred with the introduction of the Poor Law Amendment Act in 1834 which centralized the system of workhouses nationwide under the Poor Law Commission, based in London. Before 1834 each parish was responsible for its own poor, and any able-bodied person could expect assistance in their own home if they fell on hard times. The sick and elderly might also be looked after in a small parish workhouse which was seen as a "relatively friendly and unthreatening institution."

However, the new law changed the emphasis from community relief to

---

[115] Steinbach, *Understanding the Victorians*, 127.
[116] Ibid.
[117] Ibid., 4, 5.

deterrence and moral rehabilitation.[118] This change was enacted partly due to widespread abuse and the financial burden of the previous system, which was brought to breaking point due to advances in agricultural technology and a series of poor harvests.[119] It is also consistent with a peak of C in the mid-nineteenth century.

New workhouses constructed in the 1830s and 1840s were designed to be austere and to resemble prisons, acting as a visual deterrent and also representing the state's new approach to relief provision. Their accommodation was designed to be worse than that of the poorest independent laborer, although conditions did vary from location to location.

Upon entry to the workhouse, an inmate's hair would be crudely cropped and he or she had to give up their clothing and possessions for the stay. Inmates wore the plain workhouse uniform which served to identify a workhouse inmate on the street and improved hygiene. It was also seen as a way of enforcing moral discipline.[120] Alcohol, dice and gaming cards were banned, along with all reading material deemed improper, often meaning that the bible was the only available book for adults to read.[121] Food was in sufficient quantity but plain and poor quality, often a thin oatmeal porridge.[122] Life was deliberately disciplined and monotonous with daily routines consisting of continual work, prayers twice daily and little time off.[123]

Relief outside the Workhouse was now kept to a minimum and eventually made illegal, so only those who were truly destitute would seek entry. A report by the Commission in 1834 concluded that "the great source of abuse is outdoor relief afforded to the able-bodied." In reality, many local parishes would overcome this by using the loophole that excepted sickness at home, as the cost of providing outdoor relief was about half that of

---

[118] M. Higgs, Life in the Victorian and Edwardian Workhouse (Gloucestershire: Tempus Publishing, 2007), 1.
[119] Ibid., 20.
[120] Ibid., 16.
[121] Ibid., 14.
[122] Ibid., 20.
[123] F. Driver, Power and Pauperism (Cambridge: Cambridge University Press, 2004), 59.

admitting someone into the workhouse.[124]

Most controversial of all, however, given the value that Victorians placed on family, was the breaking up of families who entered workhouses. Able-bodied men with families could not enter the workhouse alone—the whole family must apply together. Once inside, mothers and fathers slept in separate dormitories to improve their "moral fiber" and prevent the breeding of new paupers. It was also a powerful deterrent to those seeking state assistance. Exercise yards for men, women and children were separated by high walls, and children over the age of 4 slept apart from their parents.[125]

So strong was the Victorian belief in the moral value of work that meaningless tasks were often assigned to workhouse inmates when no practical work was available. In 1884 the Great Yarmouth Union put able-bodied men to work in order to prevent "loafing about and making the workhouse a club house." As a result, all able-bodied men except two promptly left. The remaining men and the tramps were put to work in a shed divided into fourteen cubicles, five and a half feet square (one man to each cubicle, which was locked) shifting shingle with a shovel through a hole ten inches square. They had to move a ton an hour. They worked a nine-hour-day with one hour allowed for lunch. The shingle was then carted back to its place of origin.[126]

The significance of these practices, grueling and meaningless as they might seem, is that they were calculated to improve the character of the inmates. Basic food, obsessive routine work, sexual restriction, bans on alcohol and an emphasis on religion all raise C, and thus theoretically the economic potential of the inmates. The Victorian notion of "raising one's moral fiber" coincides very closely with adopting C-promoting behavior.

Once again, in this the shift from an emphasis on ritual and charity based on personal sympathy to one based on principles such as honesty, hard work and a more systematic philanthropy, there is evidence of a steady rise in C to the nineteenth century.

---

[124] Higgs, Life in the Victorian and Edwardian Workhouse, 12; Steinbach, *Understanding the Victorians*, 42.
[125] Fagan, *How Climate made History*.
[126] R. C. Allen, The British Industrial Revolution in Global Perspective (Cambridge: Cambridge University Press, 2009).

Population growth

Rising C also seems to have contributed to an acceleration in population growth. We have seen that one factor affecting desire for children is V, which is the trait that encourages fast breeding in areas beset by famine or animal predators. But C has an equally large effect by increasing productivity and thus the ability of the land to support more people, as well as reducing the death-rate from violence. Higher C may also increase the birth rate by increasing interest in children, an effect seen in the better maternal care of our food-restricted rats.[127]

Records of active colonization in the Domesday Book suggest that the population might have been growing by the late eleventh century. This trend accelerated in the twelfth and thirteenth centuries. New villages were being founded and old ones were growing, marshes were being drained and towns coming into conflict over pasture rights. Then came a pause, with growth slowing in the early fourteenth century even before the onset of the Black Death in 1348.[128] The late fourteenth century, and possibly the first few decades of the fifteenth, were a time of declining population.[129] This is yet another example of the secondary population cycle to be discussed in chapter eight.

Population was growing again by the last few decades of the fifteenth century or at least by the early sixteenth, a trend which continued through the sixteenth and early seventeenth centuries. Growth seems to have been slower in the second quarter of the seventeenth century, slower still in the second half, and to have been relatively static in the early eighteenth century but increased in the second half.[130] Population growth reached an

---

[127] This is ironic in the sense that the biological function of C is to control populations in areas where food is chronically limited. But this is only one of the ways in which these biological systems have an entirely different effect when operating in human societies.

[128] M. Livi-Bacci, *A Concise History of World Population* (Malden, Mass., Oxford: Blackwell, 2001), 36–7.

[129] H. A. Cronne, The Reign of Stephen, 1135–1154: Anarchy in England (London: Weidenfeld & Nicholson, 1970), 17; W. G. Hoskins, "The English Landscape" in Medieval England, Vol. 1, edited by A. L. Poole, 10, 12, 15, 19 (Oxford: Clarendon Press, 1958); J. C. Russell, British Medieval Population (Albuquerque: University of New Mexico Press, 1948), 235, 259, 260; T. H. Hollingsworth, Historical Demography (Ithaca: Cornell University Press, 1969).

[130] L. A. Clarkson, The Pre-Industrial Economy in England: 1500–1750 (London: Batsford, 1971), 27–28; I. Blanchard, "Population Change, Enclosure, and the

all-time peak in the nineteenth century, especially in the 1800s and 1870s, as indicated in Fig. 6.3 below.[131]

In short, population growth in England seems to have taken place in three main spurts with mild periods around 1200, 1550 and 1850, separated by periods of declining or static population.[132] Very roughly, the population could be said to have doubled in the first period, tripled or quadrupled in the second period, and grown eight to ten times in the third period.

Fig. 6.3. Broad trends in population growth and decline in England: 1100–1900. There is a long-term trend to faster population growth, consistent with a rise of C to the nineteenth century, plus shorter-term fluctuations to be discussed in chapter eight.

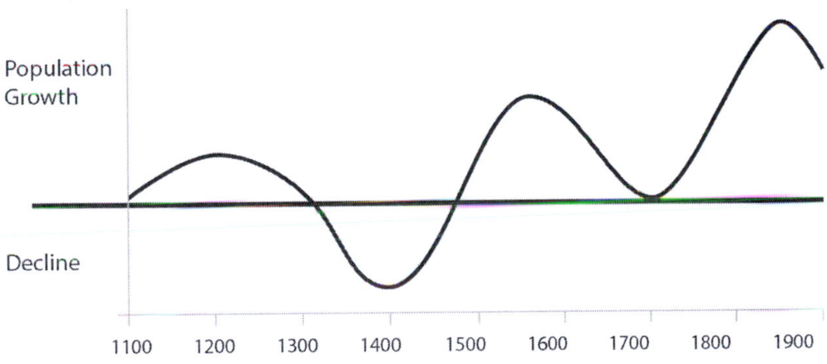

As indicated earlier, the reasons for rapid growth in the nineteenth century were a combination of better economic conditions and what seems to have been a more positive attitude to children—both aspects of high C. From the late eighteenth century onwards, childbirth came to be seen as a fulfillment of a woman's femininity, rather than as an interruption of her duties as a wife. The prestige of motherhood increased, and mothers rather

Early Tudor Economy," The Economic History Review 23(3) (1970): 427–45; J. Cornwall, "English Population in the Early Sixteenth Century," The Economic History Review 23 (1) (1970): 32–44; M. D. George, London life in the Eighteenth Century (New York: Capricorn Books, 1965); Stone, *The Family, Sex and Marriage in England.*

131 J. J. Spengler, France Faces Depopulation (New York: Greenwood Press, 1968), 53.

132 Clarkson, The Pre-Industrial Economy in England, 26; Russell, British Medieval Population, 235.

than fathers came to be seen as the center of the family.[133]

The dramatic population growth in this final stage actually understates the change, since there was substantial emigration from the seventeenth century onwards. The overall picture is of a long-term increase in the rate of population over nearly a thousand years. The peaks of growth were successively higher, and the trough around 1700 was much less severe than that around 1400. The reason for these troughs is, once again, a matter for chapter eight, but we can say at this stage that it did not represent any fall in C. Other indications suggest that C was rising quite rapidly in the late seventeenth century, even though the population did not.

Demographers commonly see the nineteenth century growth spurt as the first example of "demographic transition," in which growing prosperity leads to a fall in death rate and thus exploding population, until birth rates fall and growth slows.[134] This fits in with what has happened in a number of countries over the past half-century, including South Korea, Thailand, Taiwan and others.

But it does not fit the European pattern. English population growth took off in the late eighteenth century, well before the machine age was under way, and it continued apace for more than 150 years. Nor does it explain relatively faster growth in the sixteenth century, when wages were *lower* than in the low-growth fifteenth century. A more consistent explanation is that the trend to faster population growth reflected a rise in C from the eleventh to the nineteenth century, combined with relatively high V.

Science and technology

Original thinking—the basic principle upon which science and invention depend—has no obvious parallel in animal studies, but in the last chapter we linked it to infant C, the temperament associated with control in early childhood. Manus people, with their eagerness for change, are the archetype of such people. Thus, as parental control became more rigorous and started earlier in infancy, it is significant that we find more evidence of original thinking.

In England and more generally in Europe, the relatively conservative

---

[133] Tosh, A Man's Place, 80.

[134] W. Thompson, Encyclopedia of Population 2 (New York: Macmillan Reference, 2003), 939–40.

society of the Middle Ages began to open up during the Renaissance of the sixteenth century. This involved not only rediscovery of the knowledge of the ancient world but also the development of new ideas. Philosophical systems such as humanism and the Protestant Reformation challenged the dominance of the Catholic Church. Continental advances in art and architecture were matched in England by a unique flowering of literature, most notably the works of William Shakespeare at the end of the sixteenth century.

The sixteenth and seventeenth centuries saw the development of something even more unusual—the birth of experimental science. Though scientific thinking was known to the ancient Greeks, the unique contribution of early modern Europe was the development of theories that could be tested by experiment. Many consider Sir Isaac Newton the greatest scientist of all time. His 1687 work Principia Mathematica laid out three laws of motion and the concept of gravity as the basis for all future understanding of the physical universe. This scientific revolution continued through the eighteenth and nineteenth centuries with fundamental advances in chemistry, the understanding of electricity, geology and many other areas. A number of studies have found that Victorian Britain saw an all-time peak in per capita innovations in science and technology and in the number of scientific geniuses.[135]

Charles Darwin, in developing the theory of evolution in the late-nineteenth century, challenged conventional thought in a way that has rarely been seen before or since. Philosophical ideas began to take issue not only with conventional religion but even, for some, the very existence of God. Marxism was one of a number of ideological systems that arose in this era.

Increased creativity and openness to new ideas is another clear indication of rising C, and especially of infant C, to the nineteenth century.

---

[135] J. Huebner, "A Possible Declining Trend for Worldwide Innovation," *Technological Forecasting and Social Change* 72 (8) (2005): 980–6; C. Murray, *Human Accomplishment: The Pursuit of Excellence in the Arts and Sciences, 800 BC to 1950* (New York, NY: Harper Collins, 2003), 347; M. A. Woodley, "The Social and Scientific Temporal Correlates of Genotypic Intelligence and the Flynn Effect," Intelligence (40) (2012): 189–204; M. A. Woodley & A. Figueredo, *Historical Variability in Heritable General Intelligence: Its Evolutionary Origins and Socio-Cultural Consequences* (Buckingham: University of Buckingham Press, 2013).

Technology

The final characteristic of infant C to consider is technological aptitude. We saw in chapter six that the Manus people of the Admiralty Islands, with their intense control of very young children, were unusually good with machinery. This again has no biological equivalent, except for a general connection with impersonality, but as has been suggested it is a strong indication of high C and especially infant C. This is supported by the Japanese example where control of infants (if not at European levels) was also associated with a talent for technology. And nothing is more obviously related to the rise of Europe than accelerating improvements in this area.

Like many other aspects of C, technological progress was especially evident in the late fifteenth and sixteenth centuries with the development of sailing ships and navigational technology that allowed regular crossings of the previously impassable Atlantic. The technology of war, notably the development of the cannon and other firearms, also improved rapidly in this period. Then, from the late eighteenth century came that explosion of technology known as the Industrial Revolution. The first effective steam engine was developed by James Watt in the 1770s and within an astonishingly short time, just over a century, this was succeeded by mass industrialization, railroads, steamships, electricity and powered flight.

An interesting feature of the Victorian British that coincided with industrialization was their lack of interest in abstract theory and a tendency towards what some historians have gone so far as to call "anti-intellectualism."[136] Gentlemanly leisure and non-vocational learning, so respected in the eighteenth century, now came to be seen as idleness and uselessness.[137] The Victorians were pragmatists, relying on common sense and empirical solutions to solve the problems they were presented with—all attitudes which appear to be associated with the development of an advanced, mechanized society.[138]

A sense of this can be gained by plotting instances of major innovations in science and technology between 1455 and 2004, as developed by Bunch

---

[136] Young, *Middle-Class Culture in the Nineteenth Century*, 27.
[137] Steinbach, *Understanding the Victorians*, 134.
[138] Young, *Middle-Class Culture in the Nineteenth Century*, 27.

and Hellemans.[139] Though such an account is inevitably subjective, it correlates closely with lists drawn up by Murray,[140] Asimov[141] and with work done by Sorokin.[142] All of these show a rise of innovation to a peak in the nineteenth century. Fig. 6.4 below shows this plotted against population size.

Fig. 6.4. Key innovations in science and technology relative to population size.[143] Based on 8,583 key scientific and technological innovations, plotted against population. Science and technology skills were strongest when the West reached its peak of infant C in the nineteenth century.

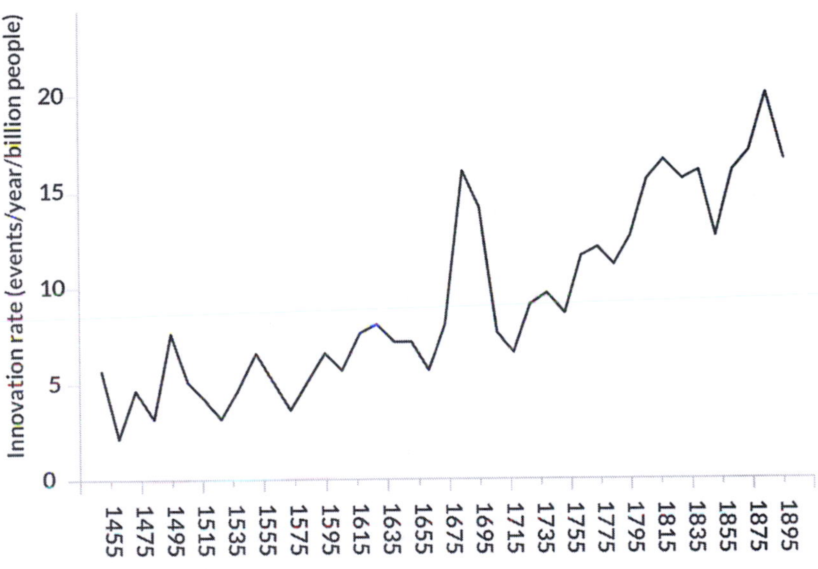

[139] B. H. Bunch & A. Hellemans, *The History of Science and Technology: A Browser's Guide to the Great Discoveries, Inventions, and the People who made them from the Dawn of Time to Today* (New York: Houghton Mifflin Harcourt, 2004).

[140] C. Murray, *Human Accomplishment: The Pursuit of Excellence in the Arts and Sciences, 800 BC to 1950* (New York, NY: Harper Collins, 2003), 347.

[141] I. Asimov, *Asimov's Chronology of Science and Discovery* (New York: Harper Collins, 1994); B. L. Gary, "A New Timescale for Placing Human Events, Derivation of Per Capita Rate of Innovation, and a Speculation on the Timing of the Demise of Humanity" (Unpublished Manuscript, 1993).

[142] P. A. Sorokin, *The Crisis of our Age: The Social and Cultural Outlook* (Boston: E.P. Dutton, 1942).

[143] Huebner, "A possible declining trend for worldwide innovation," 980–6.

The figures used in this graph are for world population, but the same pattern appears when population is restricted to Europe and North America only.[144] Technological and mechanical aptitude clearly rose to unprecedented levels in the nineteenth century, consistent with a rise in C and especially infant C.

The Industrial Revolution

The Industrial Revolution that began in Britain in the late eighteenth century, and then progressed to other European countries and Japan, can be seen as the result of all these changes. Compared to earlier centuries, nineteenth century Englishmen had become harder working, more disciplined, thriftier and more willing to invest in education and skills. They were more innovative, and better with machines. They were temperamentally inclined to the free market and less willing than before to tolerate regulations that made it less efficient. They were honest and law-abiding, supporting a stable and efficient nation state under the rule of law. Their fast-growing population and aggressive, disciplined armies helped make them masters of much of the world.

Historically, a hundred different reasons have been given for the immense changes launched in Britain in the late eighteenth century. There were incremental changes in technology going back several centuries, the rise of representative government, higher levels of literacy, scientific breakthroughs, wealth created by the discovery of new lands, and much more. But for biohistory, the real driver for the Industrial Revolution and the dominance of European peoples was a broad change which had been building for several centuries. This was above all a change in *temperament*, more specifically a dramatic rise in infant C.[145]

In the next chapter, we consider why this unprecedented rise took place.

---

[144] Woodley, "The Social and Scientific Temporal Correlates of Genotypic Intelligence and the Flynn Effect," 189–204.

[145] An implication is that individuals with epigenetic markers for high infant C should be innovative, good at business and especially skilled with technology. No other theory of the Industrial Revolution can be tested in the laboratory.

CHAPTER SEVEN

THE CIVILIZATION CYCLE

The last chapter tracked the rise of C from the Middle Ages to the nineteenth century through changes in attitudes and behavior. This chapter proposes an explanation for why this happened.

Biohistory explains rising C as the result of C-promoting traditions contained in religions and philosophical systems. Societies with religions and philosophies that are more effective at controlling behavior have, as a result, higher levels of C and V, which give them advantages such as being better organized, more hardworking or more aggressive. Thus they tend to prevail over those with less advanced traditions. The colonization of North America and Australia by Europeans is a recent example.

But C- and V-forming traditions are not enough, by themselves, to increase C and V. Christianity contains powerful C- and V-promoters, but its triumph in the late Roman Empire did not stop or even markedly slow the fall of C, leading (as we will see in a later chapter) to the collapse of that civilization. Even the barbarian Goths and Vandals were Christians at the time of their invasions. In fact, from political and other indications, C continued to fall for several centuries, and did not obviously begin to rise until eight hundred years after Christianity became the official religion of the Roman Empire. A C-forming tradition is essential to any prolonged rise in C, but it is not sufficient. Some other factor must be necessary to account for the rapid rise in C from the eleventh to the nineteenth centuries.

Biohistory proposes that this factor was an unusually high level of V and thus stress, peaking around the sixteenth century. There was a marked rise and then fall in indications of stress: harsh punishments, brutal authority, anger, treachery and suspicion. In many ways this is reminiscent of stressed social groups such as the Mundugumor and Whipsnade baboons. It is also reminiscent of societies such as the Egyptian villagers who punished children severely and had authoritarian governments (see chapter five).

In this and following chapters it is proposed that C and V have a curvilinear relationship in which a low level of C permits a rise in V and stress, high C and V cause C to rise, high C causes V and stress to fall, and falling V and stress cause C to fall. This is called the "civilization cycle" (see Fig. 7.1 below).

Fig. 7.1 The civilization cycle

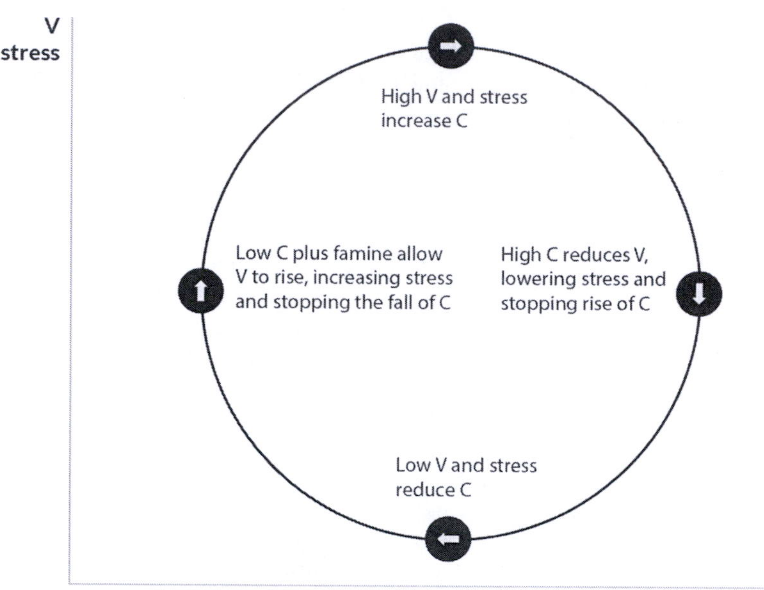

Before considering why V and stress might drive a prolonged rise in C, or why changes in C might affect V and stress, we will first consider the evidence.

Suspicion, hostility and political instability

The Elizabethan period was a glorious time in English history, one of national resurgence and creative brilliance, but there is a darker side that is less well known. In particular, Tudor England gives every indication of being a society under severe stress. In and around the sixteenth century there is increased evidence of bad temper and suspicion in personal relationships. It has been said that the men of 1485 were "reckless and ruthless, guided by temper, with higher ranks arrogant and all prone to expect evil of their fellows."[1] Letters of advice from fathers to sons between the fifteenth and seventeenth centuries show a cynical and pessimistic view of human nature, advising distrust and secrecy even within the family.[2]

The population of early seventeenth-century London has been described as volatile in the extreme, showing fear, grief or anger at the slightest cause.[3] One particular source of conflict arose from the townsfolk's habit of emptying chamber pots through open windows, which made the center of the street a perilous place to walk. Men preferred to "keep to the wall," and giving way was a sign of lower status. This often led to quarrels and violence.[4]

A characteristic of societies under stress is that status hierarchies are not only steep but unstable, something reflected in the history of English monarchs. Before 1327 English kings were several times defeated in battle but never deposed. Then Edward II was unseated by a conspiracy led by his wife and her lover, imprisoned and murdered. His great-grandson Richard II was dethroned by Henry Bolingbroke in 1399 and died in captivity at Pontefract Castle (starved to death, it is widely believed). Henry's grandson Henry VI was dethroned twice and then killed, along with his son. His successor Edward IV was deposed once. Edward's son, Edward V, was deposed after a brief reign and then murdered along with his brother (the "Princes in the Tower"). Edward V's uncle Richard took the throne but was defeated and killed at the battle of Bosworth Field in 1485 by the forces of Henry Tudor.

No monarch of the Tudor dynasty (1485–1603) was deposed but their chief ministers fared less well. Henry VII's ministers, Empson and Dudley, became so loathed for their skill as tax collectors that they were executed shortly after the king's death in 1510. Henry VIII executed two of his chancellors, Thomas More and Thomas Cromwell. Cardinal Wolsey, the chief minister of his early reign, only escaped the axe by dying on his way back to London.

---

[1] J. A. Williamson, The Tudor Age (London: Longmans, Green & Co., 1953), 8–9.
[2] L. Stone, *The Family, Sex and Marriage in England: 1500–1800* (London: Weidenfeld & Nicholson, 1977), 93–96.
[3] C. Bridenbaugh, *Vexed and Troubled Englishmen, 1590–1642: The Beginnings of the American People* (Oxford: Clarendon Press, 1968), 189.
[4] Stone, *The Family, Sex and Marriage in England*, 94.

When Henry's son Edward came to the throne as a minor in 1547, both his uncles made a play for power and died for it. John Dudley then took control of government, attempting to change the succession when Edward VI died in 1553 by placing Lady Jane Grey on the throne. He failed and was executed. The unfortunate Lady Jane soon followed. Only Elizabeth, exceptional in so many ways, was loyal to the men who loyally served her.

The relentless conspiracies and plots throughout this period, the executions, and the judicial murder of two of Henry VIII's wives, all formed part of the paranoia and suspicion of the age. Leaders had great power but lived in constant danger.

The family history of James VI of Scotland (1566–1625) gives further evidence of the perils of high rank at this time. His father was murdered by a man who shortly afterwards married his mother, Mary Queen of Scots. She was deposed and fled to England where she was imprisoned and later executed after being involved in repeated conspiracies against Queen Elizabeth. Proclaimed king of England in 1603 as James I, he was immediately faced with two serious conspiracies: the Bye plot and Main plot. In 1605 the Guy Fawkes conspiracy came close to blowing up not only the king but both Houses of Parliament. His son Charles was defeated and executed in 1649 and his grandson James II forced to flee in 1689.

But after 1689 the situation became far more stable. No British monarch or prime minister was ever again overthrown by violence (unless one counts the assassination of Spencer Perceval in 1812, which was motivated by a personal grievance), and certainly none were executed. Personal relationships also showed a similar trend. After 1700, "keeping to the wall" gave way to the convention of keeping to the right, meaning that men could pass peaceably in the street without having to assert their status, and there were other signs of calm. Nineteenth-century Englishmen were renowned for their "stiff upper lip," with disciplined self-control replacing arrogant bravado as a sign of masculinity. By the eigtheenth century, manliness was much less associated with bearing arms and fighting if insulted.[5] The gentry were less inclined to carry weapons and get into fights, and there was less violence among the working classes. These

[5] M. J. Wiener, *Men of Blood: Violence, Manliness and Criminal Justice in Victorian England* (Cambridge: Cambridge University Press, 2004), 5; P. Carter, *Men and the Emergence of Polite Society, 1660–1800* (Harlow: Longman, 2000); A. Bryson, *From Courtesy to Civility: Changing Codes of Conduct in Early Modern England* (Oxford: Oxford University Press, 1998).

trends continued into the nineteenth century as a new concept of masculinity emerged that was dignified, reasonable and restrained.[6]

These new masculine expectations of peacefulness and self-control, emulating the patterns of behavior already established by the gentry of the late eighteenth century, went through two distinct developments in the nineteenth century. One was that the new standards of restraint were applied to all men, and the other was that this new form of dignity now applied to the private as well as the public sphere, with an important emphasis on the respectful treatment of women.[7] This emphasis on emotional control came to be regarded as one of the identifiable characteristics of the British.

This pattern is quite distinct from the one noted in the previous chapter, with a relatively steady rise in political cohesion and other measures of C. There was a quite distinct change from relatively low stress in the early Middle Ages, to extremely high stress from the late fifteenth to early seventeenth centuries, with a steady fall thereafter. In other words, there was a distinct surge in stress levels just as the level of C was rising most rapidly. Could there be a connection?

Punishment

A similar pattern can be seen in the harsh treatment of children and adults. Punishment of children appears to have been on the rise in the later Middle Ages, despite the opposition of some monastic writers of the twelfth and thirteenth centuries.[8] Flogging was gaining ground in grammar schools in the fifteenth century, and in the sixteenth and early seventeenth centuries reached an unparalleled height of brutality. Many schoolmasters in this period had a reputation for ferociously whipping and "birching" schoolboys, and children could be given up to fifty strokes with an elm rod for being unable to recite lessons properly. Boys were often "pulled by the ears, lashed over the face, beaten about the head with a great end of a rod and smitten upon the lips for every offence with a rod." In 1582 William Bedell was viciously attacked by an enraged schoolmaster:

---

[6] Wiener, *Men of Blood*, 6.

[7] Ibid.

[8] L. DeMause, "The Evolution of Childhood," in *The History of Childhood*, edited by L. DeMause, 40–1 (New York: The Psychohistory Press, 1974), 132, 137.

> [He was] knocked down a flight of stairs and hit so violently … that blood gushed out of his ear, and his hearing was in consequence so impaired that he became in process of time wholly deaf on that side.[9]

Violent schoolmasters such as this were rarely dismissed, and campaigns to have them removed almost always failed, reflecting the public's acceptance of even the most serious physical violence.[10] Not only schoolchildren but even undergraduates were flogged freely between about 1450 and 1660.

In 1612, John Brisely elaborated on the bible passage "Spare the rod, spoil the child," reflecting the attitudes of the day in advocating stern punishment of children:

> God hath sanctified the rod and correction, to cure evils of their life conditions, to drive out the folly which is bound up in their hearts, [and] to save their soules from hell.[11]

With children treated so harshly, it is hardly surprising that criminals suffered even worse. This followed the same pattern of increased severity during the Middle Ages. For example, judicial procedure did not record torture before the thirteenth century, but it then became increasingly routine, applied to religious heretics from 1382.

Judicial whipping was carried out with growing enthusiasm and greater frequency in the sixteenth century, and accounts of whipping beggars and vagrants became common after 1528. In 1531, a law was passed declaring that unlicensed beggars were to be "stripped down to the waist and whipped," and vagabonds tied to the end of a cart naked and beaten with whips through the town.[12]

The fifteenth and sixteenth centuries were notorious for brutal punishments. People were tied to boats and dragged downriver, which could be fatal. They might be branded on the face with a red hot iron, something rarely seen in the past.[13] The Act of 1547 ruled that sturdy

---

[9] M. V. C. Alexander, *The Growth of English Education, 1348–1648. A Social and Cultural History* (London: The Pennsylvania State University Press, 1990), 199, 200.

[10] Ibid., 235.

[11] Wiener, *Men of Blood*, 52.

[12] Ibid., 53, 54.

[13] S. Devereaux & P. Griffiths, *Penal Practice and Culture, 1500–1900. Punishing the English* (Basingstoke: Palgrave Macmillan, 2004), 46.

beggars (those deemed fit to work) were not only to be enslaved, but also "marked with a hot iron in the breast with the mark of V." Similar forms of branding were occasionally carried out on whole groups of vagrants until the law was repealed in 1593.[14]

The death penalty was commonly applied and in an increasingly brutal form. King Henry I (1100–1135) was severe for a Medieval king but used only the rope and axe for executions. From 1241, hanging, drawing and quartering was used as a way of making death as prolonged and agonizing as possible. The victim was first choked by hanging, then cut down while still alive, castrated and his bowels cut out and burned before him. This death was most famously inflicted on William Wallace in 1305 and later became the common punishment for treason, although by Henry VIII's reign "treason" could include simply refusing to recognize the king as head of the English church—more than a hundred Catholics died in this way. These executions were popular spectacles, typically drawing large crowds, which suggests that brutal punishments reflected popular taste.

Punishments first used in the fifteenth and early sixteenth centuries included the rack, impalement and boiling alive. For religious offenses a common punishment was burning alive, as inflicted by Mary Tudor on three hundred Protestants during the 1550s. Though given the increasingly Protestant sympathies of Englishmen at the time this did little for her popularity, earning her the title "Bloody Mary."

As with other indications of stress, cruel and severe punishments became steadily less prevalent after the sixteenth century. The use of torture declined in the seventeenth century with the last recorded cases in the 1650s. The exception was "pressing" with heavy weights of people on trial who refused to plead, which was used until 1726.[15]

Capital punishment was also in decline at this time. The average number of executions in London and Middlesex fell from 140 per year in 1607–16 to only 33 in 1749–99, despite a large increase in population. An increased number of capital crimes in the eighteenth century had little effect because of the widespread use of legal loopholes, pardons and commutation.

---

[14] Ibid., 46.

[15] W. Holdsworth, *A History of English law* 3rd edn. (London: Methuen, 1945), 194; J. Lawrence, *A History of Capital Punishment: With Special Reference to Capital Punishment in Great Britain* (Port Washington: Kennikat Press, 1971), 408.

Between May 1827 and May 1830 only fifty-one people were executed in the whole of England.[16]

The same pattern can be seen in the trend away from the severe punishment of children. There was a decline in beating at most schools in the eighteenth century, with brutal flogging by schoolmasters increasingly condemned. Parents also seem to have become more lenient with their children in the seventeenth century, a trend which continued for at least the middle classes in the eighteenth century.[17] The severity of school discipline declined even more during the nineteenth century, with writers such as Hannah More advocating tighter control but with gentler and more psychological punishments. By 1867–69, half of the readers who replied to an enquiry by a domestic magazine were against physical punishment for girls.[18]

The changes of the twentieth century can be seen as no more than a continuation of these trends. By the end of the century, capital punishment was no longer used in Britain. And in recent times, not only has beating been eliminated from most schools, but the revulsion against the physical punishment of children is such that parents can even be prosecuted for it. Increasing pressure to ban all forms of smacking of children is being mounted for the Council of Europe under the assumption that it "violates their human rights.'"[19] The contrast with the widespread brutal flogging of the sixteenth century could hardly be more extreme.

Corporal punishment thus shows the same pattern as capital punishment, with a rise to the sixteenth century followed by a fall.

---

[16] L. Radzonowicz, *A History of Criminal Law and its Administration from 1750* (London: Stevens & Co., 1948), 139–142, 149–152.

[17] L. Stone, *The Crisis of the Aristocracy* (Oxford: Oxford University Press, 1965), 31–32; L. Stone, *The Family, Sex and Marriage in England 1500–1800* (London: Weidenfeld, 1977), 163–5, 168, 265, 433–4, 441; DeMause, 264–5.

[18] F. Robertson, "Home as a Nest," in *The History of Childhood*, edited by L. DeMause (New York: The Psychohistory Press, 1974), 421–2, 416.

[19] A. Hirsch, "Europe Presses UK to Introduce Total Ban on Smacking Children," *The Guardian* (April 25, 2010), http://www.guardian.co.uk/world/2010/apr/25/law-reform-smacking-europe-uk.

Formality and marked status differences

Societies under stress tend to show large status differences, with powerful individuals lording it over their inferiors. From the end of the fourteenth century there was a marked increase in status differences between social classes, including a growing emphasis on formality of relationships and rules of precedence. By the late fifteenth century the ceremony and formality of meals in noble households were extreme. Seating was strictly according to rank, often with sizeable gaps to emphasize differences. Even the people who could be spoken to were carefully defined. At some meals, conversation was forbidden. At one meal served to Elizabeth Woodville, queen consort to Edward IV, the entire court knelt in silence for several hours while she ate.[20]

This trend reached a peak in the sixteenth and early seventeenth centuries. There was an elaborate ranking system, and a constant donning and doffing of hats to indicate it. Pervasive formality applied even within the family. Children knelt to ask their parents' blessing. Middle-aged men stood in the presence of their parents until invited to sit.[21] This degree of deference and obedience goes beyond that of any society in the modern world.

There was a broad tendency to emphasize differences in rank and the rights of the more exalted. While medieval clergy had often castigated the rich, Elizabethan preachers called for humility and obedience. Gentlemen were immune from the floggings distributed so freely to their inferiors. The Crown worked to abolish the egalitarian remnants of the Middle Ages. By the fifteenth century, craft guilds had changed from associations of relative equals to very unequal groups of capitalists and poor workers.[22]

The same pattern applied to education. The growth of colleges from the fourteenth century onwards brought students under increasing control. Originally the dons were in charge, but by the sixteenth century the college heads and vice-chancellors held power. Social differences within universities were sharpened, with higher status students allowed to take degrees in a shorter time and given positions of unusual privilege and

---

[20] P. M. Rieder, *On the Purification of Women: Churching in Northern France, 1100–1500* (New York: Palgrave/Macmillan, 2006), 155–56.

[21] Stone, *The Crisis of the Aristocracy*, 21, 34–5, 591–2.

[22] J. David, *Medieval Market Morality. Life, Law and Ethics in the English Marketplace, 1200–1500* (New York: Cambridge University Press, 2012), 20, 21.

freedom.[23]

However, such favorable attitudes towards hierarchy began changing in the early seventeenth century, especially among the Puritans. The gentry began to refuse the old forms of respect to the nobility. Foremost of all with the new spirit of equality were the Quakers, who in the later seventeenth century made a point of refusing to doff their hats to men of higher status. This breach of the law was met with violence and imprisonment, but by 1689 the government had given up and no longer enforced this law. In the eighteenth century, nobility and gentry could mix with much greater ease. Formality between parents and children was also much reduced by the nineteenth century, and customs of deference had largely been abandoned by the middle class and much of the lower class. Deference survived mainly in rural communities, but even there it had slackened by 1900.[24]

One important effect of weakening deference in the nineteenth century was the rise of democracy. In the previous chapter we saw how the rise of C and especially infant C strengthened impersonal loyalties and thus the rule of law. Such loyalties help support democratic institutions but they do not necessarily produce them. Rising C may actually strengthen the power of government, as happened in Europe for much of the nineteenth century.

But in Britain, falling levels of stress were accompanied by a gradual diffusion of political power down the status hierarchy. The autocratic power of Tudor kings had been gradually weakened in the seventeenth century, culminating in the triumph of Parliament with the Glorious Revolution of 1689. But Britain was still far from a democracy, and the new system effectively gave power to the landed gentry and aristocracy.

Starting with the Reform Act of 1832 this changed with startling speed, and within a few decades all men had the vote. Similar changes took place on the continent, with male suffrage in France and Switzerland after 1848, and in the German Empire after 1871. Popular pressure for reform, and in some cases outright revolution, were the key drivers for these changes. All of this suggests a dramatic decline in willingness to accept the power of

---

[23] Stone, *The Crisis of the Aristocracy*, 29–33; A. Birnie, *An Economic History of the British Isles* (London: Methuen, 1961), 93–97.

[24] Stone, *The Crisis of the Aristocracy*, 35–6, 747–53; M. Girouard, Life in the English Country House (New Haven and London: Yale University Press, 1978), 85–6, 182–3, 231, 308; Stone, *The Family, Sex and Marriage in England*, 102.

higher status people.

We have focused mainly on Britain, but a similar pattern can be found in other European countries. French patterns of deference and treatment of children were very similar to the English, and Spaniards in the sixteenth and early seventeenth centuries were a byword for cruelty. This was the time when thousands of people were burned alive by the Inquisition.[25]

Monasticism

A long-term rise in stress up to the sixteenth century may explain why monasticism reached a peak of popularity in the Middle Ages (far higher than in the early Christian period, as noted in the previous chapter). This is despite the fact that sexual restrictions in the Middle Ages were less rigorous than they later became.

Stress is uncomfortable and unpleasant, but what seems to be especially unpleasant is a higher level of adult stress than that experienced in infancy. This can drive people to drug-taking, alcoholism and overeating, but religious asceticism is an opposite but equally valid response. In the long term it is also more adaptive, since reducing C causes people to avoid stimulus and thus lower their level of anxiety.

This last approach was strongly favored in medieval times, when the "peace of the monastery" was more than an expression. It represented an escape from inner turmoil and unhappiness. It is surely no accident that the popularity of monasticism, from the hermits of the later Roman Empire to the dawn of the Protestant Reformation, coincides exactly with an apparent long-term rise in stress to the sixteenth century.

The civilization cycle and the causes of high stress

Thus, while C rose consistently for a thousand years up to about 1850, stress shows a very different pattern. Not only did it reach a peak in the sixteenth century, it was high in the fifteenth to early seventeenth centuries. In other words, stress, after peaking in the sixteenth century, fell relatively little in the early seventeenth century, more in the late

---

[25] J. Dedieu, *L'Inquisition* (Paris: Les Editions Fides, 1987), 170–73; W. Monter, *Frontiers of Heresy. The Spanish Inquisition from the Basque Land to Sicily* (Cambridge: Cambridge University Press, (1990), 53.

seventeenth and early eighteenth centuries, and very fast in the nineteenth. What this suggests is the curvilinear relationship outlined at the beginning of this chapter, in which C rises when stress is high, and ceases to rise when stress falls past a certain point (around 1850 in Britain). In biohistory, this pattern is called the civilization cycle (see Fig. 7.2 below).

Fig. 7.2. The civilization cycle in England—initial model. A high level of stress around 1550 was accompanied by a rapid rise in C. The peak of C in the nineteenth century was associated with a rapid fall in stress.

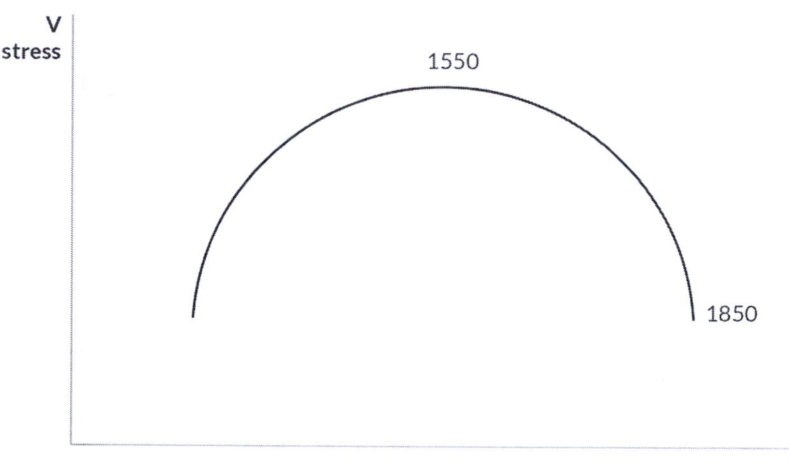

This provides a possible reason for the sustained rise in C, when combined with the powerful C-promoting traditions of medieval Christianity. When stress was only moderately high, around 900 AD, C does not seem to have been rising at all. When it was very high, around the sixteenth century, C rose rapidly. Then, when it dropped back to a moderate level, around 1850, C ceased to rise.

The studies cited in chapter two show that high levels of the stess hormone cortisol reduce testosterone, which makes it a C-promoter. So it is quite reasonable to suppose that C rose because of high stress, and the higher the level of stress the faster it rose. Not only was stress rising after 1100 but by this time Christianity had thoroughly penetrated all levels of European society. The unprecedented rise of C, which eventually resulted in the Industrial Revolution, could be explained by a combination of stress with the C-promoting traditions of Christianity.

But stress alone is not a sufficient explanation. High levels of stress among peoples such as the Mundugumor do not have anything like this effect. In fact, we saw in chapter four that the overall effect of stress in Mundugumor society was a breakdown in traditional restraints. The other point to consider is why the level of stress in Britain rose so markedly and for such a long time. The obvious place to start is with the civilization cycle itself. When C was low, as in the eleventh and twelfth centuries, stress was rising. When C was at a moderate level around the sixteenth century, it reached a peak and ceased to rise, and when C reached a peak in the nineteenth century all indications are that stress fell rapidly. This leads to the possibility that low C increases stress and high C reduces it, at least in the context of British history.

We already have a plausible motive force, with the indication that higher levels of child and adult C cause people to avoid stimuli and thus reduce their level of anxiety (see chapter five). But low C as such does not increase stress, as indicated by the fact that most hunter-gatherer peoples are quite indulgent with their children. Population density may have an impact, as indicated by problems such as alcohol abuse found in Indigenous Australian communities, which are far more densely populated than they were in pre-colonial times.

High levels of V increase stress

There is almost certainly another reason, which is that Europe in and around the sixteenth century had an unusually high level of V. One indication is the extreme patriarchal attitudes of the period around the sixteenth century, a pattern found across Europe. Orthodox theology began to downgrade women in thirteenth- and fourteenth-century Europe, and an assembly of notables in Paris in 1317 barred them from the throne. Their legal position also declined from the fourteenth century onwards, so that by the sixteenth century a Frenchwoman needed the consent of her husband for any legal act to be valid. In both England and France, the low point for women was reached in the sixteenth century.[26]

Significantly, the status of women began to improve just as the levels of stress began to decline. By the early seventeenth century preachers were

---

[26] P. Das, "Shakespeare's Representation of Women in his Tragedies," *Prime University Journal* 6 (2) (2012); S. Doran, "Elizabeth I: Gender, Power and Politics," *History Today* 53 (5) (2003); S. Broomhall, *Women and the Book Trade in Sixteenth-Century France* (Aldershot: Ashgate Publishing, 2002).

beginning to stress companionship in marriage, a feeling that seems linked to greater equality. The rise in women's status continued thereafter, especially in the nineteenth and early twentieth centuries, when in country after country they won the right to vote.[27]

Inequality between the sexes in one sense mirrors the inequality of society as a whole, but stress does not necessarily promote patriarchy. The Yanomamo as described in chapter five were intensely patriarchal but did not punish their children overmuch, while the Mundugumor were severely stressed but not especially patriarchal.

A high level of V could explain the high level of stress in sixteenth-century Europe. V reduces tolerance of crowding (see chapter four), which is a basic part of its biological function. Famine-prone animals react to population density with rising stress, making it more likely that they will migrate. This in turn makes them more likely to find and colonize new areas made available by shifting patterns of climate and rainfall, rather than building up in one place until wiped out by some local calamity. Farming populations cannot shift wholesale to new areas, so it is to be expected that a rise in V would increase the overall level of stress. By the same token, falling V after the sixteenth century would reduce stress.

Reasons for the rise of V

This, of course, leads to the obvious question of what caused this rise in V. Periodic famine and the harsh climate of Northern Europe clearly had an impact. But a more powerful influence came from Christianity. Christian teaching, which we have treated as an effective C-promoter, also contains powerful V-promoters. We have seen that restraints on sexual behavior are effective promoters of V as well as C. The bible, especially as interpreted by the early church, is a powerful support for patriarchy. And it admonishes parents to punish children, which (applied in later childhood) would tend to raise V.

Also, the Lenten fast of Orthodox and Catholic Christianity is of the annual type which tends to raise V, rather than the monthly or weekly fasting which should have more effect on C (this is because it mimics the effect of an occasional famine, rather than of an ongoing but milder food shortage). In this sense it is similar to the Ramadan fast of Islam, which is

[27] J. Tosh, A Man's Place: Masculinity and the Middle-class Home in Victorian England (New Haven and London: Yale University Press, 1999), 39.

the archetypal high-V religion. It is also noticeable that as stress fell after the sixteenth century, this annual V-promoting fast was abandoned by many Protestant Christians in Northern Europe.

If we suppose that the underlying reason for rising and falling stress was a rise and then fall in V, we can also much better explain the long-term rise in C. The first point to note is that infant C tends to undermine V for a number of reasons. The first is that infant C requires control of infants, which undermines the indulgence required by V. The second is that infant C promotes skeptical and independent thinking, which undermines V-promoting traditions. And the third is that infant C tends to make people prosperous.

In Northern Europe around 1100, C and especially infant C were low. Christianity, with its powerful V-forming traditions such as the Lenten fast, was strongly entrenched. And the chaos and poor communications of the period meant that famine was a common experience. The result was an extremely rapid rise in V and thus stress.

The combination of rising V and stress, in turn, caused a rise in C and thus infant C—the higher the level of V, the faster the rise in C. But rising infant C undermines V, causing the rise in V to first slow and then cease (around 1550). Then, as infant C continued to rise, V began to fall. Changes in C are an exact reflection of this. Around 1100 V was relatively low so C rose slowly. When V was at its peak around 1550 it increased most rapidly. And as V declined the rise in C first slowed and then ceased (around 1850). The rise and fall in stress mirrors the change in V.[28] Fig. 7.3 below illustrates this pattern.

Note that the sixteenth century in England was an age of innovation and brilliant creativity, most notably expressed in the plays of Shakespeare. Thus, increased punishment of children around the sixteenth century did not make people more traditional in their thinking. Though punishment of older children supports tradition, punishment of younger children and especially infants appears to undermine it. Once again we can refer to the Mundugumor as described in chapter four, who punished and stressed children at all ages and were fast abandoning their traditional codes of behavior.

---

[28] Alternatively, V itself may act as a C-promoter, alongside stress. Clarification of this, as with so many issues, must wait for animal studies.

Fig. 7.3. Civilization cycle, developed model. In this model it is the rise and fall in V which causes the changes in stress and is responsible for the long-term rise in C.

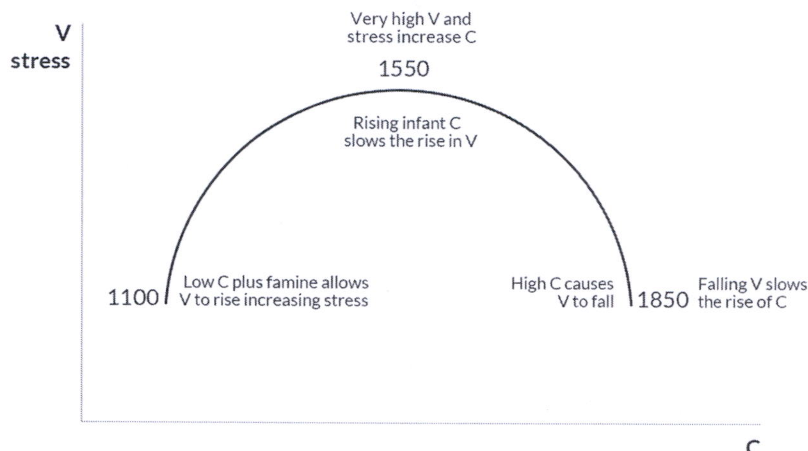

The civilization cycle in Japan

If the civilization cycle hypothesis is correct, then any other culture area experiencing a rapid and prolonged rise in C should show a similar rise and fall in stress. Fortunately, we have detailed historical records from another civilization which is about as far from Europe as possible, and so can be considered an independent case study. Europeans did not visit Japan until the closing years of the sixteenth century, and its main cultural influences were from China and, to a lesser extent, India.

Japan, like Europe, experienced a dramatic rise in C over the past thousand years. The study cited in chapter five showed Japan in the mid-twentieth century to have all the hallmarks of high C, including tight control of sexual behavior and control of children at all ages, including infancy. It also showed early success at industrialization, indicative of high infant C.

Evidence of rising C in Japan

Historical evidence shows much the same long-term pattern in Japan as in Britain. There is evidence for the growth of the nuclear family at the expense of the extended family in at least one region in the fifteenth century, which became general for the whole country in the Tokugawa period (1603–1867).[29] Not only did families at the time tend to live separately and as distinct economic units, there was also a decline in the co-operative work groups of earlier centuries.[30] As in England, the rise of C was evident first in the more economically advanced areas and the cities. A study of one city shows the nuclear family dominant since at least the eighteenth century, with little change in household size up to the twentieth century.[31] However, in remote regions of central Japan the extended family did not even begin to decline until the late nineteenth century.[32]

Monogamy versus polygyny is clearly associated with food-restricted environments and correlates with C in the cross-cultural survey. In Japan, polygyny was normal for the aristocracy at least until the twelfth century, after which monogamy gradually became the rule. Originally, there was the possibility of taking a "secondary wife" if the first did not produce an heir, but this was abandoned in favor of adoption in the fifteenth century.[33] The same situation applies to all classes in the twentieth century,[34] suggesting a peak of C at this time.

In terms of child training, Japan showed a similar pattern to England. It will be recalled that in the oldest centers of civilization, such as India and China, the key age for training is seen to be age 6 onwards. Even in medieval times the English began training children much earlier, and the same applies to Japan. The Sekyosho, written in the early fifteenth century, assumed that children under the age of 3 could not be taught but

---

[29] J. W. Hall, *Government and Local Power in Japan: 500–1700* (Princeton: Princeton University Press, 1966), 257, 291; E. O. Reischauer & J. K. Fairbank, *East Asia: The Great Tradition* (Boston: Houghton Mifflin, 1960), 628; T. C. Smith, *The Agrarian Origins of Modern Japan* (Stanford: Stanford University Press, 1959), 145–146.

[30] Smith, *The Agrarian Origins of Modern Japan*, 140–56.

[31] R. J. Smith, "Town and City in Pre-Modern Japan: Small Families, Small Households, and Residential Instability," in *Urban Anthropology*, edited by A. Southall, 180 (Oxford: Oxford University Press, 1973); C. Ueno, *The Modern Family in Japan: Its Rise and Fall* (Melbourne: Trans Pacific Press, 2009), 72.

[32] H. Befu, "Origin of Large Households and Duolocal Residence in Central Japan," *American Anthropologist* 70 (2) (1968): 309–19.

[33] L. Frédéric, *Daily Life in Japan at the Time of the Samurai* (New York: Praeger, 1972),118, 148.

[34] R. Benedict, *The Chrysanthemum and the Sword* (London: Routledge & Kegan Paul, 1967), 130; R. K. Beardsley, J. W. Hall & R. E. Ward, *Village Japan* (Chicago: The University of Chicago Press, 1959), 216–220.

that the key to character development was found between 3 and 7.[35]

But also, as in England, the commencement of training began earlier as C rose. From the seventeenth and eighteenth centuries, writers began to teach that children should be strictly and carefully trained from infancy.[36]

Standards of sexual morality also became stricter with time. The lower classes did not greatly stress the chastity of women before 1600, though the samurai were less lenient.[37] But Confucian and samurai codes with much stricter standards gained ground in the Tokugawa period (1603-1868. There is also some direct evidence of a tightening of standards since the eighteenth and nineteenth centuries, and mid-twentieth-century attitudes were fairly restrictive. Except in very backward areas, lack of virginity would severely hurt a girl's marriage chances, and—most significant of all—women often appeared to gain little pleasure from marital sex.

Folk traditons of the village of Niiike, as described in chapter six, indicate higher levels of premarital sex a few generations earlier. But by the 1959s premarital liaisons for girls were rare and marriages almost always arranged.[38] This suggests a rise of C to the mid-twentieth century.

There was also a trend for later marriage. Age of marriage in pre-Tokugawa times (before 1600) seems to have been fairly early.[39] Scattered data from the Tokugawa period suggest a somewhat later date, around the early 20s.[40] Finally, national statistics show a slight rise in age of first

---

[35] M. Ohta, "The Discovery of Childhood in Tokugawa Japan," *Wako University: Bulletin of the Faculty of Human Studies* 4 (2011).

[36] Ibid.

[37] Frédéric, *Daily Life in Japan at the Time of the Samurai*, 42, 58.

[38] Beardsley et al., *Village Japan*, 315–6, 333; Benedict, *The Chrysanthemum and the Sword*, 198; E. F. Vogel, Japan's New Middle Class: The Salary Man and his Family in a Tokyo Suburb (Berkeley: University of California Press, 1965), 107, 113, 116, 221–223; J. B. Cornell & R. J. Smith, *Two Japanese Villages* (Ann Arbor: The University of Michigan Press, 1956), 69; J. F. Embree & Suye Mura, *A Japanese Village* (Chicago: University of Chicago Press, 1964), 193–194. This relatively remote village showed substantially freer attitudes towards sexual activity in the 1930s.

[39] Frédéric, *Daily Life in Japan at the Time of the Samurai*, 42.

[40] S. B. Hanley & K. Yamamura, *Economic and demographic change in preindustrial Japan: 1600–1868* (Princeton: Princeton University Press, 1977), 247; S. B. Hanley & K. Yamamura, "Population Trends and Economic Growth in

marriage from 23 in 1910 to 24 in 1936.[41] For example, marriages in Niiike in the early 1950s were reported as later than they had been fifty to sixty years earlier, despite rising prosperity.[42] Thus, the age of marriage seems to have been latest just at the peak of C. The age of puberty was also very late until around 1910, though there is no evidence from earlier times.

A change of emphasis in religion from ritual to behavior can be traced in Japan, as it was in England. Early Japanese religion was a form of Shinto with a strong emphasis on ritual but little ethical content. Buddhism was accepted by the court aristocracy very early, but appears to have made little impact on common people.

But in the thirteenth century a number of sects with strong ethical content began to spread. The Zen sect in particular emphasized personal character and discipline, and helped its mainly warrior followers to live up to their code of courage and loyalty.[43] From the sixteenth century and through the Tokugawa period, all classes gradually became imbued with the Confucian values of duty, hard work and discipline, loyalty and respect for the social order.[44] This change from a focus on ritual to one on morality and hard work is also characteristic of a rise in C.

A long-term rise in the market economy has the same significance. There is scattered evidence of the growth of trade, manufacturing and the use of money from the twelfth century. Economic growth seems to have been particularly rapid during the fifteenth and sixteenth centuries, when the location of castles descended from the hilltops to the plains and great

---

Preindustrial Japan," in *Population and Social Change*, edited by D. V. Glass & R. Reveille, 482 (London: Edward Arnold, 1972); S. Hanley, *Everyday Things in Premodern Japan: The Hidden Legacy of Material Culture* (Berkeley: University of California Press, 1997), 141.

[41] I. Taeuber, *The Population of Japan* (Princeton, Princeton University Press, 1958), 224, 227.

[42] Beardsley et al., *Village Japan*, 312.

[43] Frédéric, *Daily Life in Japan at the Time of the Samurai*, 197, 105–6; Reischauer & Fairbank, *East Asia: The Great Tradition*, 544–9; H. P. Varley, *Japanese Culture: A Short History* (New York: Praeger, 1973), 147, 68, 70–3.

[44] Varley, *Japanese Culture*, 116, 118; Reischauer & Fairbank, *East Asia: The Great Tradition*, 614–6, 618, 657, 662–5, 643. E. O. Reischauer & J. W. Fairbank, *East Asia: The Modern Transformation* (Boston: Houghton Mifflin, 1960), 267; L. Greenfeld, *The Spirit of Capitalism, Nationalism and Economic Growth* (Cambridge Mass.: Harvard Universty Press, 2001), 227–242.

trading cities such as Sakai became virtually independent. The seventeenth century saw money starting to displace rice as a currency among the common people, and from this time onwards there was an increase in tenancy, wage labor, cash crops and rural cottage industry. A strong parallel with Europe is that Japan had largely outgrown the monopolistic craft guilds of the earlier period by the sixteenth century.[45] Japan was also like Britain in experiencing a major growth of literacy in the eighteenth and nineteenth centuries, an indication of falling time preference (greater willingness to invest for the future).[46] Then came rapid industrialization from the late-nineteenth century, which as we have seen is the clearest sign of very high C.

The growth of the money economy in Japan has an interesting sidelight. The government had tried to introduce money in the eighth and ninth centuries under Chinese influence, but the effort failed so completely that the government mint disappeared. Then, quite spontaneously, without any government action, money came into use in the twelfth century using mainly Chinese coins.[47] Biohistory suggests that the level of C was simply not high enough in earlier times to support a currency, while later on it was.

The industrialization of Japan was accompanied by fast population growth. The number of Japanese more than trebled between 1875 and 1975, compared with a mere doubling in the previous three hundred years.[48] As in Europe, a peak rate of population increase is an indication of high C. The peak rate of population growth came in the 1920s, reinforcing the idea of a peak of C around this time.

---

[45] H. P. Varley, *Imperial Restoration in Medieval Japan* (New York: Columbia University Press, 1971), 557–9; M. Takizawa, *The Penetration of the Money Economy in Japan* (New York: Amis, 1927), 34, 36; Smith, *The Agrarian Origins of Modern Japan*; Frédéric, *Daily Life in Japan at the Time of the Samurai*; Hall, *Government and Local Power in Japan*, 259; C. J. Dun, *Everyday Life in Traditional Japan* (London: Batesford, 1969), 20; W. Cole, *Kyoto in the Monoyama period* (Norman, Oklahoma: University of Oklahoma Press, 1967), 74–75; D. Landes, *The Wealth and Poverty of Nations* (Cambridge Mass.: Harvard University Press, 1998), 350–370.

[46] G. Clark, *A Farewell to Alms: A Brief Economic History of the World* (Princeton, NJ: Princeton University Press, 2007), chapter thirteen.

[47] Reischauer & Fairbank, *East Asia: The Great Tradition*, 484, 558.

[48] Taeuber, *The Population of Japan*, 14, 116, 20–2, 41.

Finally, for at least a thousand years there has been an increase in impersonal loyalties in Japan, as reflected in the political system. Formally at least, the government of seventh-century Japan had an elaborate system of law and administration based on Chinese models. But this was never effective over much of the country, and became less so with time. By the eleventh century lawlessness and warfare were endemic, with the government powerless to maintain order even in the capital.[49]

The eleventh and twelfth centuries saw a growth and consolidation in the power of certain military families, culminating in effective control of the whole country by the Minamoto in 1185. An effective judicial system was set up for the warrior class, but authority was still based on the personal loyalty of fief-holding vassals. In the following Ashikaga period (1338–1573) central authority collapsed, but this was also the time when impersonal administration replaced personal loyalties. By 1500 the stronger lords had become absolute rulers, with complex laws controlling agriculture, commerce and the private lives of their vassals. Uniform taxes and labor levies and the growth of bureaucracy also suggest a relatively impersonal administration. These trends continued with the unification of the country at the end of the sixteenth century.[50] The successful adoption of many Western legal, administrative and political institutions since 1868 also suggests a high degree of impersonality.

As in Europe, the rise of C did not always result in larger political units, which tend to reflect child V. For a time in the fifteenth and sixteenth centuries they actually became smaller. Even after the country was unified at the end of the sixteenth century, local loyalties remained strong and each fief continued to be governed by its daimyo or feudal lord. Japan was not fully unified until the late nineteenth century, at about the same time as Germany. Running across this trend was a rising sense of nationalism, expressed from earliest times by loyalty to the emperor but becoming an element of popular religious feeling in the thirteenth century, and reaching its highest pitch before the Second World War.[51]

From all these indications the peak of C in Japan came later than in Europe, in the first half of the twentieth century.

---

[49] Reischauer & Fairbank, *East Asia: The Great Tradition*, 483, 498–500.

[50] Ibid., 552–5, 571–3, 603; Varley, *Japanese Culture*, 571.

[51] Reischauer & Fairbank, *East Asia: The Modern Transformation*, 569, 662–3; Varley, *Japanese Culture*, 72, 142; Greenfeld, *The Spirit of Capitalism, Nationalism and Economic Growth*, 227–298.

Stress and the civilization cycle in Japan

In Europe the rise of C was accompanied by a rise and then fall in V and stress. Japan experienced the same rise in C, so it is highly significant that it experienced exactly the same rise and fall in V and stress. They even peaked at almost exactly the same time—the late fifteenth or sixteenth centuries. This must be a coincidence since there is no way that Europe could have influenced Japan or vice versa at this time, and China (as described in chapter fourteen) was following a very different pattern.

Early Japan seems to have been relatively low in both V and stress. The imperial law codes issued from the eighth century AD onwards were strongly influenced by Buddhism and reluctant to impose capital punishment. For example, offenses which would have led to execution in earlier times were now punished by banishment. Rulers were also relatively secure. A single imperial family presided all through this period and a single family and the Fujiwaras, was dominant at court, indicating that status hierarchies were stable. And relatively low V is suggested by the unwarlike nature and strong female figures of the court aristocracy in the Heian period, as expressed by its magnificent literature (largely written by women).

In the following centuries the provincial aristocracy grew in power, and in the late twelfth century took effective control of the country. This change to a more martial culture suggests a rise in V. But though society was more patriarchal than the Heian court culture, women still had a relatively strong position at this time, especially among the lower classes. They might, for example, control households.

There is also evidence of increasing stress. In 1232 the government established a new legal code applying the death penalty for a number of offenses, including theft. But transition of power remained relatively peaceful, after some early struggles. The Hojos kept effective control for well over a century, and great nobles at this time were rarely ousted by rebellious vassals. The important families of the fourteenth century were mostly direct descendants of provincial governors of the early thirteenth century, themselves often offshoots of the imperial and Fujiwara families who had migrated from the capital in earlier times.

A peak of stress in the fifteenth and sixteenth centuries

The following Ashikaga period (1336–1573), however, was far more brutal. Feudal warfare became ferocious and endemic, especially when central authority collapsed completely. During the Onin War of the 1460s, armies slaughtered each other in the streets of Kyoto, burning and looting as they went. After one engagement eight cartloads of heads were collected, only a fraction of those who were killed.[52] And those who fought were mainly absconding peasants rather than professional soldiers, indicating that the violence had spread through all levels of society.

Also consistent with high levels of stress is that suspicion and mistrust were pervasive. Feudal lords kept close watch over their subjects and dictated, to an extreme extent, what they could and could not do. They might be forbidden to marry, adopt a child, employ a servant, or make a journey without consent. Outsiders to these feudal domains were treated with hostility and often not permitted to enter local fiefs at all, and closely supervised at all times if they were.[53]

The late fifteenth and sixteenth centuries were also the period when local rulers were establishing absolute power in their domains, and the law codes of these new rulers were brutal in the extreme. The punishment of a man's servants and family, previously only applied for high treason, was now used for a number of crimes. A whole village might be punished for the failure of one man to pay taxes. Methods of execution came to include burning alive, boiling in oil, impalement, sawing, dismemberment and crucifixion. Even European visitors were shocked by the severity of punishment for the most trivial offenses, and we have seen how brutal Europeans were at this time. In the past, only direct vassals of the shogun had been executed for rebellion. Now, even low-ranking prisoners were often tortured to death.

This brutality could extend even to a man's own family. The Japanese ruler Hideyoshi had originally adopted a nephew as his heir. When his own wife gave birth to a son he first exiled his nephew and then ordered him to commit suicide. Family members who failed to follow his example were then murdered in Kyoto, including thirty-one women and several children.

---

[52] W. S. Morton & J. K. Olenik, *Japan. Its History and Culture* (New York: McGraw Hill, 1994), 91–2.

[53] G. B. Gibson, *Japan, A Short Cultural History* (London: The Crescent Press, 1931), 431.

Society-wide stress is also indicated by unstable hierarchies, which is of course one reason rulers acted so ruthlessly. In no more than a century leading up to the 1560s, all but one of the great families were destroyed or reduced to insignificance, often by rebellious vassals. Only the Shimazu of distant Kyushu managed to hold on. This was a far more drastic change than anything seen in Europe. It is as though every European royal and ducal family was wiped out within a hundred years. A crucial part of this change was the weakening of loyalty to feudal superiors, an extraordinary breach of the samurai code, which was meant to value loyalty above life. The new feudal lords of the period had effectively risen to power by disregarding every law, written and unwritten, of fealty and gratitude, and as such it is not surprising that they had trouble controlling their vassals.[54] A number of foreign visitors noted the prevalence of treachery during this period, supporting the idea of a high level of stress.

But, as in Europe, signs of stress must be balanced against evidence of extraordinary energy and confidence, indicative of high V. During the sixteenth century, Japanese traders moved out into Southeast Asia, founding settlements in the Philippines and elsewhere. The Japanese under Hideyoshi invaded Korea in the 1590s with the aim of conquering China. This was also a time of rapid population growth.

Furthe evidence of rising V is that the authority of husbands was growing in the fourteenth and fifteenth centuries to the point where it became theoretically absolute. Men were not even supposed to consult their wives on significant decisions relating to the family.

Declining stress from the seventeenth century

Many of these trends showed a sharp reversal in the seventeenth century. The warlike Japanese became a nation at peace, united under the Tokugawa shoguns. A single lineage ruled the country for the 265 years of the Tokugawa period (1603–1868). Not only this, but most of the great families accepted Tokugawa primacy and survived until 1868. For more than two centuries after the Shimabara revolt in 1638 there was no significant political change or warfare in Japan.

Political unity alone is not enough to explain this change in attitude. Tudor rulers held extraordinary power but still felt the need to execute many of

[54] Ibid.; J. Whitney Hall, *Japan: from Prehistory to Modern Times* (New York: Dell Publishing, 1970), 126–134.

their chief ministers. Their Stuart successors after 1603 were far more humane, and the same applied in Japan. After 1684 a number of powerful men served the shogun including Yanigizawa Yoshiyasu, Arai Hakuseki, Tanuma Okitsugu, Matsudaira Sadanobu and Mizuno Tadakuni. Not one of these was executed, and most retired peacefully. An entire culture of violence, treachery and suspicion had disappeared, indicating a major decline in stress.

There is some evidence of a similar pattern of declining stress in childrearing. The early fifteenth-century Sekyosho assumed that children would be beaten, but by the eighteenth century the emphasis was on affection and patient teaching, and parents were specifically warned against bad temper when dealing with children.[55] As seen in chapter five, the physical punishment of children was not common in twentieth-century Japan.

This is not as clear a picture as in Europe. Under the influence of Neo-Confucianism the status of women remained low, at least in theory. Punishments remained severe throughout the Tokugawa period, though painful deaths were less often inflicted in later years. Only in the 1870s did legal forms limit the number of crimes punishable by execution, reduce the use of torture and abolish flogging. Criminals were now to be beheaded or hanged. The law code of 1925 was more lenient still, and this trend has continued. In recent times, opposition to the death penalty has grown and it is normally only applied to multiple murderers.[56] This indicates a very rapid fall in stress over the past century and a half.

So while there is clear evidence of a fall in stress since the sixteenth century, a far more conservative regime and society may have slowed the decline, or perhaps reflected a slower rate of decline. This slower decline in V may perhaps be why the peak level of C came a century or so later than in Europe. With this exception, though, we see changes in C, stress and V that are strikingly similar to those in Britain (see Fig. 7.4 below).

---

[55] Ohta, "The Discovery of Childhood in Tokugawa Japan."

[56] Clark, *A Farewell to Alms*, 22; D. Cunningham, *Taiho-Jutsu, Law and Order in the Age of the Samurai* (Boston Mass,: Tuttle Publishing, 2004), 31–36

Fig. 7.4. The civilization cycle in Japan. As in Britain, a high level of stress and presumably V is associated with a rapid rise in C, and high C with falling stress and V.

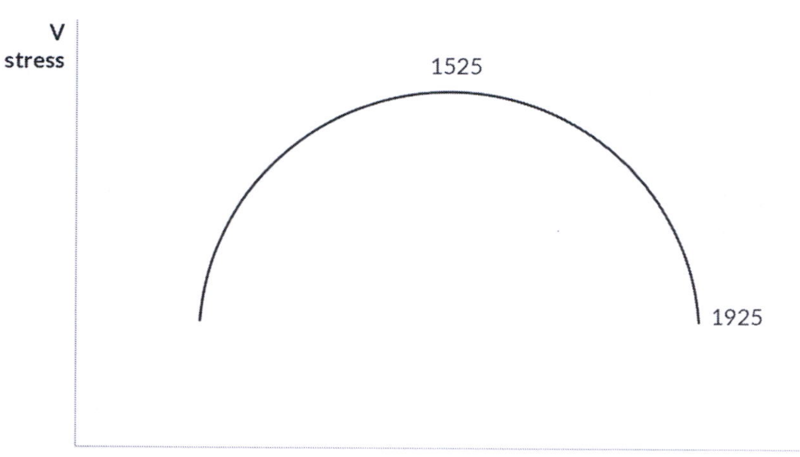

The same pattern in Europe and Japan

The similarities between Europe and Japan are striking, and difficult to explain by conventional means. In both societies there was a rapid rise in C, indicated by very similar changes in family patterns and childrearing. In both there was a significant *increase* in stress to a peak in and around the sixteenth century, and then a fall. This is in sharp contrast to the conventional view that societies become more humane as they develop. The civilization cycle provides a clear explanation for these changes and why they followed the same pattern in both areas.

However, we have repeatedly seen shorter-term fluctuations that cut across these trends. To understand this pattern we must leave history for a time and move back to biology, starting with an animal that is well known but commonly misunderstood—the lemming.

CHAPTER EIGHT

LEMMING CYCLES

We have seen how changes in C and V can both describe and explain the rise of the West up to the mid-nineteenth century. and of Japan to the mid-twentieth. In particular, the increase in C creates a society with more impersonal loyalties, strengthening first the position of rulers and then, as C rose to a much higher level, republican and democratic institutions. Among societies with higher C there was also a trend towards faster population growth and an increased orientation towards industrialization and a market-based economic system. We have also seen in both cases that the increase in C was accompanied by an initial surge in V and stress peaking around the sixteenth century, in a pattern labeled the civilization cycle.

Even a cursory knowledge of history, however, shows that the pattern of change is not this simple. There were periods such as the 1130s and 1450s in England when a previously strong monarchy dissolved into feudal anarchy. And population may have been generally on the rise, but there were long periods when it was stagnant or even declining. Japan was even more liable to periods of political fragmentation, such as happened in the aftermath of the Onin War at the end of the fifteenth century. Another example is the periodic disintegration of the Chinese state in what is termed the "dynastic cycle."

To understand why this pattern occurs we need to revisit the field of zoology. In earlier chapters we considered gibbons as exemplars of C strategy—timid, territorial, slow breeding. We have also used baboons as exemplars of V strategy—bold, migratory, fast breeding. But there are animals which seem to alternate between these two strategies. Two such species are lemmings and muskrats.

C and V strategies in lemmings and muskrats

Lemmings are small rodents living in and near the sub-arctic regions of Eurasia and North America. They are known for having a very distinctive population cycle, in which every few years their numbers expand massively and their behavior also changes.

Normally timid and shy, at the population peak lemmings lose their fear of humans and become bold and even aggressive. In a wave of mass migration they pass through every obstacle, swimming lakes and rivers and even out to sea. Huge numbers of them drown, which has led to the false idea that lemmings commit suicide. In truth the sea is simply another river too wide to cross, though numbers of lemmings have been found on offshore islands.

These cycles in lemmings are highly regular, with a period of three to four years. There is a phase of rapid growth, then migration when numbers are at a peak, then a population collapse, then renewed growth.[1]

The same cycle has been observed in several species of mice, hares, grouse and muskrats, sometimes three to four year cycles but more commonly around ten years.[2] Muskrat studies show that growing populations are distinct in a number of ways from declining populations. They are more resistant to disease, have larger litters, breed earlier and tolerate more breeding pairs per acre. In the decline periods the opposite applies. More die from disease, litters are smaller and breeding sites are further apart (see Fig. 8.1 below). They also seem to have a very poor reaction to any form of stress, commonly dying from hypoglycemic shock such as when taken into captivity.[3]

There are some factors affecting muskrat birth rates that relate more obviously to environmental conditions such as drought and population density. These include survival of late-born young and the number of litters born. Though with some evidence of cyclical effects, environmental influences are more important. But with regard to breeding in the year of birth and especially the number of young per litter, cyclical effects are significantly greater. Litters born in drought years are only a half pup

---

[1] D. Chitty, *Do Lemmings Commit Suicide? Beautiful Hypotheses and Ugly Facts* (Oxford: Oxford University Press, 1996).

[2] L. B. Keith, *Wildlife's Ten-Year Cycle* (Madison: University of Wisconsin Press, 1963), 61–2.

[3] Ibid.; P. L. Errington, *Muskrat Populations* (Ames: Iowa State University Press, 1963); J. J. Christian, "The Adreno-Pituitary System and Population Cycles in Animals," *Journal of Mammalogy* 31 (3) (1950): 247–59; J. Erb, N. C. Stenseth & M. S. Boyce "Geographic Variation in Population Cycles of Canadian Muskrats (Ondatra zibethicus)," *Canadian Journal of Zoology* 78 (6) (2000): 1009–16.

smaller on average than those born in good years, while litters born in the trough phase of the cycle are up to two pups smaller.

Fig. 8.1. Vital statistics of Iowa muskrats.[4] In the growth phase of the cycle litters are larger, disease resistance is stronger, and more animals breed in the year of their birth. All these factors work together to promote population growth.

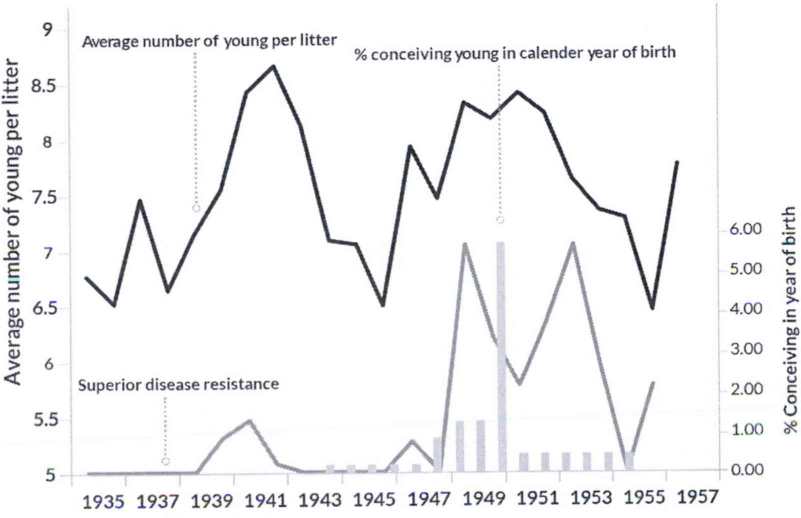

The same applies to death rate. Because disease and other forms of mortality are affected so much by environmental conditions there is no clear cyclical pattern in the total death rate, but deaths from hemorrhagic disease clearly show the effects of cycles. Muskrats in the decline phase of the cycle are far more likely to die from it than muskrats in the growth phase.

Another significant observation, in terms of biohistory, is that muskrat territory sizes are much larger in the decline phase. When large numbers of muskrats remain in an area, breeding territories are distributed with striking uniformity throughout good and poor regions alike. They prefer to maintain themselves in inferior places rather than crowd the better ones. By contrast, in growth periods muskrats pack into the most attractive cattail and bulrush stands so that much of the second-class marshes are empty, or nearly so. A decline in territoriality is also a feature of snowshoe

---

[4] Errington, *Muskrat Populations*, 56, 62, 530, 532.

hares. Most years they are intensely territorial, but in the growth phase they have been seen to migrate in huge numbers.[5] Animals in the growth phase are also a lot less wary.[6]

These cycles have been observed in a number of northern mammal and bird species including lemmings, muskrats, grouse and varying hares, plus the species preying on them (though with predators, population changes may simply depend on the availability of food).

A way of summarizing these findings is that animals in the growth phase act like high-V species—bold, social, migratory, fast breeding and resistant to disease. Those in the decline phase act to some extent like high C species—timid, slow breeding, solitary and territorial.[7]

We have seen that different species experience cycles of different lengths. As a general rule, more northerly species cycle every three to four years and show massive changes in numbers. Population peaks may be hundreds of times the crash level. In more southerly regions the highs occur only every ten years, and both build-ups and crashes are less severe. This difference is found even within some species. Animals may have three-to-four-year cycles in the north and ten-year cycles elsewhere. In other words, a harsher climate is associated with larger and more frequent cycles. Consistent with this is that captive animals of the same species, which presumably have unlimited food, do not show cyclical behavior at all.

Smaller species, and those which breed most rapidly, are more likely to show 3–4 year cycles. There is also some evidence that larger species such as elk and moose have even longer cycles, with one authority claiming that the only non-cycling species he could identify in North America was the beaver.[8]

---

[5] Keith, *Wildlife's Ten-Year Cycle*, 89.

[6] Ibid., 96.

[7] Note that the link between high C and population growth in human societies does not apply to animals, and nor should we expect it to. The two reasons given for the link in humans is that high C people create better economic conditions and have more interest in offspring. High C animals do not materially change their environments, and interest in offspring only increases the birth rate because people who lack an interest in children can limit births.

[8] C. H. D. Clarke, "Fluctuations in Populations," *Journal of Mammalogy* 30 (1) (1949): 21–5.

A further important observation is that the lengths of the cycles tend to be the same in different populations of a species over very wide areas, but local peaks and troughs occur in different years. In one part of the continent all populations of a particular species tend to peak and trough together, with the peaks becoming progressively earlier or later in more distant regions.[9]

Population cycles in animals are commonly called "ten year cycles," because that is the most common length. In fact they vary considerably—as low as 3–4 years among lemmings and some other species, while elk and moose populations appear to follow cycles that are much longer than ten years. So because lemmings are the best known of the species that go through these cycles, biohistory refers to them as "lemming cycles."

Explanations for lemming cycles

These cycles are one of the great mysteries of biology, and although various theories are proposed they have no accepted explanation.[10] Early explanations centered on population density, proposing that dense populations brought about stress which caused numbers to collapse, and that low density reduced stress and allowed populations to recover.

But repeated observations have shown that the cycles occur independently of actual numbers. For example, all local populations in the "decline" phase show the signs that typify it, such as poor disease resistance, even though in some areas the numbers are kept low by drought and have never peaked. In other words, populations tend to crash together over a wide area, even though population densities in some localities may be quite low to begin with. And crowding is considerably higher in captive animals of the same species which do not cycle at all. Exhaustion of local food supplies may be ruled out, since populations still crash when there is plentiful food. Nor do either of these explanations show why behavior should be so different in growth and decline phases, or why the cycles are so strikingly regular. What this means is that population levels seem to be the *result* of the cycle rather than the *cause*. It is as though all factors

[9] Keith, Wildlife's Ten-Year Cycle, 70.

[10] For example, theories of predator-prey relations: M. K. Oli, "Population cycles of small rodents are caused by specialist predators: or are they?" *Trends in Ecology and Evolution* 18 (3) (2003). Some other models: A. Lomnicki, "Why do Populations of Small Rodents Cycle? A New Hypothesis with Numerical Model," *Evolutionary Ecology* 9 (1) (1995): 64–81.

conspire to make rapid growth in one phase and decline in the other.[11]

The biological function of lemming cycles

Biohistory proposes that lemming cycles are a strategy to cope with environmental extremes. Small rodents in harsh climates live in a dangerous world. A bad winter can wipe out a local population entirely and the best wintering spot may be anywhere, even a place that was formerly inhospitable. It is advantageous to have a large number of very bold animals which can periodically move out and colonize new lands.

But this creates a serious problem, in that bold animals are at risk from predators and will likely get eaten long before they peak. The advantage of the "decline" phase is that it frustrates predators, which tend to breed more slowly. Predator numbers fall dramatically in the trough years, when the prey are timid and few, and they cannot breed fast enough to take advantage of the peak years when prey are numerous and bold. This allows a truly massive buildup of very bold animals, intolerant of stress and thus primed to migrate. The harsher the environment the greater the need for migration, which is why cycles are bigger and more frequent in northerly areas.

What are lemming cycles?

As discussed at the start of the chapter, biohistory proposes that lemming cycles represent fluctuations between V and C in a multi-generation cycle. All aspects of cycling behavior can be explained in this way (see Table 8.1 below)

Growing populations act like baboons, with high V. They breed fast and are bold. They are gregarious and can form dense populations, but they have an underlying intolerance for crowding which in time causes them to become stressed and thus to migrate. This seeming paradox—that an individual can prefer denser populations but be stressed by them—will be discussed in chapter ten.

---

[11] **Keith, *Wildlife's Ten-Year Cycle*; Errington, *Muskrat Populations*;** Christian, "The Adreno-Pituitary System and Population Cycles in Animals," 247–59.

Table 8.1: Behavioral characteristics of lemming cycles. Peak populations have V characteristics and trough populations C characteristics.

Growth phase	Decline phase
Higher birth rate	Lower birth rate
Lower death rate	Higher death rate
Bolder	More timid
Less Territorial	More territorial
Migratory (more stressed by crowding)	Less migratory (less stressed by crowding)

The decline populations act like gibbons, with high C. They breed slowly and are timid and territorial, all of which makes life difficult for predators. A territorial animal knows its environment very well, so that the slightest disturbance can send it dashing to safety. The only aspect of the trough phase not consistent with C is the higher death rate, since food-restricted animals tend to live longer. But this is likely the obverse of the V effect. All this does not explain what drives these cycles, but it does suggest further avenues for investigation.

First, it suggests that animals collected at different phases of the cycle should have distinct epigenetic signatures, with a V pattern in the growth phase and a C pattern in the decline phase. Second, it indicates that all mammals have the potential for lemming cycles and that they are driven by periodic severe stresses—not at any particular stage of the cycle but as a general environmental condition. Some cycling animals do not appear to cycle in the more southerly sections of their range, and non-cycling animals may cycle when placed in a cycling environment—as happened when the Hungarian partridge was introduced to western Canada.[12] In fact, biohistory proposes that most natural populations have a cycle length in proportion to their generation length, but the less pronounced cycles are dwarfed by other factors affecting mortality. As noted earlier, there is evidence for such cycles in larger species such as elk and moose.

But the main evidence for this hypothesis is that humans also have lemming cycles. Societies all over the world show clear evidence of lemming cycles at every stage in their history, quite independent of longer-term changes in V and C. In fact, lemming cycle fluctuations in V and C do not seem to affect or be influenced by the underlying levels of V and C,

---

[12] Keith, *Wildlife's Ten-Year Cycle*, 25, 116.

except in the sense (as will be seen) that increases in V and stress in the civilization cycle seem to make lemming cycles longer. Lemming cycles and civilization cycles appear to operate at quite different levels, much as waves on the surface of the sea (lemming cycles) do not influence the tides (civilization cycles).

It is also likely that changes in C driven by lemming cycles are much smaller than those associated with civilization cycles. Thus it is that a society such as Egypt with low infant C remains such even at the stage of the lemming cycle when infant C is highest. For example, it does not become noticeably better at industrialization. Confirmation of all this will need to wait for experimental evidence.

Modeling the lemming cycle

On the other hand, our understanding of V and C tells us what lemming cycles should look like in human populations, presuming they can be identified. The first point to consider is their length.

One persistent feature of lemming cycles is that they are uncannily regular. Lemmings and some other species, especially those in harsher environments, cycle every three to four years. Then there are species such as muskrats and varying hares that normally cycle at ten-year intervals. If humans show a similar cycle it would most likely be a longer one, because most human populations do not live in harsh and marginal habitats.

The lemming cycle is a multi-generation mechanism, and to compare a human cycle with that of muskrats we must factor in generation length. A muskrat matures at around six months, compared with a human around thirteen years. Taking this as a guide we might anticipate a human cycle length about 26 times a muskrat one, or 260 years. Alternatively, given that muskrats rarely breed in the calendar year of their birth and suffer very high mortality, a ten-year muskrat cycle probably represents close to ten generations, which with a human generation length of thirty years would make a three-hundred-year cycle. In fact, as we will see, human lemming cycles vary quite consistently in length depending on the stage of the civilization cycle, but most are around 300 years, and the shortest around 260 years.

In animal populations the bold, migratory behavior typically happens at the peak of population which is, by definition, the generation after the peak of growth. Thus, we might expect more aggressive behavior in the human generation born at the peak of growth. This is also consistent with the idea, as indicated earlier, that children born to larger families should have higher V because they are more likely to be supplanted at weaning age by a younger sibling. The next generation after that should show a peak of child V, because parental punishment reflects the V of the previous generation.

At the opposite or "decline" phase of the cycle we should see a low point of growth, with lower birth rate and more deaths from infectious disease. A generation after this should come a peak of C, and then a generation later a peak of infant C. This is because high-C parents tend to control their children, maximizing infant C.

Lemming cycles in human populations—initial model and predictions

To make this theory precise we need to consider exactly *when* each form of behavior should peak, in relation to the time of maximum growth. Referring to the growth year "G" and assuming a thirty-year human generation, the model is given in Fig. 8.2 below.

The lemming cycle is divided into ten sections of thirty years each, rated as to their time from the G year. The peak of V is 30 years after the G year or G+30, and child V a generation later at G+60. C peaks at the opposite end of the cycle to V at G-120, and infant C a generation later at G-90. Since we already know a great deal about these variables, the behavior and attitudes of people at different points in the cycle are as follows.

The G generation is the one where parents have the maximum number of children and death from disease is at a low point. The G+30 generation is the peak of V. High-V people are confident, aggressive, patriarchal, energetic and warlike.

The G+60 generation is the peak of child V and the low point of infant C. This is the personality profile characteristic of the Egyptian village described in chapter five, with no training of infants and the strict discipline of older children. Such peoples have weak local loyalties and accept any sufficiently powerful authority, even if alien in culture or language. As a result they tend to form large, cosmopolitan empires. They also tend to have rigidly conservative values.

Fig. 8.2. Lemming cycle initial model. A generation after the peak of growth (G) is the peak of V at G+30, then of child V at G+60. A generation after the low point of growth is the peak of C at G-120, and then the peak of infant C at G-90.

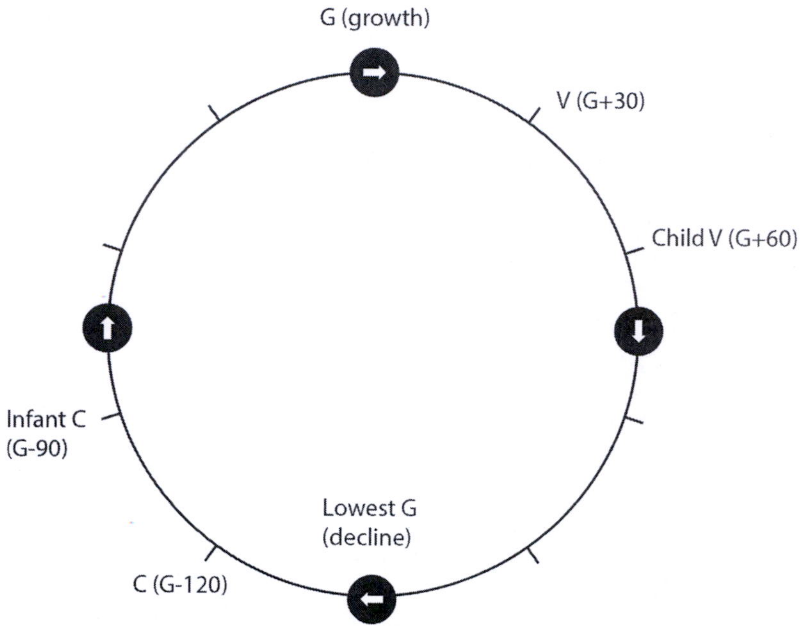

The G-150 generation is the low point of G. Birth rates should be lower and death rates higher. The G-120 generation is the high point of C and the low point of V. People should be economically skilled and least likely to engage in wars of expansion.

The G-90 generation is the obverse of G+60. Infant C is high and child V low. This means that loyalties should be intense and local, with a strong preference for leaders who are similar in culture and language. Distant authorities, no matter how powerful, will gain little obedience. Quite obviously, this should strengthen the power of local leaders. And given the strength of the hereditary principle in societies with higher infant C, as indicated by the prevalence of figurehead rulers, these local leaders may well be hereditary nobles.

These findings are summarized in Table 8.2.

Table 8.2. Predicted behavior and attitudes at different stages of the lemming cycle.

Period	Traits	Behavior and attitudes
G	G	High birth rate, low death rate
G+30	V	Aggressive warfare, migration
G+60	Child V, lowest infant C	Autocracy, weak local loyalties, tradition
G-150	Low G	Low birth rate, high death rate
G-120	C, low V	Economic growth, least warlike
G-90	Infant C, lowest child V	Local loyalties strong; autocracy weak, open to change

These predictions will now be tested against actual social and political changes in Britain over the past thousand years. Though the findings are not completely in line with the above predictions, they form a consistent pattern which is in most ways remarkably close.

Lemming Cycles in Britain

The first thing to do is chart patterns of population growth and decline, something we can do with relative certainty even though British demographic information is only approximate until the early-nineteenth century. Population seems to have been growing from the late eleventh century and especially in the twelfth and thirteenth centuries. It declined in the late fourteenth and early fifteenth centuries, and seems to have grown somewhat in the late fifteenth century but much more in the sixteenth and early-seventeenth centuries. It was static in the early-eighteenth century and then grew to the early twentieth, especially in the nineteenth century.[13]

---

[13] H. A. Cronne, *The Reign of Stephen: 1135–54* (London: Weidenfeld & Nicholson, 1970), 17; W. G. Hoskins, "The English Landscape" in *Medieval England*, edited by A. L. Poole, 10,12,15,19 (Oxford: Clarendon Press, 1958); J. C. Russell, *British Medieval Population* (Albuquerque: University of New Mexico Press, 1948), 235, 259, 260; T. H. Hollingsworth, *Historical Demography* (New York: Cornell University Press, 1969); L. A. Clarkson, *The Pre-Industrial Economy in England: 1500–1700* (London: Batesford, 1971), 27–8; I. Blanchard,

Allowing for a general trend for faster population growth, these growth periods center around 1230, 1550 and 1850, which are thus our G years. Note that these were not necessarily the exact decades when population was growing fastest. In the nineteenth century this actually happened in the 1810s and 1870s, with a mid-century dip. The "dip" is the result of a third cycle, the same one responsible for the post-war "baby boom" in Western countries which will be discussed in chapter ten. In effect, mid-nineteenth century population growth was the result of a population peak of the civilization and lemming cycles, combined with a trough of this shorter cycle. But the nineteenth century was generally a time of very rapid growth.

We can place the midpoints of slowest growth or decline at 1400 and 1700. These are roughly midway between the periods of fastest growth, which is what the lemming cycle model would suggest (see Fig. 8.3 below).

Taking each cycle from the midpoint of the period of fastest growth we get two complete cycles: 1230–1550 and 1550–1850. But, unlike lemming and muskrat cycles, which are reasonably uniform, these would seem to be different in length. The gap between 1230 and 1550 is 320 years, while that between 1550 and 1850 is only 300 years. Though only a minor change, it would be stretching the evidence to make them the same. For example, a G period around 1250 would imply that the eleventh and fourteenth centuries were demographically similar when they were not. One experienced growth, the other decline. This issue will be dealt with later.

"Population Change, Enclosure and the Early Tudor Economy," *Economic History Review* 23 (3) (1970): 427–31, 435; J. J. Spengler, *France Faces Depopulation* (New York: Greenwood Press, 1968), 53; L. Stone, *The Family, Sex and Marriage in England: 1500–1800* (London: Weidenfeld and Nicolson, 1977), 65, 67, 69, 71; M. D. George, *London Life in the Eighteenth Century* (New York: Capricorn, 1965); W. S. Thompson & P. K. Whelpton, *Population Trends in the US* (New York: Kraus Print, 1969), 266; A. J. Coale & M. Zelnik, *New Estimates of Fertility and Population in the United States: A Study of Annual White Births from 1855 to 1960 and of Completeness of Enumeration in the Censuses from 1880 to 1960* (Princeton: Princeton University Press, 1963), 21–23.

Fig. 8.3. Population growth and decline in Britain—1000–1900. Growth takes place in three distinct surges, separated by periods of static or declining population. This cuts across the trend to higher population growth associated with the long-term rise in C.

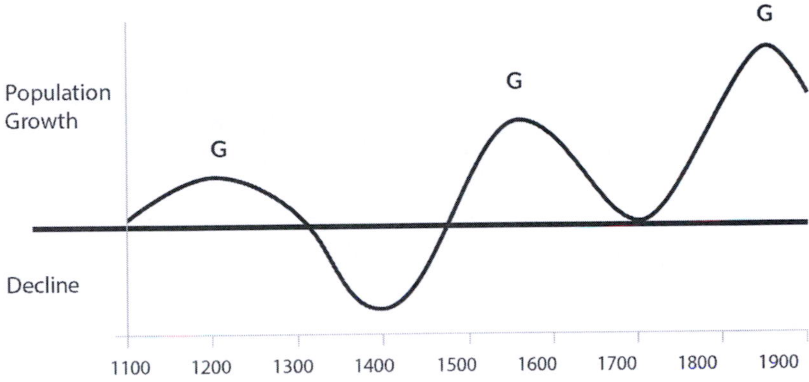

G-120—population decline or stagnation

The study of muskrats quoted earlier found that populations in the growth phase of their cycle not only bred more but had greater resistance to disease. It is thus of particular interest to see the same pattern in British populations. The birth rate was higher in the G period and lower in the G-150 period, and the death rate showed the opposite pattern.

One immediate application of this theory is that it gives us a clear and consistent explanation for that demographic catastrophe known as the Black Death. We have seen that the population fell quite markedly in the late-fourteenth and early-fifteenth centuries, surrounding the low point of V. The common explanation for this is the Black Death, which killed one third of the English population in 1348 and many more over the next century and a half. But this does not really explain the broader population trends of the time. The English population had been static or falling since at least 1300, well before the plague arrived.[14] It seems horrifying to us that one person in three should die of disease, but sanitation was poor in the fourteenth century. Towns crawled with vermin, and chamber pots were emptied into the street.

---

[14] L. R. Poos, "The Rural Population of Essex in the Later Middle Ages," *Economic History Review* 38 (1985): 515–30; E. Miller & J. Hatcher, *Medieval England: Rural Society and Economic Change, 1086–1348* (London: Longmans, 1978), 59–61.

A study of victims from a plague cemetery in East Smithfield found that they were more likely than the general population to show evidence of malnutrition or other weaknesses in their bones.[15] What this means is that although a greater number of otherwise healthy people died than would have been the case in normal years, many of the victims would likely have been short lived in any case.

Plague was not even the most common cause of death, though terrifying because of the sheer scale of the outbreaks. Most deaths at this time, especially those of children, were from other forms of infectious disease. Ten percent of undergraduates died while at university in the fifteenth century, and many noble houses died out because they were unable to rear sons past infancy. Neither of these groups were ill-fed, and the latter tended to live on country estates which were a lot healthier than the towns.[16]

Poverty cannot explain the mortality of ordinary people either. Real wages were considerably higher in the low-growth fifteenth century than before or after.[17] It is hard to escape the conclusion that people simply had less resistance to disease, just like muskrats in the decline phase of their own cycle. In fact, rather than poverty explaining the high death rate, the high death rate explains relative affluence.

People also seem to have had fewer children, just like muskrats during their decline phase. In part this was because of lower fertility. Many noble families of the fifteenth century were unable to have children. But with humans, unlike muskrats, the decline also seems to have been in part voluntary. Coitus interruptus was commonly used in the fourteenth century.[18]

The same pattern can be found in the demographic slump of the late seventeenth and early eighteenth centuries. Compared to earlier centuries this was a relatively prosperous time, and yet the population of Britain

[15] S. N. DeWitte & J. W. Wood, "Selectivity of Black Death Mortality with Respect to Pre-Existing Health," *Proceedings of the National Academy of Sciences USA* 105 (5) (2008): 1436–41.

[16] K. B. McFarlane, *The Nobility of Later Medieval England* (Oxford: Clarendon Press, 1973), 150, 168–70.

[17] G. Clark, *A Farewell to Alms: A Brief Economic History of the World* (Princeton, NJ: Princeton University Press, 2007), chapter three.

[18] P. Biller, "Birth Control in the West in the Thirteenth and Fourteenth Centuries" in Social History Society Newsletter (Spring 1980).

grew very little. Not only was the birth rate lower than in the sixteenth and nineteenth centuries, but the death rate was higher (see Fig. 8.4 below).

Fig. 8.4. Crude birth rates and death rates (per 1000) in England, 1541–1871.[19] As with muskrats, human birth rates are highest and death rates lowest in the G (growth) stage of the lemming cycle. The opposite applies in the low G stage, at G-150.

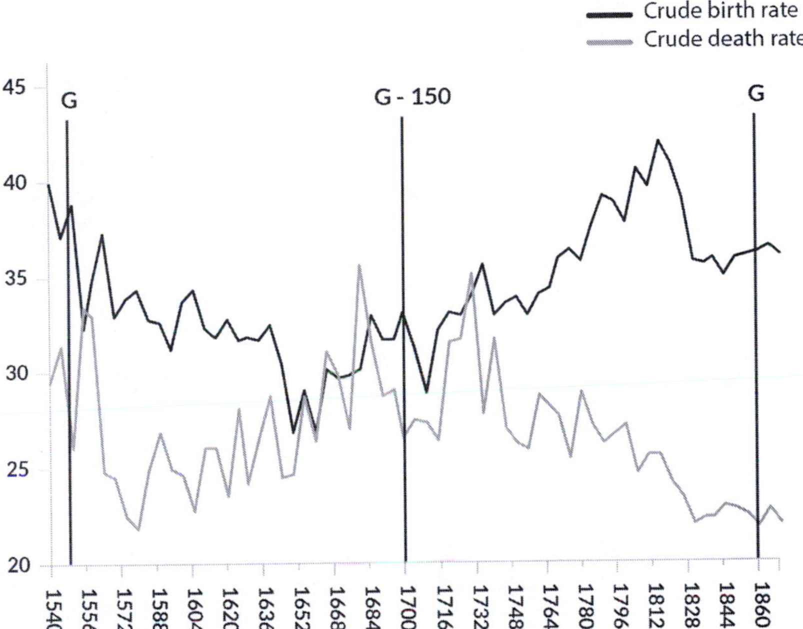

What makes this pattern even more striking, as in the earlier period, is that the prevailing economic conditions should have had the opposite effect. The late-seventeenth century was a relatively prosperous time in British history and yet the birth rate fell and the death rate rose. It is equally striking that the same pattern applies even to wealthy landowners and the sons of peers. Despite generally living in the countryside and with ample food, their death rate rose and life expectancy fell in the late seventeenth century (see Figs. 8.5 and 8.6 below).

[19] Wrigley & Schofield, *The Population History of England 1541–1871*, 528, 529.

Fig. 8.5. Median age of father's death of squires and above.[20] Even for wealthy landowners, the death rate is lower in the G (growth) period and higher in the G-150 (lowest growth) period of the lemming cycle.

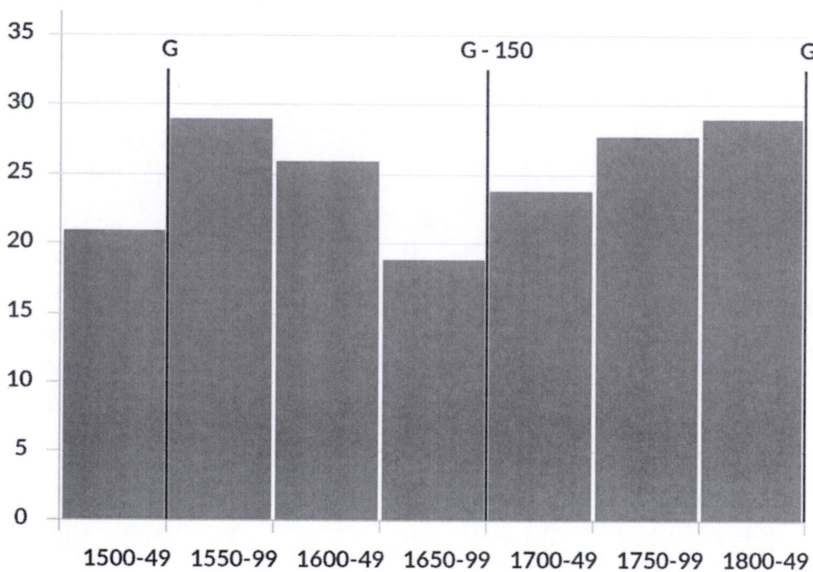

In the late-sixteenth century, 26% of the children of peers died before the age of 15 (Fig. 8.7). Between 1630 and 1730 more than 30% of them failed to reach this age, with a peak of 37% in the late-seventeenth century. This is extraordinarily high for the healthiest and best-fed families in England, living to a large extent in the country. From then on there was a steady drop to only 14% in the mid-nineteenth century, still high by modern standards but much reduced. The inhabitants of the village of Coylton in Devon showed a similar pattern, with a peak of mortality in the late-seventeenth century.

[20] Stone, *The Family, Sex and Marriage in England: 1500–1800*, 59

Fig. 8.6. Expectation of life at birth for sons of peers (noblemen).[21] Even the wealthiest people in the country die younger in the lowest G section of the lemming cycle.

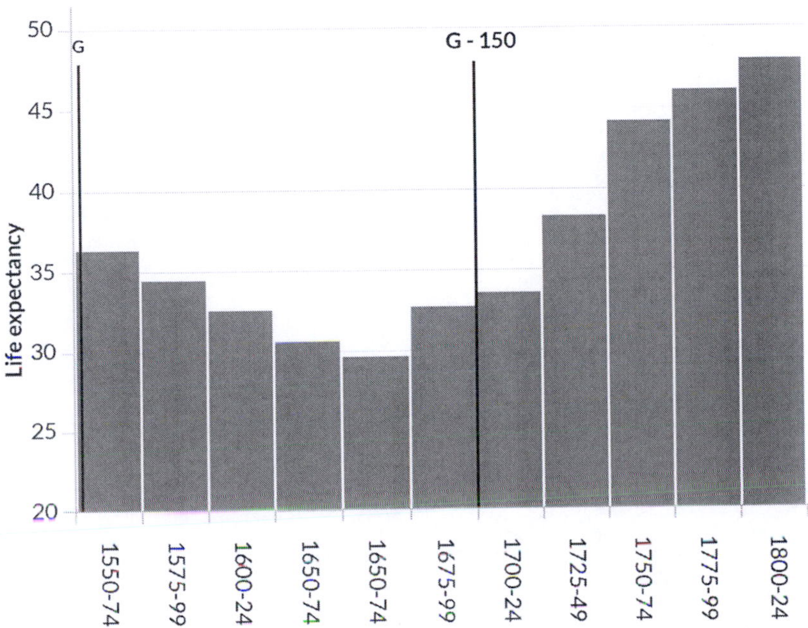

The only reasonable conclusion is that people in the late-seventeenth century seem to have been more susceptible to disease than those living before or after, regardless of economic trends.[22] This explains why the last great pandemic of the plague struck England in 1665, just thirty-five years before the G-150 year.

Birth rates were also affected. Peers and their families also had slightly fewer children in the late seventeenth and early eighteenth centuries than in the sixteenth and nineteenth centuries (see Fig. 8.7 below).

A similar pattern was found for squires, but with the lowest birth rate at the end of the seventeenth century. All of these patterns defy conventional explanations. Why should infectious disease be more of a problem in prosperous times and in people who had every opportunity to escape it?

---

[21] **Stone, *The Family, Sex and Marriage in England*, 71.**
[22] Ibid., 69.

Fig. 8.7. Mean number of children born to sons of peers.[23] There is a slight dip even in the number of children born to peers in the lowest G (growth) section of the lemming cycle.

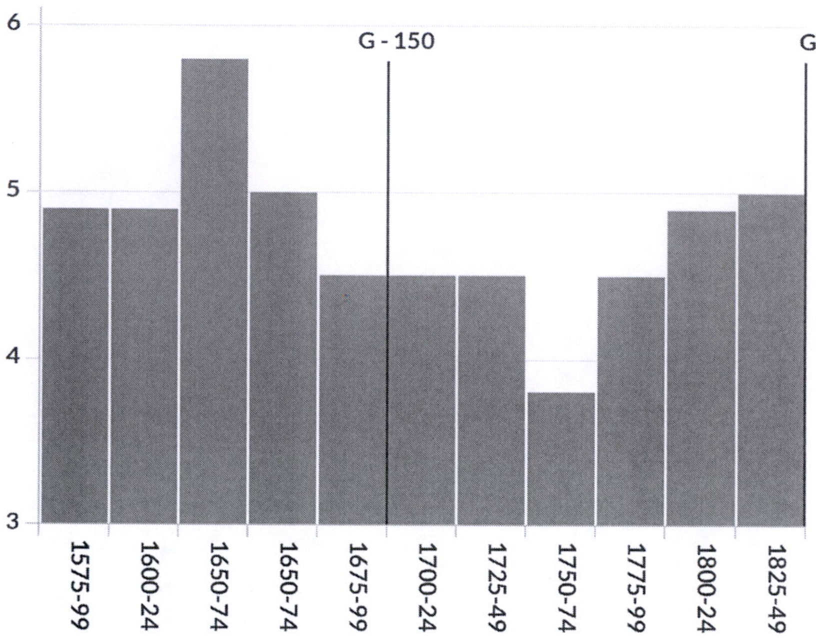

And while later marriage and contraception may explain a lower birth rate in the early eighteenth century, it does not explain the subsequent rise. But these are exactly what we would expect from a lemming-cycle pattern.

G period behavior—national unity and economic growth

Now we must turn to the behavioral aspect of the G period, which is somewhat different to what our initial theory predicts. High-G periods in British history show the confidence and energy we associate with V, but they are also characterized by something else—a strong and vibrant sense of national unity. This is not based on cowering submission to stronger authority but what appears to be positive patriotism. The best way we can explain it is that people in the G periods seem to want to support their rulers.

---

23 : Stone, *The Family, Sex and Marriage in England: 1500–1800*, 65.

The first G period in English history, based on political rather than demographic evidence, can be found in the late ninth century in the reign of Alfred the Great, around the year 880. After a long period of chaos when the various Anglo-Saxon kingdoms had been unable to fight off either Viking marauders or Danish invaders, the English were reduced to the kingdom of Wessex in the south-west. But within only a few decades they rallied, formed a powerful and (by the standard of the time) centralized resistance, defeated the Danes, and united the English kingdoms under Alfred's son Edward the Elder and his grandson Athelstan. Even given Alfred's remarkable character and ability, this was an astonishing achievement.

It is also characteristic of the lemming cycle that a *rapid rise* from anarchy to unity was followed by a *slow decline* into disunity and weakness, a pattern we will see repeatedly. The effect of the decline was to allow first Danes and then Normans to gain control. The first Norman kings were indeed powerful, as might be expected when almost the entire ruling class was replaced, but even this regime descended into anarchy within seventy years of the conquest.[24] This is the G-90 period of 1140, which we will discuss in more detail in the next section.

As always with lemming cycles, recovery of central authority after the G-90 period was swift. In the 140 years centered on the G year 1230, English monarchs such as Henry II, Richard I and Edward I were once again powerful and effective. Though rebellions still took place, this was in general a time of rapid advance for the rule of law. Partly as a direct consequence of political stability, it was also a time of economic growth.

But then, in the fourteenth century and especially after mid-century, central authority gradually weakened. There was a rise in outlaw gangs and growing disorder, growing worse under Henry VI (1421–61). Such disorder, and the population decline of the late-fourteenth and fifteenth centuries, took its toll on the economy. This came to a peak in the 1450s, consistent with a G-90 year of 1460.

Then began a swift return to central authority. Henry VII took power in 1485 and crushed further rebellions. He also enforced taxes and put the monarchy on a sound financial footing. His son Henry VIII (1491–1547)

---

[24] The same pattern can be seen when the Mongols conquered China in the thirteenth century—an initially strong regime of conquest collapsing into anarchy within two generations. See chapter fourteen.

was the most powerful monarch in English history, and Henry's daughter Elizabeth (1558–1603) one of the greatest. The Tudor period was also a time of economic resurgence, added to which were ardent nationalism and voyages of discovery.

The monarchy weakened in the seventeenth century as King and Parliament struggled for control, and the eighteenth century was dominated by landed aristocrats and the government was relatively inactive. This time there was no feudal chaos since C was at a much higher level, but aristocratic dominance is exactly what we should expect from a G-90 period around 1760.

And then came the nineteenth century, centered on the G year 1850, when Britain rose to become a world power. Government became more powerful than ever before, and the economy expanded rapidly. In chapter six it was suggested that the key reason for the Industrial Revolution was an unprecedented level of C and especially infant C, but the vigor found in societies at times of high G could also have been a contributing factor. In any case, the confidence and energy of the Victorian industrial age has been a byword ever since. The twentieth century did not bring about a weakening of government in any sense, but it certainly saw a decline in self-confident patriotism.

Thus two further elements must be added to the original model. The first is that the period around the G year is one of patriotism and national unity. And the second is that it is also the time when economic growth is at a maximum—not at the time of highest C which we would otherwise expect.

A way to understand both of these findings is that the rise and fall of C in the lemming cycle seems to be relatively small by comparison to its movement in the civilization cycle, while the rise and fall of G characteristics (unity and economic growth) are relatively large. Thus, C may be rising in a civilization cycle, yet its rate of increase will hasten or slow with lemming cycle changes. And when the two traits with opposite effects compete, so to speak, the G influence wins out in lemming cycles. Thus it is that the positive energy associated with G, plus restored national unity, more than make up for any loss of C.

Local loyalties and feudal disorder at G-90

The predicted G-90 points for Britain are 1140, 1460, and 1760.

Now we come to look in more detail at the phase of the lemming cycle when infant C is at a peak and child V lowest. As described earlier, people with high infant C have intense hereditary loyalties and prefer rulers similar to themselves in culture and language. At a time when regional dialects were much stronger and less mutually comprehensible than today, this gave a powerful advantage to local leaders. It is also the time when child V is lowest, meaning that people are least likely to accept powerful authority. This makes kings weak and local aristocrats strong—the exact opposite to the G+60 year.

The most common effect of powerful barons and weak kings is anarchy. Between 1100 and 1135 England had a powerful king in Henry I, youngest son of the conqueror. But in 1135 he died without a male heir and his nephew Stephen seized the throne, overriding the claims of Henry's daughter Matilda. For seventeen years armies marched back and forth, towns were sacked and crops laid waste while "Christ and his saints slept."[25] But this was not a single war where two large armies faced each other in the field. It was a maze of local conflicts in which each side would align themselves with a royal faction. Nobles were free to defy central authority because local people followed them rather than the king.

This period of feudal chaos exactly brackets the G-90 year of 1140. The breakdown of authority was sudden, rather than gradual as we might expect if it simply reflected the changing temperament of people in the early eleventh century. But the entire ruling class had been replaced only seventy years earlier after the Norman invasion of 1066, so perhaps not fully influenced by the local lemming cycle pattern.[26] Also, Henry I was an effective and brutal king, even going so far as to blind his own grandchildren to punish their father's treachery. We will find other examples of a ruler's personality affecting the exact timing of political events, especially in Chinese history, though not their eventual occurrence.

---

[25] J.A. Giles & J. Ingram, *The Anglo-Saxon Chronicle* , http://www.gutenberg.org/ebooks/657 (retrieved 13 September 2014)

[26] A similar pattern will be applied to the Mongol conquest of China in the thirteenth century (see chapter fourteen). Taking power in the decline phase of a lemming cycle, the Mongol rulers were initially powerful. But within a couple of generations they succumbed to the typical chaos of the Chinese G-90 period.

The next cycle fits the lemming cycle model exactly. The next G-90 year is 1460. This is near the midpoint of decades of disorder, starting when Henry VI came to the throne in 1422 and concluding with the defeat of Richard III at the battle of Bosworth Field in 1485. Rival branches of the English royal house spent several decades fighting for the throne in what became known as the Wars of the Roses, with the worst turmoil occurring in the late 1450s.

The common explanation for such periods of disorder is political. For example, Henry I died without a male heir and, in a patriarchal society, there were objections to his daughter succeeding him. Henry VI came to the throne as a child and was a weak ruler, unable to control his turbulent barons. But this is not an adequate explanation. Henry III in the mid-thirteenth century was a weak character briefly deprived of power by Simon de Montfort. This involved armed conflict, but not the violent chaos of the mid-fifteenth century. Similarly in the mid-sixteenth century, Henry VIII was succeeded by a child but no insurgency occurred. His successor Mary was deeply unpopular as a Catholic, and became notorious as "Bloody Mary" for burning Protestants, but no effective rebellion was raised against her. An attempt to enthrone the Protestant Lady Jane Gray to forestall Mary's coronation simply collapsed.

Close examination of these times of feudal disorder show the strength of local loyalties, which is exactly what biohistory predicts. The Paston letters, written by a landowning family in fifteenth-century Norfolk, give a vivid picture of society at this time. The fourth of these letters concerns events in 1423, decades before the national conflict began. William Paston had become involved in a lawsuit against one Walter Aslak, who was notorious for having kidnapped and butchered a local man and members of his household. To silence his enemy, Aslak posted notices in prominent places—including cathedrals—threatening to kill Paston in a similar manner if he stirred from home. Paston tried both legal remedy and an appeal to the Duke of Norfolk but failed totally. Aslak was protected by a powerful patron.

This was no dispute between claimants to the throne, because there was none at this time. In fact, royal power was irrelevant to the dispute. And this is precisely the point. When people are loyal to those they know, local leaders are stronger and the central government weaker. This is why Paston appealed to the Duke of Norfolk, and Aslak to his own patron. The victor was the man with the stronger patron, or in this case the more interested one.

Looking beneath the big events of such times, we see a profusion of local conflicts. Ambitious men took the opportunity to seize neighboring lands. To gain support, they might align themselves with one or other royal faction. Often they changed sides in a bewildering series of turnarounds. The most famous of such characters in the fifteenth century was the Earl of Warwick, already discussed in chapter seven. He was the most powerful supporter of the house of York but later abandoned the Yorkists and (briefly) restored the Lancastrians to power, earning his title of "Kingmaker." In a similar way, Lord Stanley decided the battle of Bosworth Field in 1485 by switching his support to Henry Tudor. This happened during the actual battle, and it was never doubted that his followers would fight for whichever claimant he chose. In the event, the crucial factor in Lord Stanley's decision was probably his marriage to Henry's mother.

In such times, ambitious men turn against their monarch if it suits their interests, and they are able to do so because their followers are more loyal to them than to the ruler. As long as a royal pretender has some ghost of a legitimate claim, a local baron can switch sides and be confident of taking his vassals with him, which will depend in turn on him supporting his followers in their own conflicts. Tenants and retainers followed their lord both for support in times of need, and because there was a deep-rooted assumption in English culture that they should do so.[27]

This attitude of service and support was not, of course, unique to the fifteenth century—it was characteristic of the entire Middle Ages and still a factor in the English Civil War of 1642–49. What made the fifteenth century different was that the relative strengths of loyalties had shifted towards local leaders, so that even an effective and capable king like Edward IV could be removed (temporarily) from his throne. No state regime can be totally secure when primary loyalties are local.

National peace returns only when primary loyalties begin shifting once more towards the center, a change of which Henry Tudor was the fortunate beneficiary. In our own day, of course, this movement of loyalties to the center has happened to an extreme degree. If the local authority in a modern English town tried to raise a rebellion against the government, the leaders would very likely end up in a mental hospital, such is our complete acceptance of government authority. When it comes to rebellion the

[27] R. Horrox, Fifteenth-Century Attitudes: Perceptions of Society in Late Medieval England (Cambridge: Cambridge University Press, 1994), 50.

crucial point is not what a leader chooses to do but whether others choose to follow.

We have already noted that while the monarchy only gradually weakened during the fourteenth and fifteenth centuries to a low point in the 1450s, the restoration of central power was swift. Henry Tudor was only one of a series of claimants to the English throne, and his claim was remarkably weak.[28] Yet within a very short time he had crushed any remaining opposition and enforced taxes (against bitter objections) that put the monarchy on a sound footing.

The next G-90 year is 1760. This was a time of peace and stability and so, on the surface, quite distinct from the 1130s or 1450s. But the mid-eighteenth century was a time when the English landed aristocracy and gentry were serenely in control of both their regional estates and, through Parliament, the national government. In 1760, the last shred of the monarch's power was removed with the surrender of the Crown Estate revenues to the government. Among many other measures which showed the power of the aristocracy, a ban on imported wheat kept farm profits (and food prices) high. The same web of patronage and loyalty existed as in the Wars of the Roses, but exercised more peacefully given the higher levels of C at the time.

Biohistory proposes that the strength of any person or institution of government reflects the loyalties of ordinary people. When loyalties focus on the king, the king is powerful. When they focus on a parliament, the parliament tends to be in control. And when they focus on local aristocrats, the aristocracy is strong. Thus the strength of the eighteenth-century aristocracy was made possible by the loyalty and respect of ordinary people, just as it had been in the 1450s and 1140s. But because C was now considerably higher, this did not result in feudal chaos; instead, the aristocratic classes formed the national government in Parliament.

This pattern of feudal disorder at G-90 is by far the most obvious sign of the lemming cycle, and we will use it later to track these cycles over thousands of years in many different societies. As we have seen repeatedly, lemming cycles typically appear as a slow descent into disorder, followed by a rapid resurgence of central authority with the

---

[28] Being descended from a mistress of Edward III, whose children were legitimized with the sole stipulation that neither they nor their descendants have any claim to the throne.

exuberant nationalism of the G period. The time gap between feudal anarchy and unified nationalism is strikingly uniform, even when cycles are longer. The G-90 point—the time of disorder ninety years before the peak of restored authority—is by far the best indication of lemming cycles we will find. Intellectual trends and attitudes to religion may not always be evident, but historians can hardly miss a society falling into feudal chaos.

G+30—aggressive warfare, migration

The predicted G+30 points for Britain are 1260, 1580 and 1880.

One characteristic of the lemming cycle is mass migrations, in which otherwise timid animals become bold and spread out in search of new places to live. Lemmings do this just after the peak of population growth—the G period. In humans, this boldness and migration urge is expressed largely by wars of conquest or expansion into new territory, which as we will see in the next chapter are typically a generation after the peak of population growth.

This approximately fits our pattern. Most medieval English kings fought aggressive wars, up to and including Henry V in the early fifteenth century. But by far the most successful was Edward I (1272–1307) who conquered Wales in the 1280s and even briefly subdued Scotland in the 1290s. This is sixty years after the G year rather than thirty, but Edward was a far more forceful figure than his father and, more importantly, benefited through a sharp decline in local loyalties as we will discuss in the next section.

English territorial ambitions, or at least their effectiveness, ebbed after the death of Henry V in 1422. For more than a century the English made no serious attempts to conquer anyone, and gradually lost their possessions in France. English territorial policy became more aggressive in the Elizabethan period, with privateers such as Francis Drake leading attacks on Spanish shipping and colonies. In 1588 this policy provoked the Spanish to launch their Armada (unsuccessfully) against England. It is also just after this period that the mass migrations began which would turn North America into a British colony.

The next G+30 period, in the late nineteenth century, will be the subject of the following chapter.

G+60—weak local loyalties, acceptance of powerful authority

The predicted G+60 points for Britain are 1290, 1610 and 1910.

We have seen that in the G-90 period, when infant C is high and child V low, local aristocrats are strongest and central government weakest. The G+60 period is the opposite situation, when infant C is low and child V high. We would expect local loyalties to be weakest and acceptance of powerful authority strongest at this time, and this is what we find.

The reign of Edward I (1272–1307) fits this pattern exactly. Edward's successes against Wales and Scotland undoubtedly owed much to his desire for imperial glory, but there was another and probably more important factor—a decline in local loyalties. Llywelyn ap Gruffudd had been recognized as Prince of Wales in 1267, but in 1274 his own brother Dafydd and another Welsh prince defected to the English. When Edward invaded Wales in 1277, more than half his force was Welsh. It was only the imposition of English law that roused widespread resistance, crushed with the death of Llywelyn and his brother by 1283.

This is even clearer in the case of Scotland, the original "conquest" of which was achieved less by armed force than by making claimants to the Scots throne promise fealty in return for English support. Robert the Bruce, who eventually made himself king of an independent Scotland, was actually an English supporter for most of Edward's reign.

Another factor behind Edward's strength was his ability to cow the English nobility, which allowed him to raise much heavier taxes than in earlier times. This was achieved partly by giving the Commons their own representatives in Parliament, but the English nobility probably had less clout in his reign than in any period between the Norman Conquest and the Tudor age. All of this evidence suggests a low point of local loyalties around the G+60 year of 1290.

The same pattern can be seen in the next cycle. In 1603, just before the G+60 year of 1610, the English and Scottish crowns were united when James VI of Scotland was crowned James I of England. The acceptance of this change by both countries, after centuries of warfare, is a powerful indication of reduced local loyalties. It is also characteristic of the lemming cycle that the opposition to the early Stuart kings was based not on local barons but on a competing center of power, which was Parliament. The absolutism of the G+60 year can be powerful but is potentially less stable, because it lacks the positive nationalism of the G period. Lacking a standing army to enforce his rule, the king soon became fatally weakened.

A decline in local loyalties is also a factor at the G+60 point in the next cycle, though in this case it applied beyond national borders. The period immediately after the G year of 1850 was one of rising nationalism, but only nine years after the G+60 year of 1910, the League of Nations was founded, a clear indication of a reverse direction. The growing internationalism and rise of the European Economic Community in the second half of the twentieth century will be discussed later as a product of the civilization cycle, but the start of this trend can be attributed at least partly to a lemming cycle pattern.

Overall, the G+60 year shows clear signs of weaker local loyalties, and of what may be termed "unstable absolutism"—that is, absolutism that is strong only when the ruler is clearly the most powerful source of authority.

Religious independence at G-60; orthodoxy at G+90

Next we will turn to religious and intellectual trends, which also show strong lemming cycle patterns. There are two factors that determine independence of thought versus tradition. One is infant C, which peaks at the G-90 period and should promote innovation, and the other is child V which peaks at G+60 and should promote tradition. From this we would expect the most independent thinking to be at G-90, but in fact it can better be placed a generation later around G-60. Similarly, the peak of religious orthodoxy tends to be found at G+90, rather than at G-60 the peak of child V. The reason may be that G itself has an effect on thinking, just as it does on politics. Perhaps the energy and optimism of the G period also promotes independent thinking, or the more settled state of society contributes to intellectual endeavor.

But in any case, the pattern is quite consistent. At G-60 people tend to be independent in their religious thinking, able to change rapidly and with little concern for orthodox theology. Priests tend to have a lower status. The epitome of such attitudes would be non-conformist Protestantism. At G+90 theology becomes more rigid and entrenched, religious divisions more bitter, and priests gain status. This favors high church Anglicans, or Catholics.

Within the lemming cycles of British history, the peaks produced are:

Religious independence: (G-60): 1170, 1490 and 1790.

Orthodoxy (G+90): 1310, 1640, and 1940.

In twelfth-century England, the doctrine of the church was only vaguely defined. A large number of schools flourished and competed, and there was even some support for the idea of religious toleration. The founding of the Inquisition in the thirteenth century, which followed the G year 1200, marked the beginning of a change. Open debate was no longer to be allowed, and a massive structure of orthodox theology was developed. Without being too specific about dates, this is consistent with a change from unorthodoxy around 1170 to orthodoxy around 1290.

Moving towards the next G-60 year of 1490, the tide began to turn. Fifteenth-century Englishmen were religious in terms of attendance at mass and donations to chantries but there was a new mood of anti-clericalism, reflected in the Lollard movement and more generally in changed attitudes among the upper classes. Priests were considered something like the property of those who owned them, and not of especially high status. Sir John Falstolf, a knight of the Hundred Years War and a model for Shakespeare's Falstaff, paid his chaplain less than his cook.[29] Also, monasteries were in decline. All this suggests a turning away from orthodox religion consistent with a G-60 period around 1490, though this change was not to bear fruit until a few decades later. A study of Tudor wills in the early sixteenth century shows a decline in bequests to the church and to religious corporations such as monasteries and friars.[30]

The most striking evidence of religious indifference was the ease with which Henry VIII dissolved the monasteries in 1536–41, plundered their wealth and made himself head of the English church. This was some four decades after the G-60 peak, but Henry had an advantage of power that earlier kings could only dream of. It is hard to imagine Henry VII taking such action in 1490, his feeble claim to the throne made good only by the doubtful loyalty of feudal lords.

Attitudes to religious dogma were strikingly "flexible" in the period leading up to the G year of 1550. Until the 1550s most people were content to follow abrupt reversals in religion depending on the whims of

---

[29] Horrox, *Fifteenth-Century Attitudes*, 197.

[30] A. D. Brown, *Popular Piety in Late Medieval England* (Oxford: Clarendon Press, 1995), 226.

the monarch. Henry VIII broke with the Papacy in the 1520s but continued to enforce Catholic doctrine. After his death in 1547 his children steered the nation Protestant, then Catholic, then Protestant with some Catholic elements after 1558 (the "Elizabethan compromise"). But after the G year of 1550, people began to take religion a lot more seriously and their positions hardened. Catholics were seen increasingly as potential traitors, a prejudice not much helped when Catholic plotters tried to blow up the Houses of Parliament in 1605.

The early seventeenth century was a time of bitter religious division between Puritans and proto-Catholic followers of Archbishop Laud, which became one of the major drivers of the Civil War that broke out in 1642. This was also the time of the Counter-Reformation and religious war on the European continent, most notably the near-genocidal Thirty Years War (1618–48) that destroyed perhaps 30% of the population of Germany. All of this is consistent with a G+90 year in 1640.

In England, in the decades after the Restoration of 1660, there was a gradual slackening of religious fervor and the first stirrings of religious tolerance in attitudes towards the Quakers and others, leading to the markedly irreligious attitudes of the early eighteenth century. An index of religiosity can be found in editions of the Bible and New Testament during this period. There is an overall increase in the number and variety of editions, which can be explained by a growing and increasingly literate population. But within that trend is a clear lemming-cycle pattern. Bible editions peaked in the 1620s and 1630s, consistent with a G+90 year of 1640. The low point relative to this trend came in the 1740s and 1750s, some decades *before* the predicted low point at the G-60 year of 1790 (see Fig. 8.8 below).

But the religious fervor around the G-90 year was largely associated with Methodism, an unorthodox religious movement which often broke free from the established church. This is consistent with unorthodox thinking at such times. And as we pass the G year of 1850, Methodism declines and there is a steady increase in communicants to the Church of England, reflecting a trend towards increasing orthodoxy (see Fig. 8.9 below).

Fig. 8.8. Editions of the Bible and New Testament in England per twenty-year period: 1520–1798.[31] Religious fervor changes to relative religious indifference as the G-60 year approaches.

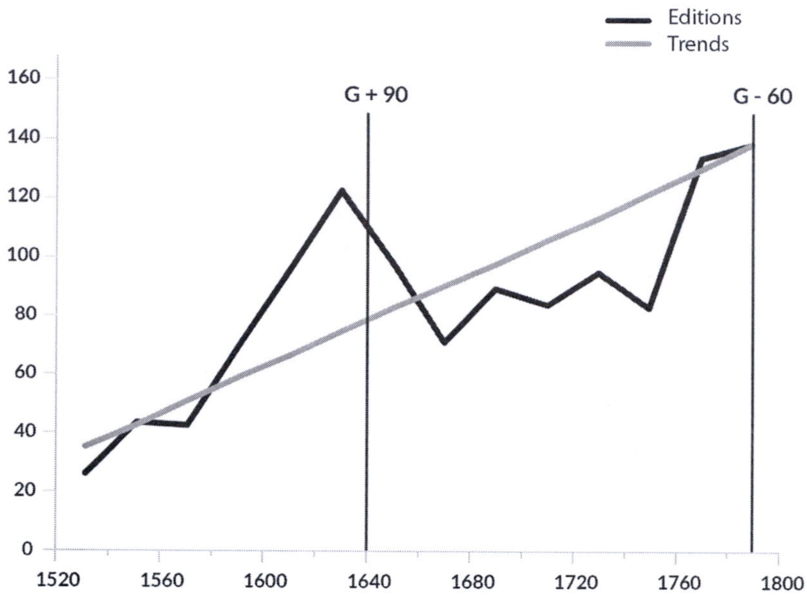

As with most aspects of the lemming cycle, this pattern was swept away by the massive changes of the twentieth century. In this case there was a rise in secularism associated with the civilization cycle, something we will examine in a later chapter.

In another trend reflecting the move towards orthodoxy, the Church of England itself became more tradition-minded over the course of the nineteenth century. Starting in the 1830s the Oxford Movement began to place more emphasis on liturgy and priestly authority, and clergy influenced by them were in the majority by mid-century. This was largely at the expense of evangelicals, who put a greater stress on personal morality and individual scripture reading.[32]

---

[31] H. Kearney, *Gentlemen and Scholars* (London: Faber & Faber, 1970), 40–1.

[32] S. Mitchell, *Daily Life in Victorian England* (Westport: Greenwood Press, 1937), 246.

Fig. 8.9. Percentage of over-15 population attending Anglican Easter Communion, and registered Methodists.[33] Moving from the G-60 year of 1890 to the G+90 year of 1940, attendance at the "established" Church of England increased relative to Methodism.

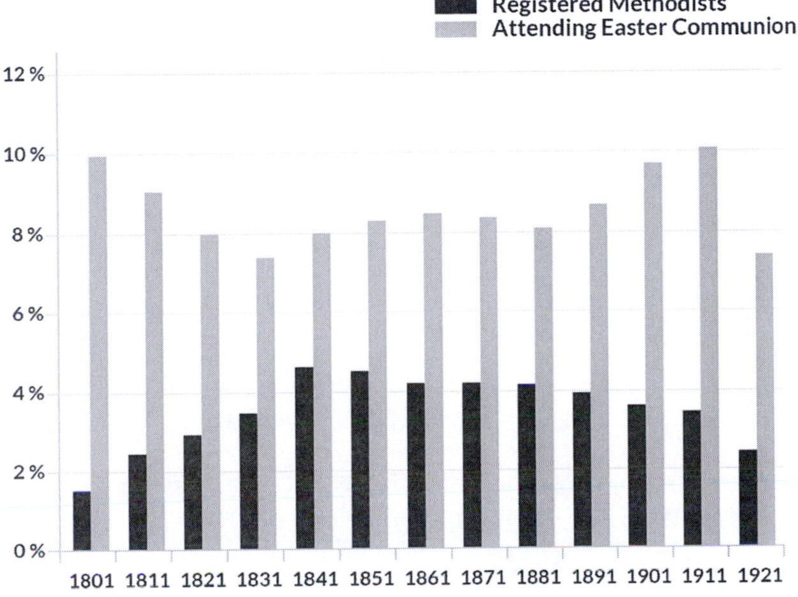

Renaissance and Enlightenment at G-60

The unorthodox thinking of the G-60 period applies to areas other than religion. The term "Renaissance" has been applied to two periods of cultural vitality which occurred in the twelfth and early-sixteenth centuries in England. These were eras of intellectual ferment, of contending ideas and change, of the triumph of reason over tradition. This phenomenon fits quite well into the G-60 years of 1140 and 1490, though the full flowering of the English Renaissance was somewhat later than 1490, perhaps due to greater wealth and security as England prospered under the Tudors.

The "Enlightenment" of the late eighteenth century was another movement in thought which fits this pattern and is appropriate to a G-60 year of 1790. Traditional ideas were being challenged by radical thinking, which was

---

[33] C. E. Gilbert, *Religion and Society in Industrial England* (London: Longmans, 1976), 20.

best exemplified by the Revolution in France.

Science in the G+90 period

There is one important exception to the association of the G+90 period with opposition to new ideas, and that relates to science. Though this period tends to be rigid and orthodox in the fields of theology, it does seem to favor science. Roger Bacon (1214–94) is often considered the founder of the empirical method, with an emphasis on testing theories through observation. William of Ockham (1288–1348) was a noted philosopher and the originator of "Ockham's Razor," the principle that we should prefer the simplest explanation for the observable facts. These two principles are among the foundation stones of modern science, and their originators bracket the G+90 year of 1310.

The seventeenth century was even more extraordinary, especially in the period around the G+90 year of 1640. William Harvey (1578–1657) first described the circulation of blood and the role of the heart. Robert Hooke (1635–1703) deduced the wave theory of light, and was the first to recognize that matter expands when heated and that air consists of small particles separated by relatively large distances. Edmund Halley (1656–1742) was an astronomer, geophysicist, mathematician, meteorologist and physicist best known for working out the orbit of the comet named after him. Most of all, Isaac Newton (1642–1727) is often considered the greatest scientist in history, developing the first great systematic theory with his laws of motion and theory of gravity, and making important contributions to optics and mathematics. His greatest work was done in the 1660s. The next G+90 period (1940 in Britain) was also a great period for science. Developments of the early-twentieth century included relativity, quantum mechanics, atomic physics, genetics and much more. If this pattern is genuine it is a little difficult, on theoretical grounds, to understand why it might be so. Science is one of foremost expressions of high C and especially high infant C, so could be expected to reach a low point at G+90.

Technical education at G-90; humanities at G+60

The level of interest in technical versus humanities education also follows a lemming cycle pattern, and this time a more predictable one. People around G-90, the peak of infant C and low point of child V, tend to prefer technical and practical subjects. At G+60, the low point of infant C and peak of child V, people tend to favor humanities subjects such as (in England) Latin, Greek, theology and history.

This can be most easily tracked by counting the number of students at Oxford and Cambridge, the main source for a humanities education. Enrollments of students at Oxford may have reached as high as three thousand in the early-fourteenth century (G+90 at 1290), falling to a low point of about 1,000 in the mid- to late-fifteenth century (G-60 at 1490).[34]

Fig. 8.10 below gives exact figures for Oxford and Cambridge from the sixteenth century. Enrollments were highest in the 1630s (G+90 year 1640) and lowest around 1760 (G-60 year 1790).

Fig. 8.10. Students entering Oxford and Cambridge per twenty-year period: 1560–1840.[35] The early seventeenth century (highest child V, lowest infant C) shows a positive attitude to classical education, the late-eighteenth century (lowest child V, highest infant C) a more negative attitude.

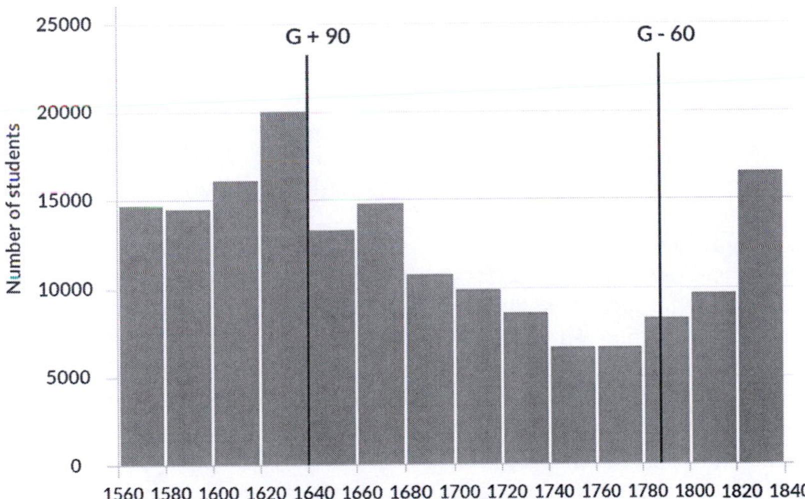

Education was highly valued in the early-seventeenth century, especially by the Puritans, with an Oxbridge degree considered an important measure of social status. This changed after 1660. Fewer people attended

[34] J. I. Catto & T. Evans, *The History of Oxford University*, Vol. 2 (Oxford: Clarendon Press, 1992), 487–495.

[35] Kearney, *Gentlemen and Scholars*, 40–1.

university, new endowments of grammar schools declined, and many village schools disappeared. Only in the nineteenth century did an Oxbridge degree and a good public school education again became the hallmark of the elite.

This is consistent with the pattern from cross-cultural studies. The one people showing the G-90 pattern of high infant C and low child V are the Manus, with their intense practical bent and passion for machines. Margaret Mead noted that Manus children showed a striking lack of interest in storytelling and imagination.[36]

Englishmen in G-60 periods also seem to be far more interested in technical and practical subjects such as engineering. These were the kinds of people who launched the Industrial Revolution in England in the late eighteenth century (G-90 at 1760). They were not typically Oxbridge graduates. For example Thomas Newcomen, who developed the first practical steam engine, was an ironmonger by trade. James Watt, whose far more advanced engine was perhaps the single greatest driver of the Industrial Revolution, was an instrument maker.

At G+60, when child V is high and infant C low, there is more interest in humanities subjects such as were taught at Oxford and Cambridge. Societies showing this pattern are those of India, China and the Middle East, and all these cultures show a bias towards the arts and humanities.

In Bangladesh one study found a 44% unemployment rate among high school and university graduates, with another 17% under-employed. One reason was the subjects studied. Only 24.3% of social science graduates in India and Bangladesh found their studies useful for getting a job, compared with 50% for agriculture, 52% for science and 62.3% for engineering.[37] China also has a problem with graduate unemployment, while there is a huge and growing demand for the kind of low level technical qualifications required by industry.[38] Unemployment among graduates can be a problem for governments, playing a large role in the

---

[36] M. Mead, *Growing up in New Guinea* (New York: Perennial Classics, 2001), 92–3.

[37] A. K. Das, "Unemployment of Educated Youth in Asia: a Comparative Analysis of the Situation in India, Bangladesh and the Philippines." IEEP Occasional Papers no 60, Unesco: International Institute for Educational Planning. 1981.

[38] Chinese Bureau of Statistics.

overthrow of regimes in Egypt and Tunisia during the Arab Spring.[39] The result is a fierce competition for government jobs, for which arts graduates are suited, while industry struggles for qualified workers.[40]

Reverence for scholarship was a marked feature of traditional China, where prestigious and lucrative jobs in the imperial bureaucracy were assigned mainly by success in examinations on the Confucian classics.

G-120—Later age of marriage

Finally, we may consider another implication of lemming cycle theory, which is that higher C at the G-120 period might delay the age of marriage. Information is only available for the G-120 year of 1730, but this may help explain a curious feature noted in the last chapter. While later age of marriage is characteristic of higher C in the civilization cycle, the latest age of marriage was in the late seventeenth and early eighteenth centuries rather than in the nineteenth century, at the peak of C. This can now be attributed to a lemming cycle effect, with the age of marriage latest at the G-120 year of 1730. It then fell slightly, though the lowest level was in the first half of the nineteenth century and not, as would be predicted, around the G+30 year of 1880 (see Fig. 8.11 below).

Revised lemming cycle

A study of British history over the past thousand years shows many of our predictions borne out, but some major revisions are required. The first is that the G period can be characterized as a period of national unity, something not necessarily predicted by the theory. The second is that economic growth seems to be strongest at the G period rather than at maximum infant C. We can only suppose that the energy and optimism of high G promotes economic activity, which is also aided by public order and population growth.

---

[39] B. Van Niekerk, K. Pillay & M. Majaraj, "Analyzing the Role of ICTs in the Tunisian and Egyptian Unrest from an Information Warfare Perspective," *International Journal of Communications* 5 (2011): 1406–1416

[40] G. Perkovich, "Is India a Major Power?" *Washington Quarterly* 27 (1) (2003): 129–44; "How India got its Funk," *The Economist*, (August 24, 2013).

Fig. 8.11. Average age of first female marriage, 1551–1901, England and Wales.[41] Even though there was a general increase in age of marriage up to the nineteenth century consistent with rising C in the civilization cycle, the latest age of marriage for ordinary people was in the early eighteenth century. This is a lemming cycle pattern consistent with higher C in the G-120 period.

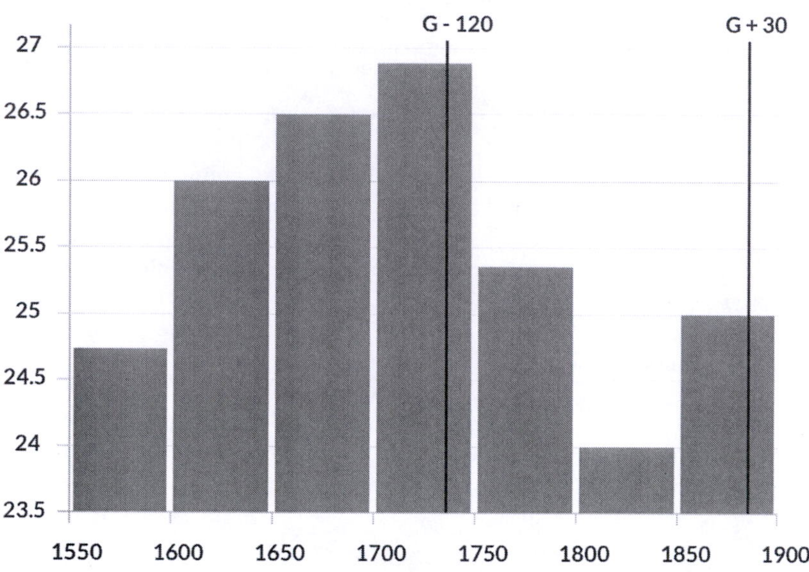

Another difference is that the "new ideas" aspect of the cycle peaks somewhat later than predicted, around G-60 rather than G-90, perhaps also because new thinking is promoted by the energy and optimism of high G. All these findings are summarized in Table 8.3 below.

[41] Source for 1551–1837: R. Woods, *The Demography of Victorian England and Wales* (Cambridge: Cambridge University Press, 2000), 108. Variation in mean age of first marriage among 10 English parishes, 1551–1837, E. A. Wrigley, "Variation in Mean Age of First Marriage Among 10 English Parishes, 1551–1837," in *The Population History of England 1541–1871: A Reconstruction*, edited by R. S. Schofield (London: Edward Arnold, 1984); Source for 1850–1901: E. A Wrigley & R. S Schofield, *The Population History of England 1541–1871* (London: Edward Arnold, 1981), 437, and averaged out between six figures ranging from 1851–1901 to fit the format of the previous data. Data comprise England only, minus Monmouth.

Table 8.3. Behavior and attitudes at different stages of the lemming cycle.

Period	Characteristics
G	Population growth, national unity
G+30	Migration, war
G+60	Autocracy, weak local loyalties, humanities education
G+90	Orthodoxy in religion and thought, science?
G-120	Population decline or stagnation, later age of marriage
G-90	Local and hereditary loyalties, technical education
G-60	Unorthodox religion, renaissance and new ideas

Lemming cycles in France

Having refined the lemming cycle pattern in England it can now be applied to other countries. As mentioned earlier, one characteristic of animals such as muskrats is that length of cycles in neighboring populations tends to be the same, but their timing varies from region to region. For example, a species may show ten-year cycles over a wide range, but one population is in the growth phase while the other is in decline.

Humans show the same pattern, and the closer they are geographically the more similar the cycles tend to be. For example, French cycles are almost the same length as English cycles, but with G years just a few decades earlier. This apparently minor difference has had a serious impact on relations between the two countries.

One advantage of studying French cycles is that they can be traced back all the way to Roman Gaul. Not only is historical information better for France than Britain after the fall of Rome, but the original population largely remained in place, which in Britain it did not. This is reflected in the fact that French derives from Latin and English from German.

As we will see when dealing with the Roman Empire in chapter twelve, Rome had a G year around 340 AD when powerful emperors restored order after the chaos of the mid-third century (G-90 at 250 AD). This seems to have applied to the Empire as a whole. In the fifth century, Germanic tribes invaded but took over the Roman administration until most of Gaul was united under the Frankish Merovingian dynasty at the close of the century. But from then on the nobility grew in power, the last strong ruler being Dagobert I who died in 639. His successors were given the rather derisory nickname of "*rois fainéant*," or "do-nothing kings." By the end of the seventh century the Frankish state had effectively disintegrated, corresponding to a G-90 year of 690. All this amounts to a fairly steady decline from Imperial unity in the early fourth century to complete anarchy by the end of the seventh, with the fall of Rome being only an incident along the way.

Central authority was restored by Charles Martel who took power in 718 and ruled until 741, though nominally as a servant of the Merovingian king. He was known as Charles the Hammer for a creditable series of victories, most notably the Battle of Tours which drove the Muslims out of France. The Frankish state reached its zenith in the reign of his grandson Charlemagne (768–814), who formed an empire covering not only modern France but most of Germany and northern Italy. This is consistent with a G year around 780. We see again here the typical lemming cycle pattern of a slow decline in order and a rapid restoration, and the 90-year gap fits quite well—even though the G periods 340 and 780 are 440 years apart.

It is also consistent with the disease resistance of the G year that a terrible series of plagues which had afflicted France (and much of the world) since 541–542, were no longer recorded after 750.

There was also at this time a "Carolingian renaissance" in literature, art and architecture, but around the G year rather than (as is more typical) before it. The best explanation is that this brief flowering depended heavily on royal wealth which had not been available in earlier reigns. Charlemagne was not only rich but a great admirer of the arts, even though himself illiterate.

The religious pattern fits well, also. It will be remembered that anti-clerical attitudes are prevalent before the G year and orthodox religion strongest after it. Charles Martel, whose reign covered the G-60 year of 720, was able (like Henry VIII) to despoil the church on a massive scale, while Charlemagne's son and successor "Louis the Pious" (814–840) ruled in the years approaching the G+90 year of 870. As the nickname suggests, he was respectful of the church and relied on it to buttress his power, and he certainly made no attempt to seize its property. This is all the more striking given that he faced a number of serious rebellions and was at one stage briefly deposed, so he had ample motivation.

The decline of royal authority continued under his descendants until they died out, and Hugh Capet was elected king in 987. But this did nothing to halt the decline. Eleventh-century French kings had very little power beyond the Île de France, and even here they were at constant war with the local barons. The lords of other provinces such as Burgundy and Aquitaine had similar problems. The low point of royal power can perhaps be set in the reign of Henry I (1031–1060), just before the expected G-90 year of 1070.

Attitudes of the time show an intense localism. The fief rather than the kingdom was known as a "patria" or fatherland, and outsiders were seen as strangers or even enemies. Even the defeat of an enemy did not bring lasting gains. When Henry gained control of Burgundy he passed it to his brother instead of adding to his own domain. Nothing more clearly indicates the limits of royal power at this time. We can only presume that Henry knew very well that people could not and would not obey a distant ruler, even the king of France, so much did loyalty depend on an immediate personal presence. Once again, regardless of the change of dynasty, there was a slow and steady decline in central authority from the Carolingian peak to the chaos of the eleventh century.

As in all lemming cycles, the slow decline in central authority was followed by a much faster rise. Henry's grandson, Louis the Fat (1108–1137), began to restore royal power as the G year of 1160 approached. A point of difference with British history is that the curbing of rebellious barons and the growth of central authority first took place at the provincial level. As Louis gained control of the Île de France, the provincial dukes and counts also established authority in their respective areas, such as Burgundy and Aquitaine. This is, in fact, one of the strongest arguments for broad changes in temperament driving changes in political order. If the character of individual rulers was the crucial factor then each Province should show distinct patterns. In fact, they all experienced much the same changes at much the same time.

It was Louis' grandson Philip Augustus (1180–1223) who established royal authority throughout France, which is attributable to declining localism around the G+60 year of 1220, similar to that which allowed Edward I of England to subdue Wales later in the century. It must be admitted that a G year in 1160 conflicts with demographic data showing an especially high rate of population increase between 1150 and 1250, suggesting a G year around 1200. Our prediction is that further demographic research will show a peak period of growth somewhat earlier, in the late twelfth century.

We have seen that the English state lost power in the early fourteenth century, but that the French decline started earlier, as is consistent with French cycles being some eighty years in advance of England. The financial position of the French monarchy became worse after the death of Louis IX in 1270, and virtually collapsed after the battle of Poitiers in 1356. Indeed, one reason the English won that battle and gained control of much of France is that the French crown had largely run out of money. The low point of French state power can be placed in the early fifteenth century, when the successes of Henry V owed as much to French internal chaos as to the famed longbow. This was not only the well-known struggle of Burgundians versus Armagnacs, but a series of local conflicts all over the country.

Population figures for the fourteenth and fifteenth centuries show that the demographic crisis may have come earlier than in England, as is consistent with the idea that French lemming cycles were several decades in advance. This is based on evidence from Burgundy, one of the few areas where medieval population data are available. In England there was a decline of population in the late fourteenth and early fifteenth centuries, and only a modest rise in the late fifteenth. In Burgundy, the population fell dramatically in the late fourteenth century, was static in the early fifteenth, and grew quite fast in the late fifteenth.[42] That would put the greatest decline of population and thus the G-150 period in 1375, which is about tweny-five years earlier than in England. This is roughly consistent with a G-90 year of 1420. France was so feeble at this time that after his victory at the battle of Agincourt in 1415, Henry V of England was able to have the French king accept Henry's male issue as heir to the French throne.

From then on, as always with lemming cycles, the recovery of power was swift. French nationalism resurged with the victories inspired by Joan of Arc. Louis XI, who reigned from 1461–83, played a similar role to Henry VII of England in the next generation. He was known as the "Universal Spider" for the cunning and intrigue with which he restored royal power.

[42] C. L. Crumley, http://www.clas.ufl.edu/users/caycedo (accessed September 5, 2014).

This was helped, of course, by the collapse of England into its own G-90 crisis around 1460. By the G year 1510, France was once more a powerful and unified country.

A major divergence from England in political terms is that the French monarchy survived the challenges of the 1640s to become increasingly absolute, especially during the reign of Louis XIV (1643–1715). It only eased with his death, which was after the G-90 period around 1685. This is the one really striking exception we find in French history to the pattern of local and personal loyalties being stronger at G-90, for which biohistory currently has no explanation.

When it comes to lemming cycles, being in advance of one's neighbors is no great advantage. The horrors suffered by France during the Hundred Years War can be attributed at least partly to France being eight decades ahead of England in the slow decline of royal authority, leading to the nadir of the French G-90 year in 1420.

Countries at or soon after their G year can be at an advantage, however, when faced with neighbors at different stages of their cycle. As will be suggested in the next chapter, demographic information tells us that France reached its peak of population growth in the 1770s, well in advance of any other European country, which explains the fervent nationalism and imperial aggression of France under Napoleon.

Acceptance of radical change is characteristic of the period before the G year, though in France it only found political expression after the Revolution. In terms of radical innovations such as the metric system and the sweeping away of feudal privileges with the Napoleonic Code, the French were well in advance of the rest of Europe. The slogan "liberty, equality, fraternity" is an expression of people relating directly to the state and each other, rather than through traditional hierarchies. This was reinforced by very high C which, as we have seen, also causes individuals to relate strongly to central authority.

But as France passed beyond its G year and other countries rose towards theirs, the balance of power shifted. In the Franco-Prussian War of 1870–71 the French were seventy years after their G year and the Germans only twenty-five or so before theirs. Decades earlier, France had fought Prussia, Russia and Austria-Hungary to a standstill. In this new war it yielded to Prussia alone. Then in 1914–18 it took four bloody years to defeat Germany, despite being part of a much larger allied coalition.

Finally, in 1940 France reached its low point of G, as clearly indicated by demography. France in the 1930s was the first country in Europe where births fell below replacement level, recovering somewhat after 1945. The result was the rapid collapse of an army which was, on paper at least, fully equal to that of Germany. But the Germans were only fifty years from their own G year, and thus only shortly after their peak of V. French G years are summarized in Table 8.4 below.

Table 8.4. Lemming cycles in France

G-90 year	G year	Length in years
250 CE	340	
690	780	440
1070	1160	380
1420	1510	350
1685	1775	265

Putting the English and French cycles together, there is the same trend from longer to shorter cycles as the Dark Ages give way to civilization (see Fig. 8.12 below)

Lemming cycles in Japan

An uncannily similar pattern of lemming cycles can also be found at the other end of the world. It is no surprise that Japanese lemming cycles have no direct connection with European ones, since the Japanese had very little interaction with Europeans until the nineteenth century. It is more striking that there is no connection between Japanese and Chinese cycles either. The reason is presumably that Japan and China had much less interaction than, say, England and France.

But although further apart, China had a profound influence on early Japan. Seventh-century Japan went through a series of major reforms which meant, in effect, the wholesale adoption of Chinese culture, religion and administrative ideas. The high point of change came with the Taika reform after 646 A.D. This openness to new ideas is a feature of the G-60 period. If we place this at 660 then the G period becomes 720, and the eighth century was in fact the high point of imperial power.

Fig. 8.12. Length of lemming cycles in England and France. Cycles become shorter in the falling-V section of the civilization cycle.

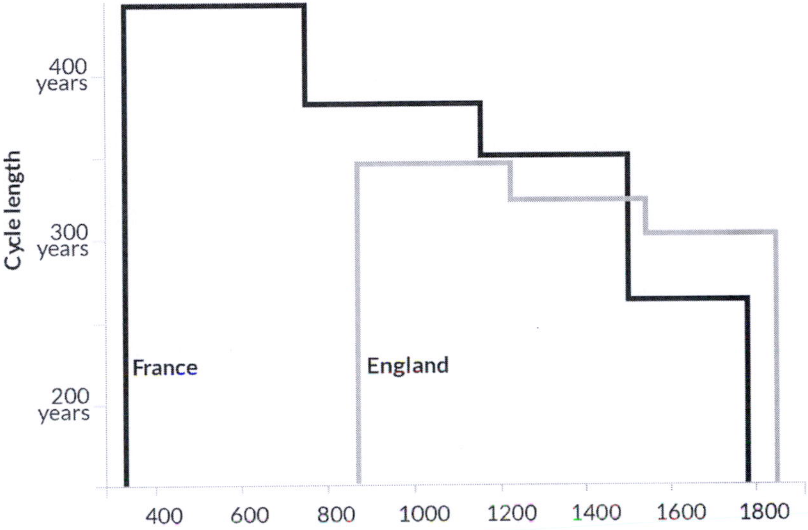

Central power ebbed gradually from this time, as is typical of lemming cycles, with local military leaders becoming increasingly independent. It reached a low point in the eleventh and early twelfth centuries as warfare broke out all over the country and armed bands appeared in the capital (G-90 at 1110). But as always, the recovery was much faster than the decline, with the Taira faction gaining control of the government by 1160, and their Minamoto rivals controlling the country by 1185. This is consistent with a G year around 1200. Also consistent with the lemming cycle is an unusual passion for Chinese culture between 1160 and 1180, reflecting the openness to new ideas around the G-60 year of 1140.

For a century or so the new military government was in effective control. But again after this came a steady decline, barely affected by the passing of the role of shogun (military ruler) to the Ashikaga family in 1368. Local lords or Daimyos took more and more authority, until central authority completely collapsed after the start of the Onin War in 1467. Though ostensibly a succession dispute over the shogunate, the Onin War became a trigger for local power struggles all over the country. This is exactly the same pattern as we saw in England during the Wars of the Roses, with the royal struggle no more than an excuse for local conflicts. The underlying reason in all such cases is the power of local loyalties, which give local leaders the power to act as they wished.

It was at this time that most of the great daimyos were overthrown by smaller-scale daimyos with closer ties to their own people. This is a sign of high stress in the civilization cycle but also a reflection of extreme local loyalties. There are many other signs of localism at the time. Large-scale peasant rebellions were often able to overawe governments and force the cancellation of debts. And in 1485 the peasants of southern Yamashiro rose under local samurai and expelled two contending armies from the region as undesirable "foreigners." All this is consistent with a G-90 year of 1500.

Once again, central power was restored much faster than it had declined. Tight central authority was first established by the daimyos in their own realms, as in twelfth-century France, and then by Hideyoshi over all of Japan in 1585.

All signs point to a G year around 1590. There is the establishment of central authority, following the chaos at the start of the century. Wars were fought by mass citizen armies, rather than the warrior elites of earlier times, and two massive armies were sent to conquer Korea in 1592 and 1597, consistent with a G+30 year around 1620. Overseas trade exploded, with colonies set up in much of Southeast Asia. Japanese of the late sixteenth century were also remarkably open to new ideas, as is common before a G year. There was a passion for wearing Western clothes and millions of people accepted Christianity.

Then, after the G year, there was a sharp reaction and a rejection of the West as Japan turned back to isolation. Christianity was outlawed and foreigners confined to the tiny enclave of Nagasaki. This again is typical of the more orthodox and conservative thinking of a G+90 period around 1680.

From this time we also have demographic evidence. The Japanese population grew in the period up to the seventeenth century but was relatively static in the eighteenth and early nineteenth centuries. From then on it began growing rapidly, with a high point in the late 1920s and a postwar surge. This suggests a G-150 year in 1775 and a G year in 1925.

As in England the political aspects of the lemming cycle were completely consistent with the demographic ones. The G-90 year of 1835 is when we would expect local loyalties to be weakest, and it was just eighteen years later that Commodore Perry arrived with his gunboats and a letter of demand for the emperor. For the next two decades the great fiefs acted with an independence they had not shown for more than 250 years, until the forces of Choshu and Satsuma overthrew the shogun in 1868. But from that time, thirty-three years after the G-90 year, strong central government was restored. By way of comparison, the Tudor dynasty was founded twenty-five years after the English G-90 year of 1460, and Henry II restored order thirteen years after the G-90 year of 1140.

This date was also just three years after the G-60 year (1865 in Japan), which is the time when people are most open to new ideas. The Japanese, who had been so adamant about rejecting all foreign influences for two and half centuries, became enamored of Western technology, science, political institutions and even clothing.

Finally, on and immediately after the 1925 G year came fervent nationalism, militarization and wars of conquest. In the 1930s and early 1940s, Japanese forces occupied vast areas of East and Southeast Asia, much as they had tried to do in the 1590s. Once again they were unsuccessful and there was an abrupt return to isolationism, though this time only in military terms. Japan continued to engage with the world in economic and cultural terms, as would be expected from a rapid fall in C. As in the West, the orthodox/religious/conservative aspect of the G+90 period (2015), which depends on child V being at a maximum, has been swamped by falling C. Japanese lemming cycles are summarized in Table 8.5 below.

Table 8.5. Lemming cycles in Japan

G-90	G	Length in years
630	720	
1110	1200	480
1500	1590	390
1835	1925	335

Once again, there is a clear pattern of movement to shorter cycles as C rises. Japanese cycles are also slightly longer than European cycles, for reasons not currently understood.

The last chapter introduced the idea of the civilization cycle, which describes the rise and fall of civilizations in terms of change in C and V (the latter being identified largely by high levels of stress). Low levels of C, in conjunction with powerful V-forming traditions, allow V and stress to rise. High V and stress in conjunction with C-forming traditions increase C, and high C causes V and stress to fall, which in turn reduces C.

Lemming cycles in England, France and Japan all show the same pattern of longer cycles when V and stress are rising in the civilization cycle, and shorter cycles when they are falling. Chinese lemming cycles, as described in chapter thirteen, will show the exact same pattern.

An initial clue can be found in a key observation. Lemming cycles do not stretch evenly, in that the period between the time of feudal chaos and the G year always appears to be around ninety years. For example, if the 340–780 French cycle were stretched evenly, the peak of feudal chaos would be put back 132 years (compared with ninety years in a three-hundred year cycle), which would make it 648 AD, almost to the reign of Dagobert, the last strong Merovingian king. Alternatively, if we accept the year of greatest chaos around 690 and go forward 132 years we are well past the reign of Charlemagne and into a time of growing disorder.

Since the gap between G and the behavior typical of G+60 (absolutism, weak local loyalties) also seems to be consistent in length (around sixty years), this implies that the "stretched" phase of the cycle must be somewhere between G+60 and G-90. This is the period of the cycle when V is falling.

Thus, the *falling V* section of the lemming cycle is lengthened by the *rising V* section of the civilization cycle. That is the observed pattern, but it must be admitted that on theoretical grounds it makes little sense. Lemming cycles in animals tend to be longer in southerly areas with less extreme climates where V would be lower. A solution to this problem must wait until lemming cycles can be modelled in animal populations, or at least proper sampling done of cycling animals in the wild.

Culture shock

The lemming cycle theory also helps explain the phenomenon of "culture shock." This is what commonly happens when a society is overrun by a powerful and alien civilization, such as when Europeans overran Mexico in the sixteenth century or Pacific islands in the nineteenth century, even past what can be explained by introduced diseases. Added to this are demoralization and loss of confidence, and an inability of people to organize and work together.

Culture shock can be explained as a massive fall in V. It will be recalled that V is associated with energy, confidence, resistance to disease and an interest in having children. It is reinforced by high adult status and undermined by low status, which tends to result in chronic stress (see chapter four). The *effect feedback cycle* suggests that a severe cultural shock would reduce V, and this is a fair description of conquest by an alien power which challenged and largely obliterated the native culture.

The islanders of the Marquesas in the Pacific, contacted by Europeans at the close of the eighteenth century, provide an extreme example. The population fell by an estimated 98% in a little over a century, and the people became totally demoralized, lethargic, suffered from alcohol addiction, and died easily of infection. The position of men declined, in that women in the 1930s were described as having effective power in the household, something not noted in the 1950s when recovery had begun.[43]

Obviously, imported disease was a significant factor. Isolated populations may have little resistance to such common ailments as measles and flu. But it was probably not the major reason. The Marquesan decline owed much to a decision not to have children. Even in the 1930s, birth control and abortion were common. Another reason for discounting imported disease is that existing *native* diseases were among the most devastating. As late as the 1950s the Marquesans were highly susceptible to illness, but of current major health problems, only venereal disease was clearly a European import.

Also, European diseases had much less impact on some other Polynesian peoples. In Tikopia, another Pacific island, records show no signs of

---

[43] R. C. Suggs, *Marquesan Sexual Behavior* (New York: Harcourt, Brace & World, 1966);
R. Linton, "Marquesan Culture," *American Anthropologist* 27 (3) (1925): 474–8.

demoralization or drastic population loss. In fact, a major current problem is that the population is more than double the original level and rising, threatening to overcrowd the island.[44]

The same applies to indigenous populations in Central and South America after the Spanish conquest in the early 1500s. The worst effects of diseases such as smallpox ought to have been felt in the first two or three decades, but populations fell for more than a century. Also, some peoples were far less affected than others. The mountainous state of Tlaxcala helped the invaders against their Aztec foes and was rewarded with a special status. The Tlaxcalan population was halved in the following decade, presumably because of disease, but then stabilized for almost a century while the Aztecs and others suffered catastrophic decline.

Further, a small group of Tlaxcalans who migrated north actually *increased* in number during the period of catastrophic decline. Tlaxcalans had as much exposure to European diseases as other Mexicans, so their greater resistance must be based on other factors.[45] As relatively high-status colonists in a less advanced area, their psychological situation would have been quite different.

To the inhabitants of Pacific islands, or those of the New World, the arrival of Europeans left them feeling helpless and unable to control events. The certainties and beliefs of their own cultures were swept away, resulting in a drastic fall of V. This would clearly be more of a problem when V was already low, as for example indicated by fewer restraints on sexual behavior. Societies with the strictest mores, such as Arabs and Indians, suffered no obvious culture shock when taken over by Europeans. But even in the Pacific, there were major differences in response. Samoans and Tikopians, who put significant controls on premarital sex, suffered only modest losses of population.[46] For the Samoans it was about 50% and

---

[44] R. Firth, *We the Tikopia* (London: George Allen & Unwin, 1961); R. Firth, *Rank and Religion in Tikopia* (Boston: Beacon Press, 1970).

[45] C. Gibson, *Tlaxcala in the Sixteenth Century* (Palo Alto: Stanford University Press, 1967), 1, 14–15, 26, 143–45, 161, 169–73, 271.

[46] Firth, *We the Tikopia*; M. Mead, *Coming of Age in Samoa* (New York: William Morrow, 1964); D. Freeman, *Margaret Mead and Samoa: The Making and Unmaking of an Anthropological Myth* (Cambridge Mass.: Harvard University Press, 1983).

[46] I. Goldman, *Ancient Polynesian Society* (Chicago: University of Chicago Press Chicago, 1970), 20–21, 580; J. L. Rallu, "Population of the French Overseas

the Tikopians little, if any. On the other hand, the sexually freer Tahitians and Marquesans suffered losses of around 90% and 98% respectively.[47]

But the real test of this hypothesis is the way in which population recovery happened. If biohistory is correct, the onset of culture shock should act like the launch of a lemming cycle at the time when V is falling most rapidly (see Fig. 8.13 below).

Fig. 8.13. Lemming cycle triggered by culture shock. The lemming cycle theory suggests the fastest decline is likely seventy years after impact, apart from the effect of infectious disease.

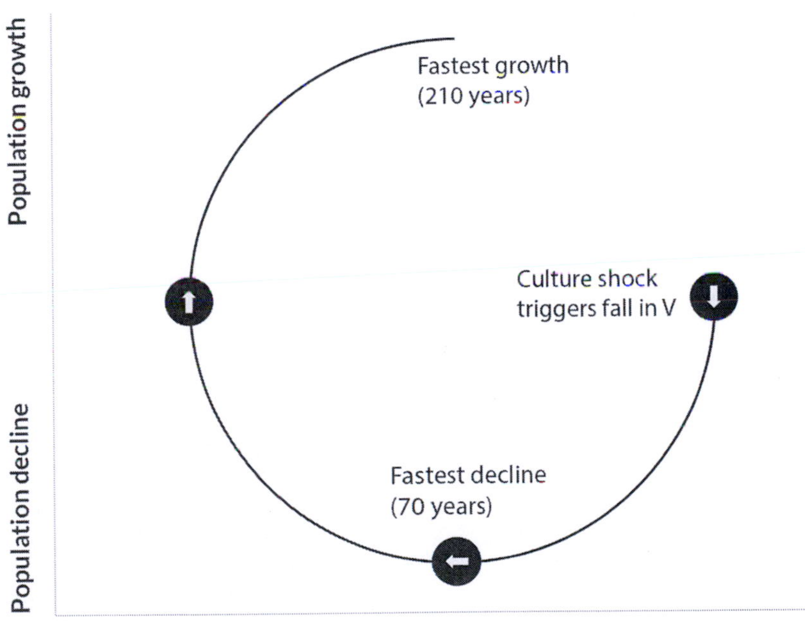

We have seen that lemming cycles lengthen when V is rising, so a society with falling V should have a shorter lemming cycle, at around 280 years. What this means is the biggest effect on the population, leaving aside the effects of disease, should be seventy years after the impact. We can see this pattern in the Marquesas Islands, first visited by Europeans in 1791

Territories in the Pacific, Past, Present and Projected," *The Journal of Pacific History* 26 (2) (1991): 169–86.
[47] Ibid.

but conquered by the French in 1842 (see Fig. 8.14 below).

Fig. 8.14. Marquesas Islands birth rate and death rate per 1,000 population. The death rate (but not the birth rate) shows a clear lemming cycle pattern by peaking after seventy years, not at the time of conquest as might be expected.

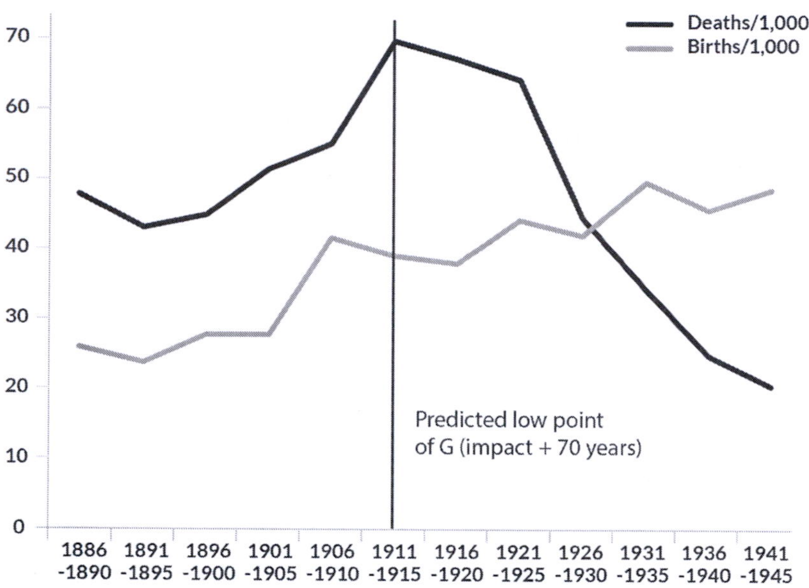

Though evidence from the early years is missing, the birth rate started rising fifty years after the French took over, a little early for our model. But the death-rate peaked exactly seventy years after this. If disease were the crucial factor then the death rate should have been higher immediately after the occupation, or even thirty years earlier when the first European ships arrived. But it fits very well with an expected trough of G occurring seventy years after the psychological shock of foreign occupation.

Recovery from culture shock

Another implication of this theory is that if culture shock launches a lemming cycle, a new G period might be expected after 210 years, which is three quarters the length of a standard cycle. There has not been enough time to observe this in the Pacific, but two Mexican populations show evidence of such a pattern.

The onset of culture shock can be dated precisely for Mexico in the year 1520 when Hernan Cortes took Tenochtitlan and overthrew the mighty Aztec Empire. This implies that there should be a renewed peak of G in 1730, and this is exactly what we find (see Fig. 8.15 below).

Fig. 8.15. Annual rate of population decline and recovery on west-central Mexico and the Mixteca Alta, 1520–1960[48], plotted against a 280-year lemming cycle precipitated by the Spanish Conquest in 1520. Population decline seems to have been greatest immediately after the conquest, but the eighteenth-century recovery and the Mexican Revolution of the 1920s show a clear lemming cycle pattern.

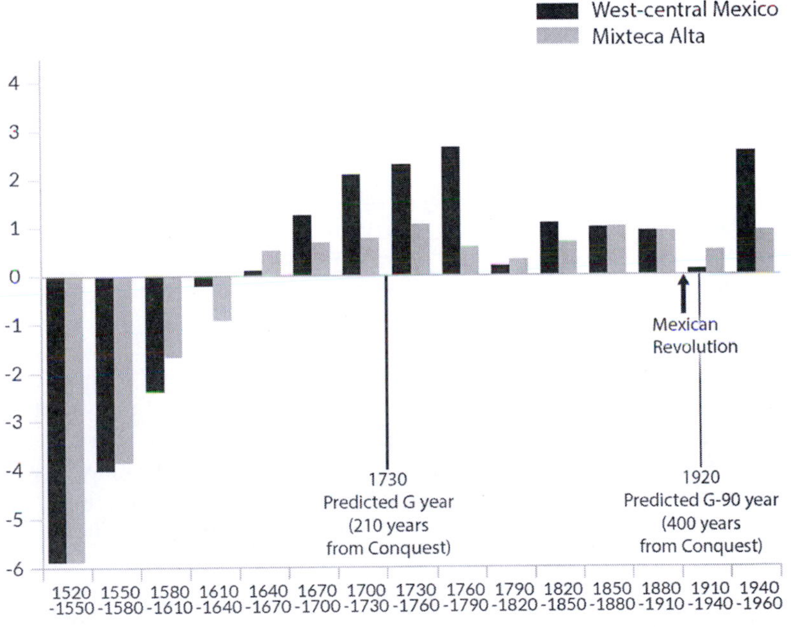

In fact, both populations show a peak of growth within a few decades of the predicted peak. This is especially striking when we consider that the mid-eighteenth century was a time of relatively slow population growth in Europe, so there is no possibility that the Mexican lemming cycle was "imported" from Europe.

---

48 S. F. Cook & W. Borah, *Essays in Population History: Mexico & the Caribbean*, (Berkeley: University of California Press, 1971), 107, 312.

A continuation of this lemming cycle would predict a further G year in 2010, which would place the prior G-90 year in 1920. Again, this exactly coincides with the bloodthirsty and chaotic Mexican Revolution of 1920–29, which is entirely characteristic of G-90 years. All of this is a simple forward projection of a lemming cycle launched by the Spanish Conquest four hundred years earlier.

Tasmanian Aboriginals

In all these examples, the populations eventually recovered from culture shock. But another people, the Tasmanian aboriginals, died out completely. After prolonged conflict with European settlers and enduring the ravages of imported diseases, a population of many thousands was reduced to two hundred demoralized survivors on Flinders Island, where the last full-blood Tasmanians died within a few decades.

In previous chapters, hunter-gatherers have been considered as generically low C, but Tasmanians may have had significantly lower C than mainland aboriginal populations.[49] For example, their political units were tiny, of only a few hundred people, compared with the much larger confederations found on the mainland, where the size of an armed party was typically limited by the ability to find enough food in one place.[50]

Another point is that Tasmanian technology was simple. Technologies such as the making of bone fishhooks had been lost, with the result that they made little use of the abundant fishing potential of Tasmanian waters. This is significant when considering that higher C people are more likely to spend time learning skills at the expense of current consumption (see chapter six). Low C might thus make it harder to pass on certain skills.

How the mainland peoples might have maintained higher C is not certain. The climate may have been tougher, at least in some areas, and there may have been effects from customs such as incest taboos which defined many eligible women as unavailable. But a lower level of C could explain why Tasmanians fought European settlers almost to their own extinction, even when they knew it was hopeless. The survivors were relocated to Flinders Island because experience had shown that the two groups could not live together.

---

[49] N. Clements, *The Black War: Fear, Sex and Resistance in Tasmania* (Chicago: University of Queensland Press, 2014).

[50] H. Reynolds, *The Other Side of the Frontier: Aboriginal Resistance to the European invasion of Australia* (Ringwood, Harmondsworth: Penguin, 1982), 97.

By contrast, mainland aboriginals typically resisted Europeans for a limited period and with relatively little bloodshed, especially when settlers were numerous and there were no inaccessible areas into which they could retreat. They tended to accept European control, once the weakness of their own position became clear.[51] After this they commonly integrated to some extent with white society, such as taking work on pastoral stations where they made able and dedicated stockmen. People with higher C tend to be more hardworking, disciplined, law-abiding and able to relate to impersonal institutions. So even the modest-C levels of mainlanders would have helped them integrate.

Lower C would thus explain why the Tasmanians died out rather than declined and then recovered, as did the mainland tribes. As discussed earlier, lower C peoples are more vulnerable to culture shock.

Summary and conclusions

The lemming cycle can be seen as an alternation between V and C, driven by the rise and fall of the variable labeled "G." G is strongly linked to population growth, to the extent that it dwarfs the effect of C in the lemming cycle and causes a significant perturbation in long-term population growth trends in the civilization cycle.

The lemming cycle seems to be a naturally occurring pattern, with shorter and more intense cycles found among species in harsher environments. Its likely function is to create a large and bold generation that can colonize areas where existing populations may have been wiped out, an increase made possible by the obliteration of predators by the previous crash. Given that this cycle is clearly found in humans, though at less extreme levels than in lemmings, the capacity to cycle is probably common to mammals and birds.

Changes in political and intellectual attitudes can be reliably linked to different stages of the lemming cycle in humans. These can be explained by relative levels of V, infant C and stress, though G itself seems to have significant and independent effects. The G year is a time of confidence, energy and high morale. Birth rates are high and resistance to disease strong. Human populations tend to be nationalistic and united. Energy plus strong government allow the economy to grow. People born at this time

[51] Reynolds, *The Other Side of the Frontier*, 62, 116.

have maximum V.

The G+30 period is the time when adults have maximum V, associated with mass migrations in animals, and to some extent with wars of conquest in humans.

The G+60 period is the time when infant C is weakest and child V strongest. As a result, local loyalties are weak and people are more likely to accept arbitrary authority. Rulers can seem especially strong at such times, but can also be readily overthrown when another power arrives on the scene, since obedience is based on fear rather than loyalty. It is just this attitude that we found in Egyptian villagers, who tend to accept any sufficiently strong authority. It is also associated with a taste for education in the humanities.

The G+90 is a high point of religious orthodoxy, with powerful clergy, which can perhaps be attributed to high child V combined with less of the vigorous and consensual "G" attitudes.

The G-150 period is minimum G, a time of population decline or at least minimum growth. Plagues tend to be especially virulent at such times.

The G-90 period sees local loyalties strongest and acceptance of authority weakest. It is the exact obverse of the G+60 period, in that infant C is highest and child V lowest. This is usually a time of chaos and disorder. It provides the clearest and most universal sign of the lemming cycle, which is a slow decline into anarchy or disorder (the G-90 period) and a rapid restoration of order towards the G year. It is also the time when technical education is favored over education in the humanities.

The G-60 period is a time of cultural renaissance and new ideas, perhaps from a combination of high-G energy and the flexible thinking associated with infant C. It is opposite in character to the G+90 period.

The falling V section of the lemming cycle becomes stronger in the rising-V section of the civilization cycle, for reasons not currently clear.

In the next chapter, we examine in more detail how lemming cycles drive wars.

CHAPTER NINE

WAR

In the previous chapter we considered the occurrence of large-scale war as equivalent to a mass migration, occurring in the lemming cycle after the period of most rapid population growth. We found some evidence of this in human history, most especially in the world wars of the twentieth century, which occurred after rapid population growth in the previous century. This chapter examines these wars more closely, refining the model to explain why these wars broke out when they did, and why certain countries were involved and not others. Lemming cycles are a factor, but other aggression-related variables are involved too, such as maternal anxiety and testosterone.

In historical terms the wars of 1914–18 and 1939–45 can be seen as particularly odd because they followed a century of peace in Europe, broken only by relatively minor and short-lived conflicts, such as the Franco-Prussian War of 1870–71. Yet within only three decades, two vast conflicts took tens of millions of lives in a frenzy of destruction and territorial conquest. Understanding these wars will also help explain other major wars in the historical record, as well as some internal conflicts such as the Chinese Cultural Revolution. Biohistory suggests they all have their origins in the treatment of infants.

Biohistory and the origins of war

In terms of biohistory, warfare becomes more likely when political units are small due to low C. Hunter-gatherers and small-scale agricultural societies, such as those of the Amazon or the highlands of Papua New Guinea, tend to engage in almost incessant warfare which causes a high proportion of deaths, especially among males.[1] Among the Yanomamo, one of the few tribal societies studied closely while still engaging in warfare, around 45% of all living adult males had participated in the

[1] L. H. Keely, *War Before Civilization: the Myth of the Peaceful Savage* (Oxford: Oxford University Press, 1996).

killing of at least one person. Two thirds of people over the age of 40 had lost a close relative (a father, brother, husband or son) to violence.[2]

But within societies above a certain level of complexity, the main factor that determines the prevalence of war is V, which is maximized by a harsh environment where famines alternate with better conditions that allow populations to grow rapidly. V is also promoted by cultural traditions, most notably patriarchy. Patriarchal peoples subordinate women and thus make them anxious—an anxiety which they transmit to their infant sons. V is also higher when the bond between mother and son is broken abruptly about the age of two, such as by the birth of another child, since close contact with an anxious mother after age 2 seems to increase anxiety in later life. In such conditions, boys grow up to be higher in status and less anxious; they become men whose high V makes them both aggressive and well organized in small groups—a perfect combination for warriors. Girls also experience this combination of anxious mothering and early weaning but, growing up to have lower status, they would not acquire the same level of V.

Thus it is that for thousands of years the patriarchal, warlike tribes of desert, steppe and mountain, where life was harsh and unpredictable, preyed on and conquered the settled lands. These include the Elamites, Aramaeans, Parthians, Huns, Magyars, Arabs and Mongols, among many others.[3]

Testosterone surges from C being lower than infant C

Aggression is also increased by testosterone, and there are various conditions which boost testosterone in men. One is tight control of women, giving them higher C which they transmit to their infant sons as

---

[2] N. A. Chagnon, *Noble Savages: my Life among Two Dangerous Tribes—the Yanomamo and the Anthropologists* (New York: Simon & Schuster, 2013), 274.

[3] A. Gat, *War in Human Civilization* (Oxford: Oxford University Press,2006), 189–221; E. Hildinger, *Warriors of the Steppes, A Military History of Central Asia, 500 B.C. to 1700 A.D.* (Boston Mass.: Da Capo Press, 1997); R. Amitai & M. Biran, (eds.) *Mongols, Turks and Others: Eurasian Nomads and the Sedentary World* (Leiden, Boston: Brill, 2005); I. Askold, "Early Eurasian Nomads and the Civilizations of the Ancien Near East (Eighth-Seventh Centuries BCE," in *Mongols, Turks and Others: Eurasian Nomads and the Sedentary World*, edited by Amitai & Biran (Leiden, Boston: Brill, 2005).

infant C. As men grow up to a life of relative freedom which lowers their C, the consequence is high testosterone. This is what we saw in rats who were not calorie restricted, but whose mothers were (see chapter two).

Another cause of higher testosterone is a general decline in C. People are born with a high infant C which reflects the C of their parents. Growing up with lower C, the result is a surge of testosterone throughout the whole society. Several generations of falling C can deliver several generations of high testosterone.

Chapter six presented evidence that C rose to a peak in mid-nineteenth century Europe. But starting in the late-nineteenth century came a steady fall of C in all Western nations. In later chapters we will look at several indicators of falling C, but for now we will consider just one—the declining age of puberty. This was a feature of all Western populations from the late nineteenth century, but Fig. 9.1 (below) gives the figures for Britain, France, Germany, Sweden and the United States.

Such a rapid decline in the age of puberty (and thus presumably C) suggests that all these populations should have experienced a surge of testosterone in people born from the late-nineteenth century onwards. And the biggest surge would be expected in Germany, which experienced a decline of nearly three years in half a century. The decline in the age of puberty was far less marked in Sweden, which stayed neutral through both world wars, and in the United States, which entered both wars late and reluctantly. What was the result?

The Great War

Rising tensions can be observed towards the end of the nineteenth century with a frenzied "land grab" for colonies that saw most of Africa carved up by European powers. But even allowing for this, what happened next was shocking in its scale and brutality.

Europe in 1914 had been broadly at peace for almost a century since the end of the Napoleonic Wars. Conflicts such as the Crimean and Franco-Prussian wars were limited in scope and inflicted relatively few casualties. The Great War was on a completely different level. What is most striking about the events at the beginning of August 1914 is the mass enthusiasm for war in all social classes. It was seen as an exciting adventure, and young men rushed to enlist for fear that it would be over too soon and they would miss out. The fervent nationalism and enthusiasm can hardly be overstated. This is how one German viewed the outbreak of war:

Fig. 9.1. Age of menarche in Germany, Great Britain, France, Sweden and the United States, by year of menarche.[4] Falling C, as indicated by decline in the age of menarche, reduces C below infant C and thus causes a surge of testosterone. The decline was fastest in Germany, which on this basis would expect to experience the biggest testosterone surge.

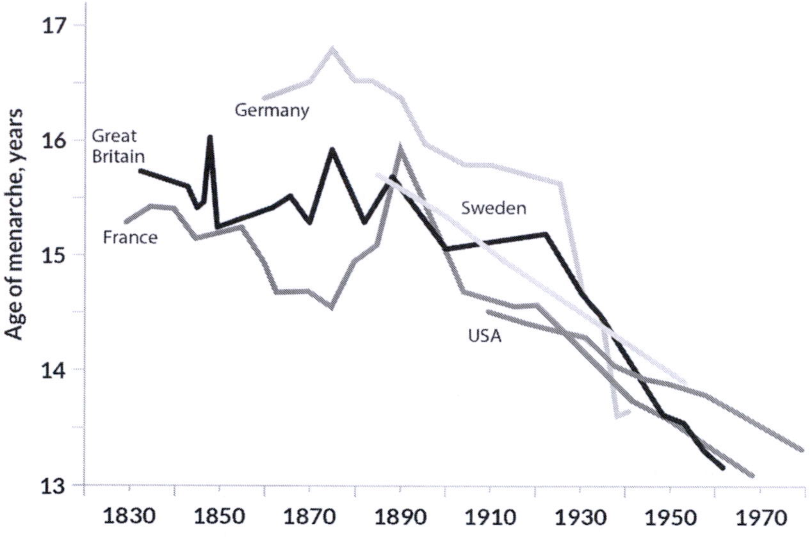

> A gigantic wave of fiery hot feeling passed through our country flaming up like a beautiful sacrificial pyre. It was no longer a duty to offer one's self and one's life—it was supreme bliss. That might easily sound like a hollow phrase. But there is a proof, which is more genuine than words, than songs, and cheers. That is the expression in the faces of the people, their uncontrolled spontaneous movements. I saw the eyes light up of an old woman who had sent four sons into battle and exclaimed: "It is

---

[4] J.M. Tanner, (1962): http://www.breastcancerfund.org/assets/pdfs/publications/falling-age-of-puberty.pdf (accessed September 6, 2014); P. E. Brown, "The Age at Menarche," *British Journal of Preventive & Social Medicine* 20 (1) (1966): 9–14. (i) Figures for Germany and have been taken from this source due to superior detail compared to the Journal of Epidemiol Community Health, and Tanner, (ii) Figures from Britain taken from this source up to 1950, then filled in from Tanner in the later part of the 20th century, (iii) France was taken from these figures up until 1920 due to superior detail, then from the *Journal of Epidemiol Community Health.*

> glorious to be allowed to give the Fatherland so much!"[5]

Or this by Ernst Junger, who was to be seriously wounded seven times and fought throughout the entire war:

> We had left our lecture-room, class-room, and bench behind us. We had been welded by a few weeks' training into one corporate mass inspired by the enthusiasm of one thought … to carry forward the German ideals of '70. We had grown up in a material age, and in each one of us there was the yearning for great experience, such as we had never known. The war had entered into us like wine. We had set out in a rain of flowers to seek the death of heroes. The war was our dream of greatness, power, and glory. It was a man's work, a duel on fields whose flowers would be stained with blood. There is no lovelier death in the world … anything rather than stay at home, anything to make one with the rest.[6]

Millions more responded to conscription with little protest or resistance. They became the soldiers who surged out of the trenches in vast human waves, tangled in barbed wire and shot down by enemy machine guns. Casualties, especially among junior officers, were at levels no Western army would tolerate today.

Yet amazingly, especially for many on the German side, even this could not wholly extinguish the lust for battle. This from Ernst Junger on the March 1918 German offensive:

> The turmoil of our felling was called forth by rage, alcohol, and the thirst for blood as we stepped out, heavily and yet irresistibly, for the enemy's lines. And therewith beat the pulse of heroism—the godlike and the bestial inextricably mingled. I was far in front of the company, followed by my batman and a man of one year's service called Haake. In my right had I gripped my revolver, in my left a bamboo riding-cane. I was boiling with a fury now utterly inconceivable to me. The overpowering desire to kill winged my feet. Rage squeezed bitter tears from my eyes.[7]

---

[5] G. Reuter, "The German Religion of Duty," *The New York Times Current History* 1 (1915): 170–173.
[6] E. Jünger, *The Storm of Steel: From the Diary of a German Storm-troop Officer on the Western Front* (New York: Howard Fertig, Inc., 1996), 1.
[7] Ibid., 253–5.

Over nine-million soldiers were killed, including more than a million from Britain and the Commonwealth.[8] Millions of civilians died from starvation, disease and other war-related causes.

What made this war so appalling, aside from the availability of weapons of mass destruction, was the intense discipline resulting from high C. Peace came in 1918, but lasted only twenty-one years before international conflict broke out again. This time the results were even more catastrophic. During the global war that followed more than sixty-million people died from combat, bombing, disease, starvation and genocide.

A striking point about both world wars is the apparent lack of rational calculation. Germany and Austria had been aggressive, militaristic societies well before this time. Germany had been united under the leadership of Prussia following a short, sharp war with France in 1871. The campaign was swift and victorious, France humiliated and two border provinces seized. Germany gained immensely in terms of territory, power and prestige with minimal casualties and risk.

By contrast, in 1914–18 and 1939–45 Germany took on nations with vastly greater resources of manpower and wealth. It is as if warlike ferocity blew away any sensible calculation of national interest. The result for Germany was catastrophic, with two generations of young men slaughtered, civilians dying, cities firebombed, national humiliation, occupation and territorial dismemberment.

The same could be said of Japan, which in the decades since the Meiji Restoration had built itself from feudal backwater to industrial powerhouse. After crushing the Russian navy in 1905 it had seized Korea and Taiwan, becoming the dominant military power in the area. All this was at immense cost to the conquered peoples, but with little risk or danger to the Japanese. Then, after about 1910, the Japanese age of puberty began to fall (see Fig. 9.2 below). The decline was much less than in Germany or Great Britain, but Japan was a far more patriarchal (and thus higher V) society to begin with, so the underlying potential for aggression was greater.

---

[8] M. Heffernan, "Forever England: The Western Front and the Politics of Remembrance in Britain," *Cultural Geographies* 2 (3) (1995): 293–323.

Fig. 9.2 Age of menarche in Japan, by year of menarche.[9] Declining C from the 1920s increases testosterone and thus aggression, making aggressive war more attractive for a few decades.

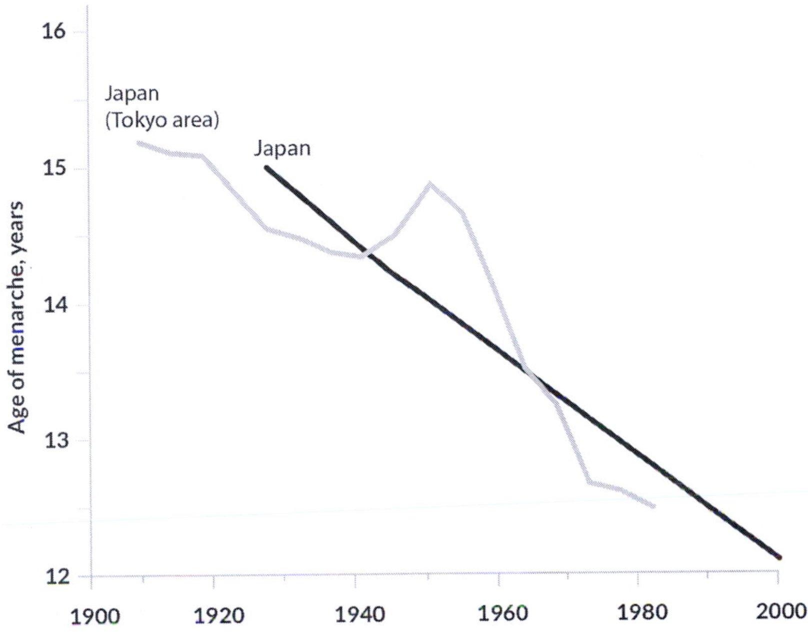

In 1931 the Japanese army occupied Manchuria. In 1937 they launched an all-out assault on the rest of China, and in 1941 declared war on the United States with the attack on Pearl Harbor and began a swift occupation of Southeast Asia, taking on the might of the British Empire. As with Germany in 1914 and 1939, the fervor for war was powerful enough to override all objections. Many of those who spoke out against the war were assassinated by nationalist fanatics. As with Germany, the Japanese took on nations with far larger populations and, in the case of Britain and the US, with greater (or potentially greater) capacity for industrial production. The result, as in Germany, was millions of deaths, mass destruction, occupation and the loss of empire.

---

[9]I. Nakamura, M. Shimura, K. Nonaka & T. Miura, "Changes of Recollected Menarcheal Age and Month among Women in Tokyo over a Period of 90 Years," *Annals of Human Biology* 13 (6) (1986): 547–54.

According to biohistory, then, it seems quite plausible that declining C, as indicated in a falling age of puberty, should be responsible for a surge of testosterone that fueled warlike aggression. But what accounts for the timing? Why did Germany launch into war in 1914 and 1939, instead of (say) a decade earlier or later? In this case it is an understanding of V which provides an answer.

Surges in V after population growth peaks

One of the main features of V, as described in chapter four, is its link with rapid population increase. High-V societies tend to have high birth rates, and population growth in itself tends to increase V. But the lemming cycle pattern, as detailed in the previous chapter, indicates that the highest V generation is the one born *at the peak of population increase*, so it is this generation that should be the most aggressive.

The figures used in all cases will be rate of natural increase, or birth rate minus death rate. It was proposed in chapter four that high birth rate increases V when a new baby forces mothers to reject their next oldest child. When death rates are high they are typically highest for infants, so older children may receive more care. Similarly, rate of population increase is a less-valid measure because it involves the movement of entire families, which does not affect the number of children raised per family.

Taking this into account, several European nations and Japan experienced rapid population growth during the nineteenth and twentieth centuries, but with peak growth (G) at very different times. To a remarkable extent, each of these peaks was followed by gradually rising militancy, resulting in the outbreak of war twenty to twenty-five years later.

France

The first European country to reach a growth peak was France. Though national statistics are not available prior to 1800, local studies provide a relatively consistent picture. These show that after very little growth during the early eighteenth century, French population grew rapidly between 1750 and 1790. The peak of growth seems to have been in the 1770s (see Table 9.1 and Fig. 9.3 below).

For example, records of nine rural departments near Paris show substantial growth between 1750 and 1790, with a clear peak in the rate of natural increase (birth rates minus death rates) in the 1770s. Paris itself shows much the same pattern. Though deaths were generally higher than births, something normal for most pre-modern cities,[10] there was significant growth between 1750 and 1780. And once again, there was a clear peak in the 1770s.

Table 9.1. Rate of natural increase in nine departments of the Paris basin, and in Paris, 1740–89[11]

	1740–9	1750–9	1760–9	1770–9	1780–9
Paris basin	0.3	8.7	6.5	11.0	7.0
Paris	-2.6	0.6	-0.8	1.9	0.0

Fig. 9.3 Rate of natural increase per 1,000 (5 year rolling average)—France 1740–2000.[12] The French Revolution and Napoleonic Wars took place when the aggressive and warlike high-V generation, born around the peak of population increase (G), came of age

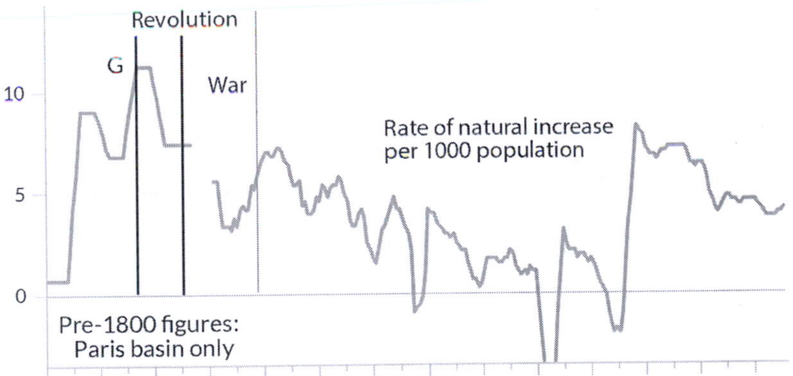

[10] Urban populations were normally maintained by migration from the countryside.

[11] Source: Louis Henry, "The Population of France in the Eighteenth Century," in *Population in History*, edited by D. V. Glass & D. E. C. Eversley (London: Edward Arnold, 1965), 443–4

[12] Ibid., 443–4; B. R. Mitchell, *International Historical Statistics: Europe, 1750–2000* (Basingstoke: Palgrave Macmillan, 2003), 95–6; Council of Europe 2002, "Recent Demographic Developments in Europe." Belgium: Council of Europe Publishing, December 2002, 34.

This provides a demographic pattern different from that of any other European country, with a peak rate of natural increase in the 1770s and a quite steady decline from then until around 1940. One likely effect of this rapid population growth was the French Revolution of 1789, reflecting the aggression of a generation born just before the peak. The other effect was war, on a massive scale.

Taking 1775 as the G year, it was exactly twenty-one years later, in 1796, that French armies invaded Italy. In the next two decades the French occupied Spain, defeated Russia, Prussia and Austria, and advanced as far as Moscow. The French appetite for war and conquest at that time was almost insatiable. In twenty years France lost more than a million men in action and from disease, especially in the invasion of Russia, along with hundreds of thousands of civilians. Yet Napoleon's escape from Elba in 1815 was greeted with widespread enthusiasm and a resurgent volunteer army, which came close to defeating the allies in the Waterloo campaign.

The fighting spirit of the French armies was remarkable. In the immediate aftermath of the Revolution the French were at war with Austria, Prussia, England and Spain, while also facing internal revolt and with an army deprived of almost all its experienced officers. Yet they drove back the invaders. It was this victory which was celebrated with the song that later became known as the Marseillaise, now the French national anthem.

When Napoleon first took over the army of northern Italy he found a force made up of misfits and rejects from the armies in the north, ragged, ill-paid and often barefoot. That such an unpromising force could seize control of northern Italy is testament to his skill as a commander, but also to the troops he led. The swiftness of his movements, in particular, reflected the extraordinary toughness of his men.

Even when Napoleon blundered, fighting spirit could save the day. At the battle of Marengo in 1800 he failed to recognize the Austrian main assault for what it was and even withdrew troops from that location. Napoleon himself considered the battle lost by 3 o'clock. It was one last counter-attack by exhausted troops that turned near-defeat into victory and drove the Austrians out of Italy.[13] Though Napoleon was a brilliant commander, his successes are easier to explain if we appreciate that he had exceptional soldiers. This is the effect of exceptionally high V, especially in a population which also had the discipline of very high C.

---

[13] A. Schom, *Napoleon Bonaparte* (New York: HarperCollins, 1997), 301–302.

In or around the 1770s, the French population was growing faster than it ever had before (or has since). The generation growing up at this time also showed an unprecedented martial vigor and appetite for war. Rapid population growth increases V, and V makes men effective and enthusiastic soldiers. But from that time onwards, the growth slowed steadily to a low point in the 1930s, when France became the first Western nation to experience fertility falling below replacement level.

As population growth declined, so did the French martial spirit. The Franco-Prussian War of 1870–71 was a crushing defeat, even though the French began it with a larger and more experienced army. More dramatic still was the collapse of French military forces in 1940, which happened within weeks, again despite an army that was (on paper) well-armed and formidable. Lower population growth reflects a decline in V, and lower V reduces the aptitude for war.

Great Britain

Britain shows a very different demographic pattern (see Fig. 9.4 below). Population growth first peaked in 1800–10. On this count alone we might expect a peak of militancy in young men reaching military age in the late 1820s and 1830s. But there is no sign of it. There were no mass citizen armies setting out on campaigns of conquest, and no clamor for war. Foreign policy was largely pragmatic. The Crown did not even bother to take over direct control of the vast domain of the East India Company until the Indian Mutiny of 1857–58 made it unavoidable.

The obvious reason is that there was at this time no decline in C and so no surge of testosterone. In fact, indications are that C was still rising to a peak it achieved in early Victorian times, so testosterone should have been substantially lower than it became at the end of the century when C began to fall.

This did not mean that young men of the 1820s and 1830s lacked courage. The Crimean War of 1853–56 saw feats of extraordinary bravery—not only the suicidal charge of the Light Brigade but the less famous charge of the Heavy Brigade, in which British cavalry charged uphill into a superior Russian force and routed them in hand-to-hand combat. This is perhaps an indication of what we might expect from courage and the energy of high V, without the quest for imperial glory driven by high testosterone.

Fig. 9.4. Rate of natural increase per 1,000 (five-year rolling average)—England and Wales 1800–2000.[14] When the generation born at the peak of population growth came of age the Boer War broke out, indicating a peak level of aggression. This was reinforced by the surge of testosterone resulting from a fall in C.

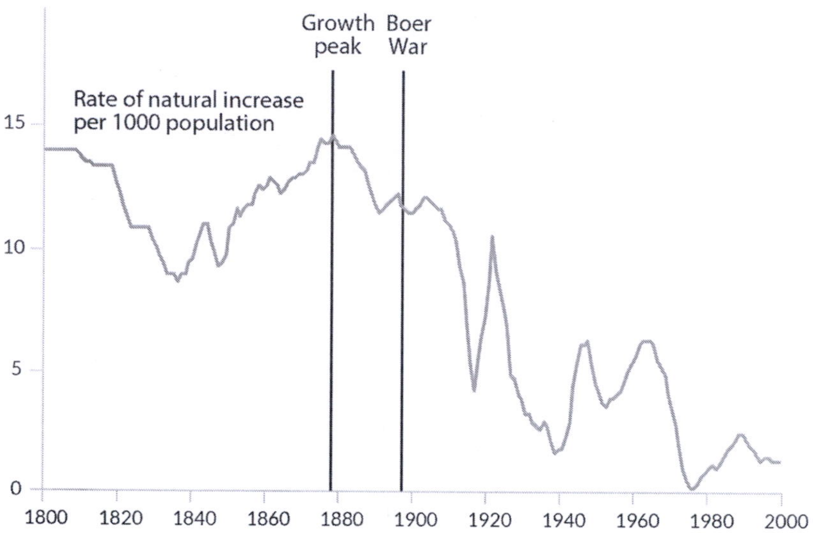

After a trough in the 1830s the birth rate and the rate of population increase began to rise steadily to a second peak around 1875, the reasons for which will be discussed in chapter ten. In chapter eight the British G year was placed in 1850, which fits other lemming-cycle patterns such as the strength of aristocracy at G-90. But the most warlike generation was not the one born at the G year but at the second peak in 1875, reinforced by the testosterone surge from falling C.

In fact it is exactly from the 1870s, when growth was most rapid, that imperialist sentiment started to rise in Britain. There was a growing enthusiasm for empire during the last decades of the century, culminating in a peak of jingoistic fervor during the Boer Wars of 1899–1902. Men rushed to volunteer, and when Mafeking was relieved after a 217-day siege there were scenes of hysterical jubilation across the country which

---

[14] Source: J. J. Spengler, *France Faces Depopulation* (New York: Greenwood Press, 1968), 53; Mitchell, *International Historical Statistics*, 95.

shocked many observers.[15] British enthusiasm for war was still strong when hostilities broke out with Germany in 1914.

This pattern is exactly what we would expect if the taste for war depended on when men of the peak military generation (early 20s) were born. Population growth was highest between 1860 and 1890 with an absolute peak around 1875. Therefore, the fervor for war was highest between 1885 and 1915, with an absolute peak around 1900.

One of the less-remembered aspects of this period is that the drive for imperial glory was accompanied by a deep sense of insecurity. There was a feeling that Britain was being surpassed by rising powers, and stories of Germany's military armament and colonial ventures were widely read. In many circles, war with Germany was seen as inevitable. It seems likely that such fears are the flipside of high-V confidence and morale. People tend to see others as having motivations similar to their own, and so aggressive people are more likely to see others as similarly aggressive, which can be a self-fulfilling prophecy. Britain in the 1930s was markedly different, with the national mood less bellicose and the dangers of Nazi Germany underestimated. Chamberlain's "peace in our time" after the Munich conference was the triumph of hope over evidence.

Germany and Austria

Next to consider are Germany, for which figures are given in Fig. 9.5 below. German population growth reached its highest level later than Britain, with a clear peak in the 1890s. Thus, purely from demographic evidence and especially considering the rapid fall in C at this time, the generation coming of age around 1914 would have been highly aggressive. A similar pattern can be seen in Austria (see Fig. 9.6 below), which was allied with Germany and whose conflict with Serbia was the trigger for war. In Austria's case the peak of population growth was slightly later—just after 1900—so peak aggression might have been expected a little later. However, V and aggression in 1914 would still have been relatively high.

---

[15] R. Hammal, "How Long Before the Sunset? British Attitudes to War, 1871–1914," *History Today* (April 28, 2013).

Fig. 9.5. Rate of natural increase per 1,000 (five-year rolling average)—Germany 1820–2000.[16] The high-V generation born at the peak of population growth (G) came of age in 1914.

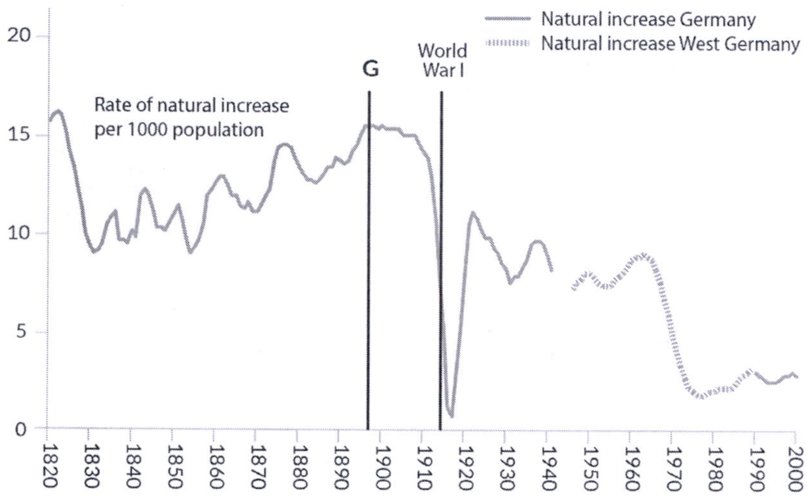

Fig. 9.6. Rate of natural increase per 1,000 (five-year rolling average)—Austria 1820–2000.[17] The Austrian peak of growth was just after that of Germany.

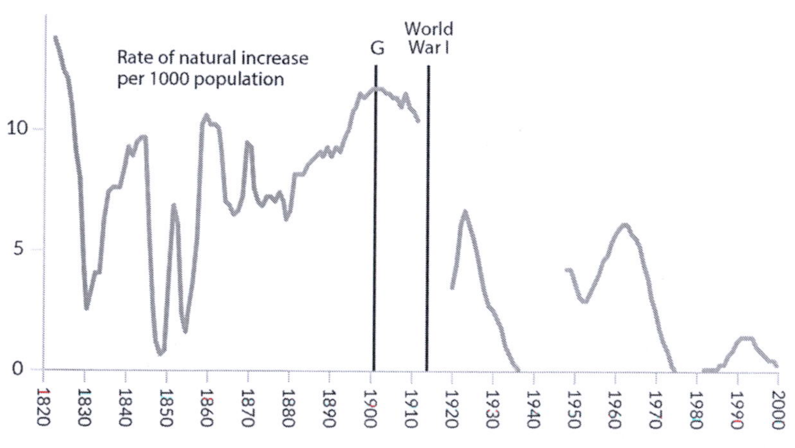

[16] Council of Europe, "Recent Demographic Developments in Europe"; Mitchell, *International Historical Statistics*, 95–112.

[17] Mitchell, *International Historical Statistics*, 95–112; Council of Europe, "Recent Demographic Developments in Europe," 34.

Russia

The next country to consider is Russia, which showed a pattern of growth similar to that of Austria with a peak just after 1900 (see Fig. 9.7 below).

Fig. 9.7. Rate of natural increase (five-year rolling average)—Russia 1820–2000.[18] The high-V generation born at the peak of increase (G) was the one that launched the Revolution and the civil war that followed.

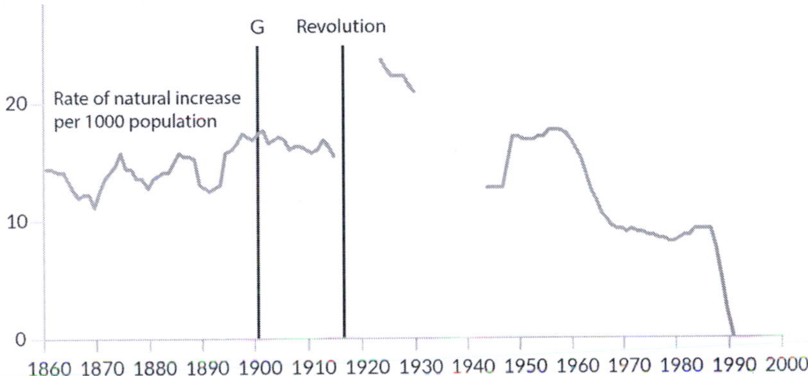

The Russian rate of natural increase remained high from the late 1890s to the outbreak of war in 1914. The influence of the super-aggressive cohort born at this time may have expressed itself through an initial enthusiasm for war in 1914:

> War hysteria quickly gripped the country and the danger of defeat was hardly discussed. The emperor's action was widely applauded. Worker and owner, peasant and landlord, civil servant, lawyer and aristocrat: all sections of Imperial society joined in the military enthusiasm. Plans for anti-governmental strikes and demonstrations were abandoned.[19]

Although the enthusiasm for war in 1914 was strong, it was perhaps more fully felt in the Revolution of 1917 and the bloody civil war that followed. This is the same pattern of enthusiastic but then brutal revolution and civil war that occurred in France after 1789, following the French peak rate of natural increase in the 1770s.

---

[18] Mitchell, *International Historical Statistics*, 100–116; Council of Europe, "Recent Demographic Developments in Europe," 34.

[19] R. Service, *The Russian Revolution, 1900–1927* (Hampshire: Palgrave Macmillan, 2009), 42.

Italy

The next country to consider is Italy, which reached its own peak of growth around 1915 (see Fig. 9.8 below). In 1935, just twenty years later, Mussolini invaded Ethiopia. In 1939 he occupied Albania, and in 1940 he formed an alliance with Nazi Germany and invaded Greece. The existence of a Fascist regime is not sufficient cause for these actions. Spain was Fascist in 1939 but stayed carefully neutral in the war, despite the help given by Germany and Italy to Franco's forces during the Spanish Civil War.

Fig. 9.8. Rate of natural increase (five-year rolling average)—Italy 1860–2010.[20] When the generation born at the peak of growth (G) around 1915 came of age, Italy launched its invasion of Ethiopia and then joined in an aggressive alliance with Germany, rather than with the Allies in the First World War.

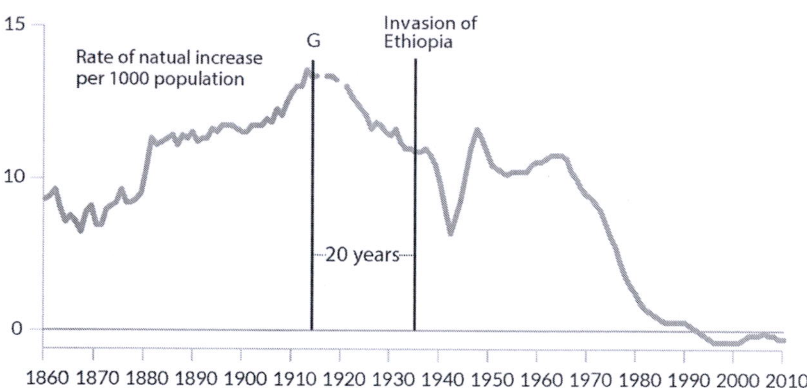

Japan

The next nation to consider is Japan, which reached its own peak of population growth much later than any European country (see Fig. 9.9 below).

[20] Source: Mitchell, *International Historical Statistics*, 106; Source: National Institute for Statistics, Ricostruzione della popolazione residente e del bilancio demografico, Table 2.3, http://timeseries.istat.it/fileadmin/allegati/Popolazione/tavole_inglese/Table_2.3.xls (accessed September 6, 2014).

Fig. 9.9. Rate of natural increase per 1,000 (five-year rolling average)—Japan 1870–2000.[21] Japan reached a peak of population growth (G) in the 1920s. The high-V generation born at that time was the one that fought in the Pacific War, their aggression reinforced by the testosterone surge from falling C.

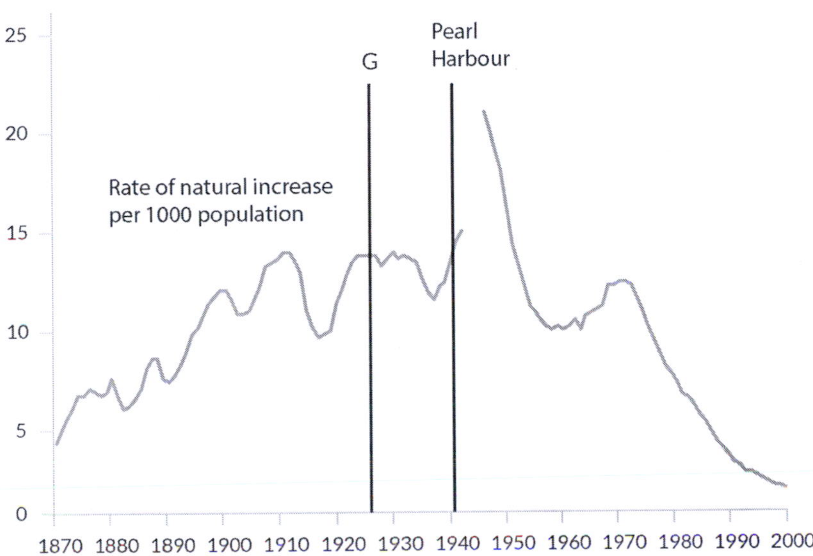

The Japanese rate of natural increase peaked in the late 1920s and early 1930s. In the 1930s there was a marked increase in warlike fervor, similar to that seen in England after the 1875 peak, but moving far more quickly into war. Japan was a much more patriarchal society than Britain at this time, which would contribute to more aggressive attitudes. The absolute peak of militancy can perhaps be seen in the kamikaze pilots who steered their planes into Allied ships in the last stage of the war. The kamikazes were of the peak-V generation born in the 1920s, when the birth rate was at an all-time high.[22]

---

[21] B. R. Mitchell, *International Historical Statistics: Africa, Asia & Oceania, 1750–2000* (London: Macmillan, 2003), 71–4; Statistical Handbook of Japan (English version). Edited by Statistical Research and Training Institute, published by Statistics Bureau of Japan, Ministry of Internal Affairs and Communications, chapter two—Population, http://www.stat.go.jp/english/data/handbook /c0117.htm#c02 (accessed September 6, 2014).

[22] M. Sasaki, *Who Became Kamikaze Pilots and how did they Feel Towards their Suicide Mission*? (Concord Review, 1997), 175–209.

The American Civil War

The final case study in this series relates to the American Civil War of the 1860s. Although demographic information is less certain than for Europe, the same pattern can be discerned of a peak growth of natural increase followed by a surge of aggressive sentiment as the peak generation came of age. In this case, though, Americans seem to have experienced two separate peaks—first in the north and then in the south. It was the generation born at the southern peak that launched the South into its war of independence.

The British North American colonies experienced rapid population growth in the late colonial period, increasing by as much as 700% between 1689 and 1760.[23] Much of this was a result of immigration, but the colonists were also comparatively well-off compared to people in Europe. Land was freely available and food was cheap, and there were none of the famines common in Europe.[24] Americans were an average 7 centimeters taller than the English by the mid-eighteenth century, a strong indication of better diet.[25]

Between 1720 and 1760 there was a demographic explosion on the coastal areas due to the extraordinary reproduction rate among the descendants of the early English colonists. During this period Connecticut's population rose from 60,000 to 140,000; Maryland's from 60,000 to 160,000; and Virginia's from 130,000 to 310,000.[26] There are no accurate demographic statistics, but from parish records and other sources, the rate of natural increase between 1720 and the end of the century seems to have been around 25 per thousand, close to the peak levels found in Europe.[27]

Such a fast-growing population suggests a very high level of V, which is supported by the extraordinary willingness of ordinary people to take up arms during the American War of Independence. In September 1774, upon hearing a rumor that Boston had been bombarded and fearing the seizer of

---

[23] R. Middleton & A. Lombard, *Colonial America, A History to 1763*, 4th Edition (United Kingdom: Wiley-Blackwell, 2011), 255

[24] Ibid.

[25] H. S. Klein, *A Population History of the United States* (Cambridge: Cambridge University Press, 2004), 55.

[26] Middleton & Lombard, *Colonial America*, 435.

[27] M. R. Haines & R. H. Steckel, *A Population History of North America* (Cambridge: Cambridge University Press, 2000), 178.

powder stores, a vast gathering of New Englanders (estimates range from twenty thousand to sixty thousand) left their homes and headed towards the action.[28] According to one eyewitness:

> For about fifty miles each way round there was an almost universal Ferment, Rising, seizing of Arms & actual March into Cambridge … [T]hey scarcely left half a dozen Men in Town, unless Old and Decrepit.[29]

Although this proved to be a false alarm, it was only a sign of things to come. The population was also eager to go to war. According to one young farmer: "The People seemed rather disappointed, when the News was contradicted." Another spontaneous uprising occurred in December 1774, four months before the British marched on Concord when armed Yankees, operating without any order from above, took control of Fort William and Mary, securing cannons and powder.[30]

While tending his fields in eastern Connecticut, Israel Putman heard news of the war and, without even returning to his house, promptly jumped on his horse and rode to Cambridge to enlist in the continental army.[31] Whether true or not, this story helps illustrate the remarkable fighting spirit of the revolutionary soldiers.

Supporting the war effort, Connecticut's committee of correspondence declared: "The ardor of our people is as such, that they can't be kept back." Within a week, 3,716 men from Connecticut were marching to the aid of their neighboring colony and by early summer an estimated twenty thousand revolutionaries had surrounded Boston, trapping the British soldiers in the confines of the city.[32]

The rate of natural increase remained strong into the early nineteenth century, perhaps even rising slightly, but from the 1820s it began a steady fall. This suggests a G period around 1825 (see Fig. 9.10 below)

---

[28] R. Raphael, *A People's History of the American Revolution: How Common People Shaped the Fight for Independence* (New York: New York Press, 2001), 47.

[29] Ibid., 47.

[30] Ibid., 48.

[31] Ibid., 50.

[32] Ibid., 50.

Fig. 9.10. Rate of Natural Increase per thousand—United States 1790–1990.[33] Americans went to war with Mexico in a giant land grab, just twenty years after the growth peak.

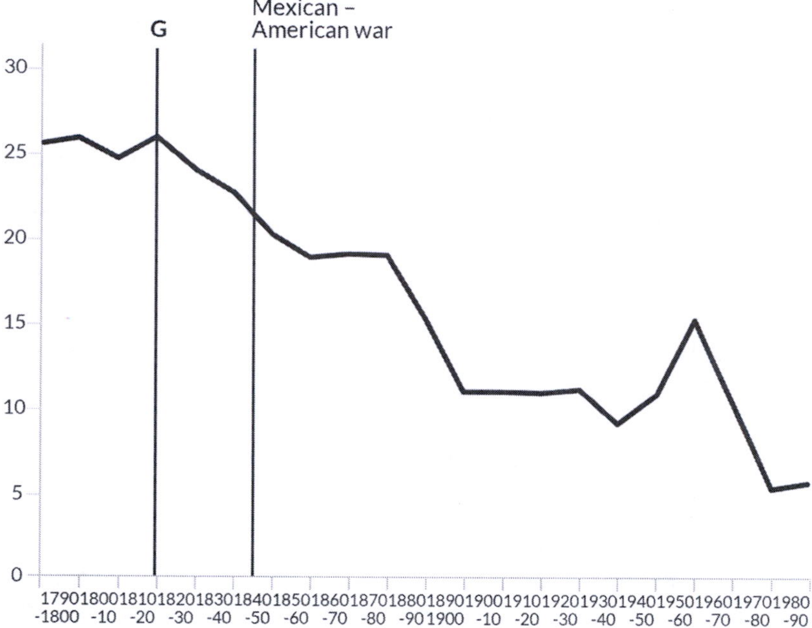

Fuelled by rapid population growth, the period from the conclusion of the war of 1812 to the outbreak of the civil war in 1861 was marked by continual territorial expansion, a concept which was exemplified by the concept of Manifest Destiny, as popularised by John O'Sullivan in 1845:

> … that claim is the right of our manifest destiny to overspread and to possess the whole continent which providence has given us for the development of the great experiment of liberty and federated self-government entrusted to us.[34]

Even those who were opposed to American expansion held a sort of fatalistic resignation that some higher power was directing the nation. In 1837 Boston preacher William Channing wrote an open letter to Henry Clay, criticizing "destinarian" justifications towards national policy.

---

[33] Haines & Steckel, *A Population History of North America*, 315.

[34] A. Stephanson, *Manifest Destiny, American Expansion and the Empire of Right* (New York: Harper Collins, 2005), 42.

> We are destined to overspread North America; and, intoxicated with the idea, it matters little to us how we accomplish our fate. The spread, to supplant others, to cover a boundless space, this seems our ambition, no matter what influence we spread with us. Why cannot we rise to noble conceptions of our destiny?[35]

Many saw the expansionistic zeal of this period as part of a universal plan in which Providence was working behind the scenes to lead the nation towards its intended destiny.[36]

In the northern states, this sentiment reached a peak in the 1840s, finding expression in the Mexican-American war of 1846–48. This, of course, is twenty to twenty-five years after the G period, the same pattern as found in Europe and Japan.[37]

But there is an exception to this pattern. While birth rates in the northern states fell rapidly after the 1820s, those in the southern regions fell more slowly or, in the case of the West-South-Central region, actually rose. They only began falling rapidly after 1840.[38] Although these figures do not take account of death rates they are suggestive of a G year around 1840 in the south, which is significantly later than the north.

If this is the case we would expect to see rising expansionist sentiment in the 1840s and 1850s, with war most likely twenty to twenty-five years after the G year, meaning the early 1860s. And this is exactly what happened. Southern notions of expansionism flavored by the language of Manifest Destiny only became popular in the 1850s, associated with the sectionalism that would eventually lead to the Civil War. There was a powerful sentiment for spreading slavery and the South's version of Americanism into the tropics, now reclassified as a paradise. As commented by one Southern congressman on Southern expansion into the tropics: "With swelling hearts and suppressed impatience they await our coming, and with joyous shouts of 'Welcome! Welcome!' they will receive us.'[39] And so it was that twenty-one years after our proposed southern G year of 1840, at an absolute peak of V in what was already a very high V society, the south launched its war for independence.

---

[35] Ibid., 50, 51.
[36] Ibid., 52.
[37] Stephanson, *Manifest Destiny*, 56.
[38] Y. Yasuba, *Birth Rates of the White Population in the United States, 1800–1860: An Economic Study* (Baltimore: Johns Hopkins Press 1962).
[39] Stephanson, *Manifest Destiny*, 64.

Like the Napoleonic wars and the First World War, also largely fought by generations born at the peak of G, it was characterized by mass citizen armies and an extraordinary fighting spirit. The carnage was made even worse by high-C discipline and industrial technology created by high infant C. It was the deadliest war in American history, with the loss of over 750,000 soldiers and an unknown number of civilians. An estimated 8% of white males aged 13=43 died in the war, including 6% in the North and an incredible 18% in the South. This was about as many as died in all subsequent wars, despite a much smaller population.[40]

Decisions about war and peace are made at all levels of society and in all age groups, especially in a democracy. Political and military leaders tend on the whole to be much older men. And yet, historical evidence suggests that wars often do break out when a generation born at a peak of population growth reaches peak military age—normally their early 20s. Surprising as it may seem, the warlike fervor of this generation somehow affects the entire society, or at least provides an unusual opportunity for aggressive and warlike leaders.

Post-war Booms

Finally, it may be noted that most countries experience a surge in birth-rate and population increase after major wars, to some extent because of births delayed during the war. In Japan, for example, the birth rate was higher after 1945 than at any point during the 1930s and 1940s. There was also a post-war baby boom in the 1950s and 1960s in many Western countries, which in the next chapter will be attributed to the Great Depression. Yet neither of these population surges seems to result in increased aggression once the age cohort matures, which should be obvious since they are (by definition) large cohorts. For example, American baby boomers born in the 1950s and early 1960s came of age in

---

[40] "U.S. Civil War took bigger toll than previously estimated, new analysis suggests", *Science Daily* (September 22, 2011); D. J. Hacker, "Recounting the Dead," *The New York Times* (September 20, 2011); J. M. McPherson, Battle Cry of Freedom: The Civil War Era (Oxford, New York: Oxford University Press, 1988) xix; M. Vinovskis, Toward a Social History of the American Civil War: Exploratory Essays (Cambridge: Cambridge University Press, 1990), 7; R. Wightman Fox, "National Life After Death: Civil War carnage and the quest for American identity" *Slate* (January 7, 2008) https://web.archive.org/web /20110716083839/http://www.slate.com/toolbar.aspx?action=read&id=2180856 (accessed September 12, 2014).

the 1970s and early 1980s, when student activism was in decline and there was no taste for military adventure. Thus it seems that the V which drives these major wars is more a product of the lemming cycle than of population growth as such. Population growth stemming from other causes does not appear to increase V.

Aggression arising from trauma experienced in infancy

The First World War took place when a super-aggressive generation born at a peak of population growth came of age. This was especially the case for Germany but applied also to Austria, Russia and Britain. Combined with the testosterone surge generated by falling C, a passion for war swept over the entire society. But only twenty-one years later another vast conflict broke out. It is true that the testosterone surge would have continued, as indicated by the continued fall in the age of puberty (especially in Germany), but what could account for the timing of this conflict?

To understand this we must return to the origins of V as discussed in chapters four and five. V is maximized by the influence of a highly anxious but indulgent mother in infancy, with an abrupt separation around age 2 when a new baby is born. This separation, although traumatic in itself, prevents an overly close bond to the mother maintaining high anxiety into adult life—a kind of "mama's boy syndrome."

The most typical cause of maternal anxiety is patriarchy, which makes women more anxious by subordinating them to men, though living with a critical and fault-finding mother-in-law serves much the same function in many societies. But any kind of maternal stressor could and should have the same effect, especially if it abruptly ceases at around age 2.

Full-scale war is a hugely stressful event for most people. Families are subject to violent intimidation by occupying soldiers, or flee their homes in terror. People cower in cellars and shelters from bombardment by artillery or aircraft. Women wait while their husbands and sons go off to war, dreading the telegram or knock on the door which tells that their worst fears are realized. In many cases, famine and privation follow as resources are diverted to the war effort or as fighting blocks off supplies.

Then the war ends, and in conflicts such as the Great War it is usually quite sudden. The fighting stops, people celebrate, the men return home and life even for the defeated becomes more normal. It follows from this that the age cohort with the highest V will be those born in the last two years of the war and who grow up during the subsequent peace. The more stressful the war and the more absolute the peace, the more aggressive should be the rising generation of young men.

There is some direct evidence of such an effect. The Six Day War of June 1967 was enormously stressful to Israeli women, with absent husbands adding to the threat of attack by the vastly larger armies of neighboring countries. A study of children born between before the war in early 1967, compared with a comparison group born between January 1969 and March 1970, showed that by school age the "war children" were found to be more tense and anxious, more often shameless liars, and more likely to run away from home. They were more socially withdrawn, irritable, resentful, hyperactive and lacking in consideration for others. Children born in the months after the war, whose mothers had experienced wartime stresses during pregnancy, were very little affected.[41] The key point about this study—unlike others showing that direct experience of war violence makes children more aggressive and anti-social[42]—is that these children would have absolutely no memory of the war. This is almost certainly an epigenetic effect.

Of course, Israeli society after the war was not immune from the stresses imposed by terrorism and hostile neighbors, but the important fact is that these stresses were far less than during the war itself when the nation was threatened with annihilation. It is the reduction of stress after infancy that maximizes V.

There is good historical evidence that wars commonly break out within twenty to twenty-five years of the end of a previous major conflict. In China, which has detailed historical records for almost three thousand years, it is common for a war of unification to be followed by a second a short generation later. China was first united under the Qin in 221 BC and widespread rebellion broke out in 209 BC. This is a gap of only twelve years, of course, but the final stages of unification seem to have involved

---

[41] A. Meijer, "Child Psychiatric Sequelae of Maternal War Stress," *Acta Psychiatrica Scandinavica* 72 (6) (1985): 505–511.

[42] R. Feldman & A. Vengrober, "Infants Remember: War Exposure, Trauma, and Attachment in Young Children and Their Mothers," *Journal of the American Academy of Child & Adolescent Psychiatry* 50 (7) (2011): 640–641; G. Chimienti, J. A. Nasr & I. Khalifeh, "Children's Reactions to War-Related Stress," *Social Psychiatry and Psychiatric Epidemiology* 24 (6) (1989): 282–287.

relatively little violence.[43] The exact timing can be further explained by the death of the Qin founder, a brutal autocrat, in 210 BC.

Another major unification was completed in 590 AD with the crushing of the south by the new Sui dynasty. By this victory they came to control all of China except the most extreme southerly province of Yunnan. Rebellion broke out in 614 on a massive scale. Between 614 and 624, some two hundred mutinies and rebellions seem to have affected virtually every province and army unit in the empire.[44] This is a gap of 24 years, which is exactly the time when the generation born at the end of the previous war came of age.

Another war of unification was concluded in 1368–71 with the triumph of the Ming dynasty. In 1398 the empire was again plunged into a violent civil war that lasted three years.[45] This is a gap of twenty-eight years, a delay which can be attributed to the influence of the Ming founder who crushed all dissent with a merciless reign of terror. The revolt took place within a year of his death.

Widespread rebellions in the 1630s allowed the invading Manchus to take Beijing in 1644 and declare a new dynasty. The south, again with the exception of Yunnan, was pacified by 1651. In 1673 the massive "Revolt of the Three Feudatories" broke out, bringing about another five years of civil war.[46] The gap was twenty-two years—just enough time for the generation born at the end of the previous war to come of age.

In 1911 the Qing dynasty collapsed, leading to decades of civil war and foreign invasion. Local warlords seized power and fought for supremacy. The nationalist government based in Nanjing was never in effective control of most of the country, and it soon retreated before the Japanese invasion. Only in 1949 was China again at peace, until the massive unrest of the Cultural Revolution in 1966–76 threatened a return to civil war.

The gap this time was only seventeen years, but it is the exception that proves the rule. Following the economic disaster of the Great Leap Forward at the end of the 1950s, Chairman Mao Zedong needed to reinforce his own authority, which had been harmed by the failure of the

---

[43] J. Keay, *China, a History* (London: Harper Collins, 2008), 87, 110–111.
[44] Ibid., 233.
[45] Ibid., 372–5.
[46] Ibid., 417, 432, 436.

plan, and wanted to impose what he saw as a purer form of Communism. His call to action was taken up by mass demonstrations of high-school students, starting in Beijing and then spreading to the rest of the country. This was exactly the cohort which had been born at the end of the civil war in 1948–49 and had grown up in peacetime,[47] the experience that creates the highest level of V. Characteristically, the demonstrations descended into chaos, which threatened to get out of hand. The army was called in to restore power, and many students were sent to the countryside. Although Mao was able to stir up and use the young generation's militant fervor for his own purposes, he was not the creator of it.

The Cultural Revolution was not a war but it was highly traumatic for those who lived through it. The student unrest of 1989, including the occupation of Tiananmen Square in Beijing, largely involved men and women born during the Cultural Revolution. This can be considered as a kind of double echo of the end of the Civil War.

The Second World War

In all the cases we have looked at thus far, there is evidence of a super-aggressive generation born at the end of the previous war. In normal circumstances another war breaks out just when this generation comes of age, between twenty and twenty-five years later. When a powerful and brutal autocrat is in control this generation's rebellion may be delayed until their death. Alternatively, when the autocrat is the one stirring up and harnessing the violence, as in Mao's case, it can be advanced by empowering this generation prior to full adulthood. But in any case, the impact is profound.

And it is this pattern which can account for the Second World War—not only its occurrence, but also its timing. The Great War was a massive trauma for the nations which fought in it, perhaps most of all for defeated Germany and Austria. Boys born between about 1916 and 1918 would have felt the full effects of this anxiety transmitted through their mothers in infancy, while experiencing lower anxiety in the post-war years. Their level of V and thus aggression would be exceptionally high. This was the generation which reached their early 20s between 1936 and 1939. Women of the same age would also have been highly aggressive, helping to provide support for the Nazi program of conquest.

---

[47] Keay, *China, a History*, 499–500, 508, 521–3.

Although the surge of aggression was experienced most strongly in Germany, it is likely to have affected other countries as well, in a variety of ways. In 1933 the Oxford Union resolved by a large majority "That this House will in no circumstances fight for its King and Country." Similar motions were also adopted by the student bodies of Manchester and Glasgow universities. When Randolph Churchill visited the Oxford Union to try and get the resolution deleted he was met with hisses and stink bombs. Yet, a mere six years later, when the War Office set up a recruiting board at Oxford, 2,632 out of a potential three thousand volunteered.[48] Such a change can be explained by the arrival of a new and more militant generation, experiencing infant anxiety between 1916 and 1918 and lower anxiety thereafter. This would give them a level of V and thus aggression distinctly higher than those born before or after. Hence the radical difference between the student cohorts of 1933 and 1939.

It must be emphasized that it is the combination of infant anxiety with lower anxiety immediately after age 2 that creates this most aggressive generation. The youngest students in 1933 would have been born during the war in 1914 and 1915, but did not experience the fall in anxiety until they were 3 or 4. Most of the 1933 students were born before the war.

Looking at the half-century following the Second World War, we might expect another surge of V and thus aggression in those born between 1943 and 1945, who would reach their early 20s between 1963 and 1966. These dates cover not only the escalation of the Vietnam War but lead into the widespread rioting and student unrest of 1968 which affected most of the combatant nations.

As a final example, Iran experienced both a peak birth rate and a savage war against Iraq in the 1980s, which biohistory suggests would produce high V and a peak level of aggression in those coming of age in the years around 2005. In that year, Mahmoud Ahmadinejad was elected president, the only candidate to adopt a pronounced anti-American line. In the years that followed Iranian militancy frightened its Arab neighbors as well as Israel, and its drive for nuclear capability alarmed the rest of the world. But biohistory suggests that these bellicose attitudes should die down over the next decade or so, especially given the falling birth rate. The election of a more moderate president in 2013 is perhaps a sign of such a trend.

---

[48] P. Addison, "Oxford and the Second World War," in *The History of the University of Oxford: The Twentieth-Century*, edited by B. Harrison, 167 (Oxford: Oxford University Press, 1994).

The idea that a whole society's attitude to war can be largely set by young men (or women) in their early 20s is quite extraordinary. But it can be proposed that the energy and enthusiasm of a youth cohort has an effect far beyond its numbers, especially if it is combined with the confidence and fervor of high V. This may have been assisted by the fact that members of the hyper-aggressive generation which fought in the 1914–18 war were now in positions of leadership—not only men such as Hitler and Goering, but the commanders on both sides.

Put together, biohistory proposes that the aggressive militarism in the first half of the twentieth century resulted from a "perfect storm" of conditions coming together at the same time. Family patterns that were still patriarchal combined with peaks of birth rate and the aftereffects of the Great War to maximize V. Added to this was a surge of testosterone as C declined rapidly from the late nineteenth century (or the early twentieth century in Japan).

Exceptions

Such a combination of factors makes war more likely, but certainly not inevitable. The obvious reason Iran did not launch a war in the early twenty-first century is that its leaders were aware of their relative weakness. Though driven to challenge American dominance, Iranians were all too aware that the "Great Satan" had sharp teeth—complete air and sea supremacy, a large and well-trained military, numerous regional allies and, of course, nuclear weapons. A high-V cohort makes war and civil disturbance more likely, but other factors—including rational calculation—often play a part.

Sweden had much the same demographic experience as Germany in the late-nineteenth century without going to war in 1914. One likely reason is that C fell less rapidly in Sweden than in Germany, as indicated by slower decline in the age of puberty. Swedes were also less patriarchal than Germans, as reflected in legal changes in the previous century. In this period a number of professions were opened to women including post and telegraph, gymnastics and the railway office. Unmarried women were granted the same rights as men in trade and commerce. Married women were granted control over their own income, and husbands were forbidden to abuse their wives. In Germany, barely any liberal rights were granted to women at this time.

The Japanese born in 1944–45 were aggressive enough, as indicated by the anti-war violence and protests of the late 1960s. Biohistory proposes that the contemporary protests in Europe and America had a similar origin, though in these cases it is not surprising that they did not result in wars of conquest. Western nations were less militant in the late 1960s than they had been in 1939, since V had been declining with the birth rate between 1918 and 1945. But the Japanese birth rate was still high, and Japan was still close to its peak level of V. Also, the Japanese were more patriarchal than most Europeans and Americans. Perhaps the rational lessons of the defeat in 1945 were strong enough to overcome the epigenetic effects in this case. America was, after all, so much more powerful by this time that any military resurgence would have been suicidal.

Germans may have been no more aggressive in 1939 than Iranians are today, but Germany was in a far more powerful position relative to its neighbors. It had a large and disciplined population, productive industries, and the most advanced industrial technology in the world. Had it not been for Hitler's insane racial policies, brilliant Jewish scientists might have given him the atom bomb in 1944 instead of helping America to achieve it a year later. In that case, the outcome of the war could have been quite different.

The sheer scale of the Second World War was made worse by what may be termed an unfortunate coincidence. This is that the Japanese experienced a surge of V and testosterone just as Europe descended into chaos. Japan might have been less inclined to attack Pearl Harbor had not the Soviet armies been shattered by Operation Barbarossa a few months earlier. Thus again we must emphasize that while a surge of epigenetically-based aggression makes war far more likely, considerations of national strength and advantage play an important role in determining whether it actually breaks out.

Conclusion

The propensity of a society to make war depends on several factors. One of these is harsh living conditions, especially experience of famine. In complex civilizations, aggression is maximized when men experience anxiety in infancy but much lower anxiety after age 2. This can be brought about by patriarchy (which subordinates women), an especially high birth rate, and the ending of a major war. All of these factors increase V.

Wars are especially likely when the generation with the highest V reaches their early 20s, and also experiences a surge of testosterone as a result of falling C. The wars of the early twentieth century, and a number of wars in Chinese history, can be explained in this way.

This hypothesis leads to the conclusion that the taste of nations for war is largely determined by the attitudes of people in their early 20s. This is not especially plausible, but a scientific theory need not be plausible. What it must be is testable.

CHAPTER TEN

RECESSION AND TERROR

So far we have seen the effects of two major cycles: civilization cycles and lemming cycles, both of which are linked to population growth and decline. Civilization cycles show rapid population growth in the high-C period, and lemming cycles show maximum growth around the "G" year, a generation before the peak of V.

But there is another type of fluctuation in population growth that cannot be explained by either of these two cycles. One example is a period of slower growth in mid-nineteenth century Britain. The growth of British population between the late eighteenth and early twentieth centuries, and most of all in the nineteenth century, can be explained by a lemming cycle with a G year around 1850 (see Fig. 10.1 below).

Fig. 10.1. England and Wales—Rate of Natural Increase 1800–1900[1]

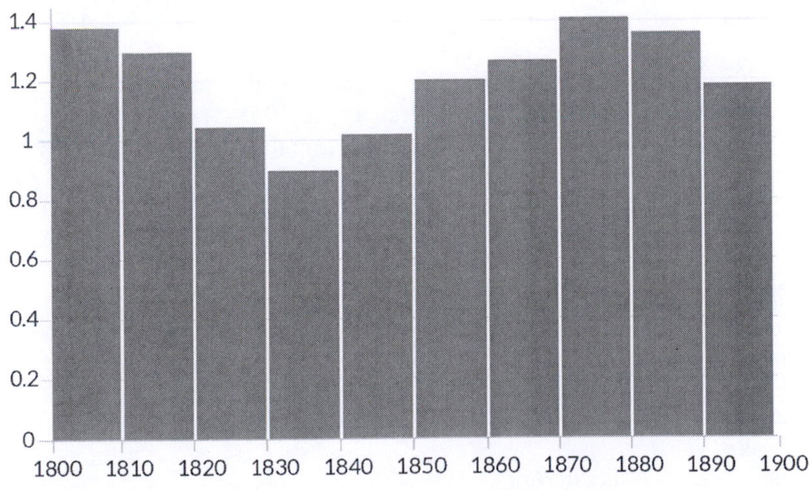

[1] J.J. Spengler, France Faces Depopulation (New York, Greenwood Press, 1968): 53

But, in fact, peak rates of natural increase are found in 1800–1810 and in the 1870s, with a slight drop to a low point in the 1830s and 1840s Another example is the "baby boom" which followed the Second World War in many Western countries, and was preceded by a "baby bust" in the 1930s (see Fig. 10.2 below).

Fig. 10.2. US crude birth rate 1910–2010 (births per 10,000 population)[2]

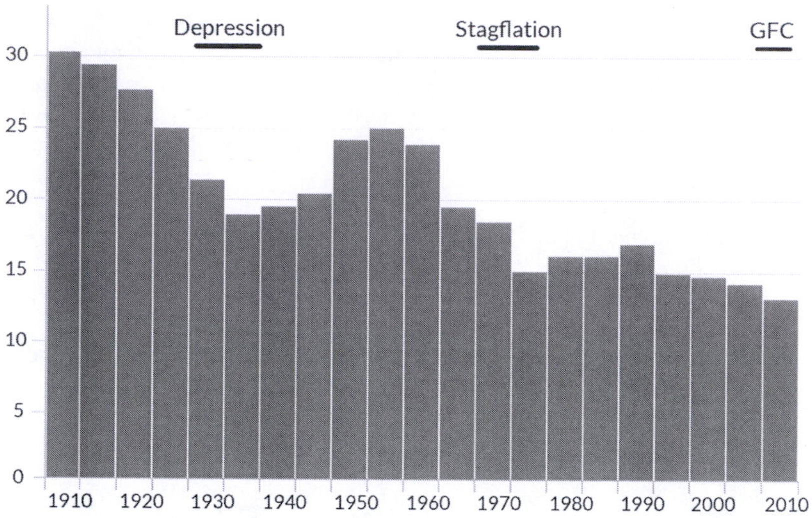

Throughout the twentieth century there has been an overall decline in the US birth-rate, but there was a sharp recovery in 1935–55 and a milder one in 1975–90. It is easy to see that the "troughs" are associated with economic recession, such as the Great Depression of the 1930s and the Stagflation of the 1970s. However, the greatest *fall* in the birth rate happened not during recessions but in periods of prosperity—the 1920s and 1960s. These were also periods of hedonism, when traditional standards of sexual behavior were loosening. In terms of biohistory, these were times of rapidly falling C. The periods of rising birth rate, especially the late 1940s and early 1950s, were more conservative.

---

[2] Department of Health and Human Services, National Center for Health Statistics, www.dhhs.gov
National Vital Statistics Reports 61(1)
" U.S. Census Bureau Announces 2010 Census Population Counts – Apportionment Counts Delivered to President" (Press release). *United States Census Bureau*, (December 21, 2010).

A similar pattern applies to nineteenth-century England. The period of falling birth rate in the 1810s and 1820s was morally conservative by the standard of the twentieth century, but the onset of the Regency marked the beginning of a wilder and more frivolous social scene, compared to the preceding, more sober era. This was exemplified by the behavior of the Prince Regent, later George IV.[3] The period of rising birth rate in the 1840s to 1860s, by contrast, was Victorian England at its most straight-laced, with a loosening at the end of the century as the birth rate began falling.

Applying biohistory to this pattern helps us to understand why these fluctuations took place, and also in particular why major recessions and political turmoil are more likely to occur in the birth-rate troughs. This applies not only to the 1930s and 1970s but also to the recent Global Financial Crisis. Because recessions and political instability are so characteristic of the trough periods of these cycles, we will refer to these cycles as "recession cycles." This does not mean, of course, that they provide an explanation for all recessions or for short-term market fluctuations.

Why Recessions Occur

There is an immense literature on the reasons for economic recessions such as the Great Depression of the 1930s. Keynes attributed it to a fall in demand. Monetarist explanations include the failure of the Federal Reserve System to prevent bank failure and maintain a large enough supply of money. Other approaches emphasize the stock market crash of 1929, price-fixing by the Federal Government, and measures of protectionism that slowed global trade.[4]

Biohistory takes a completely different view, one based on physiology rather than economics. Most people might expect the level of stress to rise as a *result* of recession, but biohistory suggests that the most significant increase in stress occurs *before* the recession and in fact causes the

---

[3] E. A. Smith, *George IV* (New Haven and London: Yale UP, 1999).

[4] J. M. Keynes, *The General Theory of Employment, Interest and Money* (Edison Martin, 2013); J. K. Galbraith, *The Great Crash 1929* (London: Penguin, 2009); S. Bernanke, *Essays on the Great Depression*, (Princeton: Princeton University Press, 2000); M. Friedman, *A Monetary History of the United States, 1867–1960* (Princeton: Princeton University Press, 1971).

recession to occur. This is a surge of stress above all in the influential youth generation. In other words, the chronic level of stress hormones such as cortisol and ACTH rises, eventually causing a sense of anxiety which involves fear and loss of confidence. To understand how this happens, we need to consider the two key factors which determine the level of stress.

The first of these is the level of V, which, as we saw in chapter four, makes people intolerant of crowding. This is part of the adaptation that drives migration when food supplies are unstable. Thus it is that *at any given level of population density*, higher V people experience greater stress.

The second factor is the level of C, which does not so much affect the actual population density as the way it is perceived, which in turn depends on whether other people are sought out or avoided. People with lower C and thus higher testosterone are more hedonistic and pleasure seeking and tend to seek out stimulus *in general*. This includes social activities such as discos and large sporting events, but also activities that induce excitement and a sense of danger such as vigorous sports, rollercoaster rides, hunting or fighting. Anything that sets the heart thumping with adrenaline is a stressor, even when eagerly sought after. Such people tend to be less anxious than others, at least partly because high testosterone reduces the level of anxiety.[5] Nevertheless, the net effect of their lifestyles is to raise the level of stress.

People with high C and lower testosterone, by contrast, are more likely to be hard working, disciplined and religious. They prefer to avoid not only crowds but other forms of stimulus, all of which has the effect of lowering their level of stress.

To illustrate how this works consider two people, John and Brad, who have the same level of V, which means their tolerance for crowds is roughly the same. If they were both in exactly the same environment, seeing as many people each day and with the same type of recreations,

---

[5] J. L. Aikey, J. G. Nyby, D. M. Anmuth & P. J. James, "Testosterone Rapidly Reduces Anxiety in Male House Mice (Mus musculus)," *Hormones and Behavior* 42 (4) (2002): 448–460; J. Van Honk, J. S. Peper, D. J. L. G. Schutter, "Testosterone Reduces Unconscious Fear but Not Consciously Experienced Anxiety: Implications for the Disorders of Fear and Anxiety," *Biological Psychiatry* 58 (3) (2005): 218–225.

they would have the same levels of stress. But John has high C, which means he prefers the quiet life. He is married and spends most evenings at home, or visiting with a small circle of close friends. He works as an accountant. He goes to church and is happy with a quiet life.

Brad, on the other hand has lower C, which also means he has higher levels of testosterone. He is a trader in a securities firm, a high-pressure job with the potential for big bonuses but also big losses. He is separated from his wife and has a stormy relationship with a co-worker. They have just had a big fight over one of his "flings." He goes to nightclubs two or three nights a week and has a huge circle of friends and contacts. He goes drag racing at weekends. He loves the challenge and excitement of his life, the constant change, the new people.

Both of them live in the same city and could even be neighbors, but their environments are completely different. Based on the number of people he sees and the background noise, psychologically John might as well live in a small country town. Brad's environment, on the other hand, has the level of stimulus of a crowded, noisy city. Though the actual population density for both men is similar, Brad's *perception* is of much higher population density and greater danger or challenge. Presuming that both men have a similar level of V, their lifetime experiences would increase Brad's stress levels and lower John's.

In other words, John's high levels of testosterone and lower levels of C lead him to seek out exhilarating stimulus such as wild parties, late nights, drugs and extreme sports which act to raise his levels of stress. It might seem strange that people can choose activities that increase stress, but this is a well-known facet of human psychology. A common reaction to excitement, novelty and danger is a rise in adrenaline and cortisol, both stress hormones. At moderate levels this can (for some people) be pleasant, even euphoric. It is exactly the thrill involved in extreme sports such as bungee jumping and mountain climbing. The danger is what makes it attractive. The same can be said of the crowds, flashing lights and loud, thumping music at a nightclub or disco, as well as the thrill experienced by elite traders in a fast-moving market during a boom.

But the fact that a particular kind of stress may be welcome does not mean it is safe. A study by two psychiatrists found that people who experienced divorce, the death of a spouse or financial crisis suffered increased risk of illness and death. But surprisingly, they found that positive events such as marriage, a change of job, the birth of a child or outstanding personal success were also stressors and increased the chance of illness or death.[6] Thus it is that people like Brad actively seek experiences which raise their underlying stress levels. We can now apply these insights to explaining boom and bust.

The Boom

Both C and V have been falling in the past century, as described in chapters seven and eighty, V somewhat faster than C for most of the time. This means that the *preference* for high levels of stimulus has been increasing as C declines, but *tolerance* for stimulus has also increased even faster because of the decline in V. For most of the twentieth century, this means that the level of stress has been falling.

But, as indicated earlier, there have been periods when the fall of C has been especially rapid. In such periods the desire for stimulus (set by C) increases faster than the tolerance for it (set by V), leading to a rise in stress. When it falls less rapidly or stabilizes the desire for stimulus increases more slowly than the tolerance for it, reducing stress.

C can change to some extent in later life but the epigenetic effects of early experience are far more powerful. Early experience means the period from age 5 or 6 through adolescence. The rat experiments described in chapter two show that CR promoters such as food restriction in infancy either do not affect or may even raise the level of testosterone, and also that C-promoters immediately after puberty have a strong and lasting effect on levels of C.

Thus it is that people growing up in tough times tend to have higher C, and those growing up in prosperity have lower C. The most conservative generation will be those experiencing these tough times between 5 and perhaps 18, which raises C and child V. It will be recalled from chapter five that Manus children going through late childhood and adolescence in the turbulent war years were more conservative and serious-minded than those born earlier or later.

The other factor to consider is that the youth generation appears to have an outsized effect on society. We saw in the last chapter that wars tend to break out when men experiencing stress in early infancy reach their early

---

[6] T. H. Holmes & R. H. Rahe, "The Social Readjustment Rating Scale," *Journal of Psychosomatic Research* 11(1967): 213–218.

20s. So it is that the behavior of people in their 20s affects not only their levels of stress but that of society as a whole.

Bringing all this together, Fig. 10.3 below shows how changes in the level of C in the younger generation might explain the various financial crises of the twentieth century.

Fig. 10.3. Uneven decline in C as an explanation for surges in stress levels. As C drops faster than V in hedonistic eras such as the 1920s and 1960s, the search for stimulus (set by C) increases faster than the tolerance for it (set by V), causing a rise in stress. This results in economic recession and political extremism.

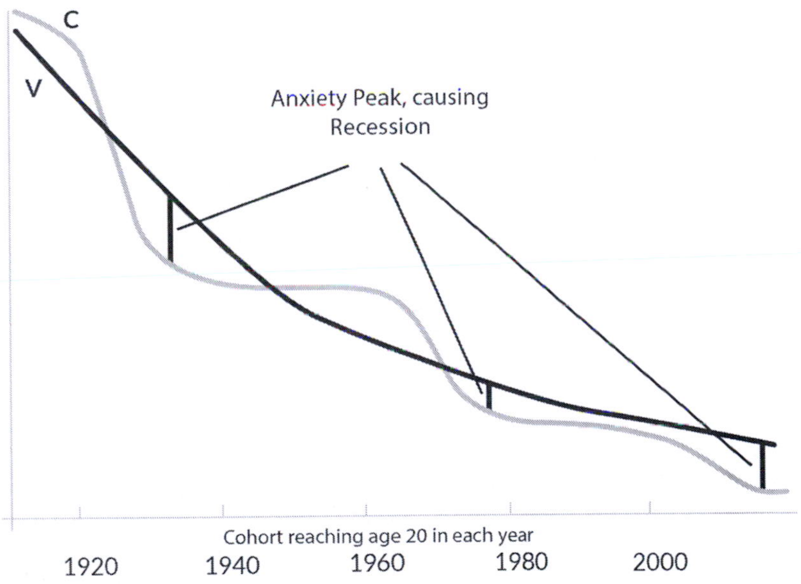

Let us see how this works out in practice.

The 1890s was a time of economic recession and uncertainty, which would have increased or at least stabilized the C of people born after about 1895. When this generation came of age from 1905 onwards they would prefer to avoid stimulus, which would reduce their stress levels and that of society as a whole. Thus the recession eased and prosperity returned.

But greater prosperity after 1900 meant that the people coming of age in the 1920s had much lower C. The result was a surge in hedonistic and pleasure-seeking behavior, both collectively (such as in the culture of flappers, wild parties and illegal liquor during the inter-war period) and in individual personal behavior.[7] For instance, casual dating replaced the earlier chaperoned courtship of the Victorian era, and popular dancing became increasingly sexualized. The *Kinsey Report* shows that women born after 1900 were far more likely to indulge in early sexual activity than women of previous generations.

One obvious effect was that the birth rate plummeted. Both V and C are associated with the desire for children, which is why populations increase so much during the high C phase of the civilization cycle and the high V phase of the lemming cycle. As V and C fell in the early twentieth century so did the birth rate, but the fall was especially rapid in the 1920s when the decline of C was most rapid. This was particularly so given that people in their 20s and 30s are the ones most likely to have children.

Then there were the economic effects. People with lower C tend to have higher testosterone, and testosterone-fuelled optimism is no hindrance to economic growth. The 1920s saw an unprecedented boom.

An initial rise in stress may even have contributed to the boom. As stress rises people can become increasingly anxious, even though they might still hold a positive outlook due to the general exuberance of the time. Numerous studies have shown anxiety to be associated with problem gambling and poor judgment.[8] Added to the risk-taking exuberance of high testosterone, segments of the population could have been experiencing a rise in anxiety, leading to gambling on a society-wide scale. An example of this is people mortgaging themselves to buy stocks on the assumption that values would continue to rise indefinitely. The dramatic rise of companies like the Radio Corporation of America (RCA) during the 1920s helps explain the attraction of the stock market at that time. In 1921 RCA stock sold for $1.50 a share, rising to $4.75 in 1923,

---

[7] F. A. Allen, *Only Yesterday: An Informal History of the 1920s* (New York: Harper Perennial, 1920), chapter five.

[8] A. P. Blaszczynski & N. McConaghy, "Anxiety and/or Depression in the Pathogenesis of Addictive Gambling," *Substance Abuse* 24 (4) (1989): 337–350; G. J. Coman, G. D. Burrows & B. J. Evans, "Stress and Anxiety as Factors in the Onset of Problem Gambling: Implications for Treatment," *Stress Medicine* 13 (4) (1997): 235–44; A. C. Miu, R. M. Heilman & D. Houser, "Anxiety Impairs Decision-Making: Psychophysiological Evidence from an Iowa Gambling Task," *Biological Psychology* 77 (3) (2008): 353–8; A. P. Blaszczynski, A. C. Wilson & N. McConaghy, "Sensation Seeking and Pathological Gambling," *British Journal of Addiction* 81(1) (1986): 113–7.

then $66.97 the following year, and then to $101 in 1927 after the establishment of the National Broadcasting Company. By 1928 RCA stock had reached a staggering $420.[9]

Thus, as the excitement of the stock market grew during this period, so did the level of speculative borrowing. For example, instead of purchasing stock outright, an investor could give his broker his money as security on a loan of up to 90% of the price of the stock, thus allowing him to effectively buy $100 of stock for $10 and get ten times as much value from any increase in price. If stock prices fell, a broker would ask an investor for a "margin call" to put up more money as collateral, or he would sell the stock to recoup the loss. This system worked, but only so long as stock prices kept rising. As practices such as this caught on, more and more investors borrowed to reap the rewards, and broker loans rose from $1 billion in 1921 to $8.5 billion by October 1929, more than the entire amount of currency circulating in the US at the time.[10] Stocks became a national mania, and people became so caught up with the excitement that they failed to notice the slowing of the economy.[11]

The Crash

By 1929 a number of factors had come together. The first and most obvious was that stock prices had gone up far beyond the level that would have been justified by their earnings. But the other factor was the level of stress. Alongside the exuberant optimism in some circles, many people in the 1920s referred to their own time as the "Age of Anxiety" for the sense of despair, bitterness and meaninglessness that they experienced, anxiety being a common response to rising stress.[12] As stress continued to rise the balance gradually shifted towards fear. As enough individuals crossed that line the market faltered and began to fall, which intensified the level of stress, and which caused the market to crash. If economic fundamentals were all that mattered, stock prices would then have stabilized and the

---

[9] D. Kyvig, *Daily Life in the United States, 1920—1940. How Americans Lived through the Roaring Twenties and the Great Depression* (Chicago: Ivan R. Dee, 2002), 214–215.
[10] Ibid., 214–215.
[11] Ibid., 215.
[12] "The Age of Anxiety: Europe in the 1920's", *The History Guide: Lectures on Twentieth Century Europe*, http://www.historyguide.org/europe/lecture8.html (accessed September 6, 2014).

economy gone back to normal. But a society-wide stress attack, itself brought to a higher pitch by the crash, now made that impossible. Confidence gave way to fear, and thus was the Great Depression born.

Franklin D. Roosevelt understood this fully, as stated at his 1932 inaugural:

> So, first of all, let me assert my firm belief that the only thing we have to fear is … fear itself—nameless, unreasoning, unjustified terror which paralyses needed efforts to convert retreat into advance.

A sense of unreasoning fear and loss of confidence is exactly what would be expected from a society-wide rise in stress, and this was something no government policy could heal. It was a product of the extroverted exuberance of the 1920s, which in turn reflected the lower C of people born after 1900.

We are accustomed to seeing exuberant prosperity as the opposite of anxious recession, but if biohistory is correct then the prosperity itself may be the behavioral cause of the recession. Prosperity reduces C, which increases testosterone and stimulus-seeking behavior, which leads to a rise in stress, which is transmitted from the youth generation to the broader society, which is amplified by the crash into a society-wide stress reaction.

Political terror

Economic recessions can cause severe financial and personal hardships, but there is an even grimmer effect when C falls faster than V. High levels of stress are deeply unpleasant, and people experiencing them tend to be angry or fearful or both, and to suspect others of ill will. We saw in chapter four how the severely stressed Mundugumor were filled with anger and mutual suspicion. High levels of stress also make people anxious and ready to seek out extreme solutions of many kinds to alleviate feelings of unpleasantness.

Thus, it is no surprise that in periods of high anxiety in the twentieth century, people were more likely to be attracted to groups such as the Nazi Party, which preached hostility and suspicion towards Jews and other minority groups. Nor it is surprising that support for such groups increased dramatically after highly stressful events such as the economic collapse of 1929. The Nazi's thrived partly because they gave young men license to vent their anger by attacking political rivals and smashing shop windows, and by 1934 Nazi thugs were beating up strikebreakers and even passers-by. Historian Richard Evans describes the innate aggression of the SA, the Nazis' private army, after the party came to power:

> As the young brownshirts found their violent energies deprived of an overtly political outlet, they became involved in increasing numbers of brawls and fights all over Germany, often without any obvious political motive. Gangs of storm troopers got drunk, caused disturbances late at night, beat up innocent passers-by, and attacked the police if they tried to stop them.[13]

Such hostility clearly did not need to be specifically targeted at outsiders. Indeed, Hitler himself demonstrated an extreme sense of insecurity. Having already established himself as Chancellor of Germany in 1933 and sensing a possible threat to his power, in 1934 he ordered various potential rivals to be dispatched in the "night of the long knives." Hitler was quite aware of the appeal of Nazi ideology to the anxious and discontented, as he stated in Mein Kampf:

> The fact that millions of our people yearn at heart for a radical change in our present conditions is proved by the profound discontent which exists among them. This feeling is manifested in a thousand ways. Some express it in a form of discouragement and despair. Others show it in resentment and anger and indignation. Among some the profound discontent calls forth an attitude of indifference, while it urges others to violent manifestations of wrath. Another indication of this feeling may be seen on the one hand in the attitude of those who abstain from voting at elections and, on the other, in the large numbers of those who side with the fanatical extremists of the left wing.
>
> To these latter people our young movement had to appeal first of all. It was not meant to be an organization for contented and satisfied people, but was meant to gather in all those who were suffering from profound anxiety and could find no peace, those who were unhappy and discontented.[14]

Highly stressed and anxious young people were also attracted to Communism which for this reason flourished at that time, especially in its more extreme and violent forms. Though ideologically opposed to one

---

[13] R. J. Evans, *The Third Reich in Power* (New York: The Penguin Press, 2005), 23.

[14] A. Hitler, *Mein Kampf*, http://www.greatwar.nl/books/meinkampf/meinkampf.pdf (accessed September 6, 2014).

another, both Communism and Nazism were psychologically similar in that both legitimized hostility and anger. Both were therefore more readily accepted by people who were anxious and highly stressed in the first place.

It must be emphasized that while the prevailing temperament made Germans more accepting of a brutal and authoritarian regime, and of the aggressive warfare that followed, it does not account for all aspects of Nazi policy. The peculiar insanity of the Holocaust in particular, which the government took pains to hide from the general population, can be better explained by the psychology of individual leaders.

The same situation applied in Spain in the 1930s, where both Communism and Fascism flourished, eventually bringing about the civil war of 1936–39. In their brutality and commitment to government control of the economy, the victorious Fascists had much in common with their opponents, differing most markedly in their attitude towards religion. Once again, the explanation for the appeal of both Communism and Fascism in Spain is that Spaniards also experienced a surge of stress in the 1930s, along with the rest of Europe and North America.[15]

The same analysis can be applied to the situation in Russia after the 1917 Revolution, with the major difference being that Russians were already highly stressed. Writing in the late 1890s, Olga Semyonova Tian-Shanskaia gives us a vivid account of the treatment of children among the Russian peasantry:

> Up to the time Ivan takes his first steps, he is looked after by his sister, a girl of nine or ten years of age. She has difficultly carrying him and drops him … Sometimes Ivan tumbles down a hillock. When he cries, his baby-sister uses her free hand to slap him on the face or head, saying "keep quiet, you son of a bitch." Sometimes his sister leaves him on the ground … For an hour or more, the child crawls around in the mud, wet, covered with dirt, and crying.[16]

At that time Russia had one of the highest infant mortality rates in Europe, with nearly half of all babies failing to reach the age of five. One of the main contributors to infant deaths was "lying over," where a sleeping

---

[15] E. Lucie-Smith, *Art of the 1930s: the Age of Anxiety* (London: Weidenfeld & Nicolson, 1985).

[16] O. Semoyonova Tian-Shanskaia, *Village Life in Late Tsarist Russia* (Bloomington: Indiana University Press, 1993), 25, 27.

mother would roll over and smother an infant to death. Deaths in this manner were so common that it was believed that some were intentional. The prevalence of infanticide itself is an indicator of a society under extreme stress. Consistent with this trend we find that young children at all ages were treated roughly, with six-year-olds being beaten with a variety of implements including whips for such offenses as screaming, getting muddy or stealing food. At the same time, they were apparently not punished for lying, swearing or fighting, which they were encouraged to do as soon as they were able. Domestic violence was also endemic, with husbands often severely beating their wives. Deaths from domestic violence were not uncommon.[17]

This indicates that Russians were already far more stressed than other Europeans or Americans. But Russia at the close of the nineteenth century also seems to have been experiencing a decline in C. This was especially evident in the cities, but sexual mores seem to have been weakening even among the peasantry.[18]

Prior to this, villagers did not tolerate premarital relations among engaged couples, but by the close of the century a young woman in a monogamous premarital relationship was no longer regarded as "wayward."[19] Urban values had spread to the country and nowhere did traditional values decline faster than in the cities themselves, where the Bolsheviks enjoyed most of their initial support.

All of these factors create a profile of a society under considerable stress, made worse by a decline in C and (from our analysis) a consequent rise in stress.

Thus, when the Kaiser was deposed in 1918 the Germans adopted the liberal democracy of the Weimar Republic, only turning to brutal authoritarianism when stress rose in the early 1930s. When the Russian Tsar was overthrown in 1917, the stressed and anxious Russians were psychologically primed to accept a more brutal and authoritarian form of government, allowing the Bolsheviks to triumph over the more liberal Mensheviks. A bizarre but commonly expressed desire among workers and peasants during the 1917 revolutions was for a "democratic"

---

[17] Ibid., 6, 7, 30.
[18] Ibid., 50, 51.
[19] Ibid., 51.

government but with a "strong Tsar" at the head of it.[20] This does not mean they would have voted for dictatorship, if given the chance, but they were more ready to accept its dictates once in power. A government is strong because it is obeyed, whether that obedience arises from loyalty or fear.

Thus it is that when Russia experienced a surge of stress going into the 1930s, it did not turn from liberal democracy to brutal dictatorship as in Spain or Germany. It turned from an already brutal autocracy to a far more brutal and even murderous one with show trials, gulags and mass murders. Fear and paranoid suspicion were rampant at all levels of society, making people ready to accept accusations that this or that person was a traitor or saboteur. A chain of labor camps were set up in Siberia, swelled to bursting with millions of prisoners. It has been estimated that more than three quarters of a million people were executed under Stalin, with maybe three million more dying in the camps.[21]

To us looking back, this sort of behavior makes no sense, but we are not psychologically attuned to see treachery around every corner. Nor are we primed to submit to brutal authority. But it was the very brutality of the government which caused the highly anxious Russians of that time to obey it, while the same brutality would cause less anxious people to resist it. Thus it is that government tends to reflect the prevailing attitudes of the population, even when many or most of the people hate the government. Anxious people tend to accept and obey powerful authority, whether elected or imposed. The Russian Terror was not a product of one murderous dictator, but of a whole society driven mad by stress.

This is the same society-wide attitude of hostility and suspicion seen in the baboon colony of Whipsnade Zoo described in chapter four. In normal circumstances, a dominant male baboon acts in a socially beneficial manner, protecting the weak against the strong and maintaining social order. But at Whipsnade the dominant male, known as Henry, was liable to attack any animal without provocation. He killed his own consort and several of the young. The same applied at all levels of this highly stressed baboon community. Each animal was likely to be attacked and threatened by those of higher status, and to attack and threaten those with lower status. Animals with the lowest status of all, the vulnerable females and young, were almost all killed. The baboon autocrat did not create the

[20] R. Bova, *Russia and Western Civilization* (New York: M.E. Sharpe, 2003), 271.
[21] Evans, *The Third Reich in Power*, 66.

stress; the stress created the baboon autocrat.

The exact same analysis applies to Russia in the 1930s, with Stalin the product of stress rather than the cause of it. And as stress ebbed during the 1940s and early 1950s so also did the terror, without any change of regime being necessary. What happened among the Whipsnade baboons was stress brought on by overcrowding. What happened in Europe in the 1930s was stress brought on by the *perception* of crowding when humans with lower C sought stimulus by crowding into cities and seeking out challenge and excitement. Even though this was the lifestyle they sought, it was greater than their underlying tolerance (set by V) and thus served to increase their level of stress.

Stressed animals and people tend to give higher status individuals more power. In humans, this translates into governments taking or being granted more economic power. The forced collectivization of the Soviet economy made little economic sense and led to shortages that cost millions of lives, but it buttressed government power. Fascist governments also intervened quite strongly in the economy, as did the Roosevelt administration in the United States. American businessmen were prosecuted for cutting the price of food at a time when many people were hungry—one of the policies that many economists believed extended the Depression and made it worse.[22] But stressed people favor powerful and controlling leaders. When Roosevelt said "There is nothing to fear but fear itself" and proceeded to give his government unprecedented power, he was responding to a psychological need which was different in degree, but not in kind, from that which allowed Stalin and Hitler to flourish.

Once again, it was the low-C generation reaching maturity in the 1920s that experienced the highest level of stress. They would also be more likely to be without work, to be attracted to any ideology preaching hatred and suspicion, and to both revere and seek out an especially potent leader. They would also be more likely to obey a brutal and authoritarian leader who was already in power, thus consolidating the position of such leaders even in societies with no democratic electoral processes, such as Bolshevik Russia. Thus, a government reflects popular attitudes, even in a dictatorship.

Fig. 10.4 below illustrates how rising stress in Russia, Germany and the

---

[22] B. W. Folsom, New Deal or Raw Deal; how FDR's Economic Legacy has Damaged America (New York: Threshold Editions, 2008).

US had different effects, because each society started with stress at a different level.

Fig. 10.4. 1930's surge of stress in Russia, Germany and the US. Rising stress causes political systems to become more autocratic and brutal, as stressed people are more likely to accept brutal authority. The level of autocracy and brutality depends on the initial level of stress.

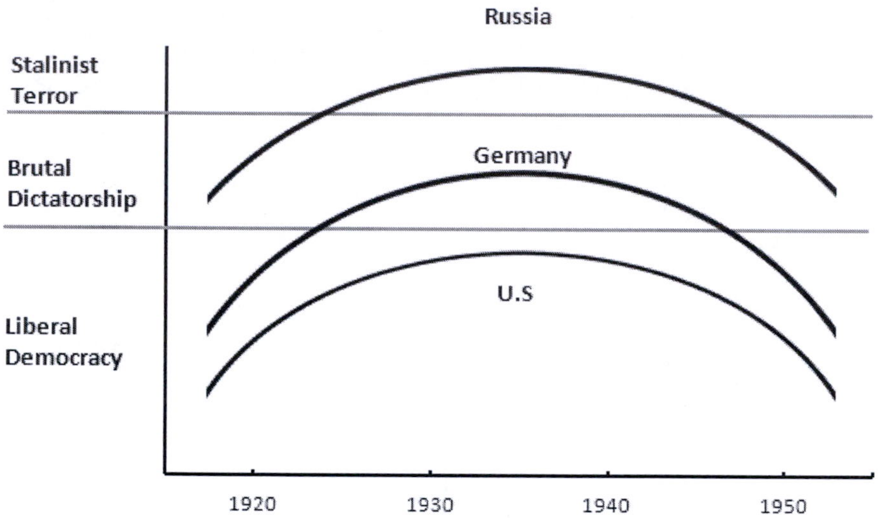

The "baby bust" of the 1930s is usually attributed to the Depression, but in fact economic factors are a very poor explanation for changes in the birth rate. Birth rates drop most rapidly in prosperous times, and in recessions they typically fall and then rise. The low point of the birth rate in 1935 can be better explained by low C than by hard times. Assuming people have children at an average age of 30, the lowest birth rate would be found in people born in 1900–1910 and who thus passed their entire childhood and adolescence in times of prosperity. Though no longer the culturally dominant generation, as they had been in the 1920s, they were the people who now chose to have fewest children.

It is not just in the 1930s that trough periods are associated with political extremism. The 1830s and 1840s, also a low point for rate of natural increase, brought England to the brink of revolution that may only have been averted by the passing of the Great Reform Act of 1832. American politics shows a similar pattern. The early 1970s saw bitter political disputes that led to the resignation of Richard Nixon, and the increasing polarization of American politics in recent years has at times almost brought the government to a halt.

Recovery

Prosperity between 1900 and 1920 reduced C in people maturing after 1920, thus increasing stress in the 1920s and 1930s. But the resulting recession had the reverse effect. Children born after about 1925 spent their childhood and adolescence in the economic gloom of the 1930s, increasing their C (or at least halting its decline).

As they reached maturity in the 1940s and became socially dominant, this cohort preferred to avoid stimulus so the level of stress and extremism in the larger society began to fall. These people became the sober, family-loving parents who gave birth to the baby boom of the 1950s—all traits typical of higher C. They would also have higher child V and thus be more traditional in outlook. So also in Russia, Stalinist terror ebbed in the 1940s and early 1950s without any change in regime. (War as such seems to have little impact on this, recalling that Western birth rates dropped after the First World War but rose after the Second.) The highest levels of C and child V would be found in the age cohort born in 1925 which spent its entire late childhood and early adolescence during the economic gloom of the 1930s. In 1925 they reached the peak childrearing age of 30, accounting for the peak birth rate at that time.

But then, once again, renewed prosperity had its impact. The cohort born after 1940 knew only prosperity, so that as they matured after 1960 their lower C fuelled the hedonism and extroversion of the 1960s—miniskirts, the Pill, "sex and drugs and rock 'n' roll." Once again, as in the 1920s, the birth rate plummeted.

And once again, as stress rose, the rosy optimism of the early 1960s turned to anger and pessimism. Spearheaded by the aggressive generation born at the end of the Second World War, civil rights and anti-war marches spawned violence and confrontation. Political divisions became bitter and an American President was forced to resign. Then followed the stagflation of the 1970s.

Fortunately, the level of stress was less pronounced than in the 1930s, making the political environment less toxic and the recession less extreme. But even so, the effect of this milder crisis was enough—once again—to at least slow the fall in C. Stress fell, the American birth rate rose slightly and the economic situation improved, leading to the "Great Moderation" when governments seemed to have learned the key to combining low inflation with low unemployment.

The Global Financial Crisis

But, of course, government policy could do nothing to halt the underlying forces in play. By the end of the twentieth century still another generation came of age which had known nothing but prosperity, increasing testosterone and driving the urge to speculate. This time the boom was not so much in stocks as in the 1920s, but in property. People borrowed more than they could afford to repay in order to buy houses bigger than they needed, with the assumption that ever-increasing property prices would make them rich. The US government supported this by arranging loans to marginal borrowers through government-backed institutions such as the US Federal National Mortgage Association ("Fannie Mae") and Federal Home Loan Mortgage Corporation ("Freddie Mack"). Trading houses bought up billions in mortgages without adequately checking the security behind them, again with the assumption of ever-rising prices. Governments added to the bubble by borrowing heavily and plunging themselves into debt, also with the assumption that ever-rising tax revenues would pay for it all.

This environment can itself be productive of stress. In The Hour Between Dog and Wolf, John Coates gives a vivid picture of traders in the financial markets in the time leading up to the Global Financial Crisis.[23] Financial traders require split-second decision-making and an appetite for risk, and their profession is thus one of the few that benefit from high testosterone. Success can bring enormous prestige and multi-million dollar bonuses, but such an environment also tends to increase stress.

Coates' study also tracked the level of hormones as the market crashed.[24] Traders' testosterone fell and cortisol rose, creating problems opposite to those of the boom. Instead of being irrationally optimistic, stressed traders became so risk-averse that they failed to see clear opportunities for gain, causing a lack of buyers and making the crunch worse than it should have been. Even a drastic cut in interest rates by the US Federal Reserve had no

[23] J. Coates, *The Hour Between Dog and Wolf: Risk-taking, Gut Feelings and the Biology of Boom and Bust* (London: Harper Collins, 2012).
[24] Ibid.

great impact, since the problem was hormonal and psychological rather than rational.

Biohistory suggests that the same hormonal and psychological change would have occurred in the general population, but it also predicts that cortisol would have been rising *before* the crash. Stress levels, which were already high, were pushed still higher by the collapse of the property markets, so that irrational exuberance was followed by irrational gloom, even in countries like Australia which were little affected by the economic fallout. People focused on paying down their debts, further reducing consumption. This recession, like that of the 1930s, can be seen as a nationwide or even worldwide stress attack. And that is why government actions, no matter how soundly guided by economic principles, have so little effect. The basic problem is not economic or even cultural but physiological.

Questions of how to test the physiological aspects of this mechanism can best be left to the final chapter, but it may be noted that this aspect of the theory could be very easily confirmed or refuted. Once there are clear epigenetic tests for C and child V there should be clear differences in cohorts born during each decade. It should also be possible to predict the onset of future recessions, and the likelihood of recovery, by tracking cortisol levels in the general population.[25] Cortisol can be reliably and cheaply measured from saliva samples.

Conclusion

The effect of prosperity on C and of C on stress can be used to explain the third of the biohistorical cycles, along with civilization cycles and lemming cycles. Recession cycles account for short-term fluctuations in population growth, economic prosperity and political attitudes. There has been no attempt to describe recession cycles in other periods but they do appear to be universal though normally much longer than the forty years characteristic of twentieth-century America. For example, recession cycles in nineteenth century Britain had troughs around 1840 and in the 1890s, which are fifty-five years apart. Eighteenth-century Japan shifted twice between hedonistic excess and sober reaction: in 1709–1716 under Arai Hakuseki and in 1787–1793 under Matsudaira Sadonobu, nearly eighty years apart. These greater lengths are entirely consistent with recession

[25] Cortisol can be measured quite cheaply and quickly from saliva.

cycle theory if it is understood that in most societies the culturally dominant generation tends to be older—perhaps 30–40 in Victorian England and 40–50 in the intensely conservative society of Tokugawa Japan.

The notion that stimulus-seeking behavior causes a rise in stress also has relevance to the understanding of drug addiction, alcoholism and similar behavior. A chemical C-promoter should be an effective treatment for all such conditions. This is again something to be discussed in the final chapter.

CHAPTER ELEVEN

WHY REGIMES FALL AND CIVILIZATIONS COLLAPSE

Civilization has been made possible by the development of physical and economic technologies such as agriculture, metalworking, writing, mercantile trade, and (in more recent times) industrial manufacturing and mechanization. Equally important, however, are the cultural technologies that adjust attitudes and behavior to the circumstances and needs of civilization by maintaining artificially high levels of C and V—systems of temperament based on physiology. Most of the cultural technologies underpinning C and V are in the form of religious and philosophical ideals and practices.

These cultural technologies have been developed by a process of competition within and between societies. Over thousands of years the religions and philosophies that bring the greatest benefits have spread widely, replacing less effective cultural technologies. Socially and politically speaking, C brings economic advantages such as a capacity for routine work, while V brings military success and a higher birth-rate. Both traits contribute to larger and more stable political units. C makes people law-abiding and infant C promotes impersonal political loyalties, such as to codes of law. V, in its child form stemming from experience of authority in late childhood, promotes acceptance of authority and of tradition.

In some societies, such as Japan and Europe over the past thousand years, the rise of C has been much faster than can be accounted for by the slow process of cultural evolution. In both areas this rapid rise in C was accompanied by a rise and then fall in V and stress, suggesting that the rise of C was driven at least in part by a high level of V and stress. As C rose still further the level of V and stress began to fall, suggesting that high C undermined V. And as V and stress rose, C increased to a peak. This pattern is known as the "civilization cycle."

So far we have examined the civilization cycle in Japan and Europe as far as this peak of C, which in Europe occurred in 1850, and Japan around 1920. In this chapter the study is taken further, bringing biohistory to bear on the decline and fall of civilizations. But first we must examine the general subject of regime change.

Regime change

Up to this stage we have considered the rise of civilization as if it were a more or less linear process. People gradually develop and adopt more and more powerful technologies, moving from farming to bronze-working to writing to iron to gunpowder to machines to lasers and nuclear weapons. At the same time states become larger and more sophisticated. The very way in which we refer to poorer countries as “developing” is an indication that we see them as just a little behind us on the same basic road.

Human history is, of course, far more convoluted than this. Regimes rarely last for more than a few centuries before breaking down. Sometimes they are overthrown by their own subjects, sometimes by foreign invaders, and occasionally they collapse completely of their own accord.

Theories of Decline

Theories of civilization collapse can be traced back to ancient Greece. According to the historian Polybius (200–118 BC) all polities pass through an inexorable cycle of growth, zenith and decay. In the Middle Ages, Ibn Khaldun (1332–1406) observed a regular lifecycle in the history of civilizations through three stages: a first stage of growth characterized by cultures with a strong sense of solidarity nurtured by hard environmental or nomadic conditions of life; a second "mature" stage characterized by a more urbane, sedentary existence and growing abundance, but leading to a weakening of group bonds; and a final third stage of senility, characterized by the complete collapse of group ties, social breakdown, and the rise of new stronger elites.[1]

In the modern age, Giambattista Vico, in his much discussed book, *New Science* (1725), also argued that the characteristic historical cycle followed three stages: first savagery and anarchy, then order, civility and industry,

[1] Ibn Khaldūn, *The Muqaddimah : An Introduction to History*, translated from the Arabic by Franz Rosenthal, 3 vols (New York: Princeton University Press, 1958).

and finally decay of civilization to a new barbarism. Vico's theory was quite novel in his insistence that the regularities of these cycles could be explained in a scientific manner because they were the result of human action and motivation. He believed that the underlying mechanism of these cycles was psychological, or accountable by motivations and attitudes in human nature.

The best-known historian of cycles is Oswald Spengler (1880–1936). In *The Decline of the West* (1918) he rejected the linear idea of progress which dominated the minds of European historians since the Enlightenment, arguing that each civilization was a unique, self-contained organism, with its own languages, arts, sciences and polities. Each civilization was similar only in following a lifecycle of childhood, youth, manhood and old age. He emphasized the deleterious effect that prosperous, cosmopolitan cities had on the character of its inhabitants, replacing blood ties, folk values and motherhood with loose urban relationships, sex and divorce, and mass politics.

Arnold Toynbee (1889–1975) wrote extensively about the life cycles of numerous civilizations, but other than writing about the creative role that minorities played in the rise of civilizations, in response to external challenges, and of the eventual inability of elites to meet new challenges, he did not add much to our understanding of the underlying logic of historical cycles. The Russian Pitirim Sorokin, in *Social and Cultural Dynamics* (1937), added some vivid images forecasting the moral degeneration of Western civilization, the loss of moral consensus and coming chaos of opinions, the denigration of freedom into empty slogans, the disruption of the family, and the replacement of higher culture by the lowest denominator of mass culture.[2]

More recent theories take a neo-Malthusian approach, focusing on population growth and the carrying capacity of the land.[3] This is the starting point for Peter Turchin and Sergey Nefodov's recent *Secular Cycles*, so-called because it dispenses with older "moralistic" theories and

[2] Bruce Mazlish, *The Riddle of History: The Great Speculators from Vico to Freud* (New York: Minerva Press, 1968); Piotr Sztompka, *The Sociology of Social Change* (Oxford: Blackwell, 1993).

[3] E. A. Wrigley & R. S. Schofield, *The Population History of England 1541–1871* (London: Edward Arnold, 1981).

attempts to follow a rigorously scientific approach.[4]

Biohistory has a great deal in common with Spengler and Ibn Khaldun, as well as theorists including Max Weber who link the rise of the West to a

---

[4] Peter Turchin & Sergey A. Nefedov, *Secular Cycles* (Princeton: Princeton University Press, 2009) contains the foremost current theory of cycles. Drawing on prior neo-Malthusian interpretations of demographic cycles in the history of Europe, Turchin and Nefodov, begin their analysis with the well-known insight of Thomas Malthus that population tends to grow in excess of productivity gains, leading to such social effects as land scarcity, declining living standards, overpopulation and famine. They add to existing literature the observation that landowning elites benefit (due to higher rents) from the overpopulation, becoming relatively richer and more numerous. They apply this model to four historical societies: medieval/modern England and France, Republican and Imperial Rome, and Russia under Muscovy and the Romanovs. In each of these societies they observe this pattern. First there is a period of agrarian expansion and increased productivity, leading to more surviving children combined with stable prices and wages. Second comes a period in which population growth starts to reach the limits of the carrying capacity of the land, with the price of the land increasing, and a larger and more affluent elite intensifying exploitation of the peasantry despite declining rates of agrarian output. This is associated with increased competition within the elite and with the state for shrinking revenue. Third comes a period of crisis, addressed through military expansion and conquest or resolved through famine, state collapse and civil war. This is followed by a depression phase, smaller populations, and the possible beginning of a new cycle if the ranks of the elite are sufficiently reduced.

With backgrounds in biology and mathematics (Turchin) and archaeology (Nefedov), these authors believe that their theory of cycles establishes the first truly scientific approach to history, "the same methods physicists and biologists used to study natural systems" (312). There is no denying that they offer a rigorously developed theoretical model with empirical time series for four countries for each of the four phases. The neo-Malthusian model has been slowly developing for many decades and includes a number of major historians such as the influential French Annales School historians and the demographic school of E. A. Wrigley and R. S. Schofield, as represented in their major work, *Population History of England, 1541–1871* (1981). However, there are a number of basic flaws with their model, starting with the obvious fact that Malthusian propositions do not hold in a state of industrial innovation. Their model oversimplifies the nature of human motivation, dealing only with the mere reality of sexual reproduction beyond the carrying capacity of the land. Biohistory provides a far richer understanding of human motivations based on actual experimental case studies.

temperamental change driven by Protestantism.[5] Its contribution is to explain their observations in terms of *biology*. In essence, religion creates the physiological basis of the temperament that supports civilization, and wealth and urbanisation destroys it.

Regime Change

But biohistory is also concerned with less catastrophic regime changes. The G-90 phase of the lemming cycle, when loyalties briefly become far more personal, is one obvious factor. No Chinese dynasty has ever survived the G-90 period, including the most recent one (ca 1920) when the Qing dynasty collapsed into warlord anarchy, to be followed by civil war and Communist victory. The English G-90 period around 1460 saw the ruling house replaced not once but twice in less than three decades.

Rather less commonly, the period immediately after the G year can also see regimes fall, the French and Russian revolutions being obvious examples.[6] Here, the causes would seem to be almost the opposite. As impersonal attitudes strengthen, loyalty to an aristocratic ruling class weakens. Ordinary people become ready to accept a more powerful and effective national regime, and the rise in aggression at the G+30 period gives them the energy to overthrow one that no longer reflects their interests.

There are situations where a regime collapses because an unstoppable new force appears on the scene, though even here lemming cycles can have a significant impact. In the Mongol invasion of China, the power of the invaders was clearly the most important factor. The process had two separate phases, the first being the loss of north China to the Jurcheds in the early twelfth century. The second was the conquest of south China by the Mongols themselves, more than a century later. The Mongols ruled a vast empire and had an army of awesome martial vigor and sophistication, but even then it took decades of bitter fighting to overcome Chinese resistance.

---

[5] M. Weber, *The Protestant Ethic and the Spirit of Capitalism* (USA: BN Publishing, 1979); P. Greven, *The Protestant Temperament: Patterns of Child-Rearing, Religious Experience, and the Self in Early America* (New York: Plume, 1979).

[6] Jack Goldstone observes the link between population growth and revolution, though providing a different explanation: J. Goldstone, *Revolution and Rebellion in the Early Modern World* (Berkeley: University of California Press, 1991).

Lemming cycles provide a partial explanation for this. The preceding G year was 1020, when the Sung dynasty was at the height of its power. The common lemming cycle pattern is of a gradual decline in government power to the G-90 period. Thus, north China was lost a century after the G year, and south China 250 years after it. Ironically, the anarchy of the G-90 period around 1330 brought down the Mongol regime established just a few decades earlier.

The Manchu invaders in the seventeenth century showed much better timing. Though having only a fraction of the Mongols' power and strength, they overran China with relative ease. A major reason was Chinese disunity, with several Chinese generals defecting to the Manchu side, including one who effectively handed them the capital. The Manchu invasion was made feasible by a G-90 period around 1630 when loyalty to imperial authority would be at a low point, and the new regime benefited from the ensuing rise in impersonal loyalties.

These examples help explain the timing of regime collapse, but equally crucial is an understanding of who is likely to replace them. Here the key principle is that groups with higher C and especially V tend to overcome groups with lower C and V. This applies as much to the fall of Rome as to the Mongol or Manchu conquest of China. It also covers the French and Russian revolutions, when lower V elites were displaced by higher V commoners.

In fact, the fall of almost any regime can best be understood as the result of falling C and V in the society affected—or at least in its elite classes, which become weakened to the point that they collapse or are forced out by another regime.[7] Any decline of C and V weakens a society or ruling class and makes it vulnerable to takeover. As a general rule, the greater and more widespread the decline of C and V, the less military force is needed to change rulers and the more total the collapse.

Seen in this light, the fall of China to the Manchus, and of Rome to the barbarians, represent exactly the same process. The difference is that in China only the ruling classes were affected, and thus partially replaced by a regime with higher V. The Chinese culture, languages, political structure and ethnic composition was little affected. The same thing happened again in 1949 when the Chinese ruling class was replaced by a higher V segment of Chinese society.

---

[7] An observation made by Vilfredo Pareto; L. Coser, *Masters of Sociological Thought: Ideas and Social Context* (New York: Harcourt Brace, 1971), 396–400.

In the case of Rome, the whole of society suffered a catastrophic fall in C and V leading to population decline and political disintegration. In areas such as Britain the native population was largely replaced, and in other areas heavily settled by incoming barbarians. Fig. 11.1 below puts this in perspective.

Fig. 11.1. Declining C and V causes regimes to collapse or be overrun. The greater the decline, the more serious the collapse.

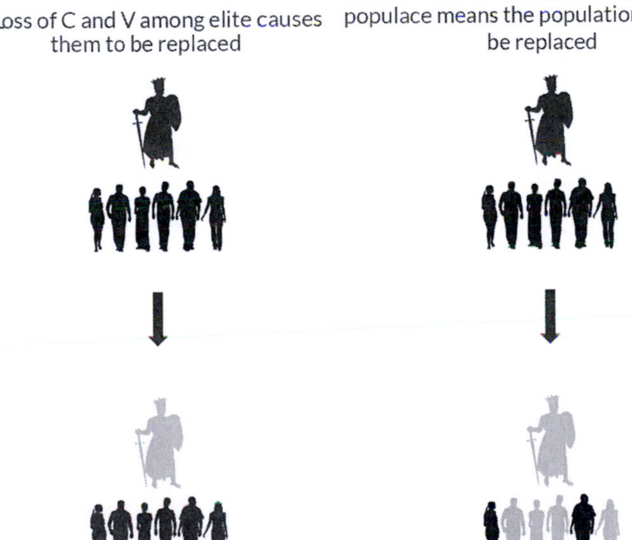

Explaining Collapse

The analysis in previous chapters explains why this fall in C and V take place. Both require humans to act in a highly unnatural manner, resisting deep-seated urges such as that for sexual pleasure. These changes in behavior cause changes in temperament that help people to survive and succeed.

But C and V are also biological systems that respond to food shortage, chronic in the case of C and occasional but severe in the case of V. A successful civilization tends to become wealthy, which works to reduce C and V and especially applies to the elite. The effect is not immediate, since temperament is set largely in childhood and people tend to bring up their children with their own levels of C and V, but over several generations these levels fall. The people first become less warlike, then less hard-working and enterprising. The downfall of a civilization is thus a direct consequence of its success.

An important point to note is that V falls first. We have already witnessed this in the civilization cycle, where high V plus stress drives the rise in C, and then begins falling centuries before C reaches its peak and then begins to fall. In other scenarios this happens faster, even in a generation or two, but V appears to be more quickly affected by wealth and population density. The characteristic result is military weakness, even while the civilization remains wealthy and united because of relatively high C.

Some time in this process a group with higher V displaces the ruling elite, or in some cases much of the population. This new elite becomes wealthy in its turn so that its own V and then C are undermined. This undermining can be slowed if the new rulers have very strong C- and V-promoting cultural practices to counter the effect of prosperity, but it can never be stopped absolutely.

This is why civilizations collapse even when there are no obvious barbarians to sweep them away. It also shows, for instance, why Chinese scholar-gentry families rarely prospered for more than a few generations, and further explains why particular regions are fertile for the development of Communist revolution, and some curious aspects of the behavior of pre-literate peoples.

This chapter provides general principles that explain civilization collapse or regime change, and examines a number of historical case studies which illustrate how these play out. In later chapters we will consider in detail the collapse of ancient civilizations, including Rome.

Prosperity counters V-promoters and C-promoters

Human beings have naturally low set points for C and V. Our ancestors were adapted to life as hunter-gatherers, living in small bands and with little need for constant routine work. Chronic food shortage would cause C to rise and famines or predator threat might increase V, but none of these changes would be very large or last beyond the environmental conditions which gave rise to them.

It is likely that the artificial raising of C and especially V began a very long time ago. Certain Indigenous Australian tribes had elaborate incest taboos which effectively put most women beyond the reach of the men in their immediate communities, which must have reduced sexual activity to some extent. And we have seen that the Yanomamo of the Amazon brutally suppress women, with the effect (through the influence on infants) of making their society more warlike.

But once farming technology becomes available, much greater changes start to take place. People vary widely in their temperament and values, and those who just happen to be a little more industrious would have more food and thus generally more surviving children than their neighbors. Not only do they expand as a proportion of the population, but by being healthier and (perhaps) having more surplus food to give away they achieve higher status. People have a tendency to imitate those who are more successful, so C-promoting ideas start spreading through the community.

The same process applies in the competition between communities. For instance, if a local shaman decrees a form of ritual behavior which happens to raise C or V, the band might flourish and subsequently expand at the expense of its neighbors, who might in turn be inclined to copy the ritual and other practices of the more successful group. This happens even though they would have no idea of the process involved, perhaps believing simply that certain actions pleased the spirits who would then bring prosperity. Those who did not adopt C-promoting behaviors are more likely to dwindle in numbers and decline in prosperity and power. Geographically they are pushed out to less productive regions and might even become extinct.

The process would proceed much faster, of course, once higher C farming societies develope in areas such as the Middle East. Immigrants carrying not only seeds and stock but a greater capacity for work prosper and expand, leading to imitation by their neighbors. The effect would be even greater if V-promoting customs give them the temperament that allows larger clans or tribes to form, or to become more warlike.

Once the process began there would be a tendency for C-promoters and V-promoters to become stronger. Most farming societies throughout history have lived at the edge of subsistence. Children are especially vulnerable to hunger and disease, with malnutrition weakening resistance to infection. Thus an industrious group which grows more food has a better chance of its children growing up than those of its less efficient neighbors.

All of this makes the increase of C and V seem like an inexorable process, but this was far from the case. Early C-promoters and V-promoters would have been supported by naturally occurring food shortages and even famines. But any long-term increase in prosperity, even the absence of famine, tended to counter the cultural influences supporting C and V.

Turning to metaphor for a moment, the raising of C and V over thousands of years might be depicted as a band of men, with great effort, slowly pushing a heavy boulder up a hill. The effort required is close to the limit of their strength, but they are forced on by the fact that their very survival depends on it. Then a stranger walks down the hill towards them and starts pushing the boulder downhill. It slows to a halt and starts to slip backward, slowly at first and then faster as the team's feet scrabble for purchase. That stranger is prosperity.

What makes this problem of maintaining C and V even more serious is that not only wealth but population density tends to undermine V. It will be recalled that V is highest when someone experiences a lower level of anxiety as an adult than they did in infancy. For example, a male in a patriarchal society spends his infancy with a subordinate and therefore anxious mother, but his adult life as a higher status man. Since higher population density increases stress and thus makes adults more anxious, it must also reduce V.

Bear in mind also that this effect is relative and applies at any level. For example, other things being equal, living in a city creates more crowding stress than living in the suburbs, the suburbs more than living in settled farmland, and settled farmland more than in the wilderness. But even in the wilderness there are different degrees of population density. If neighbors are hostile, which is commonly the case with preliterate peoples,[8] having them five kilometers away is more stressful than when

---

[8] Most preliterate peoples seem to have been in a state of chronic conflict with their neighbours, including the Yanomamo as described by Napoleon Chagnon, as well as the Fore and Gadsup of the New Guinea highlands, and the Mundugumor of the Sepik Valley. Even the "peaceful" highland Arapesh skirmished with neighboring hamlets, and the Tchambuli were forced out of their lakeside home. A. Gat, *War in Human Civilization* (Oxford: Oxford University Press, 2006); R. M. Berndt, *Excess an Restraint* (Chicago: University of Chicago Press, 1962); B. M. DuToit, *Akuna: a New Guinea Village Community* (Rotterdam: A. A. Balkema,

they are twenty kilometers away. Thus, prosperity and higher population density both work, in the long-term, to undermine V and then C. And the most obvious effect of declining V is that people become less warlike.

Prosperity may increase aggression in the short-term

This process is complicated by the fact that, in the short-term, prosperity can make people *more* aggressive. When a people on the edge of subsistence gain access to plentiful food, the common response is rapid population growth. As discussed in chapter four, population growth is a V-promoter through the workings of the *effect feedback cycle*. We saw in the previous chapter that people born at times of maximum population growth tend to be unusually aggressive.

Another reason is that rats with calorie-restricted mothers but ample food in later life had higher testosterone than those for whom food was always plentiful (see chapter two). Part of the reason is that early food restriction dampens down the effect of at least one gene which limits testosterone. And higher testosterone, other things being equal, makes people more aggressive. In our terms, testosterone is maximized when C is lower than infant C. Thus it is that declining C in the early twentieth century, as evidenced by a declining age of puberty, caused a surge of testosterone that helped precipitate two world wars.

Calorie-restricted rats in our study also had higher levels of cortisol, a key stress hormone. A rat or human experiencing food shortage in infancy but not later life will therefore have a higher level of cortisol in infancy than in later life. And having lower anxiety in adult life than in infancy is what creates V. This is why patriarchal societies are more aggressive, because stressed and subordinate women transmit higher anxiety to their infant sons, who grow up to have lower anxiety because they are high status males.

This is why prosperity undermines aggression in the long term but increases it in the short term. Someone brought up with food shortage but well fed as an adult has higher testosterone and higher V. But their children, brought up with plentiful food, have lower testosterone and lower V. Parental behavior may maintain some level of C and V, but this fades over the generations. Very high V and prosperity may also combine to

---

1975); M. Mead, *Sex and Temperament in Three Primitive Societies* (New York: Harper Perennial, 2001), 22–3.

create a population explosion, which will help to slow the rate at which V declines because rapid population growth reinforces V, just as high V causes population to increase.

Bear in mind that this fall in C and V is independent of any effects which may be brought about by culturally driven changes in behavior. It is simply environmental, a function of mammalian physiology, working on a system evolved to adapt behavior to food supplies.

We now have a model of how a sudden change from food shortage to abundance could make people more aggressive in the short term, but much less aggressive in the long term as V declines. This can now be applied to a number of different societies.

Pre-literate societies

History is replete with warlike nomads sweeping in from the wilderness to raid and conquer wealthy civilizations. The best-known examples include Aramaeans and Arabs in the Middle East, Goths and Huns in Europe at the collapse of the Roman Empire, and Mongols in China, central Asia and the Middle East. The principle is that people from the harshest environments are more aggressive and warlike than those in fertile areas.

But among hunter-gatherers and small-scale agriculturalists, this is not true at all. In fact, peoples who dwell in fertile plains are typically *more* aggressive and warlike than groups in marginal areas. This is the case among the Yanomamo of the Amazon, where groups living in the precarious highlands were distinctly less warlike than those on the fertile lowlands.[9] The highland Arapesh, a relatively peaceful people described by Margaret Mead, seem to have spoken the same language and frequently intermarried with the far more aggressive plains Arapesh.[10]

But this is not the whole picture. We have seen above that it is a sudden rise in prosperity that might be expected to increase aggression. And, in fact, the most aggressive peoples do seem to be those from impoverished areas who suddenly gain access to plentiful food. Until little more than a century before first contact, the Yanomamo were a relatively small group living in the infertile Parima Highlands, before expanding and

---

[9] N. M. Chagnon, *The Yanomamo* (Fort Worth: Harcourt Brace College Publishers, 1997), 89.
[10] Mead, *Sex and Temperament in Three Primitive Societies*, 4.

(presumably) forcing out less warlike peoples from the more fertile lands below.[11]

Until the early eighteenth century the Lakota Sioux were a small band of hunters, gathering wild foods and growing small plots of corn, squash and tobacco. Forced west to the Missouri by more powerful tribes they gained access to horses and guns, and when smallpox wiped out most of the rival tribes they were able to take up bison hunting on the Great Plains. This was so rewarding as a source of food that they abandoned agriculture.[12]

The status of women fell and they became ferociously warlike. During this period the Sioux attacked rival tribes mercilessly, either killing them or driving them from their lands. In 1873, one thousand Sioux warriors killed nearly two hundred Pawnees—men, women and children—for trespassing and hunting. There was no peacetime.[13] Along with this violence went a supreme self-confidence.[14] All this indicates a surge in V.

In the case of the Yanomamo and the Sioux, aggressive expansion was halted by European contact. But biohistory indicates that it would have faded in any case before too long. Either denser population leads to crowding stress, as seen among the Mundugumor. Or, more typically, the aggressive newcomers gradually lose V in their more prosperous living conditions until pushed aside by people from poorer lands in their turn. Because the changeover happens so quickly, the pattern of fierce lowlanders and milder highlanders is maintained.

But occasionally, the lowlanders are not immediately displaced and we see the full effects of declining V. In chapter four we encountered a remarkably unwarlike New Guinea tribe called the Tchambuli, and we can now begin to understand *why* they were so lacking in aggression. The Tchambuli had settled in a fertile lakeside area which supported a relatively dense population. They seem at one time to have been fierce

---

[11] Chagnon, *The Yanomamo*, 303.

[12] G. B. Grinnell, The Cheyenne Indians: Their History and Ways of Life (Bloomington, Ind.: World Wisdom, 2008); G. E. Hyde, *Red Cloud's Folk: A history of the Oglala Sioux Indians* (Norman: University of Oklahoma Press, 1937); G. Gibbon, *The Sioux: The Dakota and Lakota Nations* (Oxford: Blackwell Publishers, 2003), 4.

[13] Gibbon, *The Sioux: The Dakota and Lakota Nations*, 74. Royal B. Hassrick, *The Sioux: Life and Customs of a Warrior Society* (Norman: University of Oklahoma Press, 2012), 8, 62–78.

[14] Hassrick, *The Sioux: Life and Customs of a Warrior Society*, 73–74.

headhunters, as indicated by the requirement that every boy kill a man for ceremonial reasons, but at the time of contact they were distinctly low V. Mead described the men as nervous and sensitive and dependent on their wives, and they were so unaggressive that captives for sacrifice were purchased rather than captured in war.[15] Later research suggests that this picture was exaggerated and that the Tchambuli did on occasion make war, but the fact remains that they were driven from their fertile home by more aggressive peoples shortly before European contact.[16]

It is notable that of the four New Guinea societies described by Margaret Mead, including the Tchambuli, three were in a rapid state of flux at the time of contact. The Manus were rapidly expanding due to their success as traders. The Mundugumor were declining in numbers and vividly aware that their values and traditions were being undermined. And the Tchambuli had lost most of their warlike traditions and been driven from their home. Our other examples, the warlike Yanomamo and Sioux peoples, were quite recent success stories. The pre-literate world was far less stable than is sometimes supposed, and understanding C and V gives an insight into how such changes happen.

Barbarians and civilization

What happens to civilizations is essentially no different from the decline in V we have seen in these tribal societies. Intensive agriculture and state administration require cultural technologies to raise the levels of C and V; but population density and more plentiful food work to reduce first V and then C. Over time the people become less warlike and confident as V declines. Then as C also begins to fall (since V supports C) they become less hardworking and their loyalties become increasingly personal and localized, which weakens political cohesion. Eventually, more vigorous peoples from the neighboring lands move in and take over. Their influence, plus the turmoil and poverty arising from the collapse, help V and C to recover.

As we will see in a later chapter, the conquest is made more likely by the tendency of the barbarians to adopt the civilization's C- and V-promoting cultural technologies *before* they invade, giving them the ability to organize in larger groups and become even more warlike. (Such peoples

[15] Mead, *Sex and Temperament in Three Primitive Societies*, 232.
[16] D. E. Brown, *Human Universals* (New York: McGraw Hill, 1991).

are referred to here as "barbarian" but no value judgment is implied—the term simply denotes warlike peoples from tribal societies outside the territory governed by a civilization.)

A preliterate society experiencing prosperity normally loses ground quite fast, as soon as the resident members of the society become less aggressive than the newcomers. Civilizations, on the other hand, take longer to overrun because they have major advantages. The people are more numerous, since intensive farming and manufacturing, combined with heavy transport and mercantile trading, can support many times the population density of hill tribes and nomads. Civilizations are also organized into much larger groups, which can support sizeable standing armies. And they have the labor to produce fortifications and the craft skills for advanced weaponry, chariots and the like.

The result is that V can drop to a much lower level than it normally can among people such as the Yanomamo, since individuals do not need to be so aggressive when defended by professional soldiers. They may also use their wealth to hire barbarians to fight for them, a common practice through much of history.

For civilized societies, the development of more powerful cultural technologies can delay the process considerably by strengthening V and C, so that they are more resistant to collapse. An extreme example of this delay is the civilization cycle, as applied to the histories of Europe and Japan. We have seen that in both societies C increased for at least a thousand years, because of very high V and stress levels around the sixteenth century.

Inevitably, though, the combination of wealth and population density reduces V and C so much that the society is overrun by warlike invaders. This conquest may mean no more than the replacement of one elite by another. The ruling class is wealthier and more likely to live in cities, so they lose V and C faster than other segments of the population. They typically form the leadership of the army, so their loss of V is seriously detrimental to military effectiveness.

Examples of such regime change can be found in the Sinification of Chinese dynasties such as the Liau, Jin and Yuan, all established by foreign invaders.[17] Another is the Norman Conquest of England in 1066,

---

[17] P. B. Ebrey, *The Cambridge Illustrated History of China* (Cambridge: Cambridge University Press, 2010), 173–9.

when the native English kept their language and identity largely intact. They eventually absorbed the Norman incomers, partly through the process described earlier, with the Norman ruling class gradually becoming less aggressive.

By contrast, the Anglo-Saxons who colonized Britain six-hundred years earlier virtually obliterated not only the native language and culture of Romano-Celtic Britain but even the population's Y chromosome, outside the borders of Wales. In other words, the resulting population might have had British mothers but the fathers were Germanic invaders.[18] Such a wholesale replacement is more likely when the C and V of the general population has collapsed, rather than simply that of the elite.

However far the process goes, the conquest of settled peoples by warlike tribes from harsher environments is one of the great universals of history. It can be seen in the Middle East, in Europe, in India and China, in Central America and Peru. Regimes eventually collapse even when there are no obvious barbarians to do the work, such as with the decline of the Later Han dynasty in China in the second century AD. Much has been written of the "mysterious" collapse of the Mayan civilization in Guatemala, with explanations which include natural disasters, intensified warfare, famine due to over-population and civil strife.[19] But, in fact, there is nothing mysterious about it. Wealthy civilizations bear the seeds of their own collapse.

Population density, wealth, and declining V

As mentioned earlier, one reason for civilization decline is that population density tends to undermine V. Vigorous civilizations have a high level of V, which not only drives military success but supports C. The result is that population increases and population density rises, which in a high-V people causes an increase in stress. Given a culture with strong V-promoters and C-promoters, this may to some extent reinforce V by increasing the punishment of older children, as happened in Europe and Japan until the sixteenth century. But more than a certain amount of stress begins to undermine V, either by increasing infant C or by causing infants to be neglected or abused or both.

At this stage another factor steps in, which is wealth. Infant C tends to

---

[18] B. Sykes, *Saxons, Vikings, and Celts: The Genetic Roots of Britain and Ireland* (New York: W. W. Norton & Company, 2006).
[19] H. McKillop, *The Ancient Maya: New Perspectives* (New York: Norton, 2004).

create wealth, and fewer children means that population no longer presses on the food supply. Plentiful food further reduces V and then C, which makes V- and C-promoting traditions harder to maintain because they are more alien from the temperament, so these are gradually abandoned. This in turn accelerates the fall of V and C.

How quickly this happens depends on the strength of the culture behind the civilization. Europe and Japan benefited from powerful cultures with strong V- and C-promoting traditions inherited from advanced civilizations in the Middle East and China, respectively. Thus it is that V rose for several hundred years and C for a thousand or more, before starting to decline. Civilizations such as the Maya had much less advanced traditions so they declined relatively fast.

It follows from this approach that decline must be earliest in the cities and in elites, which tend to have ample food and are largely protected from cold and other environmental stresses which (unlike population density) work to promote V. The most affected groups of all are the urban elites.

Declining V and C in wealthy urban elites

To gain a better understanding of this process we will look at three urban elites which suffered a dramatic decline in V and C: court aristocrats in eleventh-century Japan, Chinese scholar-gentry families, and wealthy twentieth-century Americans. Low V should reduce energy, confidence and morale. Declining C should reduce the work ethic, self-control, adherence to rules of behavior, and interest in family and children.

The Fujiwara family, eleventh century AD

The Fujiwara were descendants of aristocrats who had had helped the Japanese imperial house gain effective control of Japan by the seventh century AD. In 645 Nakatomi no Kamatari led a coup against the dominant Soga family and in 668 was granted the surname Fujiwara by the emperor. He and his descendants came to dominate the imperial court, marrying their daughters to the emperors, with the result that for several hundred years almost all Japanese emperors had Fujiwara grandfathers, uncles and so forth.

By the eleventh century the authority of the imperial court was in decline, with local military aristocrats gradually gaining authority in the provinces.

But the court still retained wealth and prestige.

Courtiers were described as having a distaste for physical effort. They avoided such pursuits as hunting and war, and expressed contempt for the military. There was a sense of fatalism and of the futility of effort; a feeling of resignation and world-weariness. Though some courtiers held jobs, many had no interest in work and few had serious intellectual interests. Sexual behavior was notably free, to the extent that paternity was often doubtful. Unattached women in particular were highly promiscuous. Women had little interest in children, though for political reasons they seem to have borne many. Consistent with low V is the fact that women held relatively high status in the court, higher than at any time in Japanese history prior to 1945. They could inherit and keep property, and unmarried women had a great deal of freedom.[20]

As with the Tchambuli, this was a society obsessed with artistic expression. There was an intense focus on poetry, on the details of dress and personal appearance, and a deep sense of appreciation for natural beauty, which has had a marked influence on future generations of Japanese. Above all, and strikingly for a society so little focused on productive work, this period gave birth to some of the world's greatest works of literature. The Tale of Genji and other masterpieces were written by women in the imperial court.[21]

It is not possible to link artistic creativity too closely to low V, since the arts have also flourished in high V societies such as Elizabethan England and Periclean Athens. But low-V societies commonly show a distinctive artistic sensibility focused on external forms and appearances.

In most situations, an elite that lacks martial vigor or even much capacity for work would not long retain power. It is a testament to the peculiar Japanese reverence for hereditary status that the Fujiwara and the imperial court retained any measure of wealth and authority at the time. Even so, in the following century the provincial military leaders took power and

---

[20] I. Morris, *The World of the Shining Prince* (Oxford: Oxford University Press, 1964), 26, 80, 122, 143–147, 170, 177–196, 211–212.

[21] Other notable works include the Pillow Book of Sei Shonagon; M. Shikibu, *The Tale of Genji*, edited and translated by Royall Tyler (New York and London: Penguin Classic, 2002); W. Aston, *A History of Japanese Literature* (London: Heineman, 1907); D. Keene, *The Pleasures of Japanese Literature* (New York: Columbia UP, 1988).

reduced them to ritual insignificance.

Chinese scholar-gentry

The Chinese gentry, generally of south China, were the class from which the governing Chinese bureaucracy was drawn until the close of the nineteenth century. These were people who had gained wealth originally as merchants or by accumulating land, allowing them the leisure to study for the imperial examinations. Success in the examinations required an intense work ethic and high intelligence, and only a small percentage of candidates passed. Successful candidates could go on to become high officials, providing further opportunities to accumulate wealth.

A study of such prominent families found a strong pattern of decay after a few generations. Successful men seem to have had little interest in their children, presumably because of the decline in C. Their sons or grandsons tended to be far less successful, showing a lack of initiative, an attitude of resignation and a sense of the futility of effort. Declining generations were also debauched and delinquent, and infertility was a major problem.[22] These changes indicate a decline in V to a very low level, and also a loss of C.

American third generation wealthy, twentieth century

There is further evidence of declining C and V in a group of young Americans receiving psychiatric treatment.[23] They were third generation rich, brought up with the wealth generated from businesses started by their grandparents and expanded by their parents. None of this group held jobs or pursued hobbies or intellectual interests. Their lives tended to center around their social set, cars and clothes—they were the so-called

[22] H.-T. Fei, "Peasantry and Gentry: An Interpretation of Chinese Social Structure and its Changes," in *Social Structure and Personality*, edited by Y. A. Cohen, 24–35 (London: Holt, Rinehart and Winston, 1961); P.-t. Ho, *The Ladder of Success in Imperial China* (New York: Columbia University Press, 1962), 129–145, 157–160, 166; E. O. Reischauer & J. K. Fairbank, *East Asia: The Great Tradition* (Boston: Houghton Mifflin, 1960), 187, 223–228; F. Hsiao-Tung, *China's Gentry: Essays in Rural-Urban Relations*, edited by M. Park, , 205–206, 246–247, 270–272 (Chicago: University of Chicago Press, 1953).

[23] R. R. Grinker, "The Poor Rich: The Children of the Super-Rich," *American Journal of Psychiatry* 135 (1978): 913–916; R. R. Grinker, "The Poor Rich," *Psychology Today* (1977): 74–6, 81.

"beautiful people." Note once again, as with the Tchambuli and the Japanese court culture, a focus of low-V people on appearances.

The psychological attitudes of these wealthy young Americans were similar to the other groups. They complained of boredom, hopelessness and emptiness. They were described as easily angered, lacking shame and embarrassment, acting on impulse, having frequent brushes with the law, and pursuing compulsive sex lives.

The declining interest in children can be traced generation by generation. The grandparents, who built the family wealth, had sufficient contact with their children to be active role models. But the second generation was occupied with its' own activities and pleasures and left childrearing to a shifting body of servants, who were frequently fired when the parents became jealous of their attachment. Thus the third generation saw little of their parents. They had a great deal of freedom and relatively little consistent discipline, all circumstances which would minimize C.

Few of this third generation married or had children, and those that did made indifferent parents. We might say that the grandparents had relatively high C, the next generation perhaps higher infant C, and the third generation low C overall. By then, V had also fallen to an absolute minimum. It might be added that these people were not absolutely typical of wealthy American families. Those with a tradition of public service such as the Rockefellers and Kennedys have done considerably better.[24]

Overall, these three cases illustrate the kind of changes that can easily occur in wealthy, urban elites as V and then C decline. For a ruling class to lose power in most societies the change in character need not be this extreme, but these examples show how people behave when C and V fall dramatically, and also how fast this can occur. From the American example it seems to take no more than two generations of ineffective parenting for them to collapse completely.

Revolution

We have seen how wealth and urban living can reduce V and eventually C in ruling elites until they are no longer capable of holding power. Very often, regime change occurs when warlike peoples move in and take over. Through most of history such peoples remain something of a mystery,

[24] Grinker, "The Poor Rich: The Children of the Super-Rich."

since they typically did not keep written records and are known primarily by their impact on the lands they invaded. But in the twentieth century there are case studies that are far better documented in the form of revolutionaries—the equivalent of barbarians invading from within, as it were. Some have been religiously motivated, such as the Islamist revolutionaries who took power in Iran in 1979. Most revolutions, though, are political, and frequently based on Communism.

It is common to view these revolutions as victories for their respective ideologies, and there is a strong tendency to focus on the personalities and motivations of the leaders. Organized leadership is essential to the success of any large undertaking, but leaders can do nothing by themselves. Successful leaders require followers who are aggressive, self-sacrificing and disciplined. In other words, they require enough people with sufficient C and especially V to take on entrenched regimes.

From the observations thus far, it is clear that V is increased by cultural factors such as patriarchy but also by the environment. The strongest effect will be when times of hunger and famine are followed by a period of extreme prosperity, allowing populations to grow unusually fast. The experience of famine is transmitted to the next generation by the way adults treat their young, which is of course part of the value of the V mechanism as an adaptation to an unstable environment. Rapid population growth supports V, and only after some time does prosperity cause the V mechanism to fade. This is the process discussed earlier in relation to the Sioux and Yanomamo.

Thus, a population with maximum V will have experienced famine in the recent past but also at least a generation of rapid growth. Both the French and Russian revolutions were driven by mass support from ill-fed sections of the population, which in recent decades had experienced rapid growth.[25] In fact, as indicated in the previous chapter, the French and Russian Revolutions both occurred in generations born at or near a peak of population growth.

The Russian case is even more striking because there was a devastating

---

[25] This is a pattern identified by a number of scholars and known as the "J-Curve Theory of Revolutions," the idea being that Revolutions tend to occur when a long period of improvement is supplanted by a sharp reversal. James C. Davies, "Toward a Theory of Revolution," *American Sociological Review* 27 (1) (1962): 5–19.

famine in 1891–92, with an estimated half a million deaths just twenty-six years before the Bolshevik Revolution. Thus, the Russians who reached manhood in 1917 had experienced not just one but *two* major V-promoters: a population growth peak, and a major famine.

The Cuban Revolution of 1957–59 was led by urban intellectuals including the Castro brothers and Che Guevara, but the insurrection was based on Oriente Province, the most mountainous in Cuba and one with a long history of patriotic warfare. Oriente had risen in rebellion in 1868 to launch the Ten Years War, which caused terrible devastation in the area. Then, in 1895 rebellions against Spanish rule broke out all across the country, but it was only in Oriente that they were successful. Insurgents controlled the Oriente and Camaguey Provinces, as well as a few cities, until American intervention brought independence in 1898. In addition, the birth rate in Oriente was the highest in Cuba from at least the 1930s.[26] All this is an indication of high V.

So, Castro was following tradition when he based his revolution on Oriente. But even here, his only really active supporters were the poor peasants of the Sierra Maestra, the most rugged area of the province. In effect, the Cuban Revolution can be seen as the victory of the highest-V fraction of the population in the highest-V area of the highest-V province in the country. As late as 1980, more than a third of the Central Committee was from Oriente.[27]

The picture is even clearer when applied to the victory of the Chinese Communists in 1949. Mao's forces had been badly defeated in the center of the country, forcing on them the devastating Long March which brought them to Shanxi Province in the north-west. It was from here that they fought off the Nationalists and the Japanese, and eventually emerged to take over the country.

---

[26] H. L. Matthews, *Castro: A Political Biography* (London: Penguin, 1969), 67, 129; Diaz-Briquets & L. Perez, "Fertility Decline in Cuba: A Socioeconomic Interpretation," *Population and Development Review* 8 (3) (1982): 513–37; L. A. Perez, *Lords of the Mountain: Social Banditry and Peasant Protest in Cuba: 1878–1918* (Pittsburg: University of Pittsburg Press, 1989), 24–25.

[27] E. Gonzales, "After Fidel: Political Succession in Cuba," in *Cuban Communism*, edited by I. Horowitz, 499 (New Brunswick: Transaction, 1988).

Shanxi is at the center of the Loess Plateau of north China, a region of bitterly cold winters and hot summers which is notorious for famine. It is also well known as a birthplace of Chinese dynasties—the Zhou in the eleventh century BC, the Qin in the late third century BC, and the Tang at the start of the seventh century AD. Further east in Hebei province (around Beijing) the climate is almost as severe, and rebellions from this area have deposed at least two Chinese emperors: Xuanzong in 756 AD and Huidi in 1403 AD.[28]

In at least some of these cases there is evidence of a rapid increase in population over the preceding decades. The Qin had set up an irrigation system in the Wei River valley in the mid-third century BC, which is known to have allowed a rapid increase in population. But a more common cause of a population increase in a country with intense pressure on land is some kind of disaster, wiping out much of the local population in the preceding fifty years.

The background to the 756 rebellion follows this pattern. Between 696 and 720 AD the area around Beijing suffered invasion and repression by the central government, leading to prolonged lawlessness and economic disruption. About 720, troops were sent in to restore order, leading to marked prosperity. Thirty-five years later a massive rebellion broke out in this area, causing the emperor to be deposed.[29] This combines three environmental V-promoters: a harsh climate, famine thirty-five years earlier, and recent population increase.

This same area was the base for another massive rebellion in 1399, which brought the Yongle Emperor to the throne. This has been discussed in chapter nine as an after-effect of the war in the previous generation that established Ming rule. But the recent history of the province also suggests very high V. The climate, as already mentioned, was harsh. It had been devastated by the Mongols in the early thirteenth century, at least partly with the aim of turning north China into pastureland. Further massive bloodshed is recorded around the 1360s, when the resurgent Chinese murdered almost every Mongol in sight. Order was restored by the new Ming dynasty in 1368, and there would thus have been plenty of vacant

---

[28] T. R. Tregear, *A Geography of China* (London: University of London Press, 1965), 213.

[29] Reischauer & Fairbank, *East Asia: The Great Tradition*, 187, 223–228; E. G. Pulleyblank, *The Background of the Rebellion of An Lu-shan* (Oxford: Oxford University Press, 1965).

land for a fast-growing population in the decades that followed, leading up to the outbreak of rebellion in 1399.

The situation in Shanxi Province when Mao's Long March arrived there in 1935 was similar. The worst natural disaster in human history may have been the famine of 1877–78, in which between nine and thirteen million people died in Shanxi and the neighboring provinces. The population of the area had not even recovered by 1953, so there was ample scope for population increase.[30] There is thus a combination of a severe famine two generations earlier with rapid population increase since then, both factors working to raise V. Given that northern Chinese are relatively high V at any time, such a people would tend to make fierce and effective soldiers. As in Cuba, the Chinese Revolution was led by urban intellectuals. But its military triumph was as much a victory of the highest-V segment of the population, imposing their values and attitudes by military conquest.

It is not always possible to find this level of information about the background to revolution, foreign conquest or other forms of regime change, but when we do the pattern is quite consistent. Wealth and population density act strongly to undermine V and thus eventually C, leading first to military and then to economic and political weakness. Groups that triumph tend to have experienced famine and other traumas in the recent past, but also to have experienced rapid and recent population growth.

In the next chapter we turn our attention to Rome, showing how the concepts developed thus far can help us to understand how and why the Roman Empire fell.

---

[30] D. H. Perkins, *Agricultural Development in China: 1368–1968* (Cambridge, Mass.: Harvard University Press, 1959), 195–200.

CHAPTER TWELVE

ROME

The story of Rome is especially interesting because it is by far our best example of the processes that occur in a society when C declines, and of the ensuing decline in economic prosperity, social cohesion, and military and political power. Few subjects in history have fascinated people as much as the fall of the Roman Empire. That a civilization as powerful and brilliant as this should collapse so completely has spawned a hundred theories, encompassing factors as diverse as waves of barbarians and the effects of lead plumbing.[1] In fact, a proper understanding of C and V not only describes this collapse but explains how the seeds of Rome's fall were laid as early as the third century BC.

Before arriving at the rise and fall of Rome, however, it is important to examine the spread of C- and V-promoting technologies in early farming societies in the Mediterranean region.

Early farming societies and rising C

The late prehistory and early history of Eurasia is dominated by the spread of farming technology from its birth-place in the Middle East. Genetic evidence suggests that this was partly due to the migration of Middle Eastern peoples, but mainly through adoption of agriculture by indigenous hunter-gatherer populations.

The evidence of physical technology—including food-crops, tools, and domestication of animals—is plentiful and highly visible. However, the

---

[1] E. Gibbon, History of the Decline and Fall of the Roman Empire (Northpointe Classics, 2009); A. Ferrill, The Fall of the Roman Empire (New York: Thames and Hudson, 1986); J. B. Bury, A History of the Later Roman Empire from Arcadius to Ireme 395–800 AD (University of Michigan Library, 1889), Vol. 1, Chapter 9; J. N. Wilford, "Roman Empire's Fall is linked with Gout and Lead Poisoning," New York Times, (March 17, 1983); but see G. A. Drasch, "Lead Burden in Prehistorical, Historical and Modern Human Bones," The Science of the Total Environment 24 (3) (1982): 199–231.

equally vital cultural technology that went with it is less visible. Yet this is what most interests us, because it was culture and social behavior that adapted people to the routine work of farming by raising their C. Knowledge of seeds and farming tools is not enough if people have no taste for the work. This is something we see today in the Mbuti pygmies of Central Africa, who are offered work by their Bantu neighbors but quickly tire of it. Farming only becomes a way of life when people adapt to its demands. Biohistory proposes that this adaptation comes through the development or adoption of practices that raise C, such as restrictions on sexual activity.

This might also explain why the process took so long. The very earliest evidence for agriculture in the Middle East dates from around 9500 BC. Cultivation and domestication gradually developed over the succeeding two to three millennia and began to spread into Europe, but it was to be more than five thousand years before farming spread as far as the British Isles. Given the distance from Mesopotamia to Britain, that is less than 1 km per year. Not only was the spread of tools and advanced seeds slow but the change to full-time farming took much longer, since early agriculturalists continued to mix hunting, fishing and gathering with herding and crop cultivation, in a mixed pattern of subsistence that probably varied with the seasons.[2]

There is no obvious physical reason for such a slow spread. Hunter-gatherer populations have wide trading networks, as shown by finds of flint axes and other artifacts hundreds of kilometers from where they were mined. Theoretically, a person could walk the length and breadth of Europe in a matter of weeks, carrying enough seeds to launch a full-scale agricultural revolution. But agriculture requires a social structure and temperament that are different from those suited to hunting and gathering, and that means changing behavior in ways that are quite alien to our

---

[2] S. Milisauskas (ed.), *European Prehistory: A Survey* (New York: Springer, 2011), 170. There are various explanations for the adoption of agriculture including climate change, the availability of plants and animals suited to domestication, population pressures and more; L. R. Binford, "Post-Pleistocene Adaptations," in *New Perspectives in Archaeology*, edited by S. R. Binford & L. R. Binford (Chicago: Aldine, 1968); O. Bar-Yosefn & R. H. Meadows, "The Origins of Agriculture in the Near East," in *Last Hunters—First Farmers: New Perspectives on the Prehistoric Transition to Agriculture*, edited by T. D. Price & A. Gebauer (Santa Fe: School of American Research Press, 1996); P. Bellwood, *First Farmers: The Origins of Agricultural Societies* (Malden, MA., Oxford: Wiley-Blackwell, 2004).

"natural" hunter-gatherer temperament.

As farming began to develop in its original homelands in the Middle East, people developed physical technologies such as better breeds of plants, and animal domestication and husbandry. They also developed cultural technologies to create the temperament required for routine work. The spread of agricultural technologies to neighboring peoples was undoubtedly faster than the time taken to develop it. This is because better food crops had already been developed and many animals already domesticated, and because cultural technologies could be copied rather than developed from scratch. But even so, the spread of agriculture was slow in Europe. Seeds and knowledge of their use might have spread to neighboring areas with a migrant family or two, or partners brought in by marriage. Communities thus affected would tend to be better fed and thus have more surviving children. Prosperity might also give them higher status, leading to imitation by their less-influential neighbors. Over several generations the successful technologies would become entrenched, and the growing population would expand into neighboring areas.

Evidence for this process can be found in the DNA of modern Europeans. Most mitochondrial and Y-chromosome lineages stem from the existing hunter-gatherer populations of Europe, but a minority can be traced to Middle Eastern immigrants at a time consistent with the spread of agriculture. These Middle Eastern lineages are strongest in Southern Europe, indicating that this is where farming was first established.[3]

People have a tendency to imitate all aspects of the behavior of higher status people—not just their economic activities but ceremonial observances and customs. These can include, for instance, such practices as the closer guarding of unmarried girls or religious observances that require sexual abstinence. It is extremely unlikely that they would have understood why such customs were important (in terms of C), any more than people do today, except in the most indirect ways, such as the

---

[3] M. Richards, H. Corte-Real, P. Forster, V. Macaulay, H. Wilkinson-Herbots, A. Demaine, S. Papiha, R. Hedges, H.-J. Bandelt & B. Sykes, "Paleolithic and Neolithic Lineages in the European Mitochondrial Gene Pool," *American Journal of Human Genetics* 59 (1) (1996): 185–203; O. Semino, G. Passarino, P. J. Oefner, A. A. Lin, S. Arbuzova, L. E. Beckman, G. De Benedictis, P. Francalacci, A. Kouvatsi & S. Limborska, "The Genetic Legacy of Paleolithic Homo Sapiens Sapiens in Extant Europeans: AY Chromosome Perspective," *Science* 290 (5494) (2000): 1155–9.

attribution of a tribe's power to its having very potent gods, or to its having earned the gods' approval. Humans are a naturally low-C species, so C can fall very fast. But to increase C is a less natural and therefore much slower process, requiring a gradual ramping up through a process of competition between families and villages over many generations.

Other waves of physical technology were to follow with much greater speed, such as the use of writing, pottery and metalworking. These also started in the Middle East and began spreading west and north. And with them can be seen traces of the underpinning cultural technology, including formally organized religions with written texts, funerary and ritual architecture, and a specialized caste of priests.

There is also some direct evidence of the spread of C- and V-promoting customs. For example, by the Middle Assyrian period (around the fourteenth to tenth centuries BC), the women of at least some classes were rigidly secluded, an effective method of supporting both C and V, and one also practiced in Greece during the Classical period just a few centuries later. By the middle of the first millennium BC, C and V-promoters are clearly visible in the civilizations of Italy.

The rise of the Roman Republic

Women were not secluded in Rome but they were carefully guarded in the early to mid-Republic. For Romans of this period, female chastity was highly valued. The expulsion of the last Roman king in 508 BC was commonly attributed to the rape of a Roman woman by the king's son, and her subsequent suicide.[4]

This cultural technology was most likely brought into Italy by the Etruscans. According to Thucydides and Herodotus, the Etruscans came from Lydia in western Asia Minor (modern Turkey). This is supported, at least as it applies to the ruling class, by recent genetic findings which show that a substantial percentage of the modern Tuscan population, and their cattle, have mitochondrial DNA characteristic of Asia Minor.[5] It thus fits

---

[4] H. H. Scullard, *A History of the Roman World: 753 to 146 BC* (London: Routledge, 2012), 67.

[5] M. Pellecchia, R. Negrini, L. Colli, M. Patrini, E. Milanesi, A. Achilli, G. Bertorelle, L. L. Cavalli-Sforza, A. Piazza, A. Torroni & P. Ajmone-Marsan, "The Mystery of Etruscan Origins: Novel Clues from Bos taurus Mitochondrial DNA," *Proceedings of the Royal Society Biological Sciences* 274 (1614) (2007): 1175–79.

into the picture of cultural waves, including some population movements, coming out of the Middle East.

The Etruscans built fortified, autonomous cities, ruled at this time by kings. The city of Rome was most likely established by the union of several villages in the early seventh century BC, and shortly afterwards came under Etruscan domination. Roman religion was largely based on Etruscan models, with laborious and elaborate rituals. Senior Roman officials spent a great deal of their time on getting these rituals exactly right, with a single misspoken phrase or even the squeaking of a rat enough to render it invalid. On one occasion a sacrifice was conducted thirty times before it was considered correct.[6] These rituals would have been effective C-promoters.

However, according to the fourth-century Greek historian Theopompus, the Etruscans were much freer than the Romans in their attitudes to sex. Their women also seem to have been more independent.[7] Both these factors help to explain why it was the Romans, rather than their Etruscan teachers, who came to dominate Italy. The Romans may have borrowed C and V-promoters from their Etruscan mentors, but they applied them far more rigorously.

Roman cultural technology was probably also reinforced by Greek traders and settlers in southern Italy, whose culture was to make an obvious and indelible mark on Rome. Whatever the source, it enabled higher levels of C and V and thus a more complex civilization than anything previously known in Western Europe. Republican government, which the Romans maintained for nearly five hundred years, requires strong impersonal loyalties. People must be more loyal to an institution, the Senate or the Consulship, than to the individuals comprising it. They must accept that the impersonal dictates of law take precedence over the interests and ambitions of individuals. Roman skills in engineering, and the superb discipline of their armed forces and military camps, are further evidence of unusually high C and especially infant C.

All these explanations may have merit but Biohistory focuses on the temperamental adaptations needed to make farming successful.

[6] A. Everett, *The Rise of Rome: The Making of the World's Greatest Empire* (London: Head of Zeus, 2012), chapter two.

[7] Ibid., chapter three.

C was higher in Rome and Classical Greece than in the Middle East where the cultural technology developed because, for reasons to be discussed in the next chapter, the level of infant C was higher. But even this level, although well above that of previous times, was far less than would be achieved in Europe in the nineteenth century under the influence of Christianity. The Roman Republic was an impressive structure but built on personal loyalties between client and patron and was thus less stable than the regimes of nineteenth-century Europe. And though far more advanced in trade and technology than earlier peoples, the Romans never achieved an industrial or scientific revolution. Roman C was simply not high enough for such developments.

What we do see in the early Republic is a growth in C that was far more rapid than the gradual rise that would have accompanied the birth of agriculture. It was quite similar to that traced in Western Europe between the twelfth and nineteenth centuries, suggesting a civilization cycle where a rapid rise in C is driven by exceptionally high V and stress.

In the early stages Rome, like its Latin neighbors and the Etruscan states, was ruled by kings, suggesting relatively low C. Then, around 509 BC, Rome threw out its king and became a republic with two elected consuls.[8] Initially these consuls had much the same powers as the kings, but gradually a true republic evolved with a senate, a number of magistrates with different functions, and tribunes to represent the plebeians.[9] This was not an exclusively Roman development. Most Etruscans states, which covered much of northern and central Italy, became republics about this time.[10] Since republics require a higher level of impersonal loyalties than kingdoms, this suggests that the rise in C was regional rather than confined to Rome. Also indicative of rising C at the time was the creation of Rome's first written law code, the Twelve Tables, in 451–449 B.C. Among its provisions were a prohibition on making laws against individuals, and ferocious penalties for offenses against property.[11] Both of these are impersonal principles.

---

[8] Scullard, *A History of the Roman World: 753 to 146 BC*, 78.

[9] Ibid.

[10] M. A. Ward. F. M. Heichelheim & C. A. Yeo, *A History of the Roman People* (New Jersey: Prentice Hall, 2010), 19.

[11] K. Hopkins, *Conquerors and Slaves* (Cambridge: Cambridge University Press, 1978), 21.

From historical records, the peak of C can probably be placed somewhere around 250 BC. The population was growing fast in the fourth and third centuries, with forty colonies established on conquered Italian land between 384 and 218 BC.[12] Writers of the later Republic spoke of the "reserve" of the period before about 200 BC.[13] Roman citizens served willingly and ably as soldiers, just as did the citizens of Europe and Japan near their peak of C in the nineteenth and early twentieth centuries. It is also characteristic of high C that the various cities and regions of Italy kept their own identities and citizenries, even after they accepted Roman primacy by the third century BC. They were considered subordinate allies with a common foreign policy, rather than Roman subjects or citizens. Loyalty to a city or nation is characteristic of high C, just as loyalty to an individual denotes lower C.

So far, as can be judged by this, the Romans and their neighbors seem at this time to have had similar infant C levels to the Greeks in their prime. The key difference is that they would have had a higher level of child V, making them better able to build large states and also more traditional and less creative. Romans of the Republic were famously devoted to their traditions, the "*mos maiorum*."

The story of the Consul Regulus, whether true or not, gives a picture of Roman ideals of honor and integrity at this time. Regulus had been captured by the Carthaginians during the first Punic War in 255 BC and released on parole on condition that he would go back to Rome and argue the case for peace. Instead, he spoke against the proposed terms, but then felt honor-bound to return to Carthage and suffer a cruel death rather than break his word.[14]

Another example of the character of the early Romans was Cato the Censor (234–149 B.C.), as described by Plutarch.[15] Cato was famous even in his own time for personifying many of the ideals of the early Republic. Before moving to Rome to pursue politics, he served as a soldier and

---

[12] Ibid., 21.

[13] J. W. Rich, "The Supposed Roman Manpower Shortage of the Later Second Century BC," *Historia: Zeitschrift für Alte Geschichte* 32 (3) (1983): 294, 300.

[14] Scullard, *A History of the Roman World*, 437.

[15] Plutarch, "Marcus Cato," *Plutarch: Lives of the Noble Grecians and Romans*, edited by A. H. Clough, (Project Gutenberg, 1996) http://www.gutenberg.org/cache/epub/674/pg674.html (accessed September 6, 2014).

gained a reputation for military excellence. Between campaigns he worked on his inherited farm in Sabine where he dressed and acted as a simple laborer, sharing duties with his slaves. He believed that farming, unlike commerce or usury, produced virtuous citizens and brave soldiers and was a consistent source of wealth and high moral values (one of the defining traits of high C is a temperament suited to consistent, routine hard work). Later in his career as governor of Sardinia, Cato acted with strict integrity, reducing the costs of administration while himself living frugally. Not wanting to place unnecessary burdens on the treasury, he walked from city to city with a single public officer, instead of the large and impressive entourage of his predecessors.

He administered impartial justice and severely enforced the laws against usury, banishing those found guilty. He was a stern moralist. When acting as censor in his later years, Cato used his powers of office to expel many senators who he believed had acted improperly. One was a man called Manilius, who had embraced his wife in daylight and in full view of his daughter. Cato contrasted Manilius' behavior with his own modesty, claiming never to embrace his wife unless there was a loud thunderstorm.

Whilst in office he campaigned against what he saw as the growing decadence of the time, defending old laws limiting the amount of jewelry women could wear, as well as placing heavy taxes on items that were deemed "excessive luxury." Many of these changes were associated with the rise of Greek influence, exemplified by such men as Scipio Aemilianus. Cato rejected Greek culture in favor of the austere Roman tradition.

Cato ate his breakfast cold, dressed simply, lived in a humble dwelling, drank the same wine as his slaves and refused to plaster the walls of his cottages. He was seen as a classic example of the traditional Roman character—austere, frugal, harsh, self-disciplined, modest in his dress and his habits, inflexible in his integrity, disinterested and impersonal when administering justice, dogmatic and loving of order. All of this suggests that Cato had very high C and C-promoting habits. Despite being somewhat eccentric in the extremes to which he carried his asceticism, and regarded as slightly old-fashioned, he was fairly representative of Roman practices and values in the early republican period, and certainly representative of the period's ideals.

The Roman writer Sallust gives a picture of the early Republic redolent of high C and V.

> To begin with, as soon as the young men could endure the hardships of war, they were taught a soldier's duties in camp under a vigorous discipline, and they took more pleasure in handsome arms and war horses than in harlots and revelry. To such men consequently no labor was unfamiliar, no region too rough or too steep, no armed foeman was terrible; valor was all in all. Nay, their hardest struggle for glory was with one another; each man strove to be the first to strike down the foe, to scale a wall, to be seen of all while doing such a deed. This they considered riches, this fair fame and high nobility. It was praise they coveted, but they were lavish of money; their aim was unbounded renown, but only such riches as could be gained honorably.
>
> Accordingly, good morals were cultivated at home and in the field; there was the greatest harmony and little or no avarice; justice and probity prevailed among them, thanks not so much to laws as to nature. Quarrels, discord, and strife were reserved for their enemies; citizen vied with citizen only for the prize of merit. They were lavish in their offerings to the gods, frugal in the home, loyal to their friends. By practising these two qualities, boldness in warfare and justice when peace came, they watched over themselves and their country.[16]

Declining V and stress

In more recent history, the rapid rise of C in Europe and Japan was stimulated by high levels of V and thus stress in and around the sixteenth century, indicated by harsh punishments and political instability. If the Roman Republic followed the same pattern we should be able to identify a peak of stress ahead of its rapidly rising level of C.

Poor historical evidence for the pre-Republican period makes this difficult, but there is some evidence for a peak of V in the sixth century BC. Of the seven traditional kings of Rome, the first four are historically doubtful. But from then on we may be on firmer ground, and there are indications that kingship in the sixth century B.C. was both brutal and unstable. Of the last three kings, Tarquinius Priscus was assassinated in 579 BC, Servius Tullius deposed and murdered in 535 BC, and Tarquin the Proud deposed and exiled in 496 BC.[17] The Tarquins were most likely Etruscans, indicating a rule that was not only despotic but alien.

---

[16] Sallust, *The War with Cataline,* http://penelope.uchicago.edu/Thayer/E/Roman/Texts/Sallust/Bellum_Catilinae*.html (accessed September 6, 2014).

[17] Scullard, *A History of the Roman World*, 49–50.

By comparison, transfer of power under the Republic from the fifth century was less violent, a sign of lower stress. There was turbulence in plenty, but the plebeians' strongest leverage against the ruling aristocracy was for the army to go on strike.[18] All this suggests a peak of stress and thus V around 550 BC.

From this time onward there is evidence of declining stress. As early as the fifth century, ordinary people were gaining rights which protected them against the arbitrary use of government power. From the start of the Republic, a man had the right to appeal to the popular assembly against a death sentence.[19] By the mid-fifth century, popularly elected tribunes won the right to veto any act of government.[20] The Twelve Tables themselves, though harsh by later standards, shielded citizens against arbitrary power by making the law clear, and also slightly weakened the power of husbands over wives and children (patria potestas).[21] In the late fourth century BC, further legislation protected men against being seized or imprisoned for debt.[22]

If we are correct in placing the peak of V and stress around 550 BC and C around 250 BC then there was a roughly three-hundred-year gap between the peaks of V and of C, which is similar to the interval in early modern Europe between the peak of V in the sixteenth century and of C in the nineteenth. Japan also reached its peak of V in the sixteenth century but of C only in the early twentieth century, so the gap was a little longer. But for Rome, as for England and Japan, V and stress seem to have been in decline well before the peak of C.

From rising to falling C—the decline of the Republic

A rapid rise of C in the civilization cycle depends on a high level of V and stress, aided by cultural technologies which promote both V and C. As V and stress decline, partly because V is undermined by increased C, C

---

[18] R. T. Ridley, *The Gracchi, History of Rome* (Paris: l'Erma di Bretschneider, 1987), 83.

[19] Ibid., 81.

[20] Ibid., 82–5.

[21] A. Drummond, "Rome in the Fifth Century I: the Social and Economic Framework," *The Cambridge Ancient History* Volume 7, Part 2: The Rise of Rome to 220 BC, Second Edition (Cambridge: Cambridge University Publishing, 1990), 147–8.

[22] Scullard, *A History of the Roman World*, 116–7.

gradually rises to a peak and then starts to fall.

This is clearly evident in the Roman Republic from the late-third century BC. Cato the Censor was considered old-fashioned for his sternness and integrity (although still admired for them by traditionalists). Powerful aristocrats came to be known for their generosity and charming manners rather than the old Roman austerity, suggesting a switch to a more personal, lower-C attitude.[23] The decline in C was also driven by the enormous wealth brought into Rome from new conquests following the Punic Wars.[24] Sallust had no doubts on the enervating effects of wealth:

> When Carthage, the rival of Rome's sway, had perished root and branch, and all seas and lands were open, then Fortune began to grow cruel and to bring confusion into all our affairs … Hence the lust for money first, then for power, grew upon them; these were, I may say, the root of all evils. For avarice destroyed honour, integrity, and all other noble qualities; taught in their place insolence, cruelty, to neglect the gods, to set a price on everything. Ambition drove many men to become false; to have one thought locked in the breast, another ready on the tongue; to value friendships and enmities not on their merits but by the standard of self-interest, and to show a good front rather than a good heart. At first these vices grew slowly, from time to time they were punished; finally, when the disease had spread like a deadly plague, the state was changed and a government second to none in equity and excellence became cruel and intolerable.
>
> As soon as riches came to be held in honour, when glory, dominion and power followed in their train, virtue began to lose its lustre, poverty to be considered a disgrace, blamelessness to be termed malevolence. Therefore as the result of riches, luxury and greed, united with insolence, took possession of our young manhood. They pillaged, squandered; set little value on their own, coveted the goods of others; they disregarded modesty, chastity, everything human and divine; in short, they were utterly thoughtless and reckless.[25]

By the second century BC the picture is even clearer. The birth rate was falling rapidly, even in the countryside, to the concern of many. Scipio Aemilianus, who was Censor in 142–141, protested against the practice of men adopting adult heirs, so that they could gain the legal privileges of

---

[23] W.W. Fowler, Social Life at Rome in the Age of Cicero (London: MacMillan,1908), 101

[24] Ibid., 46.

[25] Sallust, The War with Cataline.

parenthood without the need to bear and raise children. Metellus Macedonicus, Censor in 131–130, urged all citizens to marry so that they could raise more children.[26] Reversing the decline in birth rate was a principal reason for the land reform schemes put forward later in the century.[27]

There is also evidence that standards of sexual behavior were loosening, as is always the case when C begins to decline. Divorce, apparently uncommon during the early Republic, grew increasingly frequent and socially acceptable from the first century BC onward. By the early imperial period it was almost routine, at least among the propertied households about which historians and poets tell us most.[28] However, fragmentary evidence from inscriptions indicates that divorce was considerably less frequent among the lower orders.[29]

Another indicataion of falling C is the growth of vast slave estates in the late Republic, reducing the number of peasant farmers who had formed the backbone of the Roman army.[30] On one hand this shows how much market thinking had come to dominate agriculture, just as booming economies accompanied declining C in the twentieth century. But it also shows that fewer people wanted to be farmers.

Slaves have to be bought and managed and controlled and are only profitable when wage labor is in short supply. This was the case in the New World in the eighteenth and nineteenth centuries, especially in the Caribbean where European workers could not survive the rigors of the climate. But slavery never dominated farming in India, China or the Middle East, including the eastern portion of the Roman Empire. For example, in China at this time slaves were used in domestic service and probably made up less than 1% of the population.[31] Farmworkers may have been low status and with little freedom, but they formed families and bred their own replacements.

---

26 Rich, "The Supposed Roman Manpower Shortage of the Later Second Century BC," 302–5.
27 Ibid., 302–5.
28 J. Carcopino, *Daily Life in Ancient Rome: The People and the City at the Height of the Empire* (Harmondsworth: Penguin, 1956), 97.
29 J. A. Brundage, *Law, Sex, and Christian Society in Medieval Europe* (Chicago: The University of Chicago Press, 1987), 23.
30 A. H. McDonald, *Republican Rome* (London: Thames and Hudson, 1966), 76.
31 E. O. Reischauer & J. K. Fairbank, *East Asia: The Great Tradition* (Boston, Houghton Mifflin, 1960), 97.

Roman farm slaves were commonly permitted to marry informally and have children, but not enough to maintain their numbers.[32] Thus they were replaced largely by purchase. Slave labor only makes sense economically where there is a labor shortage, as there clearly was in the late Republic,[33] partly due to a declining birth rate and increased demand for extended army service. Lower-C people also have less taste for the relatively tedious business of farming, causing them to migrate to urban centers—in this case, the city of Rome, where they could benefit from an economy maintained by war loot and tribute. This was an entirely different process from the urbanization of Britain during the Industrial Revolution, where people were drawn from the farms to the equally rigorous but better paid work in the new factory towns.

A careful analysis of the economics of Roman farming at this time suggests that slave labor was not obviously more profitable than the use of free labor or a sharecropping system.[34] Therefore, the rapid growth of slave estates must have been a result of rural depopulation rather than the cause of it.

Social reformers of the time were quite aware of the problem, and demanded that the urban poor be given land that would turn them into productive farmers. The most successful attempts were those of Tiberius and Gaius Gracchus from 133 BC, which gave landless plebeians much of the Roman public lands—to the deep displeasure of the rich, who had been leasing them on favorable terms.[35] Significantly, knowing his supporters, Tiberius Gracchus legislated that this land could not be sold. After the death of Gaius Gracchus in 121 BC the Senate effectively reversed the measure by permitting the sale of this land, allowing the short-term landowners to sell up and return to the city.[36] This is one of many examples showing that government edicts cannot counter changes in temperament that are based on epigenetics—in this case the intolerance that low-C people have for the drudgery of farming.

---

[32] R. M. Geer & H. N. Couch, *Rome. Classical Civilization* (Englewood Cliffs, Prentice-Hall, 1950), 155.

[33] Fowler, Social Life at Rome in the Age of Cicero, 204–5

[34] D. W. Rathbone, "The Development of Agriculture in the 'Ager Cosanus' During the Roman Republic: Problems of Evidence and Interpretation," *Journal of Roman Studies* 71 (1) (1981): 13–4.

[35] Ridley, *The Gracchi, History of Rome*, 215–225.

[36] Ibid., 236.

One obvious consequence of this change in temperament and economic behavior was a growing gap between rich and poor.[37] The elite were able to take advantage of new opportunities by buying up land, lending money, and plundering the newly conquered provinces. The great mass of people in the city of Rome, less suited than their ancestors to productive work, became increasingly dependent on government-subsidized handouts of food.[38]

Ironically, the economic decline of the plebeians was accompanied by a surge in their political influence, with a high-water mark in the reforms of Tiberius and Gaius Gracchus between 133 and 121 B.C. Though focusing largely on land distribution they increased the power of the Popular Assembly relative to the Senate, which consisted of wealthy landowners.[39] Other measures shortened the term of military service, made the state pay for soldiers' equipment and uniforms, and protected citizens against being exiled without trial.[40] Both Tiberius and Gaius were murdered by their senatorial opponents, but later leaders such as Marius and Julius Caesar relied on popular support in their rise to power.

This change was a result of a general decline in deference, which can be explained by a continued fall in V and stress. We have seen evidence of a decline in the early Republic, with ordinary citizens better protected against arbitrary power such as by the setting up of law codes. The Republic itself was, of course, some form of protection against arbitrary government. But the earliest Roman law code, the Twelve Tables, was still harsh, with many capital offenses. Ideal behavior in the early Republic was similarly harsh and overbearing, whether to slaves, freedmen or allies of the state. Cato the Censor advocated harshness to children, slaves and conquered peoples. If the European and Japanese pattern also applies to Rome, we would expect to see a significant fall in V before the peak of C in the mid-third century BC. As indicated above, there is some evidence of such a change.

The fall of V and stress is even clearer from the late third century BC, as C began to fall. By the end of this century, as already mentioned, the strict values of Cato were no longer dominant. Upper class women were gaining in status and power, one vivid sign of which was their participation in

---

[37] Geer & Couch, *Rome*, 90.

[38] Fowler, Social Life at Rome in the Age of Cicero, 36.

[39] Scullard, *A History of Rome from 133 B.C to 68 A.D.*, 25–29.

[40] Ridley, *The Gracchi, History of Rome*, 215–225.

demonstrations to repeal the Lex Oppia, which had curbed women's wealth and freedom.[41] The law was repealed in 195 BC and from this time women, who had previously been subject to men before and after marriage, came to have effective equality in married life.[42] By the late Republic women of the senatorial class were able to manage their own business and financial affairs through wide acceptance of a form of marriage, sine manu, that left them control of their own property.[43]

In the second century BC the law codes were relaxed, and army discipline became more lenient. The growing influence of ordinary citizens was part of this trend, although undermined by the decline in their economic situation, which made them susceptible to bribery.[44] One way in which the Senate defeated Gaius Gracchus was by offering still more generous terms to the urban poor.[45]

The decline in V and stress counteracted, to some extent, the effects of declining C. We have seen that C, and especially infant C, is associated with attitudes of flexibility consistent with a market economy. But punishment in late childhood results in the more rigid and traditional child V, which is less favorable to economic activity. So while declining C in the late Republic sapped the work ethic of most citizens, declining child V meant that those with money and talent found ample scope in an environment with few constraints on the scramble for wealth. It is likely that moderate C and low child V is a combination better suited to mentally demanding occupations such as banking and business, compared to high C but much higher child V. This is in contrast to occupations such as farming which require sheer hard work and tolerance for routine. Thus it is that Romans may have become better at business just as they lost their taste for farming.

Child V inhibits not only flexible economic thinking but creativity in general. This is why the first century BC, not the third century when C was

[41] A. E. Austin (ed.), *The Cambridge Ancient History* Volume 8: Rome and the Mediterranean to 133 BC, Second Edition, Roman Government and Politics, 200–134 B.C. (Cambridge: Cambridge University Publishing, 1989), 184.
[42] N. Elias, "The Changing Balance of Power between the Sexes in Ancient Rome," *Theory, Culture & Society* 4 (1987): 287–316
[43] Carcopino, *Daily Life in Ancient Rome*, 96.
[44] Scullard, *A History of Rome from 133 B.C to 68 A.D.*, 33.
[45] E. Badian, *Foreign Clientelae, 264–70 BC* (Oxford: Clarendon Press, 1958), 190–191.

at a peak, was the golden age of Latin literature. Writers and poets such as Livy, Virgil, Cicero, Catullus and Caesar have been read and revered for more than two-thousand years.

The civilization cycle model suggests that the fall in V and stress precedes and is initially faster than the fall in C. Also, the fall in C must happen first in C and only later in infant C, and infant C is the trait associated most strongly with flexible and market-oriented thinking. Thus we must expect creativity and market thinking, as well as attitudes favorable to democracy, to peak well after the high point of C, even when the work ethic and other aspects of C are strongly in decline. In the Roman Republic the high point of these attitudes came 150–200 years after the high point of C in the third century B.C.

As C fell local loyalties were also in decline, one indication being that formerly independent Italian peoples moved towards Roman citizenship. Ironically, the Social War of 91–88 BC, which precipitated the grant of citizenship to most Italians, almost certainly represents the G-90 period of 90 BC when local and personal ties were temporarily stronger.[46] In this war, Italian peoples which had been allied for centuries revolted against Roman domination, objecting to their inferior status and lack of voting rights. The G-90 period is indicated by their actions (revolt) rather than their aims, but the same could be said of any G-90 period. For example, when the House of York plunged England into civil war in the 1450s, their aim was not to fragment the state but to build a stronger state with themselves in charge. ,

Thus it is that in the early stages of declining C, some changes in Roman society can be seen as positive. Because the fall of child V was even more pronounced than the fall of C, economic activity expanded and creativity flourished. The government was less brutal and more democratic. Society was freed from the old constraints and women gained more independence. The birth rate fell but not to the extent of causing depopulation. The cultural prestige of this wealthy and creative civilization was so great that some peoples actually sought Roman protection, such as certain Greek cities and the Kingdom of Pergamon.[47] In chapter sixteen extensive parallels will be drawn with the history of the West in the late-twentieth century.

---

[46] Scullard, *A History of Rome from 133 B.C to 68 A.D.*, 64, 67.
[47] Geer & Couch, *Rome*, 99.

But as early as the second century BC there were disturbing signs for the future. Old standards of integrity were breaking down. The gap between rich and poor was growing, with Roman citizens less and less willing to do dirty and tedious jobs. People crowded into the fast-growing cities and especially Rome itself, fixated on popular entertainment.[48] These citizens used their growing political clout to gain subsidized and eventually free grain, so that many of them no longer needed to work.[49] This decline in the work ethic is a clear and inevitable consequence of falling C.

And soon the fall of infant C would have an even more ominous effect. By the beginning of the first century BC the Roman Republic was in decline. Loyalties were turning from state to family, and in the case of soldiers from state to commander.[50] This made it possible for successful generals such as Marius, Sulla and Caesar to take control.

Laws and constitutions are impersonal, and will only be strong when people's loyalties are also impersonal. When C falls below a certain level, republican forms of government can no longer be maintained. The plotters who killed Julius Caesar thought they were restoring the Republic, but killing one man did nothing to change popular attitudes. When loyalties become sufficiently personal, the only question is which man ends up in control. Fig. 12.1 illustrates this change.

The role of personal loyalties in the quest for power can be seen in the life of Mark Antony, a fairly typical politician of the first century BC. Plutarch's description of him is in marked contrast to his life of Cato:

> Antony grew up a very beautiful youth but, by the worst of misfortunes, became friends with Curio, a man abandoned to his pleasures. To make Antony dependent on him, Curio plunged him into a life of drinking and dissipation, resulting in so much extravagance that at an early age he went two hundred and fifty talents into debt.
>
> Antony's generous ways, his open and lavish gifts and favors to his friends and fellow-soldiers, did a great deal for him in his first advance to power. And after he became great, the same generosity maintained his fortunes, when a thousand follies were hastening their overthrow. One instance of his liberality I must relate. He had ordered a friend to be paid twenty-five myriads of money … and his steward, wondering at the extravagance of

---

[48] Hopkins, *Conquerors and Slaves*, 69.
[49] Ibid.
[50] Scullard, *A History of Rome from 133 B.C to 68 A.D.*, 111.

Fig. 12.1. Rise and fall of the Roman Republic plotted against changes in stress and C. The Republic fills a segment in Rome's civilization cycle. It was established when C rose past a certain point, and fell when it declined. Republican and democratic governments require high C and especially infant C.

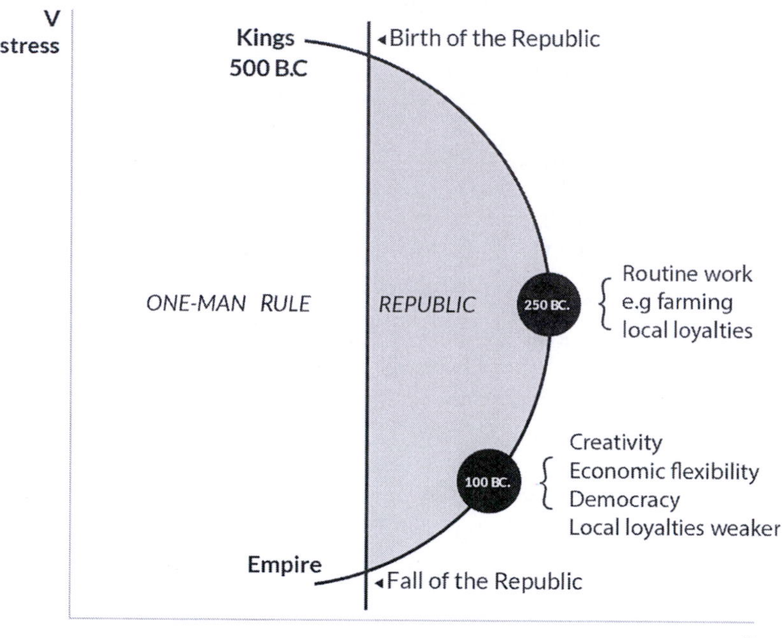

> the sum, laid all the silver in a heap, as he should pass by. Antony, seeing the heap, asked what it meant. His steward replied, "The money you have ordered to be given to your friend." So, perceiving the man's malice, Anthony said, "I thought (this amount) had been much more. It is too little—double it."
>
> Antony did not take long to earn the love of his soldiers, joining them in their exercises and for the most part living amongst them, and making them presents to the utmost of his abilities. But with all others he was unpopular enough. He was too lazy to pay attention to the complaints of injured parties. He listened impatiently to petitions, and he had an ill name for familiarity with other people's wives.[51]

In this one man can be seen all the signs of declining C: extravagance,

[51] Plutarch, "Life of Antony," in *Plutarch's Lives* vol. 2 (Digireads, 2009).

drinking, laziness and sexual indulgence; but also a personal charm that endeared him to his soldiers and came close to winning him control of the Roman world. A greater contrast to Cato can scarcely be imagined.

Changes in childrearing

If biohistory is correct, changes in political attitudes should be a direct reflection of a decline in infant C, child C and V caused by changes in childrearing patterns. Thus it is that from the late Republic onwards there is evidence of reduced harshness and a growing sentimentality towards children (lower child V), and reduced control (lower C). It may be noted that more sentimental and lenient attitudes towards children were accompanied by much less interest in having children, an apparent paradox also familiar in our time.

From at least the first century BC and over the next six hundred years, attitudes towards children in general would gradually soften, encouraged by thinkers such as Lucretius (99–55 BC), Celsus (30 BC–45 AD) and Galen (130–200 AD), whose interest in children's well-being and general popularity mark the gradual shifting of social attitudes.[52]

One way to measure this change is by looking at attitudes towards infant mortality, which was common throughout ancient times. During the Republic it was regarded as unseemly to mourn an infant's passing. In fact, it was considered inappropriate for an adult to give any attention to a child even as late as the time of Cicero (106–43 BC).[53] Yet, less than two centuries later Quintilian (35–100 AD), the noted teacher, rhetorician and consul to Emperor Vespasian, was completely devastated by the death of his two sons, aged 5 and 9.[54]

Furthermore, Quintilian wrote much on the proper education of children, preferring a syllabus of educational play in early ages, a trust in the role of nurses, and a general leniency towards the weakness of childhood. He asserted that children should not be pressed too hard or made to do "real" work, for fear of instituting a lasting bitterness in the child that would

---

[52] A. R. Colón & P. A. Colón, *A History of Children: A Socio-Cultural Survey across Millennia* (Westport: Greenwood Press, 2001), 87.

[53] Geer & Couch, *Rome*, 372–3.

[54] J. K. Evans, *War, Women and Children in Ancient Rome* (London and New York: Routledge, 1991), 173–174.

stunt their learning at a later age.[55]

This was the prevailing attitude, and fathers were charged with pressing their sons too hard in study—a far cry from the brutal discipline and scholastic expectations that parents had for their children a few generations before.[56] The happiness of the child began to take precedence over their discipline and productivity, leading to a decline in both V and C.[57]

Another indication of declining V and stress can be seen in changing attitudes towards patriarchy and deference. During the Republic, the rule of the father in the household was absolute in accordance with pater familias—the primary tenet of the fifth of the Twelve Tables (449 BC).[58] Sons were not able to purchase property, sign contracts, make career choices or perform any other action of legal validity without the consent of the patriarch, and only when a man's father died did he become legally independent.[59] This extended to the right of a patriarch to condemn his recalcitrant offspring to death, a practice which, although rarely practiced, lasted until limited by law in the first century BC.

Corporal punishment had been commonly applied in the Republic by both parents and teachers, but by the early Empire opinion favored a lighter hand in fathering. The use of violence by teachers and wet nurses to control the children of citizens was a common subject for disapproving talk, their lack of self-restraint a sign of lower class origins.

The ideal wealthy patriarch was described quite differently. He had to be represented first and foremost as a collected individual, an aristocrat who exhibited self-restraint in all aspects of emotional life and whose wisdom reflected on his entire family. The ideal of the man who is in complete control of his emotions and affects, the homo interior, became particularly popular from the imperial era. Not coincidentally, this was also a period

---

[55] B. Rawson, *Children and Childhood in Roman Italy* (Oxford: Oxford University Press, 2003), 137.

[56] P. N. Stearns, *Childhood in World History* (Abingdon and New York: Routledge, 2011), 38; Evans, War, Women and Children in Ancient Rome.

[57] Rawson, *Children and Childhood in Roman Italy*, 222.

[58] *Cambridge Ancient History* Volume 7, 147–8.

[59] Colón, *A History of Children*, 80.

when a number of writers advocated more lenient treatment of children.[60]

In practice, more lenient attitudes to children were accompanied by a reduced interest in having children, as indicated earlier. This not only led to a declining birth rate but, by the first century AD, a common practice of placing children in the care of servants and slaves, especially among the elite.[61] While this measure was rationalized by the Greek physician Soranus of Ephesus (first to second centuries AD) as a way of preventing women from becoming infertile, which he assumed could happen due to the strain of nursing, the effect was a dislocation from the traditional family structures,[62] as Tacitus (55–120 AD) lamented:

> In the good old days, every man's son, born in wedlock, was brought up not in the chamber of some hireling nurse, but in his mother's lap, and at her knee. And that mother could have no higher praise than that she managed the house and gave herself to her children. Again, some elderly relative would be selected in order that to her, as a person who had been tried and been found wanting, might be entrusted the care of all the youthful scions of the same house; in the presence of such a one no base word could be uttered without grave offence, and no wrong deed done. Religiously and with the utmost delicacy she regulated not only the serious tasks of her youthful charges, but their recreations also and their games …
>
> Nowadays, on the other hand, our children are handed over at their birth to some silly little Greek serving-maid, with a male slave, who may be anyone, to help her,—quite frequently the most worthless member of the whole establishment, incompetent for any serious service. It is from the foolish tittle-tattle of such persons that the children receive their earliest impressions, while their minds are still pliant and unformed; and there is not a soul in the whole house who cares a jot what he says or does in the presence of its lisping little lord. Yes, and the parents themselves make no effort to train their little ones in goodness and self-control; they grow up in an atmosphere of laxity and pertness, in which they come gradually to lose all sense of shame, and all respect both for themselves and for other people.[63]

---

[60] C. Laes, *Children in the Roman Empire: Outsiders Within* (Cambridge: Cambridge University Press, 2011), 145–146.

[61] Geer & Couch, *Rome*, 373.

[62] P. Garnsey, "Child Rearing in Ancient Italy," in *The Family in Italy: From Antiquity to the Present*, edited by D. I. Kertzer & R. P. Saller, 60, 48–65 (New Haven and London: Yale University Press, 1991).

[63] Dialogus de Oratoribus, 28, 29, quoted in Colón, *A History of Children*, 84–5.

Many writers echo Tacitus' concerns, especially in the late-first century AD.[64] The most obvious effect was a relaxation of control causing a decline in both C and child V, especially in the wealthy Roman elite.

These changes in childrearing patterns, themselves a result of declining C and V as a result of wealth and the civilization cycle, were to have serious political and economic consequences. The first of these, as indicated earlier, was the end of the Roman Republic. But more were to come.

The Roman Empire

The first era of the Empire began in 27 BC with the reign of Augustus Caesar. It brought political stability after decades of turmoil (though we will shortly see that this can be attributed to a G year around 0 AD). But the new regime did nothing to halt the decline in C. Augustus was vividly aware of an apparent decay in the Roman character, as in this speech given to Roman aristocrats in 9 AD over their reluctance to become fathers:

> Mine has been an astonishing experience: for though I am always doing everything to promote an increase of population among you and am now about to rebuke you, I grieve to see that there are a great many of you [who are childless]. We do not spare murderers, you know … yet, if one were to name over all the worst crimes, the others are naught in comparison with this one you are committing. Whether you consider them crime for crime or even set all of them together over against this single crime of yours. For you are committing murder in not begetting in the first place those who ought to be your descendants. You are committing sacrilege in putting an end to the names and honours of your ancestors; and you are guilty of impiety in that you are abolishing your families … overthrowing their rites and their temples. Moreover, you are destroying the State by disobeying its laws, and you are betraying your country by rendering her barren and childless. Nay more, you are laying her even with the dust by making her destitute of future inhabitants.[65]

Augustus enacted legislation to try and turn back the clock, but was no more successful than Brutus and Cassius had been in their attempt to restore the Republic through the assassination of Julius Caesar. Comments by Tacitus on the early Empire showed how old-fashioned Romans felt about the new age:

---

[64] Colón, *A History of Children.*

[65] C. Dio, *History*, translated by E. Carey (Cambridge: Harvard University Press, 1990), 4–5.

> Promiscuity and degradation thrived. Roman morals had long become impure, but never was there so favourable an environment for debauchery as among this filthy crowd. Even in good surroundings people find it hard to behave well. Here every form of immorality competed for attention, and no chastity, modesty or vestige of decency could survive.[66]

By the beginning of the first century AD, Roman C and V had been falling for several hundred years. Based on our knowledge of C, this should lead to a decline in agricultural output and trade, as people become averse to hard work and less oriented to the market.[67] The loss of impersonal attitudes and respect for authority should cause a weakening of the state, combined with military weakness as a result of falling V.

There is some evidence of this in that trade within Italy was in decline by the first century AD, but across the wider Empire it very clearly was not.[68] In fact, economically and militarily, first-century Empire was at an all-time high, with trade expanding and even some expansion of borders, such as in the conquest of Britain. So the question becomes not so much why the Roman Empire fell, but why it lasted so long, given the declining economic ability of its citizens.

The reason is that the Empire was no longer principally run by Italian-born citizens. An idea of this can be gained by looking at the most prominent writers and poets of the late Republic and early Empire, in order of the year of their birth (see Table 12.1 below).

The first three were from Rome or central Italy, with Cicero and Caesar of course being also prominent politicians. This is the cultural area for which we have traced the rise and then fall of C, which applies just as much to the Etruscan region as to Latium and Rome itself.

---

66 Tacitus (c. 55–120), Annals, 14:15.

67 W. R. Halliday, *Pagan Background of Early Christianity* (Liverpool: Liverpool University Press, 1925), 116–117.

68 E. T. Salmon, A History of the Roman World from 30 B.C. to 138 A.D. (London: Methuen and Co., 1968), 253.

Table 12.1. Birthplaces of Roman writers, second century BC to first century AD

Writer	Place of birth	Date of birth
Cicero	Central Italy	106 BC
Julius Caesar	Central Italy	100 BC
Sallust	Central Italy	86 BC
Catullus	Cisalpine Gaul	84 BC
Virgil	Cisalpine Gaul (N Italy)	70 BC
Horace	Southern Italy	65 BC
Livy	Cisalpine Gaul	59 BC
Ovid	Central Italy	43 BC
Seneca	Iberia	4 BC
Pliny the Elder	Cisalpine Gaul	23 AD
Quintillian	Iberia	35 AD
Plutarch	Greece	46 AD
Juvenal	Central Italy	55 AD
Tacitus	Cisalpine Gaul or Gaul	56 AD
Pliny the Younger	Cisalpine Gaul	61 AD
Seutonius	North Africa	69 AD
Martial	Iberia	86 AD

The next group of writers up to Pliny the Elder came mostly from Cisalpine Gaul—roughly modern Lombardy. As the name suggests, though influenced by Mediterranean culture, this was a different cultural area and had much in common with Gaul on the other side of the Alps. It was where Hannibal had recruited much of his army when he invaded Italy in 218 BC.[69] It was relatively undeveloped and was only fully incorporated

[69] H. H. Scullard, *The Cambridge Ancient History* Volume 8: Rome and the Mediterranean to 133 BC. The Carthaginians in Spain, edited by A. E. Austin, F. W. Walbank, M. W. Frederiksen and M. Ogilvie, 40 (Cambridge: Cambridge University Press, 2008).

into the Roman political sphere in the second century BC.

Most later writers came from still further afield—Iberia, Greece and North Africa. Only Juvenal was born in central Italy, and he was the son or adopted son of a freedman (ex-slave) so was unlikely to have been of Italian descent (Fig. 12.2).

Fig. 12.2. Birthplaces of Roman writers. With time, writers were drawn from further afield, as each population in turn was Romanized and thus gradually lost C and V. This same process of importing people with higher C and V kept the Empire going for centuries, despite the fall of C and V in central Italy.

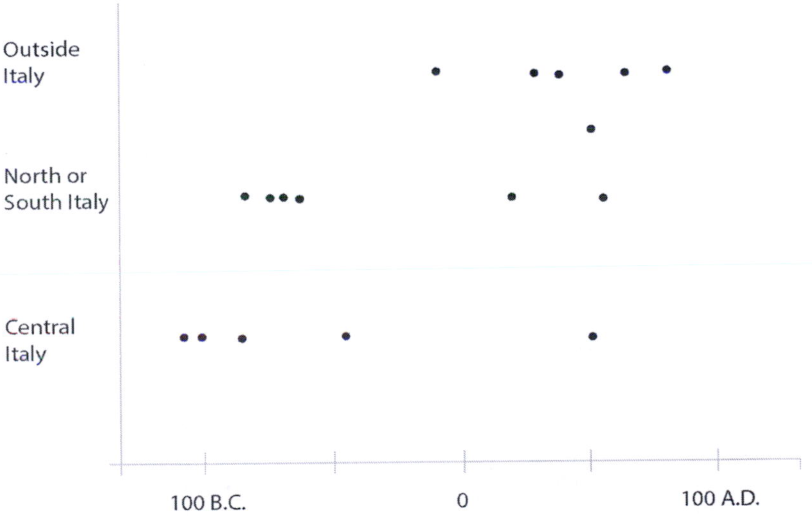

Something similar can be seen in the origins of Roman emperors, though with a significant delay. All emperors up to Nerva (born 30 AD) were from central Italy. Trajan (53 AD) and Hadrian (76 AD) were from Iberia, though Hadrian may have been born in Italy.[70] Antoninus Pius (86 AD) was born in Italy but his family was from Gaul.[71] Marcus Aurelius (121 AD) was also born in Italy but of an Iberian family.[72] A point worth noting is that the families of both Trajan and Marcus Aurelius, like that of the philosopher Seneca, are thought to have originally been from Italy.

[70] S. Perowne, *Hadrian* (London: Hodder and Stoughton, 1960), 22.
[71] H.M.D. Parker and B.H. Warmington, *A History of the Roman World from A.D. 138–337* (London: Methuen, 1958), 3.
[72] Ibid., 17

Families migrating to the provinces would have acquired the higher C of those areas (due to the cultural and biochemical effects that cause individuals to pick up the level of C of the social environment in which they live), while those who migrated to Italy would lose C within a generation or two.

Roman emperors after Commodus were largely proclaimed by the army and thus came from many parts of the Empire.[73] Septimus Severus, the most successful emperor of the early third century, was from North Africa.[74] By the late third century, imperial lineages stemmed mainly from mountainous and backward Illyria on the western Balkan coast.[75]

Of the great mass of slaves brought in by the Roman conquests, most lived miserable lives as brute labor in mines and fields but others could do very well. Many were freed by their masters, and some became quite wealthy and influential.[76] The Emperor Claudius (41–54 AD) appointed many freedmen to senior bureaucratic posts (a practice which scandalized the Senate).[77]

Few of these people would have had C at the level of natives of central Italy three or four centuries earlier. They came from regions which until the Roman conquest had limited trade and few large settlements. But C in Italy was now below that of the provinces, and possibly even below that of many barbarians. What Roman rule did was bring peace and relative prosperity, and the values and customs of a civilization with low V and falling C. Backed by Roman prestige and power, this culture was enormously attractive. Local religions and superstitions that would have maintained C to some limited degree were undermined by Roman skepticism and education.

As discussed in the last chapter, wealth and urbanization has its first effect in lowering V, and only after that does C fall significantly. For the elites in particular, an initial fall in child V meant that they became more flexible and thus more likely to succeed in business, politics and administration, though less useful as soldiers.

---

[73] Ibid., 55.
[74] Ibid., 58.
[75] Ibid., 223.
[76] Fowler, Social Life at Rome in the Age of Cicero, 225–8.
[77] Salmon, *A History of the Roman World from 30 B.C. to 138 A. D.*, 167.

But such success was temporary. Low V undermines C so their children and grandchildren had lower C and were thus, on average, less successful. Also, because of a falling birth rate, there were fewer of them. As this pattern of declining V and then C spread to more and more provincial areas, so flexibility and administration skills followed. And all this time, the C of the Empire as a whole continued to decline.

It is important to recognize that this decline was gradual, as can be traced in the continuing decline of impersonal loyalties. We have seen that these were ebbing even in the late Republic, which was what allowed leaders such as Marius, Sulla and eventually Caesar to take power.[78] But the process was far from complete in the early Empire. Augustus set up the Principate with a careful Republican façade that was not entirely a matter of form.[79] Roman magistrates were elected as they had been during the Republic, and the Senate kept significant power.[80] In 61 AD the Senate condemned four hundred slaves to die for the murder of their master, and Nero, despite his inclinations, felt unable to override the ruling.[81] Later in that same decade, Vespasian also felt the need to court the Senate to make his seizure of power respectable.[82] In a sense the early Empire remained a halfway house between a true Republic and pure autocracy.

But with time these constraints were gradually eroded. Tiberius (14–37 A.D.) ended the election of magistrates in Rome.[83] Most other cities lost this right in the second century. By the early third century the Roman state was an undisguised absolutism.[84] When Septimus Severus died in 211 AD his last words to his sons were: "Stick together, pay the soldiers, ignore everyone else."[85]

Meanwhile, through the troubled third century and even more from the late fourth, increasingly personal loyalties were beginning to undermine the unity of the Empire. The emperor himself was too distant and impersonal

---

[78] Badian, *Foreign Clientelae*, 197–211.

[79] R. V. N. Hopkins, *The Life of Alexander Severus* (Cambridge: Cambridge University Press, 1907), 87.

[80] Halliday, *Pagan Background of Early Christianity*, 47–48.

[81] M. Grant, *Nero, Emperor in Revolt* (New York: American Heritage Press, 1970), 110–111.

[82] Salmon, A History of the Roman World from 30 B.C. to 138 A. D., 215.

[83] Halliday, *Pagan Background of Early Christianity*, 48.

[84] Parker & Warmington, *A History of the Roman World from A.D. 138–337* (London: Methuen, 1958), 69.

[85] Dio, *History*, 273.

to serve as a focus for loyalties, and local leaders gained power. One result was the growth of large estates that were in many ways immune from government control. Local magnates evaded taxes and protected runaways, kept prisons, armed men and defied the law in other ways, such as by seizing debtors.[86] In some cases, even provincial governors were helpless to stop them.

As loyalties became more personal, the remaining authority of the emperor came to rely more and more on his actual presence and less on the laws and institutions of previous times. This is why the Empire was effectively split in the fourth century, with each half having an Augustus or senior emperor, and a Caesar or junior emperor.[87] Such emperors travelled a great deal, making it far easier for one to be present when needed. A distant emperor in Rome could no longer demand respect. Now he had to be physically present to exert his authority.

Population Decline

As another indication of falling C and V, the population of the Empire seems to have been in decline by the second century AD. Epitaphs show a large proportion of childless couples. Part of the reason was that women seem to have been less willing to have children.[88] For example, in the early second century Pliny wrote that his was an "era in which the rewards of childlessness make many regard even one child as a burden."[89] There might also have been an increase in infertility, as suggested by the inability of four successive Roman emperors (Nerva, Trajan, Hadrian, and Antoninus Pius) to produce sons. Some of our laboratory studies, as yet unpublished, suggests that plentiful food in early life lowers the sperm count in rats.

By the third and fourth centuries AD the fall in population was obvious and catastrophic. Roman emperors were settling barbarians in frontier areas during the third century, and even in depopulated regions of Italy and the Balkans in the fourth. But despite this the area of land under

[86] F. Lot, *The End of the Ancient World* (London, Routledge and Kegan Paul, 1953), 128–30.
[87] Parker & Warmington, *A History of the Roman World from A.D. 138–337*, 240.
[88] R. Duncan-Jones, *The Economy of the Roman Empire* (Cambridge: Cambridge University Press, 1974), 318.
[89] Plini the Younger, *Complete Letters*, translated by P. G. Walsh, (Oxford: Oxford University Press, 2006), 96.

cultivation kept shrinking, especially in the west.[90] A shortage of labor is indicated by legislation tying tenants to the land, the willingness of landlords to employ runaways in spite of severe penalties, the extreme reluctance of landlords to release men for the army, and apparently modest rents.[91] The cities suffered even greater losses due to flight to the country, and some disappeared entirely.

Plague and famine from the late-second century onward undoubtedly played a part in this, but populations in other times and places have typically recovered quickly from some afflictions.[92] It is significant that tombstones throughout the Empire show that people were not dying any earlier.[93] Even among ordinary people there seems to have been a decline in the birth rate, typical of declining C and V.

Economic decline

Economic activity flourished in the early Empire despite falling C, because the fall was counteracted by an even greater fall in child V. As in the late Republic, this made up for a declining work ethic by allowing more flexible patterns of thought. But as C continued to fall, economic growth faltered and began going backwards. We have seen that trade was in retreat in Italy in the first century AD. By the second century there was evidence of decline in the Empire as a whole, and what remained tended to be more local and less international.[94] There was a further retreat in the troubled third century and even in the more peaceful fourth, though it was far from the low ebb reached in post-Roman times.[95]

In the later Empire there was an emphasis on small-scale manufacturing for the local markets, and large estates were becoming more self-sufficient.[96] Taxes began to be levied in kind, even under a strong ruler

---

[90] Lot, *The End of the Ancient World*, 67.

[91] A. H. M. Jones, *The Late Roman Empire: 284–602* (Oxford, Basil Blackwell, 1973), Vol II, 796, 798, 803–8.

[92] Parker & Warmington, *A History of the Roman World from A.D. 138*–337, 39.

[93] Jones, *The Late Roman Empire*, 1041.

[94] Parker & Warmington, *A History of the Roman World from A.D. 138*–337, 39.

[95] Ibid., 165.

[96] J. Lindsay, *Daily Life in Roman Egypt* (London, Frederick Müller, 1963), xix–xxi.

such as Diocletian who made every effort to restore the currency.[97] However, Diocletian and some fourth-century rulers showed typical low-C attitudes by restricting the market economy. Diocletian encouraged guilds and tried to control prices and the movement of labor.[98] All this is exactly what to expect from a loss of the impersonal attitudes and flexibility associated with higher C.

Creativity followed the same pattern. We have seen that when C starts to fall there is an increase in creativity, because of an even more marked fall in child V. But as C continues to decline, creativity and flexibility are lost. Thus, Latin literature went from the golden age of the late Republic to the silver age of the first century AD to the creative wasteland of the later Empire with few if any writers of note.[99] Significant philosophers of the late Empire, including Augustine and Plotinus, were concerned largely with religion. As the skeptical infant C trait declined, religious thinking became more prevalent.[100]

The rise of Christianity

Important beneficiaries of this increased religiosity were eastern religions such as Mithraism, the cult of Isis, and Christianity.[101] All of these and especially Christianity were actually more "advanced" than the traditional Roman and Greek religions, in the sense that they had powerful C and V-promoters developed in the oldest centers of civilization. Christianity was already highly ascetic at this time, with a deep suspicion of any form of sensual pleasure, and differed radically from traditional religion in that it constrained the sexual activity of men as well as women.[102] This was also true of most mystery cults.[103]

---

[97] Parker & Warmington, *A History of the Roman World from A.D. 138*–337, 282–3.

[98] J. B. Bury, H. M. Gwatkin, J. P. Whitney, J. R. Tanner, C.W. Previte-Orton & Z. N. Brooke, *The Cambridge Medieval History. The Rise of Sacarens and the foundation of the Western Empire* Vol. 2 (New York: Macmillan, 1991) 40–41, 548–51.

[99] Lot, *The End of the Ancient World*, 151.

[100] C. Bailey, *Phases in the Religion of Ancient Rome* (Oxford: Oxford University Press, 1932), 246–8.

[101] Ibid., 177.

[102] Jones, *The Late Roman Empire*, 971–6.

[103] Halliday, *Pagan Background of Early Christianity*, 240.

Christianity also had an ethical content, and a narrative that gripped the hearts and minds of ordinary people in a way that the traditional cults no longer did. There was the image of the dying god, which had a peculiar resonance in the Roman world, the focus on charity, the hope of eternal bliss, the prospect of Jesus returning, and perhaps very soon. All this, plus an intolerance of other religions, made it far more capable of affecting behavior. Negative attitudes to sex would also be helped by lower infant C, which reduces testosterone. This is the same finding reported in chapter two, that rats with well-fed mothers had lower testosterone than those with calorie-restricted mothers.

The adoption of Christianity was not enough to reverse the enormous momentum of declining C in the Roman Empire, given the very low V at the time, but it could help individuals and families. When C is high, as in nineteenth-century Europe or Rome in the early Republic, large families are the norm and religion makes relatively little difference. Women will have about as many children as their health and economic situation allow. But when birth rates are below replacement level, highly religious people stand out. Thus in our own day Orthodox Jews, Old Order Amish, Mormons and various fundamentalists tend to have more children than their less religious neighbors. The early Christians opposed abortion and infanticide, both of which were common practices in the ancient world.[104]

Continued fall in V and stress

As C continued to decline in the early Empire, so too did V and stress. We have already seen evidence of this in the first century AD. Though emperors still executed their political opponents, the scale was nothing like that of the early-first century BC when Sulla put to death about forty Senators and nearly 1,600 Knights (though some estimates put the number as high as 9,000).[105] Claudius was responsible for the deaths of perhaps 30 Senators and 200–300 knights, Domitian (81–96 AD) rather fewer, though even this number was enough to have him vilified after his death.[106]

Laws were also becoming more humane in the first century A.D. Though Nero (54–68 AD) is a byword for cruelty, especially for his persecution of early Christians, his actions were in some ways surprisingly humane. He restricted the amount of bail and fines, stopped patrons from reducing their

---

104 Ward et al., *A History of the Roman People*, 449.
105 Ridley, *The Gracchi, History of Rome* 271.
106 Grant, *Nero, Emperor in Revolt*, 31.

freedmen clients to slavery, and acted to make taxes less oppressive to the poor.[107] After the great fire of Rome he paid for relief efforts out of his own pocket, and spent days personally searching the ruins for survivors.[108] Unimportant men who insulted him were usually no more than banished.[109] He seems to have been genuinely popular with ordinary people, and after his death there were messianic prophecies that one day he would return.[110] The affability and approachability of first-century rulers was seen as a peculiarly Roman characteristic in the first century AD, by contrast with nations such as Parthia where rulers were approached with elaborate deference.[111]

This trend would become even more pronounced by the second century. Between 98 and 180 AD a series of humane and capable emperors did still more to alleviate harshness in the law.[112] Masters were no longer allowed to kill their slaves, castrate them or sell them to brothels without good cause.[113] Care was taken for the welfare of orphans and the old.[114] The lives of citizens were guaranteed to a much greater extent, and even men guilty of conspiracy might not be executed.[115] The ideal emperor of the second century was a benevolent father who ruled by love rather than fear.[116] Civilis princeps, meaning a citizen-like emperor, was a term of high approval. And while six emperors were killed in the first century, none suffered this fate between 96 and 190 AD.

This low level of deference was a feature of all classes. The new aristocracy from outside Rome was less concerned with the old conventions of dignity. Young members would drive to race meetings or go on stage in theatre performances, both of which were considered socially inappropriate for the early aristocracy.[117]

There were other indications of low V and stress in the early Empire. The

---

[107] D. C. A. Shotter, *Nero* (London, New York: Routledge, 2005), 23–26.
[108] Grant, *Nero, Emperor in Revolt*, 151–2, 163.
[109] D. R. Dudley, *The World of Tacitus* (London: Secker and Warburg, 1968)123; Grant, *Nero, Emperor in Revolt*, 212.
[110] Grant, *Nero, Emperor in Revolt*, 250–1.
[111] Dudley, *The World of Tacitus*, 204–5.
[112] Halliday, *Pagan Background of Early Christianity*, 127.
[113] Perowne, *Hadrian*, 77–8
[114] Halliday, *Pagan Background of Early Christianity*, 139.
[115] Perowne, *Hadrian*, 49–50.
[116] Ward et al., *A History of the Roman People*, 327.
[117] Halliday, *Pagan Background of Early Christianity*, 118.

freedom and power of upper-class women seem to have been greatest in the first two centuries AD.[118] By the late first century some writers were opposing the beating of sons, and literature of the second century suggests indulgent attitudes, though schoolmasters still seem to have made considerable use of the rod.[119] The patria potestas, the total authority of a father over his child including that of life and death, was repealed by the second century.[120]

So by every measure we have so far looked at—judicial punishments, treatment of children, insecurity of rulers, deference, and status of women—levels of V and stress in the second century AD were at an all-time low.

Rising stress from the mid-second century

In the later Empire, under Christian influence, a number of laws were made to protect children.[121] But from every other indication the level of stress began to rise from the mid-second century. Even under the humane and enlightened Emperor Marcus Aurelius (161–180), torture began to be used regularly on lower-class people in criminal trials, though only after a man had been convicted, as a means of making him reveal his accomplices. Within half a century these restraints were lifted, and between the fourth and sixth centuries brutality became a feature of administration in the Eastern Empire. Torture, flogging and such cruel deaths as burning were now routine.[122]

The change in attitude was so strong that even emperors were powerless to stop the brutality. Government was losing control of officials, so that oppression became commonplace. Attempts by rulers such as Diocletian (284–305) to discourage torture and protect the weak were largely ineffective.[123] And by this time deference and status differences were becoming more marked. Etiquette became very complicated, and the pomp surrounding the ruler was used to emphasize his distance from other people, as opposed to the comparatively affable nature of second-century

---

[118] Geer & Couch, *Rome*, 372.
[119] Carcopino, *Daily Life in Ancient Rome*, 105.
[120] Geer & Couch, *Rome*, 367.
[121] Ward et al., *A History of the Roman People*, 107.
[122] Jones, *The Late Roman Empire*, 978–9.
[123] Lot, *The End of the Ancient World*, 20.

emperors.[124] There was an increase in the legal differences between rich and poor, and the corruption of justice (a sign of low C) meant that rich men were now in much stronger positions.[125]

The political system also became both more brutal and less stable. Commodus (180–92) and Caracalla (211–17) revived the reign of terror with mass executions of political opponents, something not seen for almost a century, and the position of emperor itself became more dangerous.[126] Commodus and most of his thord century successors were killed in office.

All this evidence suggests a low point of V and thus stress around the mid-second century, followed by a pronounced rise. Very low V also contributed, of course, to loss of military effectiveness, which is why soldiers increasingly came from areas such as Illyria and even from beyond the Empire.

The fall of Rome

The collapse of the Western Empire was a vast political and cultural shock, but in other ways not a sharp break. As already noted, large estates were becoming in many ways autonomous, and the Roman armies were increasingly manned by barbarians. Economic decline was far advanced, as evidenced by deposits of lead and copper pollutants in the Greenland ice sheets which show a shrinkage in metal production from the first century AD until the eighth and ninth centuries. The decline seems to have been more rapid during the Empire than after the fall (see Fig. 12.3 below).

When barbarians conquered Gaul in the early fifth century they simply took over the existing Roman administration and merged with rather than replaced the Roman upper class.[127] Political fragmentation in Gaul continued until the mid-seventh century, when royal authority was all but extinguished. Then after the Carolingian renaissance, it reached its lowest

---

[124] R. MacMullen, *The Corruption and Decline of Rome* (New Haven: Yale University Press, 1988), 60–63.
[125] Ward et al., *A History of the Roman People*, 442–3.
[126] Parker & Warmington, *A History of the Roman World from A.D. 138–337*, 29–35, 89–91.
[127] J. M. Wallace-Hadrill, *The Long-Haired Kings: And Other Studies in Frankish History* (London: Methuen, 1962), 9.

ebb in eleventh-century France, when authority was scarcely felt beyond the immediate personal presence of the ruler.[128]

Fig. 12.3. Deposits of lead and copper pollutants in the Greenland ice cap.[129] Economic decline, a consequence of falling C, started long before the Empire fell.

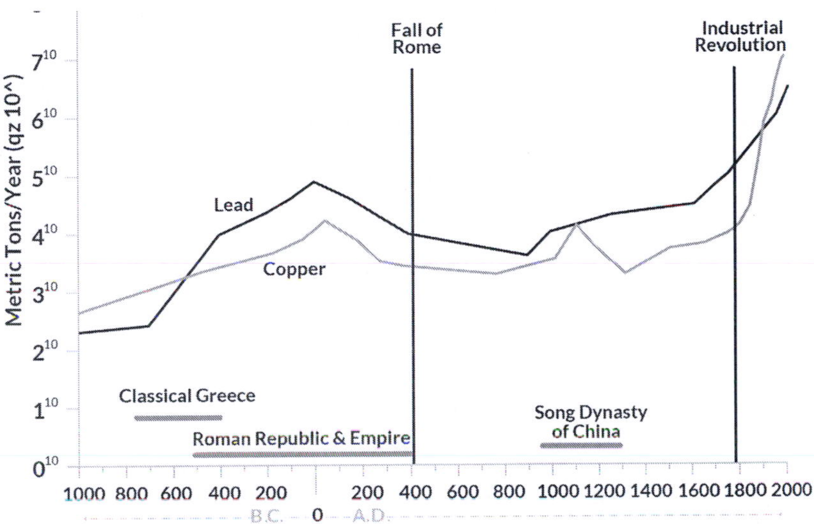

In other words, just as the fall of the Republic marked only one point in the progressive decline of personal loyalties, so the fall of Rome marked only one point in the disintegration of political unity. What mattered was not the formal changes from the Republic to Empire to barbarian kings to feudal anarchy, but the underlying and continuous decline in impersonal loyalties linked to falling C.[130] In effect, such attitudes made the fall of Rome inevitable.

The same trend can be seen in economic decline. We have already seen that trade was in decline from as early as the first-century BC in Italy, and from the first century AD for the Empire as a whole. It continued to

---

[128] P. Collins, *The Birth of the West* (New York: Public Affairs, 2013), 168.

[129] S. Hong, J. Candelone, C. C. Patterson & C. F. Boutron, "History of Ancient Copper Smelting Pollution During Roman and Medieval Times Recorded in Greenland Ice," *Science* (272) (1996): 246–249; D. M. Settle & C. C. Patterson, "Lead in Albacore: Guide to Lead Pollution in Americans," *Science* (207) (1980): 1167–1176.

[130] Wallace-Hadrill, *The Long-Haired Kings*, 10–11.

decline under the Frankish rulers of Gaul until at least the seventh century AD when the drying up of Mediterranean trade may have helped undermine the Merovingian monarchy.[131] The use of coinage, a powerful indication of C, had declined since the barbarian conquests but only disappeared in the sixth century.[132] The area of land under cultivation in the west also continued to fall until at least the end of the sixth century.[133] This suggests that C continued to fall until at least 600 AD.

It is also worth noting that the areas of the West most resistant to barbarian invaders were western Britain, Brittany and the Basque country—those areas least Romanized and thus least affected by the catastrophic fall of V that left Rome so weak militarily.[134]

The Roman civilization cycle

From the above evidence it is now possible to track changes in C, V and stress over the entire Roman period. The picture is of a cycle similar to that found later in England and Japan, but with the fall of C continuing until V and stress cease falling and start to rise (see Fig. 12.4 below). In a sense this is exactly the obverse of what happened in England and Japan over the past thousand years. Just as rising C is accompanied by a rise and then fall in stress and V, so falling C goes together with a fall and then rise in stress and V.

This is, of course, an extreme simplification of what actually happens. In reality, different sections of society change at different rates. Cato the Censor is an example of a high-C character in the Roman Republic, but even in his time he was seen as representing the values of an earlier era and the countryside. Wealth and urbanisation both reduce C, so the urban elite showed early evidence of declining C. By contrast, less affluent provincials retained higher C for a time.

---

[131] Lot, *The End of the Ancient World*, 365.
[132] Ibid., 370
[133] Ibid., 365–6; B. Ward-Perkins, *Why did Rome Fall and the End of Civilization* (Oxford: Oxford University Press, 2005), 41–2.
[134] Jones, *The Late Roman Empire: 284–602*, 1064.

Fig. 12.4. The Roman civilization cycle. As C continues to fall, the decline in V and stress slows and then V and stress begin to rise.

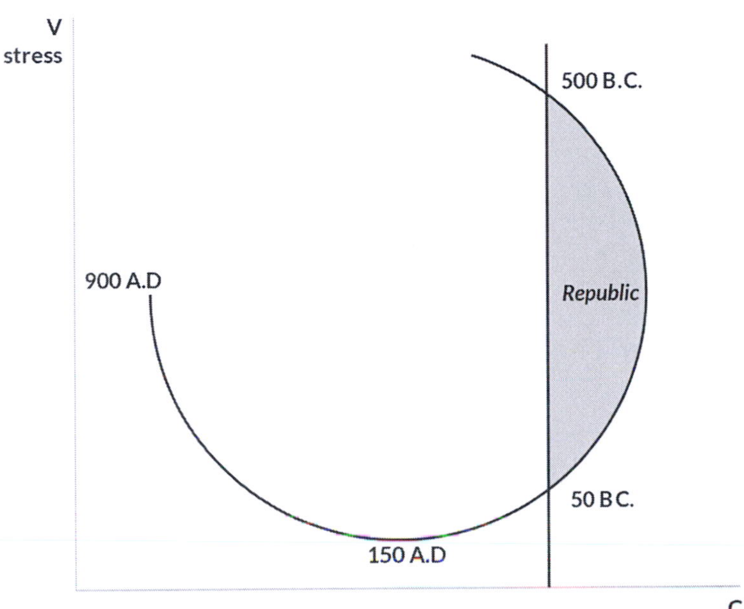

As indicated in Fig. 12.5 below, the civilization cycle can be seen as a full cycle in which stress combined with V increases C, and lower stress with lower V reduces it. By the same token, high C reduces V and thus stress, and low C allows both to rise. High levels of V and C are also, of course, dependent on powerful V and C-promoters. The civilization cycle is thus driven by both physiological and cultural factors.

But even among the provinces we have seen important differences in the rate of decline, with the effects seen first in central Italy and only later in northern Italy, Iberia, Gaul and North Africa. This clearly reflects the spread of Roman civilization, which by the time of the Empire had values that actively undermined C and V, as do the 'liberal' ideas of the modern West. As these values spread, the provinces first prospered as they lost the rigid thought patterns associated with child V, and then declined with the longer-term fall in C. In a sense, the Roman conquerors drew the provincial societies from Gaul, Iberia and elsewhere into their own civilization cycle, with disastrous results.

Fig. 12.5. Civilization cycle complete model.

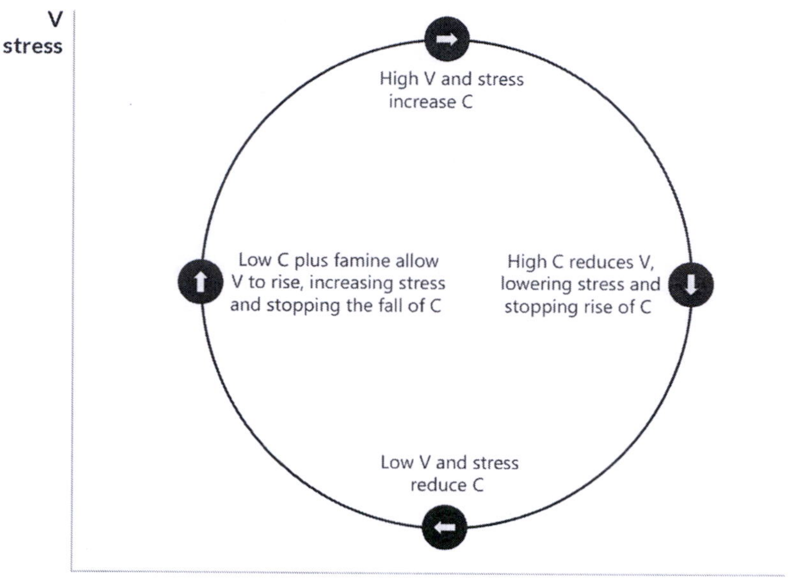

Roman lemming cycles

So far, the decline of Rome has been described as a smooth process of decay, but of course it was not. The Empire had all but disintegrated in the mid-third century AD before experiencing a revival under the Illyrian emperors.[135] To understand this, and also to appreciate why the Republic collapsed in the way it did, we must look at lemming cycles.

The civilization cycle was essentially exported from Rome to the provinces, with some delay. The falls in V and then C were experienced first in Rome and central Italy, and then in the provinces. The same can be said of lemming cycles, which the Empire as a whole seems to have inherited from Rome. The first clear G year can be placed around 280 BC. Ninety years earlier, in 370 BC, Rome was no more than the predominant city in Latium, a tiny region in west-central Italy.[136] The peoples of Italy were militarily weak, and between 390 and 329 Rome was captured and

[135] Parker & Warmington, *A History of the Roman World from A.D. 138–337*, 165.
[136] Ridley, *The Gracchi, History of Rome*, 93–98.

sacked during a series of Gallic invasions through the peninsula.[137] Such intense localism is, of course, the primary indication of a G-90 period.

But then, starting with the incorporation of Capua between 318 and 312, Roman power expanded rapidly and by 264 Rome was effectively master of Italy south of the Po.[138] Nor was this unity solely a result of military might, since most cities stayed loyal to Rome during Hannibal's invasion in 218–203, even while he rampaged unchecked over the land.[139] This rapid change from localism to centralized unity is characteristic of the transition from G-90 to the G period, especially if we consider central and northern Italy south of the Po as a single culture area (which it was in various ways, such as in the change from monarchy to republic in several states at about the same time as Rome).

From this time the Romans fought many external wars but maintained a united front within Italy. Unity first started to unravel during the reforms of the Gracchus brothers between 133 and 121 BC, both of whom were murdered by political opponents. Then, between 91 and 82 BC Italy was plunged into chaos, starting with the revolt of Italian allies in the Social War, and soon followed by a series of bloody civil wars. The key event in the fall of the Republic is often seen as the march of Sulla on Rome in 88 BC, which followed the Social War of 91–88 BC when Rome fought a bitter war against its Italian socii, or allies.[140] Sulla's march, which brought his legions within the limits of the city itself, was an unprecedented action, so much of a breach that all but one of his commanders refused to follow him.[141] In effect, their loyalties were still impersonal enough to give priority to the rule of law. But, and this is what mattered, the ordinary soldiers were now more loyal to Sulla than to the Republic. As a result, after this and a further incursion in 82 BC he was able to dominate the city, killing thousands of his political opponents.[142] Between these two events, his rival Marius was also able to dominate the city by control of the army.[143]

Then, after only three decades of calm, Julius Caesar crossed the Rubicon

---

[137] Ibid., 104–5.
[138] Ibid., 116–7.
[139] Scullard, *A History of the Roman World*, 189.
[140] Scullard, *A History of Rome from 133 B.C to 68 A.D.*, 64–68.
[141] Ridley, *The Gracchi, History of Rome*, 266.
[142] Scullard, *A History of Rome from 133 B.C to 68 A.D.*, 77–80.
[143] Ibid., 68; Ridley, *The Gracchi, History of Rome*, 266.

in 49 BC. The result was nearly twenty years of civil war until Octavian defeated Mark Anthony at the battle of Actium in 31 BC and established full control of the Empire. There followed a time of cohesion and national unity, with no form of internal conflict for almost a century. This progression from disorder to order is entirely consistent with a G-90 year in 90 BC, leading up to a G year in 0 AD. The golden age of Latin literature at the end of the Republic is also characteristic of the creativity of a G-60 period around 60 BC.

The decline of the Empire also followed an irregular course, which again can be explained by lemming cycles. There was a decline to near anarchy in the mid-third century, followed by a rapid recovery led by powerful Illyrian emperors such as Diocletian and Constantine.[144] This is consistent with a G year around 340 and a G-90 year in 250 AD. The great plague of 165–180 AD, in the reign of Marcus Aurelius, fits quite well a G-150 year of 190 AD, the G-150 year being the stage of the lemming cycle when G and thus disease resistance are lowest.[145] It was almost certainly the worst epidemic to afflict the Empire till the reign of Justinian in the mid-sixth century, which in turn came shortly before the next G-150 year of 630 AD.[146] Plotting G years at 280 BC, 0 A.D. and 340 AD also shows a gradual lengthening of lemming cycle lengths as the Dark Ages approach, from 280 to 340 years, and we have seen that lemming cycles tend to be longer in less civilized times.

Why did Christianity not save the Empire?

Before leaving the fall of Rome, there is one last question that must be asked: why did Christianity not save the Empire? The level of C in a society can be maintained or increased by cultural technologies, even when food is relatively plentiful, and one of the most effective of such cultural technologies is Christianity. We have seen that Christianity, along with a surge of V and stress around the sixteenth century, was what drove the rise in European C to unprecedented heights in the nineteenth century.

Given that Christianity was accepted as a state religion in Rome in the

---

[144] Lindsay, *Daily Life in Roman Egypt*, xix.

[145] R. J. Littman & M. L. Littman, "Galen and the Antonine Plague," *American Journal of Philology* 94 (1973): 254–55; A. Patrick, "Disease in Antiquity: Ancient Greece and Rome," in Diseases in Antiquity," edited by D. Brothwell & A. I. Sandison, 245 (Springfield, Illinois: Charles C. Thomas, 1967).

[146] Ibid., 246.

early fourth century AD, why did it not reverse the decline of C and thus prevent the fall? In fact, as the discussion above indicates, it did not even slow the decline. Diocletian, who was a notorious persecutor of Christians, did more than anyone to restore Imperial power.[147] And the decline continued full pace shortly after the Empire officially embraced Christianity.

This is all the more interesting if we consider the severely ascetic practices in the early Christian church, which was a vital part of the monastic movement spreading from the Egyptian desert to Western Europe in the fourth and fifth centuries.[148] The Egyptian monks, who led lives of extreme asceticism, separated themselves from the rest of the community, renouncing all sexual and financial activity as a form of "living martyrdom." In doing so they claimed to be practicing the purest form of Christianity.[149] These early Desert Fathers were famous for their religious zeal, including extreme piety and sexual restraint. In one popular story, the devil sends a woman to tempt a monk. She claims to be lost, and asks to stay in his cell for the night because she is afraid of wild animals:

> The monk soon had great desire for her … and he knew well that it was the devil who caused him so much anguish … And when he burned with the most passion he said "Those who do such things go into torment. This will test whether you can suffer the enteral fire where you must go." And he extended his finger and put it in the flame … But the finger did not feel the heat, because he was so filled with fleshly fire. Thus one after the other he held his fingers in the fire, so that they were all burned by daybreak.[150]

Such stories were widespread, with "heroic" asceticism of this form much admired. And such practices were not confined to monks and hermits. A monk named John Cassian, who had travelled in the deserts of Egypt and Syria but later settled in Gaul, advocated extreme sexual restraint, arguing

---

[147] G. E. M. De Ste Croix, "Why Were the Early Christians Persecuted?" in *Studies in Ancient Society. Past and Present*, edited by M. I. Finley, 27–8 (London: Routledge & Kegan Paul, 1974).

[148] W. H. C. Frend, "The Failure of the Persecutions in the Roman Empire," in *Studies in Ancient Society. Past and Present*, edited by M. I. Finley, 22–7 (London: Routledge & Kegan Paul, 1974).

[149] D. M. Hadley, *Masculinity in Medieval Europe* (New York: Longman, 1999), 109.

[150] M. R. Karras, Sexuality in Medieval Europe, Doing Unto Others (New York: Routledge, 1985), 1.

that only one with a pure heart could give pure and moral speech.[151] His ideas appealed to the aristocracy, ever anxious to preserve their power and position, and he was not the only man to advocate such ideas.[152] Cassian focused on "purifying" the body by going to extraordinary lengths to reduce nocturnal emissions:

> It takes a man six months to bring the nocturnal emission of his semen under control: he must eat two loaves a day, drink as much water as he needs, take three or four hours sleep. He must avoid any idle conversation, and curb feelings of anger. In the final stages, he should cut down on the water, and take up strapping lead plates onto his genitals at night....[153]

This is quite as extreme as anything advocated at the high point of C in Victorian England. The Penitentials, a series of guides to Christian practice produced from the sixth century onwards, were particularly strict about sex. It was forbidden on Sundays, as it was believed that the resulting pollution would make it impossible to enter the church and partake in communion, as well as on Thursdays and Fridays which were devoted to pre-communion fasting and self-denial respectively.

The early Christian calendar had not one but three Lenten periods in which sex was forbidden. The first precluded sex for forty to sixty days, the second covered the weeks before Christmas and the third related to the feast of Pentecost which again required abstinence for forty to sixty days. In addition, sex was forbidden on numerous "feast days" of the church calendar and on the days before communion on each of these.[154]

Our view of early Christianity is largely drawn from such writings, but it is important to recognize that this was almost certainly not how most Christians actually. Some church leaders did not take well to the new monastic standards of sexual purity, believing they would create a gulf between the strict standards of the monks and the less strenuous version practiced by the laity.[155] Generally, Christians in the first few centuries AD seem to have been keen to integrate into their social and cultural environments.

---

[151] Hadley, *Masculinity in Medieval Europe*, 115.
[152] Ibid.
[153] Ibid., 104.
[154] G. Hawkes, *Sex and Pleasure in Western Culture* (Cambridge: Polity Press, 2004), 67.
[155] Hadley, Masculinity in Medieval Europe, 109.

The established church did not want to see a wide gap open up between the rigorous standards of the monks on the one hand, and the less demanding version practiced by the laity on the other.[156] Church leaders at the time were attempting to popularize Christianity. In their eyes, expectations of virginity would make it look less attractive to potential converts.

Regardless, ascetic traditions continued to spread throughout the early church. Part of the reason for this was the need for a ritual separation of the priestly class from the ordinary people, as well as the zeal resulting from the monks' harsh conditioning. In an age of extreme religious feeling, to be seen as morally pure put one in a powerful position of authority.[157] Thus, in effect, sexual asceticism became the hallmark of the spiritual elite rather than something most Christians were expected to follow.

Before 590 AD popes tended to be aristocrats rather than monks, as did the bishops of Gaul such as Gregory of Tours in the late sixth century. Gregory the Great (ca. 540–604) was the first pope known to have been a monk, but even he advocated linguistic restraint rather than sexual restraint as the mark of a virtuous ruler.[158] In his handbook Pastoral Care, this "ruler" could be anyone at any level of influence or responsibility of either gender, clerical or otherwise. It was only after this time, and helped by Gregory's immense prestige, that sexual asceticism gradually came to be a universal Christian ideal. And as we saw in chapter six, it was not until the twelfth century that this ideal had taken root to the extent that priests were no longer permitted to marry.

Thus it seems that the extreme asceticism which was associated with the early church was very much a minority practice rather than standard Christian behavior. It can best be understood as a reaction to extreme anxiety in sections of the spiritual elite, resulting from a rise in stress from the late-second century AD onwards. It is likely no coincidence that the great age of monasticism from the fourth to fifteenth centuries was also the time when Europe experienced a prolonged rise in V and thus stress.

A modern parallel can be drawn with Christians in Africa, who in general have far sterner, more conservative views than those in Europe. There is,

---

[156] Jones, *The Late Roman Empire*, 979–80.
[157] Hadley, *Masculinity in Medieval Europe*, 112.
[158] Ibid., 118.

for example, a strong aversion to homosexuality and female priests in the Anglican Communion and to married priests among Catholics. Christianity is growing rapidly—there are an estimated 171 million Catholics in sub-Saharan Africa, or 16% of the population.[159] And yet the AIDS epidemic is more virulent in Africa than anywhere else in the world. The views of church leaders do not always reflect the practices of the general population, even of those identifying themselves as Christian.

The other point to make clear is that cultural technologies such as Christianity cannot raise the C of the general population without significant help. As noted earlier, the huge rise in C in Europe between about 1150 and 1850 was driven by a very high level of stress and V, peaking in the sixteenth century. The levels of V and stress in the Roman world were, according to our analysis, very low.

This is also the reason why Christians in the early twenty-first century cannot halt, much less reverse, the decline of C in the their societies, no matter how much they may oppose low-C cultural practices such as homosexuality or abortion. V has already fallen well below the level where it can act as an effective C-promoter for the society at large. This is something we will come back to in a later chapter.

The Survival of the Eastern Empire

This entire chapter has been about the decline and fall of the Roman Empire, but in fact it is not strictly true to say that the Roman Empire "fell." The Roman Empire in the West collapsed in the early-fifth century AD, but the Eastern Empire was to survive for another millennium.

Even as the Western Empire declined, archaeological evidence reveals that the Middle East and to some extent North Africa enjoyed continuing prosperity.[160] It also experienced greater political stability in the late fourth and early fifth centuries, with less of the civil conflict and usurpations that weakened the west at the time of the barbarian invasions.[161] Population even seems to have grown in the late fourth, fifth and early sixth

---

[159] *The Economist*, "Pope, CEO Management tips for the Catholic church," (March 9, 2013), http://www.economist.com/news/business/21573101-management-tips-catholic-church-pope-ceo (accessed September 6, 2014).
[160] Ward-Perkins, Why did Rome Fall and the End of Civilization, 124.
[161] Jones, *The Late Roman Empire*, 363.

centuries.[162] And judging by sophisticated ceramics and the production of new copper coins, the Levant and Egypt remained prosperous throughout the seventh and eighth centuries.[163]

Nor did the East collapse in the face of barbarian invasions.[164] Greece and the Balkans were overrun for several centuries but Anatolia, Egypt and the Levant provided a secure base which allowed the Byzantine Empire to continue and even retake territories that had been lost.[165] And when the Empire eventually came to an end, it was not by disintegration but by incorporation into another and still more powerful empire.

This is a topic that will be taken up in the next chapter, because any explanation of the fall of the Roman Empire must also explain why a substantial part of it did not fall.

---

162 P. Charanis, "Observations on the Demography of the Byzantine Empire." Thirteenth International Congress of Byzantine Studies, Main Papers XIV, (Oxford, 1966), 11–12.
163 Ward-Perkins, *Why did Rome Fall and the End of Civilization*, 124-126.
164 Jones, *The Late Roman Empire*, 1027.
165 Ibid., 1064.

CHAPTER THIRTEEN

THE STABILITY FACTOR

We have seen that the rise and fall of civilizations can be described and to a great extent explained by the rise and fall of V and C. Civilizations rise when cultural technologies create a level of V high enough to bring about a dramatic rise in C. They fall because high infant C undermines V, which eventually undermines C. Very high C also brings about wealth and urbanization, which further undermines both V and C. This is called the civilization cycle.

The potential levels of V and C depend on the strength of the cultural technology behind them. The Greeks, and then the Romans of the Republic, had strong traditions governing such matters as the chastity and subordination of women. These created levels of V and C high enough to support republican institutions for five centuries, including an effective code of law, a strong economic base, and eventually, as V and thus child V began to fall, considerable creativity. The ascetic Christian ideals of behavior which emerged in the late Roman Empire were potentially far more powerful than those they replaced, yet not adopted widely enough to halt the fall in C in the Western Roman Empire. It would, indeed, have been in practical terms impossible for most people to adopt behaviors so at odds with the prevailing low C temperament. Also, levels of V and stress were too low to permit C to stabilize.

Eventually, however, and although it took 1,400 years, Christian practices helped to gradually raise V and thus stress to levels higher than that achieved by the Roman Republic, and (as a consequence) C eventually rose far higher still. Thus, the unprecedented achievements and complexity of European civilization.

Why was infant C lower in the East?

But there is something odd here. The cultural technologies supporting C and V in Europe originated not in Europe but in the Middle East. This applies to the first wave of cultural technology involving the spread of agriculture, evidenced by archaeology and migrations of Middle-Eastern peoples whose genetic imprint can still be found in European populations.

Middle-Eastern cultural influence can also be seen in another wave of cultural technology, which contributed to the rise of Greek and Roman civilization. The development of this technology can be traced in the Middle East through changes in the position of women. Women had been accorded high status in the ancient Sumerian city states, being active in business and with goddesses prominent in the pantheon. Their status had declined by Babylonian times and reached a low point in Assyria, where by 1200 BC respectable women were required to wear veils in public and denied any right to divorce, among other disabilities. They were totally dependent on their husbands and fathers for support and were severely punished for any transgression.[1] Only a few centuries later we find Athenian women with few rights and largely confined to the home, customs that had presumably spread from the Middle East. Subordination of women is a powerful V-promoter, and both the Assyrians and ancient Greeks were ferociously warlike peoples.

As discussed in the previous chapter, the Roman civilization in Italy had two obvious antecedents. The first was the Etruscans who most likely came from Anatolia, first appearing in the ninth century BC and reaching their greatest influence in the seventh century, including control of Rome. The second influence came from Greek colonies in southern Italy and Sicily. From these two sources the strong C- and V-promoting traditions of the Middle East were brought to Rome.

The picture is even clearer on a third wave of cultural technology, most notably the spread of Christianity, which had strong roots in the very oldest centers of civilization—Egypt and Mesopotamia. The religions that competed most strongly with Christianity in the Roman Empire were also Middle Eastern in origin, including Judaism, the worship of Isis (from Egypt) and Mithraism (from Persia). A similar pattern is found in Japanese history. The cultural technologies underpinning Japanese C and V arose from the older

---

[1] B. Vivante, *Women's Roles in Ancient Civilizations: a Reference Guide* (Westport, Conn., London: Greenwood Press, 1999), 91, 95, 106, 112; N. Keddie, *Women in the Middle East: A History* (Princeton: Princeton University Press, 2006), 14, 23, http://www.truthdig.com/images/eartothegrounduploads/KEDDIE_Book-1.pdf (accessed September 6, 2014); M. Van De Mieroop, *A History of the Ancient Near East, ca 3000–323 BC* (Malden, Mass.: Blackwell, 2004), 173.

centers of civilization—Buddhism from India, and Confucianism from China. There were discrete waves of cultural influence visible here too, such as the spread of much stricter Neo-Confucian ideas from Song China.

In all these cases, the most advancec cultural technologies developed in the oldest centres of civilization. But despite this, the highest levels of infant C and thus of C arose not there but in areas with a much shorter experience of cilivized life. To understand this, let us review for a moment the characteristics of infant C as described in chapter five.

Characteristics of high infant C

The first is that infant C societies, those that control infants, tend to become wealthy. They are market-oriented, hardworking and good with machinery. The Industrial Revolution has been attributed to an ultra-high level of infant C in the late eighteenth and nineteenth centuries. They also tend to be highly creative and able to accept new ideas.

Second, societies with high infant C tend to form strong impersonal loyalties such as to republican governments and the rule of law. When combined with child V they form powerful, centralized states.

Third, societies with high infant C have an unusual respect for hereditary rank, so that noble families tend to maintain their position longer than in most societies. Ruling families may also maintain their status long after they have lost real power.

Why were Greeks higher in infant C than Persians?

By all these measures, the Greeks and Romans rated higher than the contemporary empires of Persia and Parthia in the Middle East. Both Greeks and Romans had advanced technology and were wealthy, by the standards of the time. The Greeks in particular were extraordinarily creative. Both Greeks and Romans republican government with effective law codes and, as will be discussed later, their noble families tended to be unusually long-lasting.

The advantages of high infant C were such that the tiny, squabbling states of Greece, when united under the rule of Macedon, were able to conquer and briefly control this entire cultural area. Later, and with almost casual ease, the Romans took over those parts of the "cradle" bordering the Mediterranean.

The same pattern applies to early modern Europe. The areas with highest infant C, as measured by early and successful industrialization, were in the lands never colonized by the Romans or where barbarian incursions were the strongest. These include Germany, the Netherlands and Scandinavia, plus former provinces of the Empire such as Britain and northeastern France. Within Italy, industrialization was strongest in the north, especially the area settled by and named after a Germanic tribe the Lombards. Left behind were the former Roman heartlands of central Italy and Spain, and the lands of the Eastern Empire such as modern Turkey, Syria and Egypt.

The same can be said of East Asia. If rapid industrialization is an indication of high infant C, then infant C was higher in Japan than in the older centers of civilization such as India and China, even though the cultural technologies underpinning it were imported from India and China. Japanese reverence for hereditary succession is such that a single Imperial family has been predominant for at least 1600 years.

It need hardly be said that the community studies cited in chapter six show this pattern exactly. Europeans and Japanese, with less historical experience of civilization, tend to control children in infancy. Arabs, Indians and Chinese, with the longest experience of civilization, do not.

This lower level of infant C cannot be explained by a lack of C-promoters. The Arab and Chinese villagers described in chapter five dominated women and controlled their sexuality at least as strongly as northern Europeans. But somehow these powerful C-promoters failed to be expressed as infant C. It is as though the long experience of civilization prevented this from happening.

Lower infant C, greater stability?

It is now time to bring in the observation noted at the end of the previous chapter. The collapse of the Roman Empire was largely a collapse of the West, with the East showing fewer symptoms of economic and population decline. The Eastern Empire also largely survived the barbarian invasions, with only the Balkans and Greece overrun for a time.

One reason was that the Balkans and the West were more exposed to attack across the Rhine and the Danube, though not even the Mediterranean Sea could protect North Africa from the Vandals. But it is significant that the areas that resisted were *exactly* those that had been civilized for thousands of years—Anatolia, Egypt and the Levant. These were also areas which, unlike Greece and Italy, had shown no evidence of high infant C. They had been conquered and partially colonized by Greeks in the late fourth century under Alexander, but they remained cosmopolitan empires and never formed republics or city states.

In other words, not only does C rise highest in lands never previously civilized, it also falls fastest and furthest in such lands when they become prosperous. In fact, there is an almost perfect correlation between time spent in civilization and immunity to collapse. Eastern provinces such as Syria and Egypt had built large cities and empires long before Greece and Rome, and showed no sign of the dramatic cultural collapse that occurred in the West. Next came Greece, central and southern Italy, Sicily, North Africa and coastal Iberia, with several hundred years of trade and city building. These areas still collapsed, but not as fast as other outlying provinces. Next came the rest of Iberia and Gaul, including the area of north Italy known as Cisalpine Gaul. Finally Britain, the area furthest removed from Mediterranean culture and civilization, experienced the most total collapse, with the Romano-British population largely subsumed by Germanic invaders.

Thus it seems that prior experience of civilization somehow builds immunity to the collapse of C and V when new civilizations arise in the same region, but also prevents C from subsequently rising to the highest possible level. In the Roman world, the highest levels of C and especially infant C (as judged by impersonal institutions, engineering skills and so forth) were found in Italy rather than in the longer-civilized lands of the east such as Anatolia and Egypt. But it was the eastern lands which continued to sustain a succession of powerful empires after the fifth century AD, while Italy and the west collapsed into anarchy.

Why does the late control pattern of childrearing spread so slowly?

At this point it is necessary to ask why people in the oldest areas of civilization do not control their infants or young children, since biohistory has to this point explained such control as the expression of C. People with high C are disciplined and organized in their behavior—traits which express themselves quite naturally in the control and discipline of young children.

The key to this conundrum is to observe that the patterns of behavior that support C and V, including patriarchy and restrictions on sexual behavior, seem to travel quite fast. It has been suggested that the strict seclusion of women in classical Athens was found very soon after the same behavior appeared in Assyria. Also, the Romans seem to have been strongly influenced by Greek colonies in southern Italy and by the Etruscans who most likely came from Anatolia (modern Turkey).

This is even more obvious with the third wave of cultural technologies which was carried in highly organized religions and philosophical systems—Buddhism and Confucianism in China, and Christianity in Europe. Most of the Germanic tribes had been converted to Christianity even before they settled in the disintegrating remnants of the Roman Empire.

But the tendency to indulge infants while being more severe with older children does not spread in the same way. Christianity began in the Middle East among people who had developed powerful C and V-promoters, as well as the tendency to be relatively lenient with younger children. These C and V-promoters spread along with Christianity, but indulgence of infants did not. After 1,500 years of Christianity, nineteenth-century English people maintained their early control behavior. The same can be said of the Japanese, who adopted the C and V-promoters of Buddhism and Confucianism, but not the infant indulgence of the Chinese and Indians.

Not only does infant indulgence fail to spread quickly into newly civilized lands, early control behavior appears never to spread into long-civilized lands – even when the early control culture is completely dominant. Greeks controlled the Levant for several centuries after Alexander's conquest, and many local people accepted their educational systems and even language. But there is not the slightest trace of early control behavior in the conquered peoples, such as the development of nationalism or unusual creativity. The most nationalistic people of the Middle East at this time were in fact the Jews, who were so resistant to Greek culture that they successfully rebelled against the Seleucid Empire, and twice (though less successfully) against Rome. Instead, the tendency to control older but not young children seems to develop gradually with experience of civilization.

Introducing S—a genetically based "Stability Factor"

All of this suggests that the tendency to control or indulge infants is not a cultural or even epigenetic factor, but a genetic one. Long-civilized peoples fail to control children, even when relatively high C, because they are genetically primed to be protective and nurturant of babies and very young children, similar to the way baboons react protectively to the black color of their infants. Biohistory refers to this as the Stability Factor, or "S," since its most obvious effect is to make societies more stable.

It is quite plausible that such a genetic trait could increase dramatically in a few hundred years, since other genes are known to have changed under the selection pressures of civilization. For example, the intelligence of Ashkenazi Jews appears to have risen significantly over the course of about five centuries due to natural selection.[2] There is also evidence that farming has accelerated genetic change a hundredfold, in areas such as disease resistance and the ability to digest lactose.[3]

The selection pressures driving higher S is that it makes people more likely to retain traditional values and have children in a wealthy, urbanizing society. High-S peoples are not completely immune to a loss of V and C, as this clearly happened to the Chinese scholar-gentry (see chapter ten). But as in that example, the process is largely confined to an urban elite. A similar process has happened in Middle-Eastern countries over the past century, with secular-minded elites initially gaining control in a number of countries such as Turkey under Ataturk, Iran under the Shah, Egypt under Sadat and Mubarak, and so forth. An initial decline in V and stress makes such groups more flexible and able to grasp new opportunities in occupations such as the army or civil service. But such elites tend to have fewer children, and with time the loss of C and V undermines their vigor and work ethic. Some early signs of this can be seen in the decline of secular elites and the resurgence of Islam in Iran and Turkey.

People with the high-S genetic variant have lower infant C, which makes them not only more conservative but less oriented to the market. Thus it was that the high-S peasants of the Eastern Roman Empire stayed on their

---

[2] G. Cochran, J. Hardy & H. Harpending, "Natural History of Ashkenazi Intelligence," *Journal of Biosocial Science* 38 (5) (2006): 659–93.

[3] G. Cochran & H. Harpending, *The 10,000 Year Explosion: How Civilization Accelerated Human Evolution* (New York: Basic Books, 2009).

farms and bred children, while the lower S peasants of Italy flocked to the cities and (by and large) did not. The people of the East maintained enough C and V to support strong central government and military effectiveness. Those in the West did not.

Much the same thing is happening in our age. As the low-S peoples of Europe and Japan become rich and flock to the cities, their birth rates drop well below replacement levels. Meanwhile, the peoples of the Middle East in particular have shown a remarkable ability to resist the siren call of Western civilization. Lacking infant C they are less likely to industrialize, less likely to lose V, and more likely to retain traditional values, including the subordination of women. Above all they hold firm to their archetypal high V religion, Islam. They are not immune to these trends or the effects of affluence, as shown by the falling birth rates in most Muslim countries, but they are significantly more resistant.

Just as high-S societies tend to maintain higher birth rates than low S societies, so do high-S people within a society. They are less likely to become wealthy and more likely to hold onto traditional values, so they will on the whole have more children. In twentieth-century Italy the birth rate dropped earlier and faster in the industrial north, where more low-S Germans had settled after the fall of Rome, than in the agricultural south. Once again, high S does not provide immunity. The Sicilian fertility rate is now down to 1.37 children per woman—well below replacement level though still higher than the 1.28 fertility rate of northern Italy.[4] But the process was significantly slower. In a pre-industrial society without our ability to banish famine and even hunger, the process would have been slower still.

Given that more vigorous high-S people commonly migrate into low-S societies in decline, such as Sicilians to northern Italy or Middle Eastern Muslims into modern Europe, high-S genes are widely available to be incorporated into the local gene pool and thus increased by natural selection.

This analysis also makes it clear not only how but *when* the change comes about to higher S. It is not experience of civilization as such, since nineteenth-century Britons had been living in urban societies for almost a

---

[4] Zoran Pavlovic, *Italy* (Philadelphia: Chelsea House, 2004), 49; "Birth and Fertility Rates among the Resident Population," Istituto Nazionale di Statistica (***Istat***), November 14, 2012. Statistics exclude non-Italian immigrants.

thousand years and remained low S. This, of course, is what made the, Industrial Revolution possible. In fact, up to this time the selection process would have been, if anything, to lower S. The higher infant C made possible by high S is a major contributor to economic success, and we have seen that wealthier people tended to have more surviving children before the modern era.

But as soon as affluence increases and C and V start to fall, the advantage shifts to higher S. A graph of how this might look is given in Fig. 13.1 below.

Fig. 13.1. Change from a low-S to high-S society following civilization collapse. During the civilization collapse, people with high S (and thus higher C relative to infant C) have more children, so there is a selection pressure for higher S.

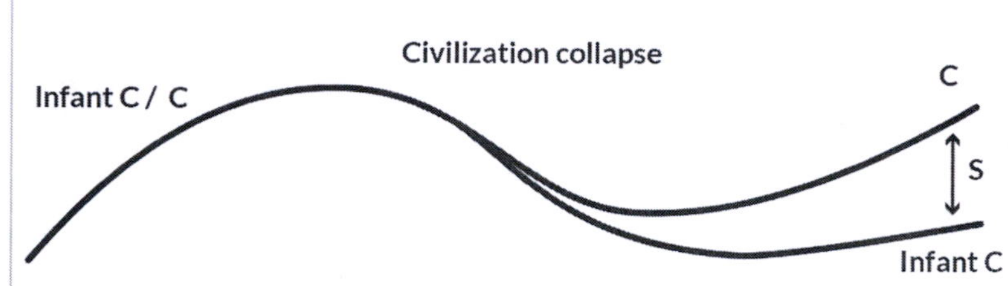

Why does this genetic change matter?

As an approach to history, biohistory is epigenetic rather than genetic. Patterns of control and punishment of children at different stages of life result in epigenetic changes which influence economic activity, creativity, attitudes towards religion and scholarship, forms of government, the size of political units, and much more. So, it might be asked, why the focus on high versus low S? If all S does is to determine how people treat their children, should we not simply focus on the actual behavior and the epigenetic changes that result from it? This was essentially the approach taken in chapters five and six, when infant C was considered as one of a number of epigenetic variables.

S matters because it is a genetic change and thus follows a completely different pattern of change. It is not affected by culture and does not spread through religions and philosophies. Rather, it is a genetic change that alters one aspect of the way people bring up their children and thus has powerful but indirect epigenetic effects. It is these epigenetic effects which explain many of the key questions of history, such as why the Eastern Roman Empire survived the fall of the West for a thousand years, and why industrial civilization took off in Northern Europe.

In certain important ways, S acts contrary to other trends in human history. A popular view is that humanity is moving towards ever-greater achievements in knowledge, power and economic activity, threatened only by ecological factors such as climate change or the exhaustion of key resources such as oil and tropical forests. But rising S puts a completely different check on human progress, by acting to curb both creativity and economic growth in the long term.[5] This means that if and when Western civilization collapses, for the reasons given in the last two chapters, the successor civilization must be poorer and less brilliant.

For all these reasons it is vital to understand what high- and low-S societies are like, and how one changes into the other. The next two chapters will follow the historic civilizations of India, China and the Middle East—societies where parents tend now to indulge their infants and young children. Their political systems in recent centuries have been cosmopolitan empires covering entire culture areas, interspersed with times of disorder. But in the distant past these societies did show signs of infant C and thus lower S, including stable nation states and unusual creativity.

Before doing so, however, we must try to understand the psychological basis of these very different patterns of childrearing.

Why do Europeans not indulge infants?

In chapter five an ethnographic study of American parents in the 1950s showed the strict treatment of infants and very young children, quite different to the indulgence of most other societies. What is of interest now is the emotional attitudes behind this practice. In particular, though these parents had no consistent preference for one age or another, they were (if anything) more inclined to favor older children. Some men were reported as being "afraid of" infants and preferring children with whom they could play. A number of women also reported that they did not care for the infant as much as for the child. At least one woman said she preferred

[5] Some historians consider the East to have been as creative as the West but see R. Duchesne, *The Uniqueness of Western Civilization* (Leiden and Boston: Brill, 2012).

older children because "they can take more care of themselves."[6] A similar ambivalence to children can be seen in the attitude of American mothers to breastfeeding from the same study:

> Some mothers said they simply could not nurse their babies, while others expressed varying degrees of distaste or revulsion towards nursing … Most mothers who nursed were not reluctant to wean their children from the breast ….[7]

These mothers demonstrate the same lack of indulgence in their attitudes to crying babies:

> Adults consider it natural and largely inevitable for babies to cry a lot, although crying is disliked and happy babies are admired … Often it is felt the baby is simply crying for companionship, as when the mother has been playing with and leaves it to do housework. Indulgence of crying for companionship is felt to be bad, for it leads to a spoiled child. Moreover, letting the baby "cry it out" is good not only for its character but also may help exercise its lungs.[8]

Later we will see how unusual such attitudes are in cross-cultural terms, especially compared with areas such as the Middle East and China.

This pattern was not confined to one community. As noted in earlier chapters, English nannies in the nineteenth century went to extraordinary lengths to discipline and control very young children, a habit taken across the Atlantic. Parental advice books of the nineteenth and early twentieth centuries advised American parents to exercise rigorous control of their infants' toilet training, feeding and sleeping schedules, even advising that playing with them could be harmful. Mothers were advised to resist emotional impulses and enforce good habits, which would lead to good character. Toilet training was to begin at two to three months, or even (according to one authority) at three to five weeks. Advice books from the 1920s and 1930s were only slightly more lenient. It may be suggested that the reason such books were popular is that most parents at this time did not feel especially indulgent towards infants, as indicated by the above attitudes. People like books which confirm and justify what they already feel.

---

[6] J. L. Fischer & A. Fischer, "The New Englanders of Orchard Town, U.S.A.," in *Six Cultures*, edited by B. B. Whitting (New York: John Wiley, 1963), 929–30.
[7] Ibid., 939.
[8] Ibid., 941.

It was not until 1946 that Dr Spock first published his book advising against such schedules.[9] But this change was not a matter of delaying discipline till a later age, which would have made parents act more like Arabs or Chinese. It was part of a general slackening of parental control of children at *all* ages. The discipline imposed by these mothers in the 1950s, even in later childhood, was far less rigorous than it had been a generation or two earlier.

Early control of children, especially infants, seems to have been a Europe-wide phenomenon. For example, the most authoritative Dutch family manual of the Victorian era was The Development of the Child by Dr Gerard Allebe. Dr Allebe's main concern was that mothers not be sentimentally over-protective. This was a serious danger, because it would deprive the child of the necessary exercise for courage and result in weakness. Mothers must suppress their inclination to indulge and spoil their children, as doing so would result in the children becoming fearful.[10]

Swedes seem to have focused even more on the early years, as judged by The Century of the Child, published in Sweden in 1909 and becoming an immediate bestseller.[11] The idea seems to have been that if discipline was firm and consistent in infancy it would hardly be needed later.

> Only during the first few years of life is a kind of drill necessary, as a pre-condition to a higher training. The child is then in such a high degree controlled by sensation, that a slight physical plain or pleasure is often the only language he fully understands. Consequently for some children discipline is an indispensable means of enforcing the practice of certain habits. For other children, the stricter methods are entirely unnecessary even at this early age, and as soon as the child can remember a blow, he is too old to receive one … The child must certainly learn obedience, and

---

[9] Internet FAQ Archive, "Parenting," (2008), http://www.faqs.org/childhood/Me-Pa/Parenting.html (accessed September 3, 2014); Internet FAQ Archive, "Discipline," (2008), http://www.faqs.org/
childhood/Co-Fa/Discipline.html (accessed September 6, 2014).
J. Wrigley, "Do Young Children Need Intellectual Stimulation? Experts' Advice to Parents, 1900–1985," *History of Education Quarterly* 29 (1) (1989): 41–75.

[10] N. Bakker, "The Meaning of Fear. Emotional Standards for Children in the Netherlands, 1850–1950: Was there a Western transformation?" *Journal of Social History* 34 (2) (2000): 371–3.

[11] E. Kloek, "Early Modern Childhood in the Dutch Context. Beyond the Century of the Child: Cultural History and Developmental Psychology," edited by W. Koops & M. Zuckerman (Philadelphia: University of Pennsylvania Press, 2012), 53.

> besides, this obedience must be absolute. If such obedience has become habitual from the tenderest age, a look, a word, an intonation, is enough to keep the child straight ... With a very small child, one should not argue, but act consistently and immediately.

The contrast between the behavior expected of European and American parents early in the twentieth century, and that of Chinese and Egyptian parents, could hardly be more extreme. In Europe, strict discipline started from the first year of life. In China and Egypt a child was barely considered able to understand right from wrong until 5 or 6.

As indicated earlier, the best way to explain this is that people's emotional reactions depend on their level of S. Low-S people can be tougher with infants and young children because they do not have the same emotional reaction to them as people with higher S.

Low S and respect for hereditary rank

So far we have considered low-S societies at times when C (and thus infant C) is very high, as in nineteenth-century Europe. Characteristics include an intense orientation to the market, aptitude for machinery, respect for the law, and republican forms of government. But low-S societies are different from high-S societies in other ways.

The first point to note is that in low-S societies training of children begins early even when C is moderate, though not early as when C is at a peak. As indicated in chapter six, discipline of children in Europe and Japan was expected to start at age 3 as early as the fifteenth century, (see chapters six and seven).

Another difference mentioned previously, and this applies especially at moderate levels of C when loyalties are more personal, is that low S peoples have a peculiar reverence for hereditary rank. Political power has tended to be hereditary in most civilizations, but high-S people favor hereditary ruling families *even when such families have lost all real power*.

An example of how this works was the peculiar situation of an English king, Henry II, ruling half of France in the mid-thirteenth century because he was the son or husband of hereditary rulers of French provinces. But at the same time, the lesser nobility of those provinces also considered themselves subjects of the relatively powerless French king. In effect, they were influenced by two different and quite incompatible loyalties which were quite independent of the realities of power.

This has a parallel in medieval Japan at the same time—another low-S society with only moderate C. The military rulers or shoguns had seized power but portrayed themselves as subjects of a powerless emperor. For a time, even the shogun became a titular ruler, with real power being held by the Hojo regents. In fact, the Japanese still revere an Emperor descended from a ruling house that has not held effective authority for a thousand years.

The same pattern exists in Europe even to the present day in the constitutional monarchs of ten European countries: Belgium, Denmark, Liechtenstein, Luxembourg, Monaco, Holland, Norway, Spain, Sweden and Britain. Significantly, all of these except Monaco and Spain are in areas of Northern Europe settled by Germanic peoples, and thus with the lowest possible level of S since none of them have gone through a civilization collapse.

Another way in which this attitude expresses itself is through loyalty to local aristocrats, which tends to make noble houses both powerful and durable, even in times of strong central authority. Rome was clearly a low-S society, and its aristocracy showed all these characteristics. Both Julius Caesar and his assassin Brutus were from patrician families which had been politically powerful for more than five-hundred years. Supreme power in the Republic was vested in the Senate, which represented such families.

Powerful and durable families are also characteristic of European history. As an example, the Howards have been Dukes of Norfolk since John Howard was granted the title in 1483 (in fact he was descended from an even earlier Duke of Norfolk through his mother). The survival of the family's wealth and power is remarkable, considering the fates of individual dukes. The first duke died at the battle of Bosworth in 1485, on the losing side. His son was imprisoned and stripped of his lands, but later restored. The next duke was stripped of his title and sentenced to death by King Henry VIII in 1547, surviving only because the king died on the day before the execution. He then lost his lands in the reign of Edward VI, but was restored by Queen Mary. His heir was executed in 1546, so the title passed to his next son, who was executed for treason in 1572, his lands and title forfeited. Again, much of the estate and the title were eventually restored to the family. A Howard in the direct line of descent is still Duke of Norfolk in the twenty-first century.

Given that Tudor monarchs were chronically short of money, and confiscated lands were a major source of wealth and patronage, it must be asked why such lands were so often returned to their former owners. After all, King Henry VIII had no scruples about plundering monasteries and stripping churches of their wealth. The key to this is enduring local loyalties which gave great families support and stability for generation after generation. Kings who tried to replace them with "new men" could face trouble and outright revolt.

For example, in 1469 there was a revolt in Yorkshire aimed at restoring the young Henry Percy to the earldom of Northumberland, forfeited by his father in 1461. He was restored soon after, and the "usurping" John Neville demoted. In 1485 Percy's son, also Henry Percy, was imprisoned by the victorious Henry VII. This led directly to Yorkshiremen joining another revolt because they feared the wrath of the king without their earl to protect them. Henry was released a few months later, his lands and titles intact.[12] The notion of a local lord as protector against royal tyranny is a frequent theme of the age.

Japanese noble houses could be equally enduring. The daimyo of sixteenth-century Japan were theoretically subject to the Emperor but in practice independent. When Tokugawa Ieyasu seized power at the end of the century he did something which would have been quite extraordinary in any other country—he confirmed most of the daimyo in power, including some of those who had been his strongest opponents, and even left them with their own private armies. This arrangement endured for nearly three centuries until two of these daimyo led the rebellion that overthrew the last Tokugawa shogun in 1867.

Examination of the ten societies in the cross-cultural survey judged as exercising significant control of infants shows the same pattern (see Table 13.1 below).

The Manus can be excluded from consideration because, not exercising authority over older children, they had little reverence for authority of any kind. But the other societies show the same pattern of respect for hereditary rank. Europeans and the Japanese have already been mentioned,

---

[12] A. Goodman, *The Wars of the Roses: Military Activity and English Society, 1452–97* (London: Routledge & Kegan Paul, 1981), 206.

and Tonga and Thailand are early control societies whose monarchs are revered and protected from criticism. They are influential in government but certainly not in control of it. Polynesians in general have a tradition of sacred chiefs.

Table 13.1. Societies with significant control of infants.

European	Asian	Polynesia	Other
Bulgaria	Japan	Tikopia	Manus
France	Thailand	Tonga	
Ireland			
United States			
Jews (from Poland)			
Holland			
Sweden			
Britain			

Low S and a preference for rulers similar in culture and appearance

There is another characteristic of low-S peoples. Just as they prefer familiar local aristocrats as leaders, they also have a strong preference for people similar to themselves in ethnicity, language and religion.

This is one aspect of high C that has obvious biological roots. Recall from chapter one that gibbons are highly discriminatory in their choice of mates, to the extent that different species of gibbons can coexist in the same section of rainforest. They are reluctant to mate with any animal that is at all different. In this, they contrast strongly with savannah baboons who mate far more indiscriminately, and for this reason form a single species over a vast area, though with local sub-species. Success in a stable, competitive environment means adapting to local conditions. An animal that does well in local conditions will reproduce most successfully with a mate that is similarly adapted, not one with variant genes that may be better suited to living somewhere else.

When C is very high, as in nineteenth-century Europe or twentieth-century Japan, this can take the quite unpleasant form of racism. Thus it is that for a long period of time, countries such as Australia and the United States strongly preferred European immigrants over those from Asia and elsewhere. The enslavement of black Africans, which was quite alien to the way European peoples treated each other, was undoubtedly made easier by such attitudes. Even today, and despite a plunging birth rate, the Japanese are reluctant to take immigrants from other nations.

Another result is that low-S people strongly prefer *rulers* similar to themselves in culture, language and appearance. In Europe, such loyalties rose to their peak in the intense nationalism of the nineteenth century but were evident much earlier.[13] In the example given earlier, the loyalty owed by Frenchmen to Henry II as hereditary ruler of Anjou and Aquitaine conflicted with the loyalty due the French king, given that Henry was English (even though he spoke French). In the fifteenth century, loyalty to their hereditary kings, combined with rising nationalism, helped drive the English out of France.

A century or so later, the Dutch fought off their Spanish overlords even though Spain was many times more powerful and wealthy. Later still, Spanish guerrillas fought Napoleon's army of occupation with equal determination and doggedness. Nation states survive because people want to be ruled by their own kind.

High-S societies—authority based on power

Hereditary monarchs have also ruled the long-civilized areas of China, India and the Middle East for thousands of years, but a key difference is that rulers were usually killed or at least dethroned when they lost power. This is clear from looking at the last emperor of each major Chinese dynasty from the ninth century onwards (see Table 13.2 below).

These were dynasties that had ruled for at least a century, and in some cases for far longer, and yet in no case was there any attempt to maintain them as figureheads. Significantly, this was *not* the case in the early history of these civilizations, such as the Zhou period which preceded the first Chinese empire, a matter that will be taken up in the next two chapters. But after this time, and in extreme contrast to Japan, no Chinese ruler reigned for

[13] A. Smith, *The Ethnic Origins of Nations* (New York: Wiley-Blackwell, 1991).

long after he ceased to have effective power, suggesting a far lesser attachment to hereditary rank. In other words, a Chinese ruler was revered only so long as he was believed to be powerful. When the power left him, so did the reverence.

Table 13.2. Fate of Chinese rulers who lost power since 800 AD (for dynasties lasting at least a century).

Dynasty	Date power lost	Fate of ruler
Tang	881	Fled
Northern Song	1127	Deposed and exiled
Jin	1233	Suicide 1234
Southern Song	1275	Drowned in flight 1279
Yuan (Mongols)	1368	Fled
Ming	1644	Suicide 1644
Qing (Manchus)	1911	Abdicated 1912

The same pattern applies to the history of the Middle East and the Muslim world in general. The last Ottoman Sultan lost power in the Young Turks revolt of 1908 and was deposed the following year. The Shah of Iran was deposed and fled abroad in 1979. Even supposedly "constitutional" monarchs like the kings of Jordan and Morocco hold actual executive power, albeit in conjunction with democratic institutions. In fact, no evidence of powerless but revered monarchs has been found in any society which does not control children at a relatively young age.

Such people normally have high child V and thus tend to accept powerful authority. Lacking impersonal loyalties and strong legal codes this gives local elites considerable power, but such power is less likely to be inherited. A famous example of elite formation is the Chinese scholar-gentry class, recruited by competitive examination. But their power is lost when their descendants cease to excel. In areas such as the Middle East, elite recruitment is often through the military, but even here status does not long survive the loss of military vigor.

High-S societies also lack the intense local and national loyalties of low-S peoples, which means they tend to form cosmopolitan empires that expand until stopped by geographical boundaries. China, for example, was bounded by mountains in the west, the arid steppe to the north, and the sea to the east. India was also bounded by mountains and sea. When empires collapse dynasties and political borders change rapidly. People do not tend to form stable political units within the culture area because they tend to accept the authority of powerful foreign leaders over weaker local ones.

Characteristics of low-S peoples

The characteristics of high-S peoples will be dealt with in more detail in the next two chapters when we consider how these societies evolved. But for now we may summarize the characteristics of low-S peoples. This is all the more important when trying to assess the level of S without direct evidence of childrearing patterns, as is commonly the case in the ancient world.

Low-S peoples, as mentioned earlier, tend to have an unusual reverence for hereditary rank, giving status to rulers when they have lost real power. Their noble houses are commonly long lasting, and the culture area is typically split into a number of stable nation states or (as we will see) city states. People in such states are loyal to their own leaders and fight ferociously against rule by "foreigners," even when they are much stronger.

As C rises to a peak, low-S societies show the characteristics of high infant C described in earlier chapters. They are uncommonly creative, wealthy and market-oriented. Populations tend to grow rapidly. Impersonal loyalties mean they may form republics, or alternatively accept rule by centralized states with strong law codes. Even absolute rulers such as Napoleon were strong proponents of the rule of law.

Finally, low-S societies are highly vulnerable to collapse. Creative peoples are less likely to hold onto tradition, and we have seen that high infant C may itself undermine V and thus (eventually) C. Added to this is that the wealth and urbanization resulting from high infant C tend to undermine V and C. We have already seen this pattern in the fall of the Roman Empire, which continued in the high-S east while collapsing in the low-S west. An even more powerful and poignant example can be found in the rise and fall of ancient Greece.

Ancient Greece

Ancient Greece was the first European society to show signs of high infant C. Greece in the twenty-first century would be classed as a moderate S society, with levels of infant C intermediate between those of Northern Europe and the Middle East. Though under Ottoman rule for several centuries, the Greeks rebelled frequently before gaining their independence with Western help in the early nineteenth century. This kind of nationalism is characteristic of high infant C. On the other hand, Greece never became an industrial powerhouse like Britain or Germany, and the level of corruption and dysfunction in modern Greece suggests infant C well below the levels of Northern Europe.

But the Greeks of the sixth to the fourth centuries BC showed all the hallmarks of high infant C. Their city states were fiercely independent and their civilization creatively brilliant in a way which has scarcely been seen before or since. And unusually for such an ancient people, there is evidence for their treatment of very young children. This is the description Plutarch gives of Spartan childhood:

> There was much care and art, too, used by the nurses. They had no swaddling bands. The children grew up free and unconstrained in limb and form, and not dainty and fanciful about their food; not afraid in the dark, or of being left alone; without any peevishness or ill humour or crying. Upon this account, Spartan nurses were often brought up, or hired by people of other countries [meaning other Greek states], and it is recorded that she who suckled Alcibiades [an Athenian aristocrat of the late fifth century BC] was a Spartan.[14]

The reference to nurses and suckling clearly indicates that this rigorous treatment started in infancy, and the fact that other Greeks favored Spartan nurses suggests that some form of training in the first two years of life was commonplace. Plato also believed that children should be taught to obey from well before the age of 3:

> Then until the age of three has been reached by boy or girl, scrupulous and unperfunctory obedience to the instructions just given will be of the first advantage to our infantile charges. At the stage reached by the age of three, and after the ages of four, five, and six, play will be necessary, and we must relax our coddling and inflict punishments—though not such as

[14] Plutarch, *Lives of the noble Grecians and Romans* (New York: Random House, 1992).

> are degrading.[15]

Aristotle had a similar view, even though he wrote at a time when Athens was under Macedonian domination when we would expect parental control to be declining along with C:

> The legislator must mould to his will the frames of newly-born children … To accustom children to the cold from their earliest years is also an excellent practice … and children, from their natural warmth, may be easily trained to bear cold. Such care should attend them in the first stage of life.[16]

He believed their physical and moral training should commence very early:

> He must also prescribe a physical training for infants and young children. For their moral education the very young should be committed to overseers; these should select the tales which they are told, their associates, the pictures, plays and statues which they see. From five to seven years of age should be the period of preparation for intellectual training.[17]

People who do not begin serious training of children until age five or six typically believe that younger children are untrainable, so that any attempt to direct their behavior is both cruel and fruitless. The Greeks of this age evidently felt it was neither. Spartan nurses, like American, Swedish or Dutch parents of the recent past, believed that crying should not be rewarded by attention. We have also seen how Dutch parents were advised to train their infants to sleep in the dark by the end of their first year.

What made the Greeks extraordinary was the combination of low S with powerful C and V-promoters imported from the Middle East. These included practices such as the rigid seclusion of Athenian women, at least for classes that could afford it. In the early agricultural societies of the Middle East, this pattern took several thousand years to fully develop. The Greeks, however, adopted it almost as soon as they began dwelling in cities. The effect of this was not only to raise the level of C by controlling

---

[15] Plato, *Laws*, Book 7, http://hoodmuseum.dartmouth.edu/exhibitions/coa/ch_discipline.html (accessed September 6, 2014).

[16] Aristotle, *The Basic Works of Aristotle*, Politics, edited by R. McKeon (New York: The Modern Library, 1941), book seven, chapter sixteen.

[17] Ibid., chapter seventeen.

women's sexual behavior, but to raise V by subordinating them to men. This last helped to make the Greeks ferociously and incessantly warlike.

But unlike the longer-civilized peoples of the Middle East, Greece had never gone through a prolonged civilization collapse which would raise S by favoring people with lower infant C. Also, protected by seas and mountains, they were never overrun by these older civilizations. Persia came close in the early fifth century BC but was beaten back. Thus there was no mass immigration of higher S peoples, including rulers who might have an outsized genetic impact. It may be that conquest by higher S empires also tends to weed out lower S genes because high infant C inclines people to rebel. And rebels who survive the slaughter commonly end up as slaves in a foreign land with little chance to reproduce. Thus, by fighting off the Persians the Greece were able to retain their low-S genes.

The Greeks, in their brief cultural flowering, were to have a huge impact on future ages. More than any previous people they were able to break free of traditional thinking and see the world with fresh eyes, leading to immense achievements in philosophy, science, art, architecture and literature. The histories of Herodotus and Thucydides are astonishingly modern in tone, far more familiar to us than (say) the works of medieval scholastics. They also created sophisticated political systems, including the first democracies. All this is familiar because their culture influenced Western development, and they had (like us) high infant C, albeit not to the level of nineteenth-century Europeans. Such a character also made possible their remarkable commercial success, with trade goods and settlements found all over the Mediterranean.

In terms of creativity, Greece had another key advantage evident from its political systems—a low level of child V. A key aspect of infant C is that it promotes loyalty to specific people and institutions, whether hereditary rulers or nations, while people with high child V are more willing to submit to authority in general. Also, as indicated earlier, people with high infant C are less likely to overlook cultural, religious and linguistic differences with their rulers. People with high infant C and low child V should therefore be intense but narrow in their loyalties. They will want their leaders to be very similar to themselves, and they will not be overawed by powerful authority. The clearest possible expression of such an attitude is the city-state.

The loyalty of Greeks to their cities was intense and durable. Dozens of tiny states retained their independence and identity for hundreds of years.

One of the very few cases where one state overcame another was the Spartan conquest of Messenia in the eighth century BC. The difficulty of holding this conquest was such that the Spartans could only achieve it by organizing their entire society for war. And when Spartan rule was broken after four centuries, the Messenians emerged with their identity and borders intact.

Sparta was the one Greek state to maintain an empire, and it is the one state where we see evidence of high child V based on the harsh treatment of older children. Spartan training of children, especially boys, was notoriously tough. From the age of 7 they were sent to live in companies under severe discipline, fed so little that they must steal, and flogged for being discovered in the act. They were allowed only a single cloak in the bitterly cold Laconian winter, bathed rarely, swam and exercised naked. In one particular ritual where boys endured flogging to steal cheeses from an altar, the treatment could be so severe that they died.[18]

But the Spartans were notoriously different from other Greeks in ways that are exactly consistent with higher child V. They had little interest in commerce, and Spartan citizens were barred from taking part in it.[19] They were notoriously lacking in creativity, conservative, superstitious and brutal to their subject peoples. Young Spartans were encouraged to randomly murder helots to keep the subject classes cowed and in place.

Though Athenian parents had no objection to beating children, discipline seems to have been less harsh than in most high-C cultures. The playwright Euripides, writing just after what biohistory suggests as the peak of C in the mid-fifth century BC, advised that "it is sweet for children to find a mild father and not be hateful to him." He also said that "a father who is mean to his children heavily weights down his old age."[20]

A lower level of child V and thus stress is also suggested by the relatively lenient punishments inflicted on adults. Penalties included fines, imprisonment, loss of certain political rights, exile or confiscation of property. People sentenced to death were commonly allowed to flee into exile, and in the event of the sentence being carried out, hemlock could be

---

[18] Plutarch, *Lives of the noble Grecians and Romans.*

[19] D. Sacks, *A Dictionary of the Ancient Greek World* (Oxford: Oxford University Press, 1995), 232.

[20] M. Golden, *Children and Childhood in Ancient Athens* (Baltimore and London: The Johns Hopkins University Press, 1990), 103–4.

used to avoid any pain. Free citizens were not to be flogged or tortured, and the law forbade imprisonment for debt.[21] Compared to the law codes of ancient Rome or sixteenth-century England, this was mild indeed.

This combination of infant C with low child V explains why city-state civilizations are creative by their very nature, a phenomenon also seen in northern Italy during the Renaissance. Cities such as Florence and Venice were republics, commercially brilliant and with extraordinary artistic achievements. The same can be said of seventeenth-century Holland, a tiny nation which built a mighty commercial empire while giving rise to some of the world's greatest works of art. We can see direct evidence of low child V in comments by French and German visitors that the seventeenth-century Dutch were astonishingly lenient with their children, in that they declined to use corporal punishment.[22] This was seen as indicating extreme laxity, but we have already noted evidence from parental advice books that Dutch parents in the nineteenth century exercised the most rigorous control of their children from earliest infancy—again, with relatively little physical punishment. It is this combination of tight control from an early age with minimal punishment which produces this most creatively brilliant form of civilization.

Low child V in ancient Greece also explains why their greatest level of creativity came quite early in the civilization cycle, at and shortly after reaching the highest level of C. We can set the peak of C in Athens with some accuracy because of the outbreak of the Peloponnesian War in 431 BC. This conflict is a perfect example of what happens when an aggressive, high-V people experiences a testosterone surge as the result of an initial decline in C. The expedition against Syracuse in 415 BC was imperial lunacy (similar to that of Germany in 1914 as discussed in chapter nine), especially given the immense sufferings of the Athenian people earlier in the war.

Based on the outbreak of this war, a peak of both birth rate and C can be placed around 450 BC, with the subsequent fall in C creating a surge of testosterone in the generations born from that time onward. This is the

---

[21] D. S. Allen, "Punishment in Ancient Athens," in *The Stoa: A Consortium for Electronic Publication in the Humanities*, edited by A. Mahoney & R. Scaife, (2003), http://www.stoa.org/projects/demos/article_punishment?page=1&greekEncoding= (accessed September 6, 2014).

[22] Kloek, "Early Modern Childhood in the Dutch Context," 53.

same pattern responsible for the World Wars of the early twentieth century, as described in chapter nine.

But while creativity in the Roman Republic peaked a century and a half after the high point of C, fifth-century Athens was arguably the most creatively brilliant society in history. The Parthenon was built between 447 and 438 BC. Thucydides wrote and Socrates taught during its last decades, while great playwrights vied in the theaters. Athens had also been securely democratic for some decades before the presumed peak of C, something not even achieved in most of nineteenth-century Europe (though it was in America). Even Britain did not give all men the vote until 1918. Observing Greeks at this time it is hard to avoid the conclusion that they were not only more creative but also more intelligent than their ancestors, or than most people have been throughout history. It is possible that high infant C plus low stress actually raises IQ, an idea to be revisited when dealing with industrializing nations of our time.

We have seen that democracy and creativity reached their high point in ancient Rome and twentieth-century Europe well after the peak of C. The reason is that both societies had relatively high child V, which needed to decline first before the full flowering could take place. In ancient Athens the flowering took place at and immediately after the peak of C, presumably because stress was already low at this time.

Greece, in its day, was an outstanding success. But as we will see many times, low-S civilizations have a serious problem. As C increases the level of infant C also rises. What makes the society successful also makes it unstable, because high infant C eventually undermines C by reducing V (which requires indulgence of infants) and promoting free thinking, which undermines traditional values. It also creates prosperity which directly undermines both C and V. The higher the level of infant C, the faster this happens, and the more immediate the weakness.

Thus it is that ancient Greece was well past its peak in the fourth century BC. By the time Alexander came to the throne of Macedon in 336 BC, C was in obvious decline. The birth-rate was dropping and the Spartan population, in particular, was already much reduced. The relative ease with which the Greeks submitted to Alexander's father Phillip is in sharp contrast to the prickly independence of Greek city states in previous centuries and their heroic resistance to the Persians less than 140 years earlier, a clear indication of reduced local loyalties by the late fourth century. Of course, it was this very fact which made it possible for the united Greek and Macedonian armies to conquer the Persian Empire.

Greeks compared to Romans

Rome was like Greece in having high infant C, by the standards of the ancient world. The key difference is that the Romans seem to have had significantly higher child V. In Rome, as described in chapter twelve, it was not until the late-first century AD that any Latin writers spoke against the beating of children, and this was after C had been in decline for more than three hundred years. The patria potestas," which allowed fathers to kill their sons without penalty, was not repealed until the following century. This would explain Roman conservativism, as expressed by their reverence for the 'mos maiorum' (ways of the ancestors), their lesser level of creativity, and of course their greater success in building a large Empire. Rome conquered the world because it was able to unite most of Italy when C was at a peak in the third century BC, rather than a century and a half after the peak as was the case for Greece.

Sicily and southern Italy—the rise in S

It has been suggested that Europe remains a low S continent because of a huge influx of barbarians following the collapse of the Roman Empire. This meant that the tendency for S to rise in declining civilizations was genetically swamped by lower S peoples. But the influx was much less in some areas than others, permitting S to rise much higher than in Northern Europe.

Greece has been part of the civilized world for almost three millennia and should, for this reason, have relatively high S. Yet, as we have seen, the signs are that it has only moderate S. The obvious reason is that between the fourth and seventh centuries AD a large number of barbarian peoples invaded and settled including Goths, Huns and in particular Slavs. Never having been civilized they would lower the general level of S in the lands where they settled.

The highest levels of S in the former Roman Empire would be expected in the areas most distant from the barbarian homelands, where fewer barbarians had the opportunity to settle. This could apply to such areas as southern Spain, but the effects of higher S can be seen most clearly in southern Italy and Sicily.

The original inhabitants, having no history of civilization and therefore being low S, were largely displaced or absorbed by Greek colonists from the eighth century onwards. This area then became known as "Magna Graecia," or Great Greece. As we have seen, Greeks at this time clearly had high infant C and were thus low S. There were also Carthaginian settlers in the west of Sicily. Originally from Phoenicia they had a much longer history of civilization and were thus presumably higher S, but genetically they were probably a small part of the mix.

By the time of the fall of Rome this area had been civilized for a thousand years, so S was presumably higher. And unlike Lombardy and areas further north with substantial Gothic populations, relatively few Germans settled in the south. When Justinian's general Belisarius invaded Italy from the south in 535–6 AD, he did not encounter resistance from a Gothic garrison anywhere south of Naples. And in southern and central Italy the aristocracy was overwhelmingly Roman until at least the sixth century, much more so than in northern Italy or Gaul.[23] The whole subsequent history of this area shows the effects of higher S and thus lower infant C, in sharp contrast to what happened further north.

One of the features of such a society is acceptance of authority in general, with a lower level of the kind of loyalty to people of similar ethnicity and language which helps local regimes to become entrenched. Thus, in place of the city-states that emerged in northern Italy, the south and especially Sicily were ruled by a succession of alien powers—Byzantines, Normans, Arabs, French, Germans, Aragonese and Austrians. The strength of the Mafia, perhaps Sicily's most famous export, derives from a culture valuing personal ties above all, seeing government as an alien entity to which no loyalty is due—characteristics linked to low infant C.

The south of Italy has remained poorer than the north, more conservative and religious and, until recently, with a higher birth rate. All this suggests lower infant C. In recent times the birth rate has fallen below replacement levels, as in the rest of Italy, which reinforces the point made earlier that high S does not stop C and V from collapsing. Higher S people tend to be poorer and more conservative. But given enough prosperity, which in biological terms means people are not going hungry, both C and V must fall.

Though Sicily and southern Italy have been used as examples of high S

---

[23] B. Ward-Perkins, *The Fall of Rome and the End of Civilization* (Oxford: Oxford University Press, 2005), 65, 67.

and thus low infant C, in a more general sense the same distinction can be made between southern and Northern Europe. The lowest S and highest infant C can be found in the northerly and Germanic-speaking areas. The distinction between Catholics and Protestants runs roughly along the same lines, the obvious exceptions being Austria and south Germany—German-settled areas that remained Catholic after the Reformation. The German-descended peoples in general also seem to be especially low S and thus suited to develop high infant C civilizations, even compared with other European peoples such as the Celts and Slavs.

Chapter seven referred to Charles Murray's meticulous study of prominent men in the arts and sciences. The geographical pattern shows a heavy concentration in England, Germany, Switzerland, Belgium, the Netherlands, northeast France and northern Italy.[24] In general, these are areas of strong German settlement and thus likely to have lowest S and highest infant C. The partial exception is northern Italy, but even this was the section of Italy most settled by Germans after the fall of Rome.

In the next two chapters we will study the history of high-S peoples including those of China, India and the Middle East. Using the characteristics outlined in this chapter it will be possible to identify that the earliest civilizations in these areas were distinctly low S, similar to those of Europe and ancient Greece. We will also be able to trace the transformation of these societies into the high-S civilizations of today, and explain why they followed a path so different from that of Europe and the West.

---

[24] C. Murray, *Human Accomplishment: The Pursuit of Excellence in the Arts and Sciences, 800 BC to 1950* (HarperCollins e-books, 2009), chapter thirteen.

CHAPTER FOURTEEN

CHINA AND INDIA

In the last chapter we saw how a high level of S, the Stability factor, is characteristic of the oldest centers of civilization including India, China and the Middle East. This genetic trait causes people to be more indulgent of infants and younger children than of older ones, thus suppressing the level of infant C below that of C. In this chapter we look more closely at the evidence for increasing S in the history of China, India and the Jews.

China

We saw in chapter five that high-S societies such as China tend to be indulgent of infants. A characteristic of such societies is that they are less resistant to rulers unlike themselves in language, culture and religion. This means empires tend to expand to fill the entire culture area, defined as a region with shared cultural traits but significant local variations in language, values and so on, such as Europe or China. They may accept rule by people from outside the culture area—in the case of China, the conquests by the Mongols and Manchus, which produced two of the last three Chinese dynasties, or four if we count the Communists.

By contrast, as discussed in the last chapter, low-S societies such as those of Europe not only strongly resist rulers from outside their culture areas, they usually resist rulers with relatively minor differences from within the culture areas, leading to a proliferation of nations and even city-states.

High-S peoples are also less loyalty to hereditary leaders, meaning that although rulers tend to be hereditary, they will not be retained as figureheads once they lose power. Also, local nobles tend to be weaker and their dynasties less enduring.

The later history of China is that of a stable, high-S civilization, but it was not always that way. Until the early centuries AD, China was low in S, and only gradually developed higher S in the centuries that followed. A study of Chinese history makes this clear, and also shows how civilization cycles play out in Chinese history, including a collapse of C and V from the second century BC followed by a steady rise of both from the sixth century onward. The combined effect was a rise in C but not in infant C, as shown in Fig. 14.1.

Fig. 14.1. C and Infant C in China. The collapse of the ancient Chinese civilization brought selective pressures for higher S, so that when C rose once more, infant C did not.

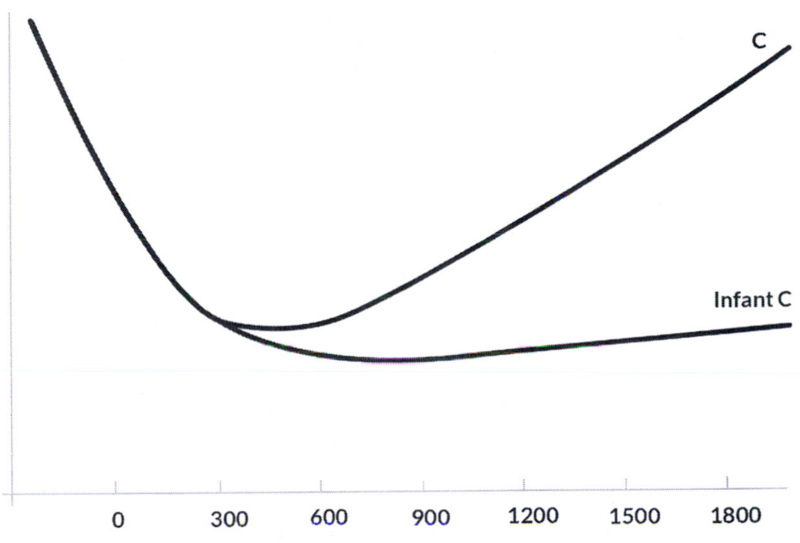

Low S—the Zhou dynasty

The earliest Chinese civilization is associated with the Shang dynasty that ruled towards the end of the second millennium BC. [1] Around 1046 BC it was overrun by invaders from the Wei river area in the northwest, who established the Zhou dynasty in control of what is now north China. In 770 BC their capital was sacked and they were forced to relocate eastwards.

Even before 770, local rulers within the Zhou sphere of influence probably had a great deal of independence, but from 770 until the end of the third century BC these states became effectively independent. However, it is a

---

[1] Chinese historical information from: J. Keay, *China, a History* (London: Harper Collins, 2008), 64; E. O. Reischauer & J. K. Fairbank, *East Asia: The Great Tradition* (Boston, Houghton Mifflin, 1960).

clear sign of low S that the Zhou emperors remained as nominal overlords for more than five-hundred years after they lost power. For example, after the Zhou ruler had been ejected from his capital by an army from the southern state of Chu in the mid-seventh century, he was restored in 635 by Chonger, ruler of the northern state of Jin. The grateful Zhou Emperor awarded him the title of ba, or "hegemon." The ceremony in which he accepted this honor, after refusing it twice as a sign of humility, shows the extreme reverence expressed for this utterly powerless ruler:

> Chonger ventures to bow twice, touching his head to the ground, and respectfully accepts and publishes abroad these illustrious and enlightened and excellent commands of the (Zhou) Son of Heaven.[2]

The emperor may have ceased to rule, except in his own small principality, but he continued to reign. This pattern, never again to be seen in China, is typical of the reverence for hereditary rulers associated with low S.

Another characteristic of low-S societies is that hereditary aristocrats tend to play an unusually prominent role when C is moderate and loyalties still personal, as compared to the more fluid and fast-changing elites in high S societies. This was certainly the case in China. Until the sixth century BC, the great majority of prominent men mentioned in the histories were from great or royal families. The great families of Jin in particular were so powerful that in the fifth century they overthrew their ruler and eventually formed three separate states.

The eras between the eighth and third centuries BC are known as the "Spring and Autumn" and "Warring States" periods, after two famous historical texts, though the latter does describe pretty well the endemic warfare of the time.[3] But this was no feudal anarchy. Though larger states tended to absorb smaller ones, most retained their identity for hundreds of years. National loyalties were so strong that they could even survive more than a century of foreign conquest.

For example, the state of Chu was overrun by its southern neighbor Wu in the late seventh century, but when Wu itself was overrun in 473 BC the exiled Chu ruler was able to take back almost all his nation's former territory. In modern Europe, the survival and rebirth of Poland is an example of the same phenomenon. Poland had been divided among more

---

[2] Keay, *China, a History*, 64.

[3] V. T.-b. Hui, *War and State Formation in Ancient China and Early Modern Europe* (Cambridge: Cambridge University Press, 2005).

powerful regimes in the eighteenth century, but the people retained their language and identity and emerged independent in the twentieth century. In a low-S society, local loyalties are intense and enduring.

During this period there were signs of rising C and especially infant C. States became increasingly centralized, with elaborate and often draconian law codes. This allowed the formation of massive armies, many times the size of armies of earlier centuries, a change which contributed much to the ferocity of warfare. The impersonal loyalties of people with high infant C make them much easier to organize and discipline, so long as it is under the authority of ethnically similar rulers, as in the Roman Republic or nineteenth-century and early twentieth-century Europe.

Political centralization went hand in hand with economic development. Until the fifth or sixth centuries BC, estates tended to be self-sufficient and trade slight. Wealth was measured in livestock and salaries were paid in grain. But by the late fourth century, farmers were buying (rather than making) their cloth, cooking pots and implements. Trade and transport conditions improved, and coins were coming into use. Population grew dramatically and the area of Chinese settlement expanded.

This was also a time of innovation and creativity without parallel in Chinese history. Early China had been a conservative society with a high regard for tradition. When Confucius offered his services as an adviser to the rulers of fifth-century states, he advocated a return to the values of the past. But his teachings were the start of what became the golden age of Chinese philosophy, known as the "Hundred Schools of Thought." These competing schools included versions of Confucianism which contained the strong ethical strands characteristic of high C. There was also the less savory school of Legalism, which supported and justified the harsh policies of the new centralizing states.

In addition, popular beliefs became organized into what became known as Taoism, a quietist philosophy advocating retreat from the world. By the fourth and third centuries BC, change was no longer seen as a thing to be feared. New customs were to be adopted where useful, without regard for the past.[4]

New ideas were springing up in other areas as well. This was the time when iron and the plough came into general use. There was some

---

[4] F. Mote, *Intellectual Foundations of China* (New York: McGraw-Hill, 1989).

standardization of weights, measures and coinage. Military innovations included the crossbow, cavalry (replacing the chariots of earlier centuries), more advanced siege engines, and new weapons and armor. All of this indicates high levels of the creative infant C. Added to long-lasting local states, strong hereditary aristocracies in the earlier period, and a revered but powerless emperor, S was clearly low.

The evidence is not strong enough to pinpoint the peak of C, though it probably occurred around the fourth century BC. There are also indications of high V and stress at this time, or possibly even earlier. Rulers in the fourth and third centuries BC imposed exceptionally harsh punishments, as advocated by Legalist philosophy. Whole armies could be wiped out, something not recorded in earlier times. Political instability seems to have been highest in the fifth century, when the rulers of Jin, Qi and possibly Sung were deposed by leading families within their state. Rulers throughout the Warring States period (464–222 BC) also demanded much greater deference than in the past.

The Shang: low S but low infant C

At this point it must be asked why the earlier Shang dynasty showed no evidence of high infant C, even though as never civilized peoples they would presumably have had low S. The Shang period was not outstandingly creative or commercially successful, never set up powerful and unified states, and never went through a prolonged collapse.

The obvious reason is that the Chinese had not yet developed the more effective cultural technologies of later times. As discussed in chapter three, though C-promoting traditions would have been developing since the domestication of plants made them useful in the Neolithic period, their development must have sped up greatly once civilization got under way. Dense farming populations, the development of trade and the need for public order would have given a strong competitive advantage to people with higher C. This would be aided by the rise of a priestly class who could codify and impose forms of behavior associated with religious practice. If these behaviors increased C and V then the people would be more successful and their deity seen as more powerful, causing other peoples to accept it. Thus there was competition between different kingdoms, as well as between families and other groups within each kingdom, which caused C and V to rise over many generations.

In the Shang period, this process cultural evolution was at a very early stage so that C and thus infant C never rose beyond very modest levels.

The First Chinese Empire

In the Zhou period, as discussed earlier, much higher levels of C and thus infant C were achieved. But by the third century BC, C was in decline. A clear sign was the weakening of local loyalties. In 256 the last Zhou emperor was deposed. Between 238 and 221 BC the state of Qin overcame the remaining great states of Han, Zhao, Wei, Chu, Yan and Qi, and the Qin ruler took the title of Shi Huang di, meaning "First Emperor." After centuries of independence, maintained at huge cost in men and material, the speed and relative ease of this final conquest is striking.

Mass revolts followed the death of the First Emperor, and after this a new dynasty known as the Han came to power. To conciliate rival generals it re-established many of the old kingdoms, an indication that the old local loyalties retained some force so a measure of local autonomy was desirable. But successive emperors gradually took back control until all were gone by the end of the second century BC. The local loyalties characteristic of high infant C were no longer strong enough to threaten the unity of the empire.

During this century, Chinese power and wealth were at an all-time high. C was falling but still high enough to support imperial authority, and without internal divisions it gained control over outlying lands. Economic growth continued, and this is considered one of the great periods of Chinese history.[5] An entirely exact comparison is the way the Greeks reached their peak of power after the conquests of Alexander, made possible only by the weakening of local loyalties in Greece itself.

But as C and child V continued to fall in the first century BC, the strong centralized state began to weaken. The tax-free estates of great families were expanding rapidly, leading to a crisis in government finance after mid-century.

The problem was not helped by a series of weak emperors, but one particular episode demonstrates that the underlying causes were social rather than political. In 9 AD a strong and capable ruler seized power.

---

[5] J. Gernet, *A History of Chinese Civilization* (Cambridge: Cambridge University Press, 1990), 138.

Wang Mang attempted, among other reforms, to break up the family estates which were robbing the central government of income. He enforced government monopolies and instituted a policy of agriculture loans to support the hard-pressed peasantry. But all this proved futile, and in 23 AD a series of uprisings swept him away.

Wang Mang failed because the problem was not a lack of political will but that as C declines, loyalties become more local and personal. In the later Roman Empire this not only led to the growth of self-contained estates but allowed generals, who could rely on the personal loyalty of their men, to rebel against imperial authority. In China the two issues were combined. The great families not only commanded local loyalties but ran their own private armies. The new Later Han dynasty which emerged in 25 AD was, effectively, a coalition of such families and never as strong as the Former Han. After 88 AD central authority was in decline, and had disappeared long before the formal end of the dynasty in 220 AD.

Declining C also affected economic activity. Trade may have increased under the united empire during the second century BC, perhaps a couple of centuries after the peak of C. This is the same pattern we saw in the Roman Republic and early Empire. Because V falls faster and earlier than C, the loss of the conservative child V compensates for the decline in C, so trade and creative thought may continue to flourish for a time. But by the first century BC this was no longer the case. Over the following centuries the great estates became more and more self-sufficient, trade declined, and coinage virtually went out of use in some areas. This, too, is exactly what happened in the Roman Empire, and for exactly the same reasons.

The decline of creativity happened even earlier, and has traditionally been put down to the infamous "burning of the books" by the First Emperor at the close of the third century BC. But again, a more fundamental explanation is the loss of infant C. Biohistory suggests that the intense creativity of the third century resulted from a marked fall in child V, even though historical accounts of the Qin Empire show no signs of slackening brutality (on the other hand, the Qin state was seen as backward and barbaric, so its policies may not have been representative of the whole). As in Rome during the "golden age" of the first century BC, this fall in the conservative child V more than made up for the initial loss of infant C.

Creativity was not totally lost in the early empire. Historians such as the great Sima Qian (145–86 BC) set a pattern for future ages to follow, though he suffered greatly for his craft. Falling foul of the brutal Emperor

Wu di, he accepted castration over exile so as to finish his work. But there was a distinct falling off from the intellectual ferment of the previous century. In the united China of the second century BC, the fall of infant C was starting to have an impact. There is a close parallel to the "silver age" of the early Roman Empire which followed the golden age of the late Republic. And by the end of the first century BC, as in the later Roman Empire, even this level of innovation was no longer seen.

There is also evidence of low V. Mass peasant levies had been characteristic of the Warring States armies, but by the close of the second century BC they were composed largely of criminals who suffered losses of up to 90% in a series of campaigns against the nomad pastoralists of the north. By the first century BC the balance had shifted to defense and a military weakness that persisted for several centuries. The harsh punishments and penalties of earlier times were also relaxed.

The end of the empire

By the end of the first century BC, as mentioned earlier, the government was losing its grip. There was a brief recovery in the first century AD under the Later Han dynasty, but the great families with their tax-free estates and private armies remained largely outside government control. By the end of the second century AD the empire collapsed into several centuries of chaos and disorder. It was by far the longest Dark Age in all of Chinese history.

The fall of most Chinese dynasties, including the former Han dynasty, can be explained in terms of lemming cycles, as will be discussed later. But the crisis that gripped China from the second century AD was far more serious. It has close parallels with the fall of Rome, which happened only a short time later. Not only political authority and military effectiveness but trade, learning and virtually every aspect of civilization were in remission for around five hundred years. In other words, what occurred was a collapse of V and C in a typical civilization cycle.

The reason for the collapse is exactly the same as the reason for the creative brilliance of the Zhou period. This was a low-S society, and low S permits high levels of infant C, which creates a civilization that is innovative but unstable. High infant C undermines V, causing a catastrophic fall of C and the collapse of that civilization.

The Chinese Dark Ages and rising S

Despite the initial similarities, the later history of China was very different to that of Europe after the fall of Rome, and the reason is that in China S rose far more. This was because, in the first place, the collapse of China was not as extreme. The population fell much less than in the Roman Empire. There were no massive slave estates with dispossessed peasants flocking to the cities. And while the population of north China declined, that of south China actually expanded during the centuries of disorder.

Also, barbarian invasions played very little part in the collapse of central power in China, largely because the nomads of the northern steppes were far less numerous than the German tribes crossing the Rhine in the late fourth and early fifth century AD. The Germans were able to flood in because the fall of C had removed the impersonal loyalties that sustained the Empire, and the fall of V had taken their military vigor. But the fact remains that when the Empire collapsed, the barbarians were waiting to take advantage, and there were a great many of them.

Chinese had settled about as far north as agriculture could reach, so there were no broad and fertile lands outside the Chinese empire with populations who had experienced no collapse in C and V. The lands to the south were thinly populated, and so largely settled by ethnic Chinese in the centuries to come.

Thus, barbarians played a surprisingly small role in the collapse of the first Chinese Empire. In fact, all the immediate successor states of the Later Han were ruled by native Chinese, though barbarian horsemen fought on all sides and it was a Xiongnu army which sacked Luoyang in 311 AD, sometimes likened to the sack of Rome by Alaric a century later. So, although barbarian dynasties fought and feuded in north China for hundreds of years, their contribution to China's ethnic makeup was far less than that of German settlers after the fall of Rome.

What this means is that the selection pressures resulting from the fall of C and V were strongly in favor of high S, and with few low S barbarians to dilute it, there was a substantial increase in S.

By the fifth century AD, C and V were once more on the rise. The barbarian northern Wei dynasty gained firm control of north China for several decades, and in 485 began distributing land to free peasants, which helped to stop the growth of large estates. But there was no return to the era of the Warring States. No stable nation states were established in north China, and there was no revered but powerless "Son of Heaven." Instead, dynasties rose and fell with startling speed. For example, the barbarian Qin dynasty erupted from the Wei valley in 351 and by 381 had taken all of north China, plus Sichuan. They sent a massive force against the south in 385 but were mysteriously defeated, after which the regime fell apart.

Eventually, the empire was reunited under the Sui (589–618) and Tang (618–907) dynasties, and from then on strong dynasties alternated with periods of chaos—the typical high-S pattern referred to in the last chapter. Rather than favoring aristocrats, as in a low-S society, these dynasties began selecting bureaucrats by competitive examination. The Chinese were changing from a low-S to a high-S people.

Rising C without infant C

Despite the fact that infant C never rose to its previous height in the Zhou period, there is evidence of a gradual increase in C. Trade and the use of money revived from the sixth and seventh centuries and grew rapidly from the eighth to the thirteenth centuries. Cities came to be based on trade, rather than administration as in the past. Political stability increased, with shorter periods of disorder, also reflecting a rise in child V.

But C- and V-promoting traditions remained relatively weak, initially. Sexual behavior in the southern courts during the centuries of disorder seems to have been very free. And there was no obvious change under the Tang dynasty, at least in the cities. It was the custom for successful examination candidates at the capital to give a feast in the brothel quarter.

Women were also relatively free under the Tang. They could gain power and wealth as Daoist priestesses, brothel owners and prominent courtesans, and are even portrayed as playing polo on horseback.[6] They played an influential role in politics, not least the Empress Wu of the late seventh century who briefly made herself China's sole female ruler. Other influential women of the early eighth century included the Empress Wei, who had women appointed to influential positions, Princess Taiping, and Yang Guifei. All of this was deeply shocking to Confucian sensibilities—the Empress Wu was deposed and the others all killed, but they at least held power for a time. The nearest China came again to a female emperor

---

[6] C. Benn, *China's Golden Age: Everyday Life in the Tang Dynasty* (Oxford: Oxford University Press, 2002), 60–66.

was in 1976 when Mao's widow Jiang Qing tried to seize power through the "Gang of Four", with equally unfortunate results.

But attitudes to sex became stricter with time, closely linked with the subordination of women and the resurgence of Confucianism. It was during the Northern Sung Dynasty (960–1126) that women began to be secluded, separated from men, have their feet bound, and be portrayed in a more fragile manner.[7] Clothing became more modest. This process accelerated with the triumph of Neo-Confucianism in the thirteenth century and under the Mongols (1239–1368). Meanwhile, the Mongol conquest and the occupation of north China by barbarian invaders in the previous century worked to raise the level of V, and thus child V.

Most of our evidence applies to the elite, but Confucian attitudes to sex seem to have been spreading to other classes by at least the Ming Dynasty (1368–1644). In the Qing period (1644–1911) the pornographic novel was suppressed and sex, according to some accounts, came to be seen as more a burden than a joy. Ethnographic studies of traditional villages in the twentieth century reveal an intensely conservative attitude to sex.

A continuing rise in S after the Tang

Along with the rise in C there are indications of a continuing rise in S from the Tang period onwards. Even though S was considerably higher after the collapse and rebuilding of Chinese civilization, it was still lower than modern levels. The Tang period did not show the intellectual ferment of the Warring States period but it was brilliant and creative in literature and the arts.[8]

Later periods, especially from the thirteenth century onward, were both less creative and more rigid and orthodox in their thinking. Charles Murray found a clear decline in important painters and especially poets from the Tang dynasty peak (eighth and ninth centuries), with few philosophers of note after the twelfth century.[9] This is completely opposite to the European pattern, as discussed in chapter six, where creative accomplishment rose to a peak in the nineteenth century.

[7] P. B. Ebrey, *Women and the Family in Chinese History* (London, New York: Routledge 2003).
[8] Benn, *China's Golden Age*.
[9] Charles Murray, *Human Accomplishment* (New York: Harper Collins, 2009), Chapter 14.

Decline of the hereditary principle

Another way to track the rise of S is to follow the decline of the hereditary principle. We have seen that in the Zhou period this was exceptionally strong, with a powerless monarch reigning for more than five hundred years. They were deposed in the third century, but the ruling Liu family of the Former Han dynasty (206 BC–8 AD) enjoyed considerable prestige even when its power was gone. The Wang family was in effective control for several decades before Wang Mang dared take the throne in 9 AD, and he did not last long. In the subsequent chaos nearly all those who vied for power claimed descent from the Han imperial family. Among these was Liu Xiu, who eventually took the throne as founder of the Later Han Dynasty.

Even after this dynasty fell in 220 AD imperial descent continued to matter. One of the three successor states was founded by Liu Bei in what is now Szechuan. Though growing up in poverty he too claimed descent from the Han imperial house, a considerable asset in his rise to power. Even in the early Tang, hereditary descent made a difference. Many if not most of the leading figures in the seventh century, including the Sui and Tang founders, came from the northwestern aristocracy.

By comparison, elites in later times tended to gain wealth initially through trade or landholding, giving them the resources to train sons for the imperial bureaucracy, and to use this as a road to greater wealth and power. But even the most successful families only tended to maintain themselves in power for two to three generations. And local loyalties to hereditary families—the essence of aristocratic power—had very little to do with it. Power was achieved not by family connections but by success in the imperial examinations. On this basis we can see a steady fall in hereditary loyalties and thus a rise in S from the Warring States period, to the Han dynasties, to the Tang, to the fourteenth century and after.

Rising S is also evident in China's military situation. Under the Han dynasty, as in the Roman Empire, infant C (made possible by low S) eventually undermined V and led to military collapse, aided of course by the breakdown of the state as a result of falling C. China also became militarily weaker during the Song period. It never even controlled the area around Beijing, and from the early twelfth century lost first north China and then (a century and a half later) the south to the barbarians.

But this was not the military weakness of the collapsing Han dynasty. The

Southern Song dynasty resisted the extraordinary might of the Mongol Empire for several decades, and later regimes were still more successful. The Ming, Qing and Communists did far better than the Song in controlling the non-Chinese areas to the north and west. This is an indication that V collapsed very badly under the Han, much less under the Song, and still less in later periods. All of this is consistent with a steady rise in S, limiting the level of infant C, which in turn allowed V to be maintained.

The Chinese Civilization Cycle

In Europe and Japan, rising C was accompanied by a high level of V and stress, and there is some evidence of such a pattern in China. Punishments seem to have been unusually severe under the Sui (581–618) and Ming (1368–1644) dynasties. Significantly, both these dynasties followed periods of "barbarian" rule, which would have worked to raise the level of V. The Tang (618–907) and Ming dynasties were also relatively strong in a military sense, another sign of higher V. The Song dynasty (960–1279) was both more humane in its laws and also militarily weak. This pattern, which is admittedly tentative, can be summarized as in Fig. 14.2 below.

Fig. 14.2. Civilization Cycle in China. V rose during the Chinese Dark Ages of the third to sixth centuries AD, and in times of barbarian incursion.

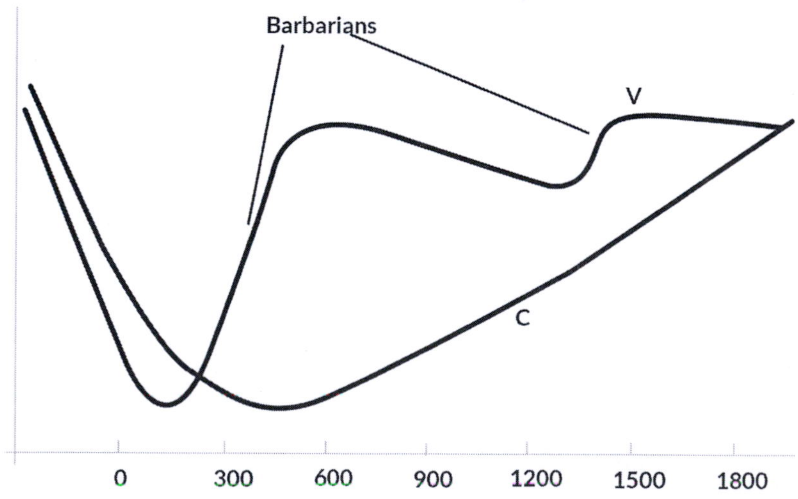

The south would show a similar pattern, except that with an easier climate and shorter periods of barbarian rule, V was lower. This is why the armies which brought dynasties to power usually came from the north (Qin, Sui, Tang, Song, Communists), as indicated in chapter eleven. Southerners, being lower V and thus with less of the conservative and anti-commercial child V, probably also had higher C and even infant C and were thus better at trade. This is also why the economic growth of China in recent decades has been strongest in the south, and why the city-states of Hong Kong and Singapore—largely settled by southern Chinese—have been so successful.

Why did C continue to rise?

One question to consider is what drove the continuing rise in C over two-thousand years of Chinese history. The civilization cycle covers this quite well up to about 700 AD, when the recovery of C coincided with a high level of stress. But after this the pattern breaks down. In fact, the strongest evidence for rising C can be found in the Song dynasty, when the level of V was relatively low and declining.

The most likely reason is that after 700 AD it was the process of competition between families and kinship groups which gradually raised C. In a stable, densely populated society on the edge of subsistence, more industrious farmers and merchants have an obvious edge over the less disciplined. A landless worker might raise one child to adulthood if he is lucky, a well-off peasant three or four, a wealthy landowner with a concubine eight or ten, unless moving to the city, in which case both V and fertility would quickly fall.

A far more difficult problem is what drove the continuing rise in S. When C is collapsing there is strong selection for people who can resist it, stay on the farm and hold to traditional ways. But China from the seventh century onwards was a stable, agrarian and increasingly conservative society. There was not the mass movement to the city found in ancient Rome or the modern west. A higher level of infant C would presumably have given economic advantages, allowing people to have more surviving children as they did in early modern Europe (see chapter six).

The most likely answer is that the reasons were political. In a low S society with generally strong feudal and then national loyalties, such loyalties were no hindrance to success and survival. But once S had risen during the collapse and the balance shifted towards cosmopolitan empires, intense local loyalties could be dangerous and even suicidal. Given the bloody vengeance inflicted by emperors on rebellious subjects, it might be wiser and safer to submit to the strongest power around. This may also be why Russia and other countries in Eastern Europe appear to have had higher S than Western Europe, having been subject to a brutal Mongol regime for two centuries.

Chinese Lemming Cycles

A survey of Chinese history would not be complete without an analysis of lemming cycles which, as mentioned previously, are especially obvious because of their connection with the rise and fall of Chinese dynasties.

We have seen that the clearest sign of a lemming cycle is a slow decline into a period of anarchy or localized conflict at the G-90 period, followed by a rapid restoration of central authority. This pattern is easy to spot in China. Chinese dynasties tended to decline gradually until they fell apart in a period of civil strife, followed by a fast resurgence under a new dynasty. To traditional Chinese historians, this is the key theme of history. In their view, the decay of central power reflects the decline in the quality of the emperors.[10] Moral strength leads to political strength. Moral weakness brings on disorder. A heavy weight to put on one man!

Thankfully, there is no need to focus on individuals. Major Chinese dynasties follow a classic lemming-cycle pattern, with central government tending to collapse in the G-90 period and then quickly recovering power under the new regime. Except for the Former Han, the G years fall in the first century of every dynasty that ruled China for more than 150 years.

We can tentatively place an early G year around 470 BC. In the early-sixth century a high proportion of men mentioned in Chinese histories were from great but not noble families, a rough index of aristocratic power, indicating a G-90 year around 560. The proportion of aristocrats mentioned fell rapidly in the early fifth century, suggesting an increase in royal authority (see Fig. 14.3 below). This occurred not in China as a whole but in the independent states into which northern China was then divided.

[10] Reischauer & Fairbank, *East Asia: The Great Tradition*, 115, 184.

Fig. 14.3. Percentage of prominent men from great but not royal families in China, 722–253 BC.[11] A high proportion of men from great families in the mid-sixth and early third centuries is consistent with aristocratic power and thus G-90 years around 560 and 310 BC.

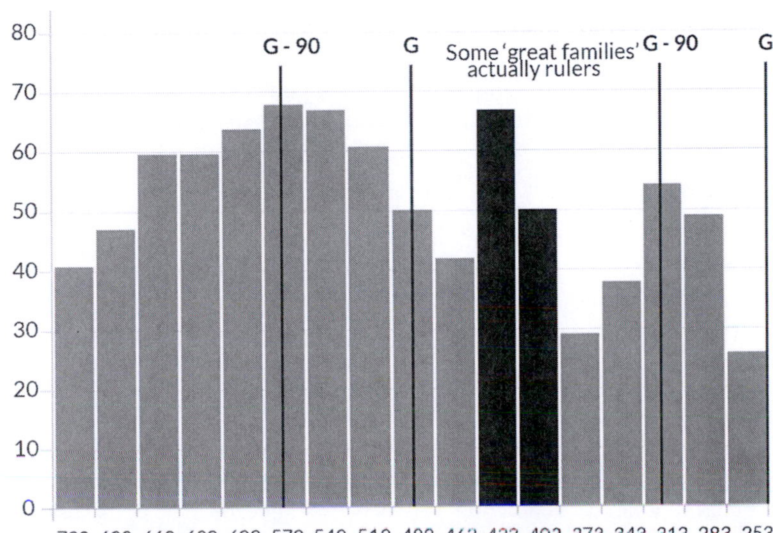

Another sign of the lemming cycle is that the late sixth and early fifth centuries BC comprised the first golden age of Chinese philosophy. Confucius lived between 551 and 479 BC and Laozi probably a bit earlier, in the sixth century. These dates are consistent with the creativity and fresh thinking of a G-60 year in 530 BC.

The next G year can be placed more reliably around 210 BC. The proportion of men from great but not royal families fell steadily from the sixth to the late third centuries BC, as Chinese states became more centrally administered and hereditary nobles had less power. But there are two major exceptions to this pattern: one in the late fifth century and another round 300 BC. The first of these (the lighter color in Fig. 14.3 above) can be largely discounted. This was the time when the rulers of Jin, a major northerly state, had been eclipsed by three great families who were in fact, though not yet in name, independent rulers. The same thing had

---

[11] Hsu Cho-yun, Ancient China in Transition (Stanford: Stanford University Press, 1965), 39.

happened in at least one other state. Remove these from the relatively small number of men mentioned in this period (15, versus an average of 44 for other periods), and we eliminate this peak.

The second exception seems to indicate a brief rise in aristocratic power. The whole trend of these centuries was to increasingly centralized power, and yet for a brief period around 300 BC most prominent men were from great rather than royal families. This was a pattern across the many states into which China was then divided, and is consistent with a G-90 period around 300 BC.

There is other evidence for a G period in the late third century. Since the eighth century BC there had been a gradual change from aristocratic armies to mass peasant levies. This development reached a peak in the third century, with armies ten times as large as in the earlier period. The relative willingness of ordinary men to bear arms is characteristic of G periods.

Another sign is an extraordinary flowering of Chinese philosophy. Over three thousand years of Chinese history, only seventeen philosophers have been given their own section in a standard text.[12] Six lived in what amounts to a single generation between 330 and 263 BC, consistent with the explosion of new ideas we typically find at G-60 (270 BC). This intellectual ferment was suddenly and notoriously brought to an end by the new ruler of united China at the end of the century, but the lemming cycle provides another explanation for this change, in that thought tends to become more orthodox after the G period.

The first few decades of the Former Han dynasty were politically stable, with evidence of declining local loyalties. The local states, which had kept some independence, were finally abolished after a revolt in 154 BC. This is consistent with a low point of local loyalties at the G+60 period of 160 BC. The reign of Wu ti (141–87) was the high point of Han military power, with massive wars of conquest into central Asia. But these armies did not have the military élan and effectiveness of the G period. They were largely composed of criminals and suffered terrible losses.

From then on the decline in government power was steady and, in a sense, went on for centuries. As discussed earlier, this was less a typical dynastic

[12] Y.-l. Fung & Y. Feng, *A Short History of Chinese Philosophy* (New York: Free Press, 1997).

decline than the collapse of Chinese civilization into a Dark Age following the collapse of V and C. But there was, significantly, a brief reversal to this decline. After a near collapse of government power at the end of the century, central authority was restored to some extent by Wang Mang (9 AD–23 AD) and more effectively by the later Han dynasty, which reached a peak around 80 AD. This is consistent with a G period around 80 AD and a G-90 period at the low point of government power around 10 BC.

From then on the decline continued until the fourth century AD, when north China was divided into a number of barbarian states. Then the recovery began. But just as a lemming cycle temporarily reversed the fall in the first century AD, so another one temporarily reversed the rise in the early sixth century. North China was united under the northern Wei dynasty in 439–524 AD, divided again between 524–77, and finally united and joined with the south by 589. There followed the resplendent Tang dynasty (618–907), usually considered the greatest in Chinese history. Placing a G year in the powerful first century of the Tang, at 650 AD, allows a G-90 year in 560, explaining the temporary setback of the mid-sixth century.

The seventh century also showed the military vigor typical of G periods. A militia system had been set up in the mid-sixth century, reaching a peak in the seventh. Military service held considerable prestige at the time and was one of the major routes to social advancement. The success of relatively small armies in the north and north-west shows their effectiveness. But enthusiasm for military service began to ebb after about 660, and the high point of military success in the 740s was achieved by professional armies including many barbarians. As is very often the case, the G+60 period around 710 saw declining national energies balanced by increasing absolutism, as local loyalties fell to a low point. Immediately following this came the last strong Tang ruler, the Emperor Xuanzong (712–756), whose reign ended in rebellion and disaster as the lemming cycle progressed past G+90.

The slow decline of the Tang into the anarchy of the early tenth century and resurgence under the Sung fits a G-90 year of 930 and a G year of 1020. It also explains the relative military strength of the early Sung, even though this was not at Tang levels. But it was not to last. Central power decayed steadily from the late-eleventh century, allowing steppe nomads to move in and take control. North China was conquered in 1126, and the whole of China by Mongols in the thirteenth century. In other words, the dynasty did not so much collapse as find itself unable to resist barbarians of unprecedented power.

The fate of the Mongol Yuan dynasty shows what happens when a regime takes power "too early." The Mongol conquest, completed in 1279, took place half a century before the expected peak of anarchy from the G-90 year (1330). The simmering discontent noted by Marco Polo was kept down only by the immensely powerful Mongol army. The last strong Mongol ruler died in 1307, to be followed by seven rulers in twenty-six years and outright civil war after 1328. In other words, a dynasty supported by the most powerful army ever lasted less than fifty years because it hit the decline phase of the lemming cycle.

The Manchus, by contrast, arrived just after the expected peak of turbulence in the next cycle (1630). Despite being equally alien and with an army vastly smaller than that of the Mongols, their Qing dynasty survived early revolts to rule China for almost three-hundred years.

Both Ming and Qing dynasties fit the lemming cycle pattern perfectly. Both rose to power immediately after the chaos of the G-90 period, both were militarily effective in the early stages, and both showed a steady decline after their first centuries. The great Ming voyages of exploration between 1403 and 1419 are also typical of the energy and dynamism of a G period around 1420. The rise of the Communist regime can be seen in much the same way. But before the Communist victory came the chaotic Warlord Era of the 1920s, which provides a valuable case study of G-90 disorder at very close hand.

The Warlord Era in China—case study of a G-90 period

A study of Chinese warlord armies in the 1920s shows how the political instability of the period can be explained by psychological attitudes.[13] There has been no hereditary landed nobility in China for well over a thousand years, for reasons discussed earlier. But warlord armies showed the same intensely local and personal loyalties as the people of fifteenth-century England. Soldiers were strikingly loyal to their own commanders. Cases of individual treachery and willful desertion were rare, except when an entire unit changed sides under direct orders of its commander.

It was this feature that made the politics of the time so confused. Common soldiers owed allegiance to their own leader, not to the warlord himself and not to any particular ideology. The local commander himself had a

---

[13] L. W. Pye, *Warlord Politics: Conflict and Coalition in the Modernization of Republican China* (New York: Praeger Publishers, 1971), 114–115.

personal tie with his own superior, the warlord, who would tend to support one of the most prominent national figures. Because of this, a defeated warlord might be quickly abandoned by his followers.

History focuses on powerful leaders because those are the people we know, but history makes better sense if we focus on the loyalty of the individual soldier, especially once we recognize that such loyalties may become more or less personal over successive generations. In a society with low infant C and high child V such as China, ordinary people normally accept the strongest locus of authority, which is the emperor or central government. Ambitious men may be willing to risk all for the sake of power but will find it hard to attract followers. What makes the chaotic periods different is that face-to-face and local loyalties are strong enough to take precedence, for a time, over acceptance of central authority. No large-scale authority is stable under such conditions, since the real focus of power is with lower-level commanders.

Warlord propaganda reflected these attitudes in that much of its focus was personal. It praised the warlord or attacked his enemies. Other themes tended to be bland statements calculated to have universal appeal such as the need for moral virtues, sound money and credit, and sympathy for the people. Nationalist and especially Communist propaganda put far more emphasis on concrete policies such as anti-imperialism and the revision of unequal treaties. Focus on policy rather than persons is a sign of more impersonal attitudes. Of course, by Western standards even Communist propaganda was relatively personal, as shown by the Mao cult and the demonization of his enemies in later times. Chinese loyalties are still more personal than in the West, but considerably *less* personal than they were ninety years ago.

The power of the warlords also depended at least in part on personal support from the broader community. A 1923 poll of students and businessmen, about the least likely groups to be supportive of warlords, found that no fewer than five of the most admired twelve living Chinese were from this group. Even people who deplored the chaos of the country often spoke of individual warlords with a great deal of respect.

But just as a shift of the younger generation towards personal loyalties and attitudes explains the collapse of central authority in the 1920s, so a gradual increase in impersonal attitudes explains the steady shift back towards central control. The Nationalist regime of the 1930s and early 1940s can be considered as intermediate between the Warlord and

Communist eras. This applies not only to its control over the country, which was improved but far from complete, but to the level of personal versus impersonal attitudes expressed in its propaganda. These were midway between the appeals of the Warlords, and those of the Communists.

Thus as each successive generation became more impersonal in its orientation, especially the young, so the balance of power shifted from warlords to Nationalists. By the late 1940s the balance was shifting still further, with the younger generation now more attuned to the Communist ideology. Thus, a political movement that had very little support in the 1920s became strong enough to take over the country by 1949.

Warlord propaganda can also be rated on its attitude to change, especially the acceptance of Western ideas. It was certainly less conservative than in the past, advocating the end of the Imperial system, for example. But there was still a focus on government by good men, a mainstay of the traditional Confucian system, and a sense of caution about Western education. The Nationalists and especially the Communists were far more committed to a radical remake of Chinese society.

In the lemming cycle, acceptance of change is most characteristic of the G-60 period, which in Chinese terms came around 1950. In this sense all the dominant systems of the twentieth century—Imperial, Warlord, Nationalist, and Communist—represent a steadily more positive attitude to social change. This means that as the broader society became more accepting of radical change, so the movement best expressing that attitude did best in the struggle for supremacy.

The acceptance of Communist ideology was the biggest single change in Chinese thought since the adoption of Buddhism nearly two-thousand years earlier, exactly fitting the G-60 year of 1950, though some changes were not without precedent from earlier times. A radical redistribution of land known as the "equal field system" was most fully applied after the Sui Dynasty came to power in 589 AD, just at the G-60 period of 590. In effect, it curbed the power of great landowners and strengthened government authority by increasing tax revenues, not dissimilar to the goals of the new Communist government in setting up communes.

The equal field system broke down in the eighth century, allowing landowners to amass large estates and reducing direct government power. This is, in a sense, not dissimilar to what has begun to happen in China in recent times, though at a much earlier stage of the lemming cycle.

Regardless of this, the transition from the Warlord Era anarchy to a new and powerful centralized Communist "dynasty" is a perfect reflection of a lemming cycle pattern. It is notable that China in the early twenty-first century has all the vigor, confidence and unity of a typical G period centered on 2010, even though population growth is relatively low as a result of the one-child policy. This is a vivid confirmation of the principle, observed in muskrat populations, that cyclical changes in attitude persist even when the realities of population are quite different. For example, local muskrat populations "peak" and "collapse" in line with the ten-year cycle, even when local populations may be quite low as a result of drought. Chapter nine showed this same pattern, in that postwar population booms did not increase V in any measurable way, unlike earlier booms resulting from lemming cycle G periods.

We now have enough information to track the length of Chinese lemming cycles (see Table 8.6 below).

Table 8.6. Lemming cycles in China. There is a smooth transition from shorter cycles in the advanced civilization of ancient China, to longer cycles in the Dark Ages that followed, to shorter cycles as civilization was restored.

G-90 year	G year	Length in years
560 BC	470 BC	
300 BC	210 BC	260
10 BC	80 AD	310
560	650	570
930	1020	370
1330	1420	400
1630	1720	300
1920	2010	290

In chapter eight we saw that lemming cycles in England, France and Japan became longer in periods of rising V and stress in the civilization cycle. Chinese history shows the same pattern even more convincingly, with a transition from short cycles in periods of falling V to about 80 AD, to a long cycle as V rose between 80 and 650 AD, to progressively shorter cycles thereafter. Significantly, the one interruption to this pattern is the slightly longer cycle of 1020–1420 AD, which was the period when steppe nomads overran the Chinese Empire.

India

The history of the Indian subcontinent provides another example of a gradual change from low to high-S.

India is now clearly a high-S society. Take, for example, the treatment of children observed in a Rajput community in northern India. Though a busy mother may occasionally leave an infant to cry for a while, in most instances a crying baby will receive instant attention. If the mother is not available, other women or a grandmother will try to calm it, and if hungry will bring it to the mother for feeding. Once slightly older it is carried so consistently that it gets little chance or opportunity to crawl.[14] This can be compared to the much sterner, harsher treatment meted out to older children. A very similar pattern can be seen in a south Indian community, except that punishment of older children was somewhat less harsh.[15]

But, as with China, there are clear signs of an early, low-S civilization, uniting into a cosmopolitan empire and then collapsing into a long Dark Age, to be succeeded by a society with increasing levels of both C but not infant C. In other words, there is a transition to higher S.[16]

The first Indian civilization was that of the Indus Valley, and like the Shang they showed no evidence of high infant C such as large states or advanced commercial organization. They did, however, collapse, after which warlike chariot-riding Aryans invaded from the north, bringing with them heroic Vedic hymns. By 600 BC these semi-nomadic peoples had

---

[14] B. B. Whiting (ed.), *Six Cultures: Studies of Child Rearing* (New York and London: John Wiley and Sons, 1963), 513–518.
[15] S. C. Dube, *Indian Village* (London: Routledge & Kegan Paul, 1965).
[16] Indian historical information from: R. Thapar, *A history of India*. Vol. 1 (Harmondsworth: Penguin Books, 1966).

settled as farmers and a new civilization emerged.[17] It comprised sixteen states covering the Ganges plain, north to the foothills of the Himalayas, and northwest into the modern Punjab. Many of these states kept their identities for several hundred years, and some had republican forms of government. As in ancient China and Greece, this is the characteristic political structure of a low-S society.

It was a time of prosperity and growth, with expanding trade and cities and the introduction of money. It was also the most creative period of Indian thought, with widespread reaction against traditional beliefs involving ritual and animal sacrifice.[18] The founders of both Buddhism and Jainism lived in the sixth century BC. Other schools of thought were deterministic (believing that the future was predestined and nothing could change it), or totally materialistic (believing in no gods at all). Buddhism and Jainism, especially the latter with its asceticism and stress on honesty, were highly ethical religions. These features also indicate high infant C made possible by low S.

But it is also indicative of low S that this civilization, like that of ancient China, proved highly unstable. At the end of the fourth century BC it was united by the Mauryan Empire, which soon came to control most of the subcontinent. By the third century BC the Empire was showing increasing signs of pacifism, as in China in the first century BC, a clear indication of falling V. The Mauryan Empire reached its peak under Ashoka Maurya, who reigned from around 269 to 232 BC, but disintegrated rapidly after his death, with the last Mauryan ruler being assassinated in 185 BC. There followed nearly six centuries of weakness and disorder, with nomadic invasions from the north. Again, as in ancient China, this strongly suggests a civilization collapsing as the result of falling C and V.

Emerging from this extended period of anarchy, the India of the fourth century AD onwards was not one of nation states like those of the Ganges civilization, but of strong empires alternating with periods of disorder. Compared to China it had less experience of empire and more of disorder, but the overall pattern was the same. Also, like China it was increasingly conservative.

---

[17] P. Olivelle, *Between the Empires: Society in India 300 BC to 400 AD* (Oxford: Oxford University Press, 2006), 36.

[18] S. K. Belvalkar & R. D. Ranade, *History of Indian Philosophy: The Creative Period* (New Delhi: South Asia Books, 1996).

All this indicates rising S, which rose for the same reason it had in China. India under the Mauryans was not radically depopulated, as was the Roman Empire, and the barbarians were relatively few. Again, as in China, the reasons can be found in geography. India is bounded by mountains and deserts in the north, and sea to the west and east. There were no dense farming populations in neighboring lands to migrate *en masse* as the Empire weakened. Thus, when selection pressures raised the level of S, as a reaction to the decline in V and C, there were few barbarians to dilute the genetic change.

There are also indications of an ongoing rise in S after this time. In terms of creativity, India in the fifth and sixth centuries AD was midway between the earlier Ganges civilization and the more conventional India of later centuries. While not as innovative in terms of philosophy and thought as the pre-Mauryan period, it was the classical age of Indian civilization in terms of architecture, sculpture, literature and the arts. In this there is a striking similarity with Tang China, also a golden age for the arts but not as fertile philosophically as the Zhou period.

The Mauryan period was also a great age for science. This was the time when Indians developed the concept of zero, a revolutionary idea which forms the basis of our decimal system. In a book written in 499 AD the astronomer Aryabhata calculated the value of pi and the length of the solar year with remarkably accuracy. He worked out that the earth was a sphere and rotated on its axis, and established the cause of eclipses.[19] All this creativity is an indication of at least moderate infant C.

But S continued to rise as the centuries passed. Instead of leading to a scientific revolution, Aryabhata's theories were opposed by later astronomers, who preferred to compromise with the demands of tradition and religion. India was becoming less open to new ideas. By the tenth century, scientific works such as medical texts were largely commentaries on earlier works, with little reference to empirical knowledge. Astronomy came to be regarded almost as a branch of astrology. The literature of this later period has been described as imitative and pedantic, an indication of a less creative society.

[19] Other notable mathematicians of the period included Varahamihira, Brahmagupta, Bhaskara I, Mahavira, Bhaskara II, Madhava of Sangamagrama and Nilakantha Somayaji; G. G. Joseph, *The Crest of the Peacock: The Non-European Roots of Mathematics* (Princeton, NJ: Princeton University Press, 2000), 372–403.

Only in the south (again as in China) was this trend less advanced. There were states with relatively stable boundaries from the sixth to the eighth centuries, and between the tenth and twelfth centuries the south was more innovative in terms of philosophy, religion and trade. But the trend was the same.

Another sign of high S is acceptance of alien rulers. From the early eleventh century, north India was overrun by waves of invaders, especially Afghans and Turks. The history of Bengal in the late fifteenth and early sixteenth centuries gives a striking example of this process. It started with a revolt by the Abyssinian palace guards, allowing their leader to take the throne. The Abyssinians were replaced by an adventurer of Arab descent, and then in 1538 by an Afghan nobleman fleeing war further west. The general population seems to have accepted these changes with apparent indifference.

In the sixteenth century most of India was conquered by the Mughals, Turkic peoples from the Central Asian steppes. After their empire collapsed it was seized with almost casual ease by the British. They overcame several serious challenges to their rule, including the Maratha Wars of 1775–1818, the Sikh Wars of 1845–49 and (the only challenge from within the British territories) the Indian Mutiny of 1857–58, all of which were defeated by armies composed largely of native troops.

Again, as in China, C and V continued to rise. An increasingly austere form of Hinduism gained ground, rolling back the Buddhist wave of earlier centuries. Brahmanical Hinduism, far more than Buddhism, lowered the status of women. And subordinating women is the single most effective way of maintaining V, and thus C, since it transmits anxiety to their infants (V) and controls their sexual behavior (C and V). Islam, the archetypal high-V religion, served the same function in much of northern India.

The success of high-S societies

People in the West tend to judge other cultures in terms of wealth, human rights and democracy. Poor and autocratic societies are seen as lagging behind. None of the high-S societies in the modern world have achieved Western levels in these terms, apart from city states such as Singapore and Hong Kong. China is growing more affluent but a dictatorship. India is democratic, but poor.

But biological success cannot be judged in such ways. Peoples that grow, spread and outcompete their neighbors must win out over those which decline in population and power. From this viewpoint, the success of the high-S, high-C cultures of India and China is beyond question. Not only have they absorbed and assimilated all invaders but their peoples have spread into many other countries, often becoming commercially successful because of the discipline arising from high C.

It should also be noted that there are important distinctions among high-S peoples. North Indians and Pakistanis appear to be higher V than south Indians, both more patriarchal and harsher in their treatment of older children.[20] This is likely to be a result of high S stemming from longer experience of civilization, plus greater exposure to warlike invaders from the harsher lands to the north. It is likely no accident that the two longest civilized areas of India, in the valleys of the Indus and Ganges, are now predominantly Muslim. As will be discussed in the next chapter, Islam is the cultural technology most supportive of V and the most antithetical to infant C.

The same can be said of the Chinese. Chinese can be intensely nationalistic, willing to accept foreign rule by the Qing but strongly resentful of it. They have also tended to be commercially successful—a sign of high C and probably also some level of infant C. Thus, Chinese, and especially southern Chinese, are also likely to rate somewhere between the low-S Japanese and the high-S Arabs, but probably closer to the latter.

High S it not incompatible with economic development. Though infant C was the key to the Industrial Revolution, high levels of C are also beneficial. And if C is high enough, even high-S people can have a substantial level of infant C. This is because S only causes people to be relatively indulgent of younger children. Thus it is that the most creative and brilliant people in the world today, whether judged by economic success or Nobel prizes per capita, are distinctly high S. These are European or Ashkenazi Jews.

---

[20] L. Minturn & J. T. Hitchcock, "The Rajputs of Khalapur, India," in *Six cultures: Studies of child rearing*, edited by B. B. Whiting (New York and London: John Wiley and Sons, 1963), 203–361; S. C. Dube, *Indian Village* (London: Routledge, 1965).

Jews and Judaism

Of all the peoples who have lived on Earth, the Jews of Eastern Europe have probably had the highest known level of C. This is the result of one of the most extraordinary pieces of cultural technology ever devised—Jewish ritual law. Jewish law as a C-promoter has already been discussed in chapter three, but it is worth emphasizing again.

Jewish law goes far beyond most other cultures in controlling sexual behavior. It applies to men as much as women, lacking the "double standard" still prevalent in nineteenth-century Europe, which gave greater sexual freedom to men. Adultery and homosexuality are abhorred. Masturbation is strongly condemned. Premarital sex is so guarded against that engaged couples may not touch each other. Even married couples are limited in their sexual activity. For the Orthodox, a strict taboo during menstruation is extended for another week until a woman has a ritual bath, thus outlawing sex for half the month. Couples sleep in separate beds. In traditional society women would shave their heads at marriage to reduce their dangerous attractiveness.

Added to this is a vast array of strictures covering diet, dress, Sabbath keeping and other requirements—a code that directs and controls behavior on a daily and even hourly basis. All of this has the effect of increasing C to the highest possible level. And it is backed by a practice of religious scholarship far more widespread than study of the Confucian classics ever was in China. Most men have some form of religious education, and full time scholarship has enormous prestige and is not a rare occupation. This focus on learning, and a sense of personal engagement with God, create a moral code far stronger than anything that could be achieved by social conformity.

C is a biological system that is a powerful aid to economic success. It makes people hard-working, disciplined, law-abiding and willing to sacrifice present consumption for future benefit. The enormous investment placed by Jews in religious observance and scholarship has been amply justified by the development of a character that has made them successful as a commercial minority among alien peoples. But there is another aspect to Jewish life that has helped to contribute to their success—they are distinctly higher S than other European peoples.

Jewish Childrearing

The first place to look for this is family patterns, and those we will look at are for the highest C Jewish societies—the traditional Jewish communities of Eastern Europe.[21] The American societies considered in chapter five exercised similar control at all ages and were notably lacking in indulgence of infants, frequently leaving them to cry. The Chinese, by contrast, were extremely indulgent to infants but much tougher with older children.

In this respect, traditional Jewish behavior was much more like that of the Egyptians and Chinese. Babies were almost never left unattended. If not carried they were rocked constantly, for hour after hour. Mothers talked to them constantly. A crying baby would be first fed and then fussed over, cuddled, comforted and attended to, never left to cry. Attention came not only from the mother but from older siblings and female relatives.

Also similar to Chinese behavior was the steady decrease in indulgence after infancy. Older children were loved but very rarely kissed or caressed. Babies were constantly praised to their face, but older children rarely. To do so would be considered as "spoiling" them. A clear impression is that these Jewish adults, at least in this society, had a much stronger emotional reaction to infants than to older children.

However, although Jews have high S and thus relatively low infant C compared with C, their very high C means that they do control infants.

> Long before [weaning], orderly toilet habits have probably been taught, for that begins some time after the 1st six months. The training is firm and even insistent ... A child is encouraged to good performance by verbal stimulus, and praised if he does well. If he misbehaves, sounds of disapproval are made. Also, if a child learns to speak before weaning, he is taught to say a blessing before taking the breast.[22]

This is nothing like the level of control exercised by nineteenth-century English nannies, or even by American parents in the early twentieth century, but it is far more than found in most cultures.

---

[21] The following treatment is derived from a study of traditional Jews in Eastern Europe: M. Zborowski & E. Herzog, *Life is with People: The Jewish Little-Town of Eastern Europe* (New York: International University Press, 1952).
[22] Ibid., 328.

The benefits of high S for Jews

In terms of our understanding of S as detailed in the last chapter, we would certainly *expect* Jews to be high S. They have been civilized, in the sense of having cities and an advanced written culture, since at least the time of King David three-thousand years ago. For 2,600 years, from the Babylonian exile in 587 BC to the foundation of Israel in 1949, they have lived under alien rulers. The last chapter suggested that one reason for a continuing rise in S, even in the absence of a civilization collapse, is that people in cosmopolitan empires with lower S and thus higher infant C have a tendency to revolt and be wiped out. The Jewish revolts against Rome in 66–73 AD, 115–117 AD and 132–136 AD, all suppressed with wholesale slaughter, are exactly the kind of events that would have this effect. The Jews who survived would tend to be those with higher S who were more tolerant of Roman rule.

But these survivors were now well suited to take on a role as a commercial minority. Judaism, which was at this time becoming far more rigorous and systematic under the leadership of the Pharisees, was a powerful C-promoter. And high S would also be advantageous if it allowed people to have high C *without* an equally high level of infant C. There are two reasons for this.

The first, as mentioned earlier, is that lower infant C makes people more tolerant of foreign rule, so less likely to riot or revolt. The second is that lower infant C makes people less open to new ideas. Control in later childhood also causes some increase in the tradition-minded child V, though not as much as punishment. Put together, lower infant C and higher child V make tradition more attractive and reduce the risk of assimilation.

Throughout European history there has been intense pressure on Jews to convert to Christianity, including mass pogroms and the threat of expulsion from the countries in which they took refuge. But enough remained Orthodox, and with a sufficiently high birth rate, for Jewish identity to persist and even to allow the recreation of their homeland after two thousand years—an achievement unparalleled in human history.

This is not the rigid traditionalism of an Egyptian village—a commercial minority needs flexibility of thought. Compared to Arabs, traditional Jews had lower child V because they were not as punitive in later childhood. They would have higher C and, as indicated earlier, a significant level of infant C. The drawback is that as V and thus child V fall in the modern

West, Jews are turning away from their faith in large numbers. Latest statistics show that a clear majority of US Jews are marrying non-Jews, casting doubt on the long-term survival of the Jewish people.[23]

Jewish temperament is thus a creation of Jewish law, as transmitted by the way a mother brings up her children. This is why Judaism is considered to be inherited from the mother—an oddity in what is otherwise a patriarchal culture.

Jews and Chinese—the parallels

Jews and Chinese, especially south Chinese, can be said to share a similar character, one based on a combination of high C and relatively high S. This accounts for a number of curious similarities between these two peoples.

The first is that both have enjoyed considerable success as commercial minorities. Not for nothing are the Chinese known as the "Jews of South-East Asia." As indicated earlier, the combination of high C with high S is ideal for this role.

The second is that the tradition of both peoples includes a reverence for scholarship—the Jews for religious studies and the Chinese for the Confucian classics. Both these disciplines have the advantage of reinforcing cultural traditions. As the levels of V and thus child V have fallen, a tradition of learning plus high C means that both groups are over represented in the ranks of university-educated professionals. For example, people with Chinese surnames were ten times over represented on the 1987 list of National Merit Scholarship semifinalists.[24] By contrast, people with higher infant C are more oriented towards technical and machine skills, which is why interest in tertiary education is lowest when infant C peaks in the lemming cycle (see chapter eight).

For Jews in particular, the fact that V falls before C in wealthy urban environments has had a particular consequence. As noted, Jews traditionally have extremely high C, moderately high infant C, and only moderate child V. The initial fall of child V, when Jews become partially

---

[23] "Pew survey: 6.8 million Jews, but majority intermarry," *The Times of Israel*, (Tuesday July 29, 2014).

[24] N. Weyl, *The Geography of American Achievement* (Washington DC: Scott Townsend, 1987).

assimilated and detached from their tradition, made possible an extraordinary flowering of creative potential.

One proposed reason for Jewish success in the modern world, at least for the Ashkenazi Jews of Europe, is that they are more intelligent as a result of specializing in occupations where intelligence is an advantage.[25] However, biohistory suggests that epigenetic factors are a more powerful explanation. Evidence for this is that in recent years the Jewish advantage has ebbed. Recent objective measures of academic performance suggest that Jews are only slightly over represented, relative to population, in the elite ranks. These include National Merit Scholarship results and admissions to Caltech (Fig. 14.4). [26]

Fig. 14.4. Jews as percentage of students winning awards in American academic competitions.

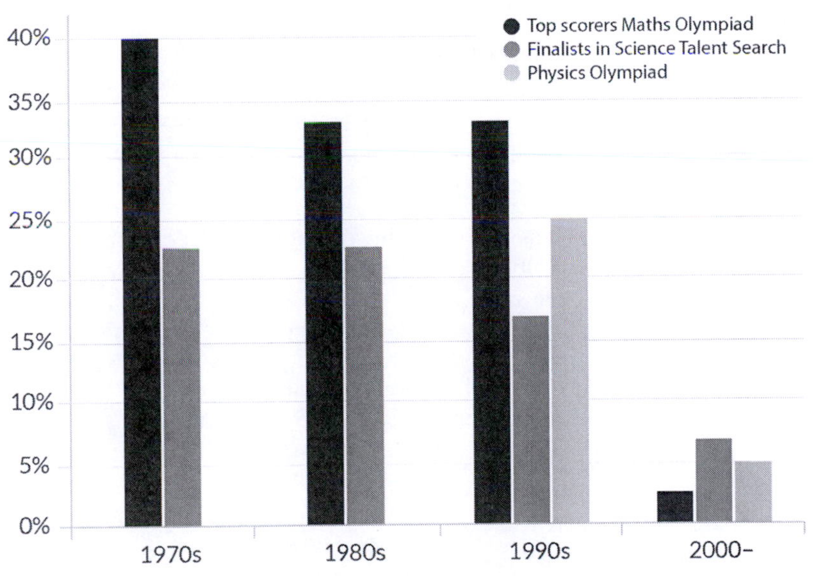

[25] G. Cochran, J. Hardy & H. Harpending, "Natural History of Ashkenazi Intelligence," *Journal of Biosocial Science* 38 (5) (2006): 659–93; C. Murray, *Jewish Genius*, Jewish Genealogy, March 2014.

[26] R. Unz, "The Myth of American Meritocracy," *The American Conservative* (2012): 14–51.

This collapse in performance brings up a point first made in the last chapter with reference to Sicilians—high S is no defense against falling V and C in a wealthy urban population. Indeed, the success of groups such as Chinese and Jews makes them unusually vulnerable. Thus, the Jewish birth rate in America is below average for the general population, and the birth-rate in Singapore has dropped to barely half the replacement level.[27]

A final point that Chinese and Jews share is an orientation towards left-wing causes, such as an active and powerful government, which is why the republic of China went Communist in 1949 and the state still plays a powerful role in economic affairs. Jews also lean towards left-wing causes. They were at the forefront of the Communist revolution in Europe. Karl Marx was Jewish, as was Trotsky and many of the early Bolsheviks. Even now, American Jews trend to the left politically, 78% voting for Barak Obama in 2008 and 69% in 2012.[28] This is especially striking given that they are generally more prosperous than average and, in that sense, natural Republicans.

This fits in with cross-cultural findings that link infant C to economic individualism. The Manus, the only people on record with high infant C but not child V, were extreme individualists with no political unit beyond the village. By contrast, lower infant C made possible the Israeli kibbutz movement, in which most property is shared.

What all this implies is that the Jewish law is not the outdated remnant of a less-rational age, as some people believe. It is the forge in which Jewish identity and success have been created, and has been the key to Jewish survival. Jews who abandon it must relatively soon cease to remain Jews.

Fig. 14.5 indicates how C and infant C may have changed over the past 2,500 years. In northern Europeans with low S, infant C has risen along with C and both remain roughly equal. For Jews with moderate S, C has

---

[27] *The Jewish Federations of North America*, “NJPS: Marriage and Fertility,” http://www.jewishfederations.org/page.aspx?id=46437 (accessed September 6, 2014);
Department of Statistics, Singapore,
http://www.singstat.gov.sg/statistics/browse_by_theme/birth.html (accessed Sept. 6, 2014).

[28] Y. Benhorin, "78% of American Jews vote Obama," YNet News, (2008), http://www.ynetnews.com/articles/0,7340,L-3618408,00.html (accessed September 6, 2014); M. Berenbaum, "Five Jewish takeaways from the 2012 Election," *Jewish Journal* http://www.jewishjournal.com/opinion/article/some_jewish_takeaways_from_the_2012_election (accessed September 6, 2012).

risen dramatically and infant C also, but not nearly as much. For Chinese with high S, C has risen while infant C may actually have fallen. The actual levels are speculative, pending physiological tests which should at least identify levels in modern communities.

Fig. 14.5. Child and infant C in northern Europeans, Jews and Chinese.

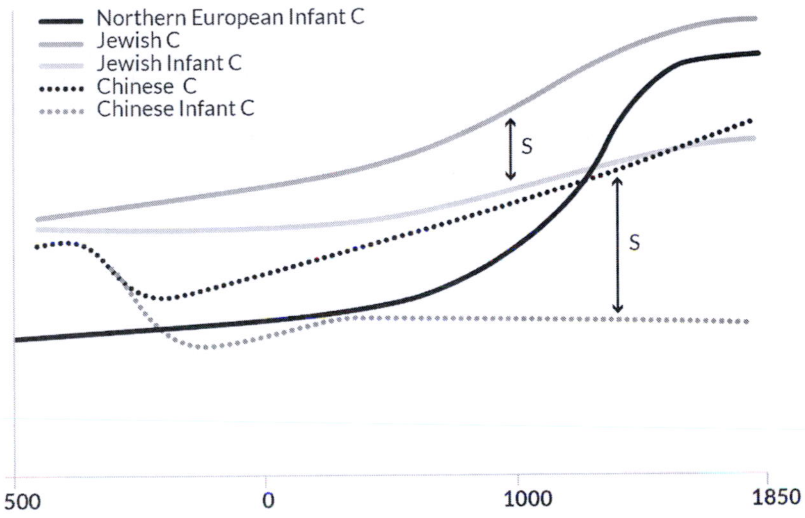

Genetics, epigenetics, and economic potential

It should be clear by now that although S is presumably a genetic factor and affects economic development, genes as such do not determine economic potential. High-S peoples can be wealthy, and low-S peoples such as Europeans of a thousand years ago desperately poor. In fact, high-S Chinese do considerably better in educational and financial terms than lower S whites, blacks and Native Americans.[29] The differences that really matter are not genetic but epigenetic.

In the next chapter we will follow changes in these factors in the Middle East, which will help us better understand the development and evolution of Judaism, Christianity and Islam, and their varying influences on the history of civilizations.

---

[29] 2010 ACS 1-year estimates http://factfinder2.census.gov/faces/nav/jsf/pages /searchresults.xhtml?refresh=t (accessed September 6, 2014).

CHAPTER FIFTEEN

THE TRIUMPH OF THE FUNDAMENTALISTS

The characteristic S (for Stability) is low when nations first become civilized, but rises when civilizations collapse as a result of a fall of V and C. Presumably genetic in nature, S appears to make people indulgent of infants but strict with older children. It thus prevents mothers with high C from giving their children an equivalent level of infant C. Lower infant C means that people tend to remain poorer and more conservative and thus retain higher V, so they have more children. Once S is high enough to allow the formation of large, cosmopolitan empires, there is a further selection for high S because low-S peoples are more likely to revolt.

The early civilizations of Greece, Rome, ancient India and China had characteristics of low S including powerful aristocracies and stable nation or city states. As C rose, these civilizations showed strong evidence of infant C such as highly centralized states and republican institutions, both indicative of impersonal loyalties. The people were also exceptionally open to new ideas and showed creative ferment, especially in the areas of philosophy and religion.

But despite their brilliance and short-term success, these low-S societies ended in collapse. One reason is that control of infants, which is what forms infant C, undermined V which eventually caused C to collapse. High infant C also tends to make societies wealthy and urban, which accelerated the process. Also, infant C caused people to be skeptical about tradition, undermining the C- and V-promoting traditions which support C and V.

With lower V they became militarily weak and their birth rates fell. With lower C, economic activity and creative thought declined, and states became less cohesive as personal loyalty to familiar local leaders replaces impersonal loyalties to more distant leaders or institutions. The result was centuries of anarchy and barbarian invasions.

In the process of collapse people with higher S, who tended to be more conservative and thus have more children, grew as a proportion of the population until the new society that rose from the collapse was of the more stable and conservative high S form, with lower infant C. These new civilizations were also aided by their adoption of new cultural systems such as Buddhism and Confucianism with more powerful C and V-promoters, which raised their child V to higher levels than before the collapse, making larger and more stable empires possible.

In the case of Europe, the new civilization that arose after the fall of Rome had the benefit of Christianity, which eventually made much higher levels of C and V possible. However, the Germanic peoples of Europe were never conquered by Rome and never went through a collapse, so their level of S remained low. Parts of the former Empire, such as north-eastern France and northern Italy, were so heavily colonized by Germanic barbarians so that their S also remained low. Thus in all of these areas, especially Northern Europe, when C rose again under the influence of Christianity it brought infant C to unprecedented heights. The result was industrialization and the worldwide predominance of nineteenth-century Europe. This civilization surpassed all others in terms of its economic and political scale and power.

In areas such as southern Italy and Sicily where the original population was less impacted by low S barbarians, S rose so that when C and V recovered, infant C did not. They thus remained poorer and more conservative and less resistant to foreign rule. In this sense they were like the Egyptians discussed in chapter five.

It is now time to return to the Middle East, where the earliest known civilizations arose. With the longest experience of civilization this area has developed what is almost certainly the highest level of S, and it has also given birth to the religion with the most powerful and effective V-promoters: Islam. This society, especially in its most extreme and fundamentalist forms, can be seem as the culmination of cultural and even genetic evolution, providing the most stable and durable form of civilization.

Egypt

Egyptian civilization including irrigation, metalworking and cities dates from the third millennium BC—the Egyptian Old Kingdom. This was notable for the pyramids of Giza, among other achievements, but lacked the cultural brilliance of later times. Also, the collapse was relatively mild. Although royal authority had been weakening during the sixth dynasty (dating from 2345 BC) it did not collapse until 2181, shortly after the death of the long-lived Pepi II. In less than seventy years, a long human lifetime, a new and powerful regime had become established. Thus again, as in India and China, there was little evidence of high infant C in this first civilization, presumably because cultural technologies were not yet powerful enough to bring C to a higher level.

But in the reunified Middle Kingdom from 2055 BC there are clear signs of high infant C. Hereditary nobles known as nomarchs retained power for two centuries, building magnificent tombs and acting as local administrators, something otherwise unknown in Egyptian history. It was not until the reign of Sunusret III in 1878–39 BC that the pharaoh gained the power to appoint his own governors.

The Middle Kingdom was considered in later times to be the classic age of Egyptian literature, including philosophical and popular works. Notable texts include the Tale of Neferty, the Instructions of Amenemhat I, the Tale of Sinuhe, the Story of the Shipwrecked Sailor and the Story of the Eloquent Peasant.[1] There were also important innovations in such fields as sculpture. Creativity is a sign of high infant C.

But so too is instability, since high infant C undermines V, and barely two centuries after its founding, on the death of Amenemhat III in 1814 BC, the Middle Kingdom was in decline. A century later it had largely collapsed into what is known as the Second Intermediate Period, permitting invasion and conquest by foreigners such as the Hyksos. This "Dark Age" was far more extensive, not ending until Ahmose I (1549–24) drove out the Hyksos and established national unity.

But the signs of high infant C were nothing like as strong as for ancient China and India. The level of creativity in the Egyptian Middle Kingdom was less spectacular and the collapse briefer than in those societies. In general, this appears to have been a moderate-S society, like Tang-dynasty China, rather than a low-S one like ancient Greece and Rome, China in the Warring States period, or the Ganges civilization of India.

The most likely reason is that Egyptians had gained some level of S during

---

[1] R. B. Parkinson, *Poetry and Culture in Middle Kingdom Egypt: A Dark Side to Perfection* (Oakville: Equinox Publishing, 2010).

the collapse of the Old Kingdom, and the influx of lower S people had not been great enough to swamp this. Being surrounded by desert, Egypt was less susceptible than other ancient civilizations to being colonized by large-scale barbarian incursions. The same principle applies to the collapse of the Middle Kingdom, which was eventually succeeded by the more stable and conservative New Kingdom.[2] Once again, barbarian invaders were not numerous enough to greatly dilute the rising S of the native Egyptians, so S could rise relatively fast. After 1550 BC there would never again be such a period of anarchy. Higher S made Egyptian civilization more stable and durable at the cost of being more resistant to any form of change.

It is a general principle that S stays low when a civilization collapses if the incoming barbarians are numerous compared to the existing population. In the case of Germany after the fall of Rome, or the Ganges area after the collapse of the Indus civilization, the new civilization was in a completely different area and thus with ultra low S. The people of the Egyptian Middle Kingdom were at the opposite extreme, largely the descendants of Old Kingdom stock with only a small admixture of newcomers, and thus moderate rather than low S.

Even so, Egypt, China and India all have in common a high point of infant C long after the earliest evidence of civilization, and after at least one significant change of regime.

Sumer and Akkad

The development of civilization in Mesopotamia shows much the same pattern.[3] The process began with the preliterate Ubaid culture, which developed advanced systems of agriculture and the first cities in the fifth millennium BC.

---

[2] William McNeill, *The Rise of the West: A History of the Human Community* (Chicago: University of Chicago Press, 1963), 84; Jan Assmann, *The Mind of Egypt: History and Meaning in the Time of the Pharaohs* (Cambridge Mass.: Harvard University Press, 2002).

[3] General sources on the ancient world: S. W. Bauer, *The History of the Ancient World: From the Earliest Accounts to the Fall of Rome* (New York & London: WW Norton, 2007); A. S. Issar & M. Zohar, *Climate Change: Environment and Civilization in the Middle East* (New York: Springer, 2004).

At some point during the fourth millennium the Sumerians, speaking a Semitic language, migrated into this region from an unknown homeland. With the Sumerians come the beginnings of recorded history, and this history shows distinct signs of high infant C. City states such as Kish, Nippur, Uruk, Lagash and Ur lasted as independent entities for more than a thousand years. And though priest-kings exercised overall leadership, assemblies of wealthy men and young warriors could at times overrule them. Long-lasting local states and republican institutions are key features of high infant C, which can only occur when S remains low.

Also consistent with high infant C is that the Sumerians were great innovators. Among much else they were responsible for the first system of writing, the first known codes of law and administration, the potter's wheel, bronze-working, wheeled vehicles, and the abacus. As mathematicians they developed arithmetic, geometry and algebra. They were the first to work out the area of a triangle and the volume of a cube. They were keen astronomers and their fascination with the numbers 6 and 60 is reflected in our system of seconds, minutes and hours, as well as the 360 degrees of the circle.[4] This creativity is even more remarkable considering how few people were responsible. The cities of Sumeria contained just a few hundred thousand people, in an area smaller than Pennsylvania.

Some time after 2500 BC the city states were becoming weaker, suggesting a fall in C. Strong kings such as Gilgamesh of Uruk were able to dominate other cities, though such regimes typically fell apart after the founder's death. And there are indications of a growing gap between rich and poor, something also characteristic of the late Roman Republic. In the twenty-fourth century BC a reforming king of Lagash, Urukagina, made laws to protect the poor against the rich but was quickly overthrown.

Falling V is indicated by military weakness, with incursions by Elamites from the mountains in the east. Then, in the late twenty-fourth century, Sumeria was conquered by Sargon of Akkad. Though Sargon came to power as a high official serving the king of Kish, he represented a people immediately to the north, speaking a quite different language. Under him the Akkadians took control and crushed Sumerian resistance. Sargon went on to form the world's first multi-ethnic empire, reaching into what is now Syria and southeastern Turkey.

---

[4] S. N. Kramer, *The Sumerians: Their History, Culture, and Character* (Chicago, University of Chicago Press, 1963).

But this empire did not last long. In the reign of Sargon's grandson a fierce and destructive mountain people known as the Gutians gradually overran the land. The king was probably killed and his successor lost control of all but the capital area, bringing on a time of anarchy. This involved far more than a change of rulers. Temples were sacked, trade routes were lost, agriculture declined and the irrigation systems were neglected. The city of Akkad was sacked in 2151 BC and so comprehensively destroyed that its site has never been firmly identified.

There was a brief restoration of political unity under the Third Dynasty of Ur in the early twenty-second century BC, but no return to the earlier spirit of creativity. Cylinder seals, rolled across wet clay as proof of the writer's identity, contain evidence of this. The rich variety of Sumerian seals gave way to renewed vigor and more realistic depictions of the Akkadian period. But, by the Third Dynasty of Ur archaeologists find only monotonous repetition of a single theme, reminiscent of the cultural aridity of the later Roman Empire.[5]

After less than eighty years there was a further collapse accompanied by an incursion of Amorites—a collection of peoples from what is now northern Iraq and Syria. Southern Mesopotamia seems to have recovered some prosperity from the end of the nineteenth century BC, with evidence of building and restored irrigation systems. But it was not until the eighteenth century that Hammurabi, an Amorite king of Babylon, reunited most of Sargon's empire under his rule.

From then on, the history of the area developed into a pattern of empires alternating with relatively brief periods of disorder—the standard high-S political pattern. And despite bigger populations and far greater wealth, never again was there the intense creativity of the Sumerian city states.

What is important to note is the length of the period of disorder. The Gutians seem to have been invaded in 2214 BC. Then, apart from the Third Dynasty of Ur, which probably represented a lemming-cycle peak, there was political chaos and economic retreat until the end of the nineteenth century, and political unity was only restored in 1763 BC. This is a gap of almost five hundred years.

In other words, the pattern is the same as the one we saw in India, China and, to a lesser extent, Egypt—an extended period of civilization

[5] J. G. Macqueen, *Babylon* (London: Robert Hale, 1964), 203–4.

culminating in a brilliant flowering with signs of high infant C; city states, immense creativity, reverence for hereditary rank (even in the absence of real power), and in later times evidence of unusually tight administrative control of individual citizens, suggesting impersonal loyalties. All these civilizations were first unified and then weakened rapidly, falling into centuries of chaos. Their successors were more stable but less-creative civilizations showing evidence of high C and child V, but lower infant C. This suggests a rise in S following the collapse of the high C civilizations.

The passing of advanced cultural technologies to the barbarians

The collapse of these civilizations all involved incoming barbarians. In the case of Sumeria, an important point to consider is where these first invaders came from. In later ages they emerged out of the vast plateaus and mountains of central Asia (modern Iran and Afghanistan, and the central Asian steppes), and from the Arabian Desert. But the Akkadians did not. They came from central Mesopotamia, a relatively flat and fertile land. If V is a product of famine, why did the conquerors come from here rather than more inhospitable regions? It was undoubtedly a poorer and more backward area than Sumeria, and may well have experienced famine, but it cannot have been as tough as the mountains and deserts to the east and west.

We can start by recognizing that a harsh environment alone is not enough to create the highest level of V. V will certainly be increased by occasional famine, especially when interspersed with times of more plentiful food. But maximum V can only be generated when famine is reinforced by cultural V-promoters such as patriarchy.

In addition, fully-fledged barbarians need organization if they are to overrun wealthy and powerful civilizations. Even the most degenerate state can draw on thousands of soldiers, many of them experienced professionals. To defeat them requires an ability to form and co-ordinate large coalitions at considerable distance from their home bases, even among people who might be more accustomed to fighting each other. It also requires warlike ferocity and willingness to suffer high casualties in a pitched battle, something rarely found among hunter-gatherers and small-scale agriculturalists such as in Australia and New Guinea. In other words, what is required is cultural technologies to increase the levels of both V and C.

The best place to find such technologies, of course, is from the civilized peoples who have been working hard to develop them. This is why the first barbarians emerged from areas relatively close to existing civilization (see Fig. 15.1 below). The Semitic Akkadians of ancient Mesopotamia had been close neighbors to the Sumerians for well over a thousand years. Even the barbaric Gutians were from the headwaters of the Diyala River, only about three hundred kilometers from the Sumerian border.

Fig. 15.1. Ancient Mesopotamia

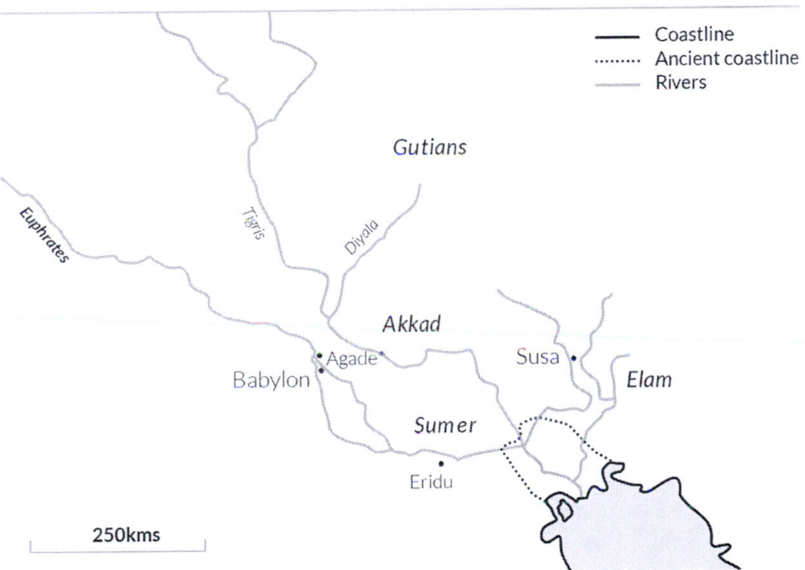

This is the same pattern as in ancient China (see chapter fourteen), which was first unified by the semi-barbarian but agricultural state of Qin from the Wei river valley. Nomad raiders from the north were a problem, which is why early versions of the Great Wall were being built at this time. But it was to be five-hundred years before true nomad horsemen from the steppe overran China, and fourteen-hundred years before the nomads reached their peak of power under Genghis Khan.

Civilizations can only arise because of changes in culture, especially control of sexual behavior, that increase V and C. Without this there is little temperamental basis for farming, and less acceptance of powerful authority makes the rise of larger political units possible.

V and C would not have been very high in ancient Sumeria compared with later times, which is one reason the states were so small, but they were crucial to the political, economic and intellectual achievements of that civilization. Sumerian cities poured immense resources into their religious establishments, which makes very good sense if it allows priests the time and energy to enforce the values and behavior that increase V and C. Competition between cities would also have promoted such efforts. If a city's god demanded behavior that increased V and C then the population would grow faster and be more effective as farmers, merchants and soldiers. It might conquer nearby areas, or its people emigrate and take up leading positions in other cities. The city itself would tend to become more successful and prestigious. All of this would encourage other cities to adopt its religious practices and values and maybe even its gods.

One of the most important C-promoting traditions is the seven-day week, with sanctions against working on one of the days. This was another Sumerian invention, including the performance of rituals and prohibition of certain kinds of work on the seventh or "holy" day. Though not as advanced or strict as it became in later Jewish and Christian tradition, any regularization of work would serve to increase C.

By the Babylonian period, which followed the collapse of Sumerian civilization, these cultural technologies seem to have become even more pronounced. The law code of Hammurabi, which drew heavily on earlier codes, contained a number of provisions relating to the control of sexual activity, with heavy religious sanctions. For example, a wife suspected of adultery had to take an oath before a god, and was drowned if she failed to do so. If accused by neighbors she could be thrown into the Euphrates, where sinking was considered proof of guilt, a somewhat more lenient version of the Medieval European ordeal where drowning was considered proof of innocence. This suggests a strong belief that the gods disapproved of sexual misbehavior. The chastity of unmarried girls was also strictly protected. The subordination of women had become more pronounced since Sumerian times. Women were considered in some sense property of their husbands and could even be seized for debt, though they were not confined to the home as done in later times by at least some classes of Assyrians.[6]

---

[6] Ibid., 73–5; K. Nernet-Nejat, "Women's Roles in Ancient Mesopotamia," in *Women's Roles in Ancient Civilizations: a Reference Guide*, edited by B. Vivante (Westport, Conn.: Greenwood Press, 1999).

Without conquest, the spread of such cultural technology can be slow, but it will happen gradually through trade and migration. The Akkadians, Gutians and other barbarians adopted Sumerian religious systems, and as a result became fiercer in war and better organized as their levels of C and V rose. These peoples also had the "benefit" of more frequent famines, the result of incessant feuds and absence of central authority, so they were especially open to the type of religious innovations that tend to promote V.

After a time, V and then C began to drop in the neighboring civilizations, as they always must, at which time these fiercer and better-organized barbarians moved in and took over. They subsequently influenced the civilization's native population. It is natural for humans, and indeed any primate, to copy the behavior of higher status individuals. As we saw in India and China, conquest by a higher V people tends to increase V in the conquered population as they imitate the behavior, values and beliefs of the conquerors. And because V combined with high population density increases stress there will also be a rise in C, since cortisol reduces testosterone.

In other words, there is a two-way process whereby a civilization and its barbarian conquerors reinforce each other's C and V. Civilized peoples are better at developing C because this is vital for farming. Barbarian invaders are more likely to develop high V because of their experience of famine and because families and tribes with high V do better in societies that are constantly at war. But key behaviors such as sexual restrictions and domination of women are helpful to both, so cultural beliefs that support one tend to support the other.

Once again, it should be emphasized that people had no idea of what they were doing any more than they do today. It is simply that societies and families whose religious beliefs and customs contained stronger C and V-promoters tended to outcompete the others.

What makes this process slow is that people do not easily accept values requiring them to act in ways that are too far removed from their current practices. For example, a man accustomed to regular sex might accept abstinence for a week, if a priest told him it would cause the gods to bless his crops. He might be less likely to do so if the period of abstinence were a month.

A religion promoting high C cannot be fully accepted by people with much lower C, so behavior in the beginning will only be slightly affected.

But as levels of C and V rise, so people are willing to accept more stringent codes which further affect behavior in a kind of ratcheting up effect. This was the problem found by Christian missionaries in Northern Europe, for whom nominal conversion of the people was only the first step. There began a centuries-long struggle to wipe out pagan superstitions and associated behavior, and impose the much stricter standards of Christianity.

The conversion of Africa

The same process can be seen in sub-Saharan Africa today, where the rapid growth of Christianity and Islam (see Fig. 15.2 below)

Fig. 15.2. The growth of Christianity and Islam in sub-Saharan Africa.[7] In the past century, most people have become either nominally Christian or Muslim.

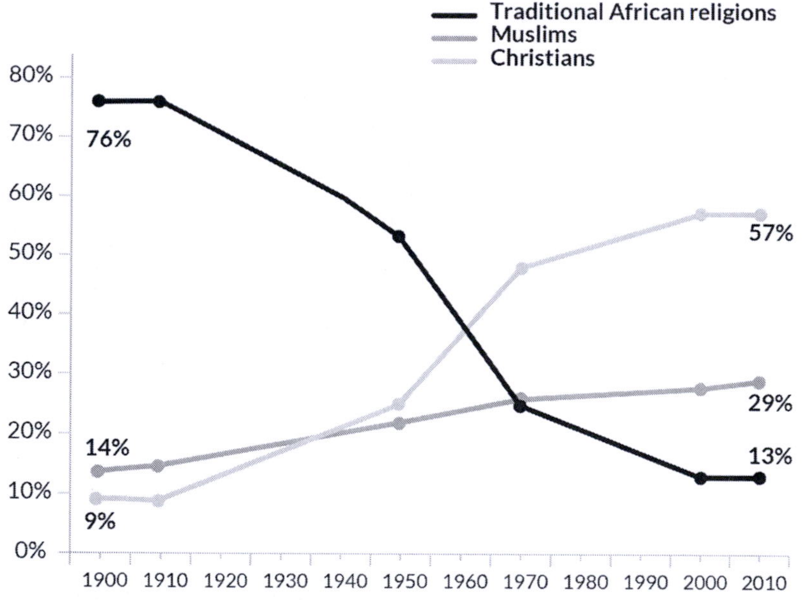

[7] *Pew Research*, "Tolerance and Tension: Islam and Christianity in Sub-Saharan Africa," (April 15, 2010) http://www.pewforum.org /executive-summary-islam-and-christianity-in-sub-saharan-africa.aspx (accessed September 7, 2014).

This has hardly dented the explosion of AIDS, showing that the adoption of a high-C religion causes only limited changes to behavior. The most common way for HIV/AIDS to spread is through sexual activity, especially if an individual has multiple partners. Christianity and Islam work hard to control sexual activity, yet their introduction has had little impact on southern Africa. There, traditional sexual practices still dominate after one-hundred years of prolonged missionary activity. The prevalence of AIDS in sub-Saharan Africa (see Fig. 15.3 below) illustrates this.

Fig. 15.3. Prevalence of AIDS in Africa.[8] Though most people are now nominally Christian or Muslim, sexual behavior largely reflects pre-conversion ideas.

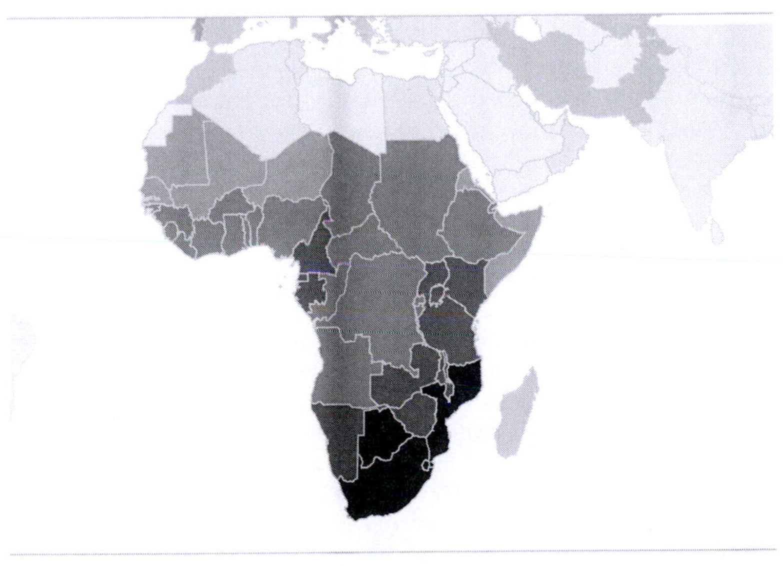

No data <0.1% 0.1% - <0.5% 0.5% - <1.0% 1.0% - <5.0% 5.0% - <15.0% 15.0% - 28.0%

[8] UNAIDS Report on the Global AIDS Epidemic 2010, Global HIV Prevalence Map.

This pattern is not seen in areas of North Africa and the Middle East which have been Muslim for more than a millennium and where stricter sexual standards have taken deep root.[9] But polygamy and promiscuous sexual behavior are far more common in sub-Saharan Africa. In Nigeria, a third of all women are involved in a polygamous union, and South Africa alone accounts for around two thirds of the world's 15–24-year-old HIV-positive people.[10]

South Africa is also one of the most religiously observant countries in the world, with about half the population attending worship every week.[11] In Edendale, a township in KwaZulu, the province with the highest rates of HIV/AIDS in South Africa, there is a large Christian population (85.5% in 2000, 10% higher than the national average), but also a wide acceptance that young men will have seven to ten sexual partners per year and most men will not be faithful once they are married. One young man commented on the phenomenon, describing the practice as "the Zulu way." A study in 1992 showed that although less promiscuous, Zulu girls were often sexually active by age 12. Other studies suggest that 70% of children born to black mothers are "illegitimate" and that mothers who had children with more than one partner were common.[12]

A study undertaken in Edendale, South Africa, provides a clear indication of the ineffectiveness and laxity of mainstream Christian sexual practices. On such issues as illegitimate conception, the age of births of first children, pre- and extramarital abstinence, mothers having children to different fathers, and support for use of condoms, the congregations of four different denominations (with the notable exception of the Pentecostal Church, accounting for 4.3% of the population of Edendale) scored little better and sometimes worse in their levels of sexual misconduct than those who did not attend church at all. Anglican, Roman Catholic, Methodist and Presbyterian churches were known by the locals to express a "realistic" attitude towards people's promiscuous nature and an acceptance of ingrained patterns of sexuality. These churches were more focused on

---

[9] R. C. Garner, "Safe Sects? Dynamic Religion and AIDS in South Africa," *The Journal of Modern African Studies* 38 (1) (2000): 46, 55, 58.

[10] *National Agency for the Control of AIDS*, Naca.gov.ng (accessed September 7, 2014); A. Harrison, J. Cleland & J. Frohlich, "Young People's Sexual Partnerships in KwaZulu-Natal, South Africa: Patterns, Contextual Influences, and HIV Risk," *Studies in Family Planning* 39 (4) (2008): 41–69.

[11] Garner, "Safe Sects?"

[12] Harrison et al., "Young People's Sexual Partnerships in KwaZulu-Natal."

attracting participation, and as such there were few obligations placed on followers at church or in youth groups and programs.[13]

The standards of sexual morality emphasized by the Christian church were not always so slack, especially when coming from Western missionaries in the early twentieth century. As one of the White Brothers states about their missionary station:

> [T]hey have got their hands on 3000 hectares of land on which about 8000 people live; they exercise the authority of a king over this property, not only judging cases but conscripting labour, ordering fatigues for construction materials, chasing out polygamists, removing amulets, demolishing the little huts for sacrifices, replacing even a chief whom they have expelled, and imposing on the chiefs catechists ....[14]

These relaxed attitudes reflect the extreme difficulty missionaries had in attracting converts. In the early days, pagan "backsliding" was common. And even when Christianity took root it did little to displace pagan sexual practices. Thus, in order to popularize the faith, moral requirements were relaxed. Standards in many places are so low that in recent years many clergymen have suggested that polygamy be permitted, and one leading Cardinal even conceded that "Christian marriage works badly in Africa."[15]

As mentioned previously, the Pentecostal movement has had more success in imposing stricter standards on its followers. Through a rigorous church life, charismatic sermons and a strong emphasis on community, they manage to enforce a far higher standard of sexual morality than competing denominations. They are often fervent for abstinence, with support provided by the AIDS epidemic. Many Pentecostals are starting to believe that the pandemic is God's punishment for non-Christian behavior, especially sexual looseness, and only by turning to God can one be "saved."[16] A blending of tribal theology with modern religious discipline has created a culture where young members are not only far more resistant

---

[13] Garner, "Safe Sects?"

[14] E. A. Isichei, *A History of Christianity in Africa: from Antiquity to the Present* (London: Society for Promoting Christian Knowledge, 1995), 136.

[15] Ibid., 98, 328–329; B. Sundkler & C. Steed, *A History of the Church in Africa* (Cambridge, Cambridge University Press, 2000), 216, 634; Garner, "Safe Sects?" 65.

[16] H. Dilger, "Healing the Wounds of Modernity: Salvation, Community and Care in a Neo-Pentecostal Church in Dar es Salaam, Tanzania," *Journal of Religion in Africa* 37 (1) (2007): 69.

to HIV but have achieved greater economic success.[17] Although African Pentecostalism may not be as effective a system of moral discipline as that which the original European missionaries sought to impose, its practices would certainly serve to raise C to some degree. Thus, we see the familiar pattern of sexual restraint improving economic success.

Israel

In the previous chapter we looked at Jews in the modern world, considering how their powerful C-promoting religion and relatively high S have contributed to their survival and success. In this chapter we will consider the origins of their religion, a well-documented example of how C- and V-promoting traditions arise and spread, aided by the interaction of civilized and barbaric peoples.

According to their own traditions, the Israelites were a desert people, at least some of whom had spent time in Egypt and who invaded and occupied Canaan, though none of this can be verified by archaeology.[18]

The book of Genesis records a family from Haran in northwestern Mesopotamia living temporarily in the city of Ur, and this is consistent with the archaeological evidence. From the twenty-second century BC the Amorites, a nomadic people from this area, were settling in and taking over various Sumerian cities following the decline of the Third Dynasty of Ur. Abram's ancestors Serug, Nachor and Terach had typically Amorite names. There were towns called Sarugi (Serug), Nakhur (Nachor), and Til-turakhi (Terach) in the vicinity of Haran.[19]

In addition, the customs described in Genesis are those of northern Mesopotamia. For example, tablets dug up from the city of Nuzu in northeastern Mesopotamia reveal customs familiar to readers of the bible. A childless couple commonly adopted an heir, but if the couple had a son of their own then the adopted son was required by law to surrender his right. This is what happened when Abraham adopted his slave Eliezer as heir, with Eliezer being disinherited when Sarah gave birth to Isaac. The

---

[17] Ibid., 66; D. Maxwell, "'Delivered from the Spirit of Poverty?': Pentecostalism, Prosperity and Modernity in Zimbabwe," *Journal of Religion in Africa* 28 (3) (1998): 350–73.

[18] N. Finkelstein & A. Silberman, *The Bible Unearthed* (New York: Free Press, 2001), 107.

[19] D. Klinghoffer, *The Discovery of God* (USA: Doubleday, 2003), 10.

tablets also show that a childless wife should give a handmaid to her husband, as Sarah did with Hagar.[20]

The names of Abram's generation, on the other hand, tend to be Sumerian. Sarai (his wife and sister) is the Akkadian name for the wife of Sin, the Sumerian moon god, and Terah's daughter Milcah was almost certainly named after Sin's daughter Malkatu.[21] All of this accords with the biohistorical picture of nomadic peoples taking high-C elements from the religions of the settled lands, and combining these with their own higher levels of V to form larger and more powerful political units.

The stories of Joseph and Moses represent an equally significant narrative in which a pastoral people spent several generations living in Egypt. While keeping elements of their high-V culture, as indicated by a high birth rate, they were clearly influenced by the customs of Egypt which was, at this time (the mid-second millennium BC), one of the world's most advanced and sophisticated cultures. It is notable that Moses, the great lawgiver, was brought up as an Egyptian prince, given that literacy and laws are key indications of high C. Modern historians may question whether such people actually existed or represent combined traditions from different individuals, but from the viewpoint of biohistory the narrative is highly plausible.

The Israelites who moved into Canaan had a religious and cultural system combining effective C-promoters and V-promoters. With relatively high C from Egypt and with V levels elevated by their harsh desert experience, as recorded in the Book of Exodus, they were both aggressive and cohesive enough to displace native peoples and eventually take over the land.

By the time of the Babylonian exile in the sixth century BC they had developed still more powerful C- and V-promoting traditions—strong enough to survive transplantation of the elite to another country. This was extraordinary in itself, since the whole purpose of mass deportations by the Assyrians and Babylonians was to break down ethnic loyalties, as was largely successful with the northern kingdom of Israel. Only the people of Judaea, strengthened by the reforms of Josiah, were able to resist assimilation.

---

[20] S. W. Bauer, *The History of the Ancient World: From the Earliest Accounts to the Fall of Rome* (New York & London: W. W. Norton, 2007), 133–35.
[21] Ibid., 128.

It must be noted that the development of this advanced religious technology involved far more than controlling behavior in ways that would increase C. Only three of the Ten Commandments (keeping the Sabbath, respect for parents and forbidding adultery) relate directly to V or C. The first three (only one God, no idols, no blasphemy) build reverence for a single deity and no other—an unusual idea at the time but increasingly important as the demands of the Jewish faith became more rigorous. Polytheists can escape from overly demanding deities while maintaining their basic worldview. Monotheists cannot.

Of particular interest is the threat of "visiting the sins of the fathers upon the children, even to the third and fourth generation." This was not only a powerful inducement to people who placed a high value on descendants and genealogy, but also an astute observation. As has been shown so far, the behavior of parents has a massive effect on future generations. Someone who acts in a way that reduces their C may be quite successful as a result of the effects of early life experience. But the workings of epigenetics means that their children, grandchildren and beyond are likely to be far less successful, and there will probably be fewer of them. We have seen this pattern in the decline of wealthy urban families described earlier—a pattern that the family of the intensely religious John D. Rockefeller did not follow.

The last five Commandments, of course, are to do with treating other people well. This makes the religious community attractive, and helps the society as a whole to be more cooperative and thus successful. Ethical concerns are a major feature of all the Abrahamic religions and others such as Buddhism. Increased C helps this by making such behavior more congenial, as has been seen in relation to testosterone. And inducements to act ethically work to increase C, as does any form of restraint on behavior.

Finally, there are the other elements including music, ritual, architecture, vestments and, most of all, literature. The Old Testament is a profoundly moving compendium of religious writings, including stories that teach fundamental values in a way that is both entertaining and easy to remember. Humans naturally tend to imitate high-status people, and revered characters from the bible form powerful role models. Along with this overt religious culture, customs and behaviors were developed, reinforced by public opinion, such as those to do with styles of dress and modes of speech, which would also serve to maintain C.

The result of all this was that the Jews achieved unusually high C and V, for the time. Living under the control of the Babylonian, Persian and Greek empires for several centuries would have worked to increase their S, as indicated in the last chapter, since people with higher infant C are more likely to rebel and thus be slaughtered. But the Jews seem to have been less affected than other peoples because they rebelled several times—once successfully against the Seleucids and twice futilely against Rome. Presumably this was because their powerful C-promoting traditions gave them higher infant C than other peoples, despite rising S.

Their V-promoting traditions were also quite effective, and Jews at the time were mercenary soldiers as much as traders. Only after the destruction of the Temple, as indicated in the previous chapter, did they begin developing the unique pacific traditions of the diaspora, promoting ever-higher C but more moderate V.

Even more significant for human history is that, starting two-thousand years ago, these potent Jewish traditions began spreading to other peoples through Christianity and then Islam. In the form of Christianity they have largely built the modern world, by creating the temperament that has made advanced industrial civilization possible.

The Strengthening of V- and C-promoting technology

While Israel was developing its religious technology in the remote backwater of Palestine, the great empires of Mesopotamia, Anatolia and Egypt were rising, clashing and falling. As mentioned earlier, these were not like Sumerian city-states or the later nations of Europe, with borders relatively constant for hundreds of years. Huge empires rose, expanded and replaced each other. Assyria alone had three distinct periods of imperial power, separated by eras of weakness and foreign domination.

This was already a complex, higher S civilization, with political structures reflecting acceptance of powerful authority rather than specific local loyalties. It was not as creative as that of the ancient Sumerians, or the lower S civilizations flourishing in India and China at the time. But never again would it experience the long period of chaos and darkness that followed the fall of these civilizations. The most extended Dark Age after the time of Hammurabi was associated with the influx of nomadic Aramaean peoples in the late twelfth and eleventh centuries BC—a time of depopulation for which few written records survive, but which was relatively shosrt.

These invaders were presumably low S but their numbers were relatively small, so their genetic contribution in most areas was minor. Where they were more numerous or in the majority, such as the Israelites in Palestine and the Aramaeans in Syria, they did form independent states with a strong sense of national loyalty. The Aramaean city-states fought long and fiercely against the resurgent might of Assyria, though they were eventually crushed. As suggested earlier, once large empires arise higher S can be a positive advantage, since higher S peoples are likely to resist and rebel.

The dangers of rebellion are illustrated by the Assyrian treatment of prisoners of war. Favored measures of execution included burning, impaling, skinning alive and embedding living victims in plaster columns. With neighbors like these, people who accept powerful authority—even that of an alien people—are more likely to survive and reproduce. Thus, there would be a strong evolutionary selection for higher S.

Meanwhile, cultural evolution was moving towards ever-stronger traditions supporting C and V. There is evidence of this in the trend towards secluding women, which increases both C and V—C by limiting their sexual activity, and V by making them more subordinate and thus anxious. The first steps were taken in Babylonian times, around 2000 BC, when women were expected to cover their bodies and faces and be chaperoned in public. The Assyrians took this further by insisting at least some women stay home most of the time, concealed behind curtains, a custom also adopted by the Persians.

Seclusion of women is a very expensive habit, because of the effort needed to segregate them and because it reduces their financial contribution. But the military success of Assyrians and Persians was ample reward for the cost involved. In the long term, biological and cultural success is based not on wealth but on the number of surviving children and the status they hold. Military prowess, in both offense and defense, is an effective way to achieve such success. It is the same process of cultural evolution traced in China, such as in the development of Neo-Confucian ideas during the Song dynasty and later, but in this case achieved more by conquest and so with a stronger element of V compared with C.

As a result of this increase in V and thus child V, combined with lower infant C as a result of rising S, there was a tendency for empires to grow larger. Sargon of Akkad ruled most of Mesopotamia. The Assyrians at their peak controlled Mesopotamia, modern Syria, Palestine, northern Arabia,

Egypt and southeastern Anatolia. The Persian Empire added to this the rest of Anatolia and their homeland in what is now Iran (see Fig. 15.4 below).

Fig. 15.4. Expansion of civilization in the Middle East.[22] Civilization expanded steadily from its Sumerian origins to successively harsher terrain as barbarians took up C- and V-promoting technologies.

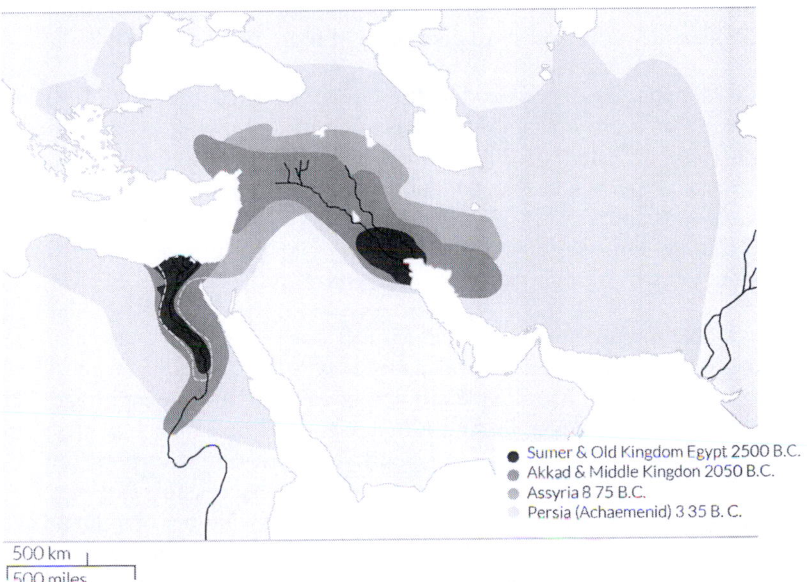

The Greeks under Alexander annexed this vast empire with ridiculous ease in less than a decade. Defeating the Persians only required three great battles, and most cities and provinces surrendered without a fight. The Macedonian conquest illustrates very clearly the effects of higher S, reducing infant C to a minimum and making ever higher levels of V possible. The result was a tough, resilient population which simply accepted rule by the strongest, no matter how alien. Even at this early period there is clear evidence of the attitudes to authority observed in modern Egypt.

---

[22] L. Woolley, *The Sumerians* (Oxford: The Clarendon Press, 1928). xii, 75; M. Brosius, *The Persians: An Introduction* (Abington: Routledge, 2006), 88, 93, 143, 153–157; J. Haywood, *The Penguin Historical Atlas of Ancient Civilizations* (London: Penguin Books Ltd., 2005), 22–53; P. K. O'Brien, *Atlas of World History* (London: Philip's, 2007), 28, 39, 41–43, 53; H. Saggs, *Civilization before Greece and Rome* (New Haven: Yale University Press, 1989), 17–18.

The Greek conquest also illustrates the ironic point that low-S civilizations reach their greatest influence when C is in decline. This applies to the Sumerians under Sargon of Akkad, the Chinese under the Qin and early Han dynasties, the Mauryan dynasty in India, the Greeks under Alexander, and of course the Roman Empire. Not only are they united and so militarily stronger, but it is easier to assimilate subject peoples to a cosmopolitan and liberal culture of sensual indulgence, rather than the prickly local loyalties and stern discipline of high infant C.

This was certainly the case for the urban elite of the new Hellenic kingdoms that arose out of Alexander's empire, following his early death from disease in Babylon. But on another level, the penetration of Greek culture was superficial. The common people of most of the conquered lands never learned to speak Greek, despite what may seem to us the overwhelming superiority of Greek culture. Nor did they in earlier times adopt the tongues of their Assyrian and Persian overlords.

Instead, the *lingua franca* of the Middle East remained Aramaic, the language of the illiterate, pastoral nomads who had brought on the Dark Age of the late twelfth and eleventh centuries BC. This is despite the fact that the Aramaeans never formed a major empire, and their city-states in the area of modern Syria were crushed by the reviving Assyrian Empire. In other words, as empire builders they were not successful. But they did have one "gift" for the peoples of the settled lands: higher V. Their very lack of unity, combined with military ferocity and what seems to have been a fast growing population, suggests that they had an unusually high level of V, even for a nomadic people.[23]

Peasant farmers have little to gain from great works of literature or art, such as the high civilization of the ancient Greeks. What helps them to survive is a high birth rate, strong local organization, and aggressive self-defense combined with a willingness to submit to any seriously powerful invader. And this is precisely the cultural technology the Aramaeans brought with them when they invaded, influencing the peoples of the Fertile Crescent to such a degree that they adopted the Aramaic language.

---

[23] E. Lipinski, *The Aramaeans: their Ancient History, Culture, Religion* (Leuven: Peeters, 2000).

Islam

We have seen how cultural and religious technologies developed in a region controlled by a civilization spread to ever more distant peoples in harsher environments, who then combine this with their own higher V to become effective invaders and conquerors.

This process was traced in China, where the first conquerors came from the Wei Valley in the northwest, and it was fifteen-hundred years before true nomads from the steppes overran the empire, in the shape of Mongol horsemen. In the Middle East, the advanced cultural technology developed first among the Sumerians in southern Iraq, spread to the nearby Akkadians and Elamites, then the Amorites and Aramaeans from northern Mesopotamia and Syria, and following them then the Medes and Persians from mountainous Iran. After the Greeks it was the Parthians, a nomadic people from the area east of the Caspian Sea, and then again by a revived Persia under the Sassanians. In the seventh century AD a new group of conquerors erupted from the harshest environment of all—the Arabian deserts. They brought with them the most powerful V-promoting cultural technology the world had yet seen—Islam.

It is not surprising that the united Arab tribes should sweep over the Middle East in scarcely more time than it had taken Alexander a thousand years earlier. The desert tribes were fierce warriors once their harsh living conditions were combined with the cultural technology for maintaining higher V, especially patriarchy and punishment of older children.[24] Child V also made them better able to accept authority and thus unite in larger groups, though against this the incessant feuding and disunity usually associated with high V must be set.

What finally united the Bedouin was almost certainly a lemming cycle G period. The period before the G year is associated with innovation and new ideas, and the Arab adoption of Islam fits this pattern.

But Islam itself was more than a rallying cry. It was a cultural technology with unique benefits not only for the Bedouin but for the peoples of the Middle East they were shortly to overrun. To pious Muslims, the Quran is the direct word of God. To other people, Muhammad clearly built on existing cultural technologies. There were sizeable Jewish and Christian

---

[24] E. Dehau & P. Bonte, *Bedouin and Nomads: Peoples of the Arabian Desert* (London: Thames & Hudson, 2007).

communities in Arabia at the time. Muhammad accepted the Jewish scriptures and saw Jesus as a prophet—nowhere is this clearer than in the Muslim fast of Ramadan.

The Christian church had established a forty-day fast before Easter by the early fourth century AD. Although customs varied, it typically involved abstinence from richer food such as meat and dairy products, with a single meal in the evening. An annual fast is a highly effective way of increasing C and especially V. Muhammad's genius was to expand it into a prohibition against eating or drinking during daylight hours, a standard of behavior that is easy to define and so less likely to be breached. He also made it one of the five pillars of Islam, and thus harder to discard in the way that most Christians later abandoned their own fasting traditions.

Among the settled peoples of the Middle East the levels of child V and thus stress had been rising for millennia, aided by the regular incursion of nomad conquerors. People with child V are primed to accept powerful political authority, so they are also likely to favor powerful divine authority. Thus the Muslim stress on submission to God (the word Islam means 'submission') met powerful psychological needs

A practical example of this is the conflict in the Byzantine Empire over reverence for icons, which can be regarded as quasi-polytheistic since it allows alternative beings to whom prayers may be addressed. The Greeks were generally in favor of icons and the Anatolians generally against, which suggests that the Anatolians had higher child V. After the Ottoman conquest the Anatolians quickly accepted Islam while the Greeks, despite centuries of occupation, did not. Thus Islam is a supremely powerful and effective V-promoting religion, with a strong appeal to high V people.

There is far more to Islam than this, of course. Praying five times a day tends to increase C, as does avoiding alcohol. The requirement to eat with the right hand, leaving the left for ablutions, is an effective health measure, and this is only one of many health measures associated with Islam, such as ritual washing. Then there is the Quran itself and the sonorous power of the Arabic language, with an attractive system of ethics including a focus on alms giving and the equality of believers. Putting all this together created a powerful religious technology which made its followers more aggressive, more confident, more united and more reproductively successful than any competitors.[25]

In the case of the Arab conquest the most striking consequence is what came after, with the conquered peoples largely adopting not only the religion but the dress and even the language of the Arabs. This was not necessarily what the original conquerors had sought. The Caliph ‘Umar (634–44 AD) is alleged to have accepted the surrender of the Syrian Christians on the condition that (among other things) they did not adopt Arab dress or speech. The Arabs were concerned to maintain their distinct identity as conquerors, and not merge with the subject peoples.

When these peoples started to convert to Islam it caused surprise and even confusion among the rulers. The tax levied on non-believers was an important source of revenue and they were reluctant to let it go, so for some eighty years they continued to levy it on converts. Only very gradually did the conversion of non-believers become an aim of Muslim policy.

But the reason the subject peoples became culturally Arabic is obvious if we follow the lessons of history. High V is advantageous to settled peoples. It makes them confident, aggressive and fertile. Those who adopt high-V customs, in imitation of their new overlords, tend to outbreed and outcompete those who do not.

Muslims also took over existing cultural technologies, such as the custom of secluding women. This was not a Bedouin practice, but was common in Middle Eastern cities at the time of the conquest, at least among certain classes. The effect of Islam was to spread the custom far more widely. It is part of a process by which Islamic culture has become more rigid and severe, and thus more effective in the one way that matters for long-term success—having more surviving children. Another factor that helped the birth-rate was the Muslim prohibition of infanticide.[26] All of this is a form of ratcheting effect, by which high-V customs increase V, making even more extreme customs possible.

---

[25] I. Lapidus, *A History of Islamic Societies* (Cambridge: Cambridge University Press, 1995); B. Lewis, *The Middle East, 2000 Years of History from the Rise of Christianity to the Present Day* (London: Phoenix Press, 2001).

[26] Sometimes seen as an attempt by Mohammed to raise the status of women, but in fact acting to raise the birth rate, a powerful V-promoter in itself. J. Esposito, *Islam: The Straight Path* (Oxford: Oxford University Press, 1998).

One of the more extreme forms of Islam can be found in Wahhabism. It is no coincidence that this very high-V version of Islam became dominant in Saudi Arabia, the area of the Arab world which until recently had the harshest climate and living conditions.[27] This is also a country where patriarchal customs remain strong. Every woman must have a male guardian whose consent is required for her to take a job or to marry, travel, access education, have elective surgery or even open a bank account. Women are forbidden to drive and must be accompanied by a male family member outside the home. They must cover up all but a small portion of their face when in public. Relatively few of them work.[28]

Another pre-Islamic custom taken up by Muslims in many areas was female circumcision—a method of reducing women's sexuality that should increase C and V. It is often said that such practices, and others that have been criticized, do not belong to the "true" religion of Islam. For example, Muhammad seems to have had a relatively positive relationship with women, listening to his wives and helping them with household chores.[29] But what is important, when it comes to cultural success or failure, is not the idealized religion of a sacred text but how people behave. This is the product of how that text relates to and combines with folk customs and pre-Muslim traditions. Put together, these have a massive effect on temperament and behavior. People reared in this way think, feel and react quite differently from non-Muslims.

Failure to recognize this has disastrous consequences. The occupations of Iraq and Afghanistan by the United States and its allies arose from a belief that human beings are fundamentally the same, and that removal of tyrannical regimes, plus some education, will turn these countries into prosperous and liberal democracies. But five-thousand years of cultural evolution have ensured that people are not the same, and political regimes tend to reflect these differences. Parliamentary democracy in developed countries reflects relatively high infant C and low child V, resulting in

---

[27] Lapidus, *A History of Islamic Societies*, 11–20; K. Kelly & R.T. Schnadelbach, *Landscaping the Saudi Arabian Desert* (Philadelphia: Delancey Press, 1976); N. DeLong-Bas, *Wahhabi Islam: From Revival and Reform to Global Jihad* (Oxford: Oxford University Press, 2004).

[28] Q. A. Ahmed, *In the Land of Invisible Women: A Female Doctor's Journey in the Saudi Kingdom* (Naperville, Ill.: Sourcebooks, 2008).

[29] A. Barlas, *Believing women in Islam* (Austin, University of Texas Press, 2002) K. Armstrong, *Muhammad: A Biography of the Prophet,* (New York, Harper Collins, 1992)

impersonal political systems based on loyalty rather than fear. Arabs, Persians, Afghans and the other Muslim ethnic groups of the Middle East and Central Asia are epigenetically primed for personal loyalties rather than impersonal institutions. Child V and stress make them accepting only of harsh authority, so they tend to rebel against rulers seen as weak (such as elected presidents). They also make people rigidly conservative so that education of women, seen by Westerners as an unambiguous good, is perceived as a threat against their culture and way of life. In this they are, of course, correct.

The Triumph of the Fundamentalists

All of this explains why Muslims in the West remain poor and distinct, especially people from the Middle East and Pakistan who have the highest levels of V. Ultra-high levels of V and child V combine with low infant C and only moderate C to form people less likely to succeed economically or to assimilate. One result has been the growth of slums with large Muslim communities around French cities and elsewhere in Europe, where unemployment and birth rates both remain high.[30] Character is set epigenetically, passed down the generations by childrearing systems and possibly direct epigenetic inheritance, and therefore is not easily changed. Part of this inheritance is aggression, a vital aid to survival in these tough and unstable societies. This is especially so for migrants from the most violent places on Earth such as Somalia, Afghanistan and Chechnya.

But the above hypothesis should not be interpreted as an attempt to portray these cultures as "backward." On the contrary, the three oldest centers of civilization were Egypt, Mesopotamia and the Indus Valley in what is now

[30] M. Viorst, "The Muslims of France," *Foreign Affairs* 75 (5) (1996): 78–96; T. Sobotka, "The Rising Importance of Migrants for Childrearing in Europe," *Demographic Research* 19 (9) (2008); Shadi Hamid, "The Major Roadblock to Muslim Assimilation in Europe," *The Atlantic*, (August 18, 2011) http://www.brookings.edu/research/opinions/2011/08/18-muslim-europe-hamid (accessed September 7, 2014); S. Kern, "European 'No-Go' Zones for Non-Muslims Proliferating Occupation Without Tanks or Soldiers," *Gatestone Institute* (August 22, 2011) http://www.gatestoneinstitute.org/2367/european-muslim-no-go-zones (accessed September 7, 2014); R. Koopmans, "Fundamentalism and out-group hostility: Muslim immigrants and Christian natives in Western Europe," *WZB Mitteilungen*, December 2013, http://www.wzb.eu/sites/default/files/u6/koopmans_englisch_ed.pdf (accessed September 7, 2014).

Pakistan. It is no accident that these areas, and most of the lands surrounding them, are now strongly and fervently Muslim. Long experience of civilization has bred a genotype (high S) and culture (Islam) which perfectly adapts people to survive and expand their numbers in dense agricultural and urban populations.

And when it comes to long-term survival, creativity has limited value. The Middle East gave birth to a creative surge in science and philosophy in the centuries after the Arab conquest, perhaps because the new elite was partly descended from low-S "barbarians," and thus may have had higher infant C for a time. But creativity faded away as populations intermarried and the low-S genes were swamped. This is the sign of an increasingly tough, vigorous and conservative culture. It is the same pattern noted in China and India, and completely opposite to that of Europe. Creativity looks good in history books and the display cases of museums, but the culture that succeeds in the long-term is the one which gives its adherents the most children and grandchildren to carry on its traditions.

The partial liberalization of some Muslim regimes reflects only the influence of a small Westernized elite, which is likely to be swallowed up within a generation or two as the hardline high-V culture of the countryside reasserts itself. Some signs of this have already been seen in Iran with the establishment of the Islamic Republic in 1979, and (less strongly) in Turkey with the election of the mildly Islamist Justice and Development Party in 2003. This is exactly what happened to the urban elites set up by Alexander the Great's conquests in the late fourth century BC, who gradually faded into the general population.[31]

If Muslims have more children than followers of other religions, it may even be asked why Christians and Jews have survived as minorities in Muslim lands. Even 5% more surviving children should have allowed Muslims to swamp these minorities in fourteen-hundred years. And the experiences of countries like Lebanon and Nigeria suggest that the demographic advantage of Muslims over Christians and Jews is far greater than this.

The answer lies in the concept of an "ecological niche," by which different species divide up a habitat. In a tropical forest some species live high in

---

[31] A. J. Dennis, *The Rise of the Islamic Empire and the Threat to the West* (Bristol, IN: Wyndham Hall Press, 2001); P. Margulies, *The Rise of Islamic Fundamentalism* (Farmington Hills, MI: Greenhaven, 2005).

the canopy, others the midlevel, and others nearer the ground. Some live on insects, others leaves, others fruit or flowers. Each species is suited to its own niche, and has the advantage over all other species in that area. Even in the poorest peasant society there is an "ecological niche" involving trade, finance and, to some extent, the professions. Christians and especially Jews have lower V but higher C, which bring greater flexibility of mind and thus tend to be overrepresented in these areas.[32] They may have fewer children than their Muslim neighbors, but even a slight advantage in wealth and parental care can allow more of their children to survive. This is consistent with the principle that V tends to increase fertility while C reduces the death rate, which accounts for the very high population growth seen in Europe and Japan when C is at a peak. Thus the numbers of a higher C minority can be maintained and even grow, though with an upper limit based roughly on the size of the "niche," meaning the available jobs.

The same reasoning applies to the success of Jews in Christian Europe. Demographic information from Poland and Germany in the nineteenth century suggests that Jews mostly had lower fertility than Gentiles. But their lower death rate, especially of infants, allowed sizeable populations to establish themselves.[33] This success is especially remarkable given the discriminatory taxation and (at times) outright hostility of the dominant culture.

In *The Son Also Rises*, Gregory Clark argues that Jews in Christian lands tend to maintain their economic status because their less successful members convert into the majority population while upwardly mobile Christians have often adapted Judaism, a belief system congenial to more complex economic roles because of its focus on literacy. He makes the same point for Christians in Muslim lands.[34] This makes even better sense once we understand that Judaism shapes character in a way that is more favorable to commerce than Christianity, and Christianity shapes character in a way that is more favorable to commerce than Islam.

---

[32] G. Clark, *The Son Also Rises* (Princeton: Princeton University Press, 2014), 228.

[33] L. Dobroszycki & P. Ritterband, "The Fertility Of Modern Polish Jewry," in *Modern Jewish Fertility*, edited by P. Ritterband, 66–70 (Leiden: Brill Archive, 1981); A. Goldstein, „Some demographic characteristics of village Jews in Germany: Nonnenweier, 1800–1931," in *Modern Jewish Fertility*, edited by P. Ritterband, 125–127 (Leiden: Brill Archive, 1981).

[34] Clark, *The Son Also Rises*, 228.

Ironically, one of the reasons Islam is doing so well today is the sheer success of Western civilization—the result of Christian values driving C and especially infant C to unprecedented heights. Even the second-hand wealth of the Industrial Revolution has brought starvation to an end over most of the world, and as a result the Muslim populations have exploded. In a world where most children survive there is little biological advantage to economic success, so in recent times the Christians and Jews in Muslim lands have shrunk to smaller proportions of the population.

For example, Lebanon once had a Christian majority but is now 54% Muslim.[35] In communist Yugoslavia, the provinces with Muslim populations grew much faster and received tax revenue from the wealthier Christian states. The population of Kosovo, the spiritual homeland of Christian Serbia, grew from 733,000 in 1948 to over two million in 1994, with the Muslim component surging from 68% to 90% and lately even higher.[36]

Meanwhile, Muslims move into a Europe where C is in decline, where the birth rate is far below replacement level and the decline of the nation state presents few obstacles. Of all immigrant groups in Europe, Muslims have the highest fertility and the lowest level of assimilation. And although there is a decline in V and C with continued residence, this is slower than for non-Muslim populations.

Even more significant is the continued migration from Muslim countries. Although birth rates in Muslim majority countries have plummeted in recent years, only 6 out of 49 have a fertility rate below the replacement level of 2.1 children per women; 17 have a fertility rate of 4 or higher, meaning that their populations can be expected to double over the next generation. Some, such as Niger and Afghanistan, still retain a fertility rate more than *three times* the replacement level.[37]

---

[35] US Department of State, "Lebanaon," International Religious Freedom Report 2010, http://www.state.gov/j/drl/rls/irf/2010/148830.htm (accessed September 7, 2014).

[36] S. P. Ramet, *Nationalism and Federalism in Yugoslavia, 1962–1991* (Bloomington, IN: Indiana University Press, 1992).

[37] N. Eberstadt & A. Shah, "Fertility Decline in the Muslim World: A Veritable Sea-Change, Still Curiously Unnoticed" *The American Enterprise Institute Working Paper Series on Development Policy*, Number 7, (December 2011), http://www.aei.org/files/2012/03/21/-fertility-decline-in-the-muslim-world-a-

If biohistory is correct, and the recent rise of Islamic governments in the Middle East and North Africa suggests it is, then this tough, resilient culture will survive the affluence and liberalizing influences of the West. This is not the triumph of Islam as such, because Islam has a range of forms. The enlightened and liberal Islam of the urban elites is strikingly different from the fundamentalism of the countryside. But it is this latter form, with all its associated customs such as female circumcision and rigid subjection of women, which is likely to triumph in the end.

The Westernized urban elites will shrivel and decline and the rural traditionalists emerge and regain dominance. Meanwhile, as indicated in the next chapter, the West can be expected to become steadily feebler and less fertile. On current trends there is little doubt that Europe will become an Islamic continent in a century or so. A fourteen-hundred-year struggle is coming to end.

---

veritable-seachange-still-curiously-unnoticed_102606337292.pdf (accessed September 7, 2014).

Chapter Sixteen

The Decline of the West

Civilization rests on physical technologies such as agriculture, writing and metalworking. But it also requires cultural technologies, especially religions, to create systems of temperament that support civilized life.

One of these systems of temperament is C, which involves hard work and self-discipline, as well as the willingness to sacrifice present consumption for future benefit. Parental control in infancy produces infant C, which involves machine skills and openness to new ideas. The other system is V, which is associated with aggression, confidence and small group cohesion. Its child-V form, the result of control and especially punishment in late childhood, makes people more conservative and accepting of authority.

These advanced religious systems were developed over many thousands of years in the oldest centers of civilization, but people in these areas also went through a genetic change which gave them a higher level of "S," making them more indulgent of infants and thus less likely to develop infant C. The resulting civilizations were more conservative but also more stable.

Meanwhile, these advanced religions spread to peoples who had never been civilized and thus had never developed the high S genetic variant. The result was a much higher level of C and especially of infant C in Northern Europe and Japan.

This now brings us to Europe and North America around 1850, and Japan a century later, when each society was at a peak of C. With their advanced technology, cohesive nation states, supreme confidence and high birth-rates, these cultures dominated the world. Having examined the history of this civilization from the ancient era to the peak of C, it is time to examine the changes experienced since then, and their likely progress into the future.

Declining V and stress

The theory of the civilization cycle, as discussed in chapter seven, proposes that the rapid increase of C in Europe and Japan was partly the result of high levels of V and stress which peaked in the sixteenth century. As V fell the rise in C slowed and then came to a halt in the nineteenth century. This section presents evidence that V continued to fall through the twentieth and early twenty-first centuries, and from all indications will continue to do so. This is exactly what happened in other ancient civilizations, including Rome.

Population decline

The first and most powerful expression of the fall of V, and also of the fall of C which has accompanied it, is a declining birth rate. As noted in previous chapters, V is the "populate or perish" trait which causes animals to breed with maximum speed in environments where famine and predators are a constant threat. It is why people from the mountains and deserts have periodically flooded into the fertile plains throughout recorded history.

Higher C also promotes interest in children, something we saw in the intensive maternal behavior of calorie-restricted rats. This is helped by the stable governments and advanced economies of high-C peoples which reduce child mortality. Thus we saw a rising trend of population growth to the nineteenth century, apart from periods associated with lemming-cycle troughs.

The combination of falling V and C over the past century has contributed to a dramatic fall in birth rates over the past forty years. Not one Western country apart from Israel is producing the minimum 2.1 children per woman needed to maintain itself (see Fig. 16.1 below).

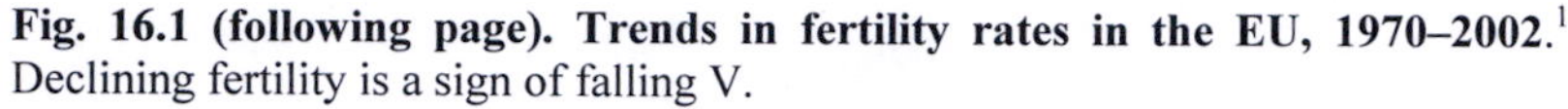

Fig. 16.1 (following page). Trends in fertility rates in the EU, 1970–2002.[1] Declining fertility is a sign of falling V.

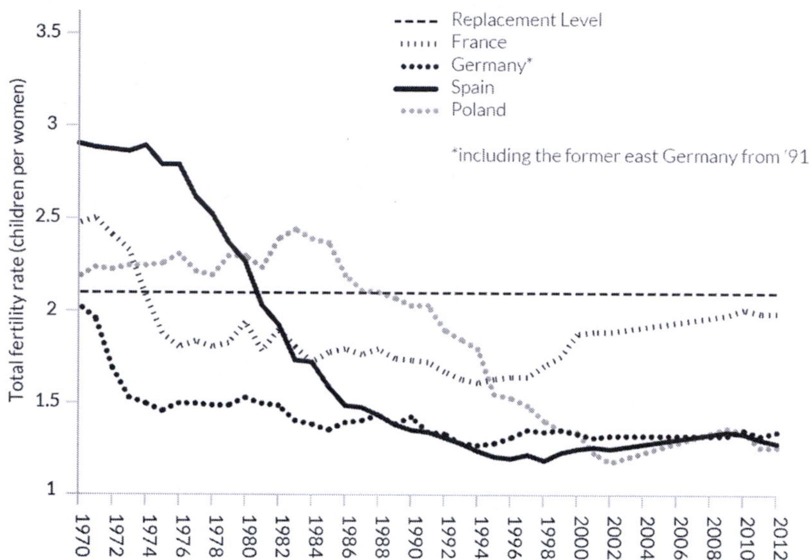

This applies not only to Europe and Japan but to other countries settled by Europeans such as the United States. Populations in these countries continue to grow only because of mass immigration from poorer regions. In Japan the decline in fertility is so severe that demographers predict the population will fall to around 45 million by the end of the twenty-first century if current trends continue (see Fig. 16.2 below).

Demographers agree that wealth and modernization lower the birth rate. Various explanations have been proposed from the cost of raising children to increased access to contraception, women in the workforce, earlier sexual activity, postponement of marriage, more divorce, high taxes, socialistic welfare systems, or even anorexia in young women.[2] Biohistory

---

[1] Council of Europe, Recent Demographic Developments in Europe, Demographic Yearbook, (2003), http://www.rand.org/pubs/research_briefs/RB9126/index1.html (accessed September 3, 2014); Eurostat, "Fertility Rates by Age," http://appsso.eurostat.ec.europa.eu/nui/show.do?dataset=demo_frate&lang=en (accessed September 7, 2014).

[2] C. B. Douglass, *Barren States: The Population Implosion in Europe* (Oxford and New York: Berg Publishers, 2005), 60;

proposes that the key reason is actually a change in *temperament*—that people with lower V and C are simply less interested in having children.

Fig. 16.2. Japanese population—historic and projected.[3] As V continues to fall, the Japanese population will cease to grow and start declining. Only large-scale immigration from poorer countries will prevent this happening throughout the West.

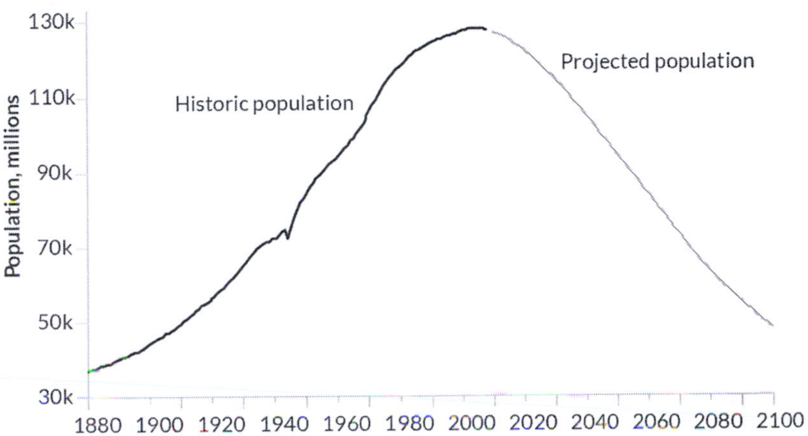

R. Lesthaeghe, "On the Social Control of Human Reproduction," *Population and Development Review* 6 (4) (1980): 527–548; R. Lesthaeghe, "A Century of Demographic and Cultural Change in Western Europe," *Population and Development Review* 9 (3) (1983): 411–435; R. Lesthaeghe, "Value Orientations, Economic Growth and Demographic Trends—Towards a Confrontation." IPD-Working Paper, 85–7, Brussel, Vrije Universiteit (1985); R. Lesthaeghe & D. J. van de Kaa, "Twee Demografische Transities?" ["Two Demographic Transitions?"], in *Bevolking: Groei en Krimp* [*Population: Growth and Decline*], edited by D. J. van de Kaa & R. Lesthaeghe, 9–24 (Deventer: Van Loghum Slaterus, 1986); D. Coleman, "New Patterns and Trends in European Fertility: International and Sub-National Comparisons," in *Europe's Population in the 1990s*, edited by D. Coleman, 1–61 (Oxford: Oxford University Press, 1996).

[3] Statistics Bureau of Japan Population by sex, population increase and population density. http://www.stat.go.jp/english/data/chouki/02.htm (accessed September 3, 2014); National Institute of Population and Social Security Research (December 2006). Population projections for Japan: 2006–2055, http://www.ipss.go.jp/index-e.asp (accessed September 3, 2014); *The Japanese Journal of Population* 6 (1) (2008): 76–114.

This also explains why groups such as Orthodox Jews in Israel are largely immune to the trend.[4] Mormons, another group which has retained powerful C- and V-promoting traditions, have also experienced declining fertility but this has steadied at three children per women since the 1990s.[5] The Amish are another group which has beaten the trend and maintain an average 6–7 children per household.[6]

Plummeting birth rates have the advantage that they reduce the danger of overpopulation and ecological collapse. But declining, aging populations are not a good basis for a prosperous future, especially if birth rates continue to fall, as based on this hypothesis we expect them to. The situation will become even worse as C continues to decline and succeeding generations lose their economic skills. Not only will there be fewer young people to support the old, but they will increasingly lack the economic skills to do so.

Governments can do nothing effective to stem population decline. Direct action to support the birth rate can have only marginal effect. Perhaps no government ever worked harder at this than the Nazi regime. They imposed harsh curbs on abortion, homosexuality and conspicuous prostitution, provided marriage loans, child subsidies and family allowances. The advertisement and display of contraceptives were banned and birth control clinics were closed down. Prolific mothers were honored with awards like front line troops, including a gold Honor Cross for women with more than eight children. Single women were encouraged to have children by "racially pure" men, and given support in various ways.

---

[4] "Israel's Ultra-Orthodox Problem," *Newsweek* (January 2, 2012), http://www.thedailybeast.com/newsweek/2012/01/01/israel-s-ultra-orthodox-problem.html (accessed September 7, 2014).

[5] "Utah fertility rate tops the U.S. charts" *Deseret News*, (November 7, 2010), http://www.deseretnews.com/article/700079435/Utah-fertility-rate-tops-the-US-charts.html (accessed September 7, 2014); Eric Kaufman, "Mormons: A Rising Force?" *Huffington Post* (March 26, 2014), http://www.huffingtonpost.co.uk/eric-kaufmann
/mormons-a-rising-force_b_1509283.html (accessed September 7, 2014).

[6] C. Totland, "Amish Enjoy Unexpected Boom in Numbers," *The Washington Times* TAFE Courses Online (August 9, 2012), http://www.washingtontimes.com/news/2012/aug/9/amish-enjoy-unexpected-boom-in-numbers/?page=all (accessed September 7, 2014).

The result of all this was a short-term rise in the birth rate, but nothing like a return to the situation of three decades before.[7] And there were also some less positive effects. By 1939 the incidence of divorce had risen by 50%, and by 1945, 23% of young Germans were infected with venereal disease—not something likely to promote long-term fertility.[8] Germany actually achieved more births per woman after the war without any major government intervention (see Fig. 16.3 below).

Fig. 16.3. Average number of children per woman in Germany, 1870–2004.[9] Even autocratic governments which do everything possible to increase birth have little success.

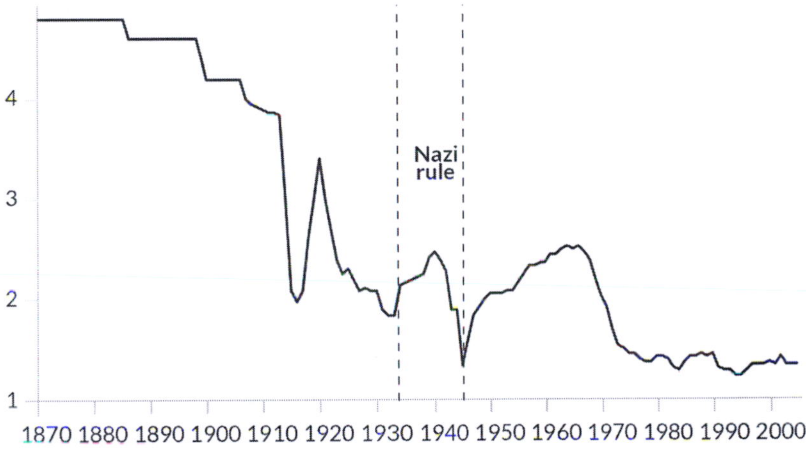

It is not even clear how much Nazi policy had to do with this increase. The Italian fascists also had pro-natal policies in the 1920s and 1930s, while their birth rate fell steadily (see Fig. 16.4 below). People have children because they want to have children, not because the government tells them to. In the long run, Nazi ideology most likely reduced the birth rate by promoting extra-marital sex and undermining Christian belief and practice, which contains powerful C-promoters.

---

[7] M. Mouton, *From Nurturing the Nation to Purifying the Volk: Weimar and Nazi Family Policy, 1918–1945* (Cambridge University Press: Cambridge, 2007),

[8] L. Pine, *Nazi Family Policy, 1933–1945* (Oxford and New York: Berg, 1997).

[9] The Berlin Institute for Population and Development, http://www.berlin-institut.org/ (accessed September 7, 2014).

Fig. 16.4. Rate of natural increase per 1,000 in Italy, five-year rolling average, 1862–1949.[10] The futility of government action is even clearer in Italy, where Mussolini's pro-natal policies did nothing to halt or even slow the plunge in birth rate.

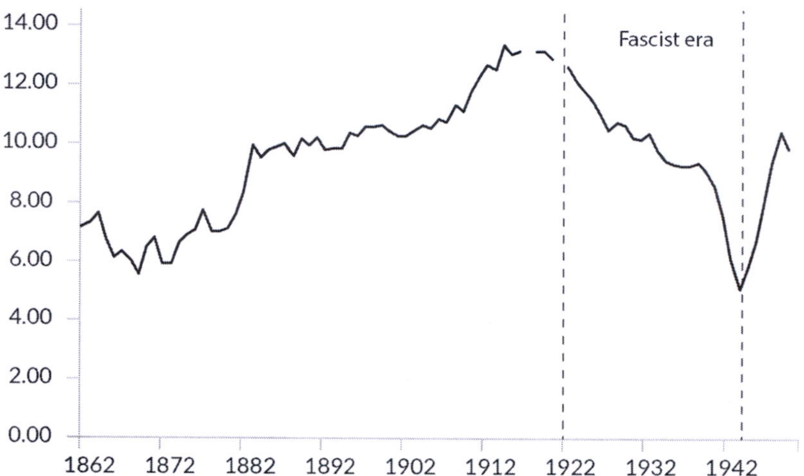

The same can be said of that other great twentieth-century totalitarian ideology, Communism. Ex-Communist countries such as Russia and the Ukraine tend to have very low birth rates, despite being far less prosperous than other European countries. Early attempts to increase the birth rate in Russia may even have been counterproductive. Kremlin youth groups set up special tents at their summer camps for young people to have sex, something that would reduce C and thus make children less desirable.

In recent years the Russian government has worked to promote Orthodox Christianity and traditional values, a policy in the right direction for increasing the birth rate but likely to be futile in the long run. Even the most autocratic governments have limited power to change popular behavior.

Recent attempts to increase the birth rate in Western countries through financial incentives have had little impact. A 2003 report for the OECD looked at the effect of government incentives in ten different countries,

---

[10] B. R. Mitchell, *European Historical Statistics, 1750–1975* (New York: Columbia University Press, 1975).

concluding that no policy was likely to reverse the decline in fertility.[11] A more recent Germany study came to the same conclusion.[12]

As with other indications of falling V, the US is significantly behind most European countries, with a birth rate only marginally below replacement level. This applies especially to the "red states" more likely to vote Republican. In 2004, George Bush won a majority in the 19 states with the highest fertility rates, while John Kerry won the 16 states with the lowest fertility.[13] This is even more obvious if minorities are excluded (see Table 16.1 and Fig. 16.5 below).

Table 16.1 States with lowest and highest Bush vote in 2003

Lowest Bush vote in 2004			Highest Bush vote in 2004		
	Babies per white woman	Bush vote		Babies per white woman	Bush vote
Vermont	1.63	38.9%	Utah	2.45	71.1%
Massachusetts	1.60	37.0%	Alaska	2.28	61.8%
Hawaii	1.59	45.3%	Idaho	2.20	68.5%
Rhode Island	1.50	38.9%	Kansas	2.06	62.2%
DC	1.11	9.3%	South Dakota	2.28	59.9%

[11] J. Sleebos, "Low Fertility Rates in OECD Countries: Facts and Policy Responses," OECD Social, Employment and Migration Working Papers 15, (2003).

[12] "Baby Blues: German Efforts to Improve Birthrate a Failure," *Spiegel Online International*, (December 18, 2012), http://www.spiegel.de/international/germany/study-german-efforts-to-increase-birthrate-a-failure-a-873635.html (accessed September 7, 2014); S. B. Westley, M. Kim Choe & R. D. Retherford, "Very Low Fertility in Asia: Is There a Problem? Can It Be Solved?" *Asia Pacific* 94, (May 2010), http://www.eastwestcenter.org/fileadmin/stored/pdfs/api094.pdf (accessed September 7, 2014).

[13] S. Sailer, "Baby Gap, " *The American Conservative*, (December 20, 2004), http://www.theamericanconservative.com/article/2004/dec/20/0004/ (accessed September 7, 2014).

Fig. 16.5. Babies per white woman in states won by Bush or Kerry in 2004.[14] Voting Republican is associated with higher V and thus more babies.

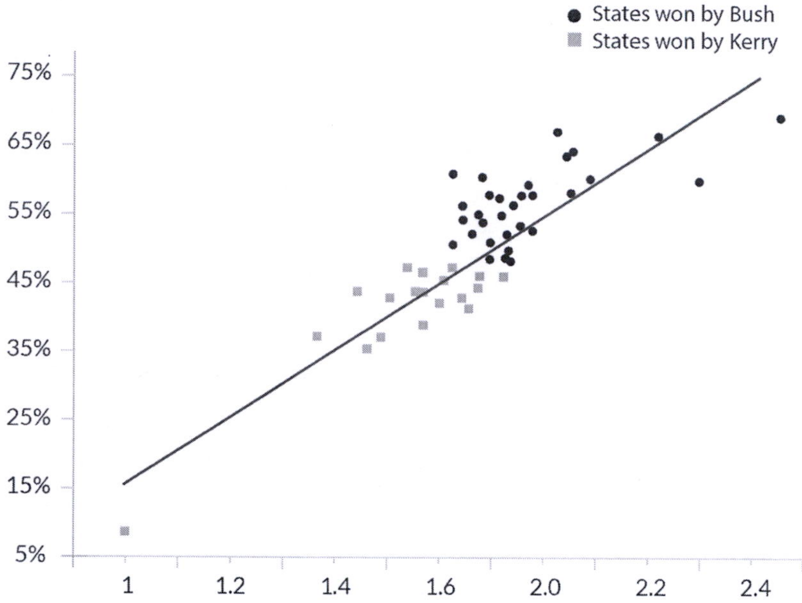

Utah, which had the highest Caucasian birth-rate in the country, also had the highest pro-Bush sentiment, with more than 70% of the electorate voting Republican. Equally dramatic, the District of Columbia is a demographic sinkhole, with women having barely half the children needed to maintain population. Not surprisingly it is a bastion of support for the Democrats. Similarly, in the 2008 election there was a correlation between lower Caucasian birth rates and voting for the Democrats.[15]

---

[14] S. Sailer, "The Baby Gap: Explaining Red and Blue," *The American Conservative*, (December 20, 2004), http://www.isteve.com/babygap.htm (accessed September 7, 2014).

[15] "Red State, Blue State, Teen Birthrate, Teen Abortion rate," *Gene Expression*, (January 27, 2010), http://www.gnxp.com/blog/2010/01/red-state-blue-state-teen-birthrate.php (accessed September 7, 2014).

Sexual equality

Patriarchy is an effective V-promoter and a reflection of high V, so trends towards sexual equality indicate declining V over the past century and a half. In the Victorian era, women lacked the vote and were effectively barred from most professions. In the absence of a marriage agreement stipulating otherwise, a married woman's entire property and money (including wages) passed to her husband, with whom she was legally considered to be one entity.[16]

The change since then has been dramatic. As early as 1840, laws in the United States and Britain began to protect women's property from their husbands and their husbands' creditors, and in 1870 the UK Parliament passed the Married Women's Property Act which gave women the right to their own property once they were married.[17]

By the late-nineteenth and early-twentieth centuries, women's suffrage movements were making headway across much of the Western world. In 1893 New Zealand gave women the full vote, followed by Australia in 1902, then Finland in 1906 and Norway soon after.[18] Britain passed the Representation of the People Act in 1918, granting the vote to women over 30 (as opposed to men, who could vote from the age of 21), but this was later amended in 1928 to include equal voting rights for both sexes. By the 1920s women had received the vote in Denmark, the United States, Austria, Germany, Canada and the Netherlands, although France did not follow suit until 1944 and Switzerland not until 1971.[19]

In the 1960s the feminist movement began to challenge conventional gender roles as well as legal and workplace inequalities.[20] In the United

[16] B. Griffen, *The Politics of Gender in Victorian Britain: Masculinity, Political Culture and the Struggle for Women's Rights* (Cambridge: Cambridge University Press, 2012), 9–10.

[17] Encylopeadia Brittanica, "Married Womens Property Acts " http://www.britannica.com/EBchecked/topic/366305/Married-Womens-Property-Acts (accessed September 7, 2014).

[18] Teachers, "Women's Suffrage," http://www.scholastic.com/teachers/article/womenx2019s-suffrage (accessed September 7, 2014).

[19] Ibid.

[20] J. Freeman, "From suffrage to Women's Liberation: Feminism in Twentieth Century America," http://www.jofreeman.com/feminism/suffrage.htm (accessed September 7, 2014); Encylopeadia Brittanica, "Women's Movement,"

States the movement grew with the passing of the Equal Pay Act of 1963 and the Equal Credit Opportunity Act of 1973.

By the twenty-first century discrimination has been reversed, in some circumstances, and there are increasing requirements for women to be given preference in political parties and on company boards to balance the male-female representation. Such significant changes in gender relations in a civilization over the course of just 150 years are unprecedented, and indicate a rapid fall in V.

Declining militarism

Further evidence of falling V is the decline of militarism in the second half of the twentieth century. Although the growing distaste for war can be partly attributed to the turning of the lemming cycle following G years (with subsequent peaks of V) in the nineteenth century, the change has been far more drastic than in past cycles.

The contrast between the fervent jingoism of 1914 and the widespread pacifism of the early twenty-first century could hardly be more marked. In 1914–18 deaths in combat included around 117,000 Americans, 887,000 Britons, 1.4 million Frenchmen, and more than two million Germans. Yet German proposals for a negotiated peace in 1916 were rejected outright.

By the late twentieth century, V had declined to the point where even small-scale wars became increasingly unpopular. A watershed moment was the loss of Vietnam. America's casualties were less than a tenth of what they had been in the Second World War, yet overwhelming public opinion turned against the war and forced a staged withdrawal. Much of the opposition to the war focused on Vietnamese rather than American casualties, and returning US troops were often abused by protesters. Ubiquitous television coverage no doubt contributed to this change, but it is also consistent with a general decline in V.

It has been indicated that the militancy of the anti-war protests reflects the relatively high V of the generation born in 1944–45, but declining V in society as a whole caused this to be expressed as an anti-war rather than a pro-war movement.

---

http://www.britannica.com/EBchecked/topic/647122/womens-movement (accessed September 7, 2014).

Although this attitude to US troops was short lived, the opposition to war became even more marked by the turn of the century. US public opinion had initially supported the invasion of Iraq in 2003, but four years later had turned decisively against it. By then, only 40% of US citizens agreed it had been the right decision. Most people wanted the troops home as soon as possible, despite a death toll significantly less than in Vietnam. The same applies to the recent withdrawal from Afghanistan, with scarcely any pretense that the stated objectives have been achieved. This is not to say that any of these wars were justified or unjustified, but only that people have less taste for conflict and far less tolerance of military casualties.

Loss of cultural confidence and morale

V is also related to high morale, so a decline should have a dramatic effect on the way people perceive their own cultures.

During the nineteenth century, Western cultural confidence was at an all-time high. European colonial empires ruled most of the world, and science and technology progressed at an unprecedented rate. This paternalistic imperialism is perhaps best summed up in Rudyard Kipling's poem "The White Man's Burden," which glorified the civilizing mission of the British Empire, whilst at the same time warning of the immense burden imposed by the divine requirement to civilize the world.

> Take up the White Man's burden—
> Send forth the best ye breed—
> Go send your sons to exile
> To serve your captives' need
> To wait in heavy harness
> On fluttered folk and wild—
> Your new-caught, sullen peoples,
> Half devil and half child[21]

Charles Kingsley wrote in a similar vein, mirroring the feeling of immense confidence of the time: "the glorious work which God seems to have laid on the English race, to replenish the earth and subdue it."[22]

---

[21] R. Kipling, "The White Man's Burden," *McClure's Magazine* 12 (February 1899).

[22] C. Kingsley, *Miscellanies*, Vol. II (Google Books, 1863), 364

By contrast, since the 1960s there has been a widespread sense of pessimism in Western nations, relating to areas such as the environment, politics and belief in science.[23] Surveys over the past forty years have shown that for every American who believes the country is going in the right direction, two believe things are getting worse. This opinion covers all fourteen areas on which questions were asked including the criminal justice system, public safety, national leaders, Americans' honesty, their work ethic, the health care system, education, standards of living, the economy, and racial issues.[24] Annual surveys of a broad range of American high-school students have also shown a steady decline in social trust since the 1970s, which is to be expected as high-V peoples tend to form tight-knit communities.[25]

There have been numerous explanations for this profound loss of morale, most focusing on the wars and upheavals of the twentieth century. But the Napoleonic wars that devastated Europe at the beginning of the nineteenth century were followed within a few decades by the supreme confidence of the Victorian era. And loss of cultural confidence did not become commonplace until the 1960s and 1970s, more than two decades after the Second World War. A far more plausible reason is a fall in V—something that is easily testable at a physiological level.

Decline in the corporal punishment of children

As always in history, the decline in V has been accompanied by a fall in stress. One indication is changing attitudes to the physical punishment of children, which have gone from unquestioned acceptance to a criminal offense in many Western nations in less than a century.

A compilation of seventy biographies of people who lived before the seventeenth century reveals that every individual was severely beaten in

---

[23] O. Bennett, *Narratives of Decline in the Postmodern World* (Edinburgh: Edinburgh University Press, 2001); A. Herman, *The Idea of Decline in Western History* (New York: Free Press, 2007).

[24] R. E. Bradley, *Upside: Surprising Good News About the State of Our World* (Grand Rapid, Michigan: Bethany House, 2011), 17–18.

[25] W. M. Rahn & J. E. Transue, "Social Trust and Value Change: The Decline of Social Capital in American Youth, 1976–1995," *Political Psychology* 19 (3) (1998): 545–65.

his or her childhood.[26] Such practices were considered not only healthy to the development of a child but an entirely necessary part of moral development. It was only in the case of a perceived injustice or severe injury that the perpetrator's actions were questioned. This continued, though with lesser severity, to the nineteenth century, when it was still seen as an indispensable means of enforcing discipline.

It was not until the mid-twentieth century that the practice became widely questioned. Following trends towards leniency, Benjamin Spock published Baby and Child Care in 1946, which advocated the abandonment of punishment and strict discipline in favor of treating children with reason and compassion. At the time of its publication the book was considered controversial and outside of the mainstream, but it soon became a bestseller, reflecting a public willingness to accept such ideas.

The latter half of the nineteenth century also saw a sharp decline of corporal punishment in French secondary schools. Though permitted to this day, the "right to correction" has become less severe and more regulated.[27]

Other countries have taken an even harder line. Physical punishment of school children was officially banned in the Netherlands in 1920,[28] Italy in 1928, Norway in 1936[29] and Austria in 1974.[30] It was widespread in German schools in the early-twentieth century, but gradually outlawed state-by-state following the Second World War, culminating in complete

---

[26] "Patterns of Child Rearing," *The International Child and Youth Care Network* 48, (January 2003) http://www.cyc-net.org/cyc-online/cycol-0103-mckerrow.html (accessed September 7, 2014).

[27] C. Heywood, *Growing Up in France: From the Ancien Régime to the Third Republic* (Cambridge: Cambridge University Press, 2007), 243; R. Slee, *Changing Theories and Practices of Discipline* (London: The Falmer Press, 1995), 44; European Committee of Social Rights, *European Social Charter (revised): Conclusions 2005, Volume 1 (Bulgaria, Cyprus, Estonia, France, Ireland, Italy, Lithuania)* (Strasbourg: Council of Europe Publishing, 2005), 241.

[28] Nederlands Juristenblad 496, (March 20 1920).

[29] Council of Europe, *Eliminating Corporal Punishment: a Human Rights Imperative for Europe's Children* (Strasbourg: Council of Europe Publishing, 2005).

[30] Article 47(3) of the School Education Act.

abolition through the 1970s and 80s.[31] In Britain caning was officially banned in state schools and some private schools in 1987, followed by a complete ban in remaining private schools in 2003.[32] In Japan it was officially banned in 1947 though still supported by public opinion in some areas.[33]

More controversially, bans on the physical punishment of children have been extended to families in many European nations. Countries that have banned all forms of physical punishment, inside and outside the home, include Germany, the Netherlands, Norway, Sweden, Austria, Spain and Denmark, although it is still permitted in France, Italy and Britain.[34] In recent times the European Court of Human Rights attempted to impose a blanket ban on all physical punishment of children on the grounds that it violates children's human rights.[35]

As in many areas, the United States is behind the trend and remains the only Western nation where physical punishment is permitted in schools, especially in the southern states. Fig. 16.6 shows which states still permit corporal punishment in schools, an indication of higher stress and thus V.

---

[31] D. Schumann, "Legislation and Liberalization: The Debate About Corporal Punishment in Schools in Postwar West Germany, 1945–1975," *German History* 25 (2) (2007): 192–218;
Council of Europe, *Eliminating Corporal Punishment*,118.
[32] M. Pate & L. A. Gould, *Corporal Punishment around the World* (Santa Barbara: Praeger, 2012), 81–2.
[33] Ibid., 83.
[34] Ibid., 55–6.
[35] Council of Europe Parliamentary Assembly (Recommendation 1666 [2004]). United Nations Convention on the Rights of the Child (Article 19).

Fig. 16.6. US States permitting physical punishment in schools, 2005–6.[36] Physical punishment in schools is a sign of higher V in southern states.

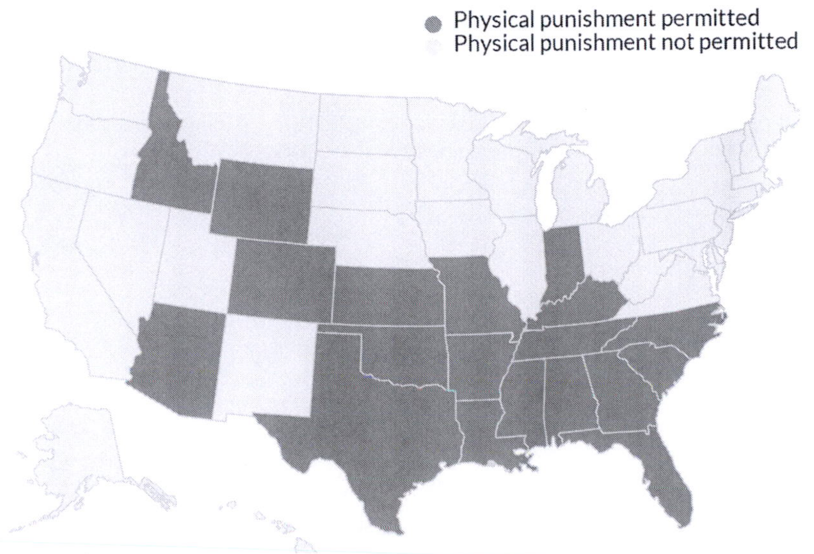

Decline of capital punishment

Another expression of stress, and thus V, is the willingness to execute or severely punish criminals. Capital punishment was reduced drastically in Western nations in the nineteenth century, and the trend continued in the twentieth.[37] In Britain the execution of juveniles under 16 was outlawed in 1908,[38] and the death penalty for murder was suspended in 1965 and then abolished in 1969. A complete ban under all circumstances occurred in 1998. Similar movements against the death penalty were underway right

---

[36] The Centre for Effective Discipline, Discipline at School—U.S.: Corporal Punishment and Paddling Statistics by State and Race http://www.stophitting.com/index.php?page=statesbanning (accessed September 7, 2014).

[37] R. Hood & C. Hoyle, *The Death Penalty: a World Wide Perspective* (Oxford: Oxford University Press, 2008), 11–15.

[38] P. Hodgkinson, "The United Kingdom and the European Union," in *Capital Punishment: Global Issues and Prospects*, edited by Peter Hodgekinson, Andrew Rutherford, 194 (Winchester: Waterside Press, 1996).

across the Western world.[39] Today, the death penalty has been banned in all Western countries except the United States,[40] with a small and perhaps temporary resurgence in recent years. Once again, America shows higher V levels than other Western countries, though the general trend of overall decline is the same (see Fig. 16.7 below).

Fig. 16.7. Persons executed in the US, 1930–2010.[41] A decline in executions is a sign of falling V, despite a slight recovery in the 1990s.

Increased IQ

The fall of stress has had effects which many people consider beneficial. It may even have raised IQ, though loss of rigid child V thinking patterns will also have had an impact. In chapter thirteen it was suggested that the success of Ashkenazi Jews could be explained in part by genetic selection for higher intelligence over the past five hundred years. But most of the achievements of such peoples (including also the ancient Greeks and Italians of the Renaissance period) were the result of an epigenetic effect in which high C combined with low stress to raise creativity and intelligence. One reason for supposing this is the startling rise in IQ scores

[39] Hood, *The Death Penalty*, 42–7,

[40] Ibid., 11–5, 49–50, 354.

[41] US Department of Justice's Bureau of Justice Statistics.

in industrializing nations over the past half-century. At the extreme, young Dutch males in 1982 scored 20 points higher than their predecessors in 1952. This means, in effect, that the average Dutch male in 1982 rated higher than 90% of those tested in 1952.

The effect has been found in fifteen European nations, the United States, China, India, Japan, South Korea, urban Brazil and Argentina, and recently even in countries such as Kenya and the Dominican Republic.[42] As near as can be assessed, these gains occur in exactly the times and places where there is evidence of falling V and stress, such as growing affluence and declining birth rates.

If this is the case we should also expect such gains to peter out and eventually reverse as the fall of stress slows and is countered by falling C. And this does seem to have been happening recently in developed countries, where IQ scores have stabilized or even declined very slightly.[43]

Implications of declining V

There are many indications of falling V and stress in Western society, most of which can be viewed as positive. There is less taste for war, more opportunities for women, possibly higher intelligence, and the end of the threat of overpopulation.

How we view these trends also depends on what is likely to happen in the future. Conventional wisdom is that the West has reached a new equilibrium, that birth rates plus immigration will maintain population, that Western nations will remain generally peaceful but capable of defending themselves, and that they will remain open and tolerant liberal democracies. An example of this thinking can be found in Francis Fukuyama's book *The End of History*:

> What we may be witnessing is not just the end of the Cold War, or the passing of a particular period of postwar history, but the end of history as such … That is, the end point of mankind's ideological evolution and the universalization of Western liberal democracy as the final form of human

[42] J. R. Flynn, *Are we Getting Smarter? Rising IQ in the Twenty-First Century* (Cambridge: Cambridge Uni Press, 2012).
[43] R. Lynn & J. Harvey, "The Decline of the World's IQ," *Intelligence* 36 (2) (2008): 112–20.

government.[44]

Biohistory indicates that this is not the case, and that such trends will continue. Some notion of the long-term implications can be seen from the Western Roman Empire, which combined depopulation with increasing military weakness. Parallels between the decline of Rome and the modern West are easy to draw and not necessarily meaningful in themselves. But once we recognize the *reason* for the decline of Rome, the parallels become more maningful. Rome fell because of dramatic collapses in first V and then C, as a result of its wealth and of the inherent instability of any civilization with high infant C.

Our civilization is in exactly the same situation. People may believe that "technology" will save them, but technology accelerates the fall in V by its power to create wealth. An even more serious problem is that the fall of V, combined with this same wealth, is rapidly undermining C.

Declining C

This section provides evidence for a widespread decline of C in the West over the past 150 years, and especially since the 1960s. Once again the decline of Rome provides the best predictive model for the long-term implications, though with the qualification that the modern West is at a relatively early stage. Given that Roman C probably peaked around 250 BC, we would now be at the rough equivalent of the year 100 BC in the Roman Republic, although machine-generated affluence means that C and V are probably declining at a far more rapid pace.

Earlier age of puberty

C is a physiological system which adjusts animals to conditions of chronic food shortage. One way in which it achieves this is by delaying the age of puberty. As indicated in chapter six, the age of puberty was around 13–14 in famine-ridden Medieval Europe, rising to as high as 17 by the relatively prosperous nineteenth century as C reached unprecedented heights. From that time it began to fall rapidly in all European countries for which figures exist (see Fig. 16.8 below).

---

[44] F. Fukuyama, *The End of History and the Last Man* (New York: Free Press, 1992).

Fig. 16.8. Declining age of menarche in Europe and the US.[45] A declining age of puberty (measured here by age of menarche) reflects falling C as much as increased prosperity.

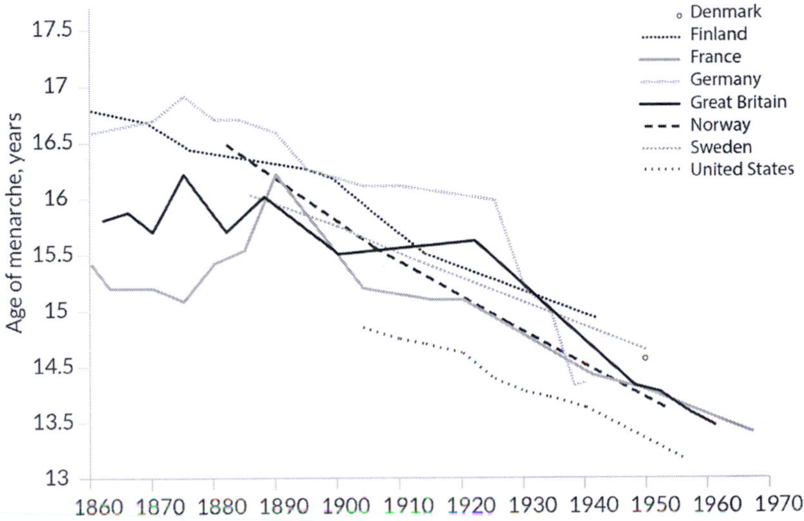

Similar patterns can be seen in Japan, though offset somewhat because Japanese C peaked later than in Europe. The age of menarche in Japan was around 15 in 1920, declined to 14.2 by 1940, rose to 15 as a result of privations during the Pacific War, and then declined to 12.2 by the present day.[46] The decline in age was even more delayed in South Korea, falling

---

[45] Tanner (1962) http://www.breastcancerfund.org/assets/pdfs/publications/falling-age-of-puberty.pdf (accessed September 3, 2014); *Journal of Epidemiol Community Health* 60 (11) (2006): 910–911; P. E. Brown, "The Age at Menarche," *British Journal of Preventive & Social Medicine* 20 (1) (1966): 9–14. Notes: (i) figures for Germany and have been taken from this source due to superior detail compared to the *Journal of Epidemiol Community Health* and Tanner; (ii) figures from Britain taken from this source up to 1950, then filled in from Tanner in the later part of the twentieth century.

[46] M. Hosokawa, S. Imazeki, H. Mizunuma, T. Kubota & K. Hayashi, "Secular Trends in Age at Menarche and Time to Establish Regular Menstrual Cycling in Japanese Women Born between 1930 and 1985," *BMC Women's Health* 12 (1) (2012): 19, http://www.biomedcentral.com/1472-6874/12/19 (accessed September 7, 2014).

from 16.8 in the 1920s to 15.23 in the 1960s and 12.6 in the 1980s.[47] The major decline, of course, coincided with post-war prosperity.

The age of puberty has fallen even further in recent decades, with an age of 8 or 9 being no longer unheard of. Significantly, this is more likely in obese children, African Americans and boys engaging in risky behaviors.[48] All these are associated with lower C—African Americans because their ancestors came from lower C societies.

Many scientists have suggested that earlier puberty simply reflects better nutrition,[49] which as an explanation is true but not sufficient. The nineteenth century was more prosperous than the famine-ravaged Europe of the Middle Ages, so the rising age of puberty cannot have been driven by food shortage. Nor, for the same reason, can the recent decline in the age of puberty be wholly explained in terms of nutrition. As we have seen, levels of C in human societies are determined more by culture than food shortage as such, so a rapid fall in the age of puberty indicates a rapid decline in C.

Obesity

Falling C is a result of overeating but it is almost certainly a cause as well, and it helps explain the rise of obesity in Western societies. Obesity has been linked to risk-taking in men and impulsivity in women, as well as

---

[47] J.-Y. Hwang, C. Shin, E. A. Frongillo, K. R. Shin & I. Jo, "Secular Trend in Age at Menarche for South Korean Women Born between 1920 and 1986: The Ansan Study," *Annals of Human Biology* 30 (4) (2003): 434–42.

[48] P. B. Kaplowitz & S. E. Oberfield, "Reexamination of the Age Limit for Defining when Puberty is Precocious in Girls in the United States: Implications for Evaluation and Treatment," *Pediatrics* 104 (4): (1999): 936–41; P. B. Kaplowitz, E. J. Slora, R. C. Wasserman, S. E. Pedlow & M. E. Herman-Giddens, "Earlier Onset of Puberty in Girls: Relation to Increased Body Mass Index and Race," *Pediatrics* 108 (2) (2001): 347–53; E. J. Susman, L. D. Dorn & V. L. Schiefelbein, "Puberty, Sexuality, and Health," *Comprehensive Handbook of Psychology*, edited by M. A. Lerner, M. A. Easterbrooks & J. Mistry (New York: Wiley, 2003); W. C. Chumlea, C. M. Schubert, A. F. Roche, H. E. Kulin, P. A. Lee, J. H. Himes & S. S. Sun, "Age at Menarche and Racial Comparisons in US Girls," *Pediatrics* 111 (1) (2003): 110–3.

[49] P. McKenna, "Childhood Obesity Brings Early Puberty for Girls," *New Scientist* (March 5, 2007), http://www.newscientist.com/article/dn11307-childhood-obesity-brings-early-puberty-for-girls.html (accessed September 7, 2014).

single parenthood and poverty,[50] all of which are indications of low C. This connection can also be inferred from the *effect feedback cycle*, whereby any behavior that increases C will also be increased by C. If eating more food decreases C, then lower C should encourage overeating.

Fig. 16.9 below shows that the proportion of overweight people is rising in all OECD nations, and the projected rise in the future. The US may score slightly higher because of large ethnic minorities. Korea, which industrialized relatively recently, has comparatively few overweight people, although this can be expected to change as C continues to decline.

Fig. 16.9. Past and projected overweight rates in selected OECD countries.[51] A steady increase in obesity could be a consequence of falling C as well as plentiful food.

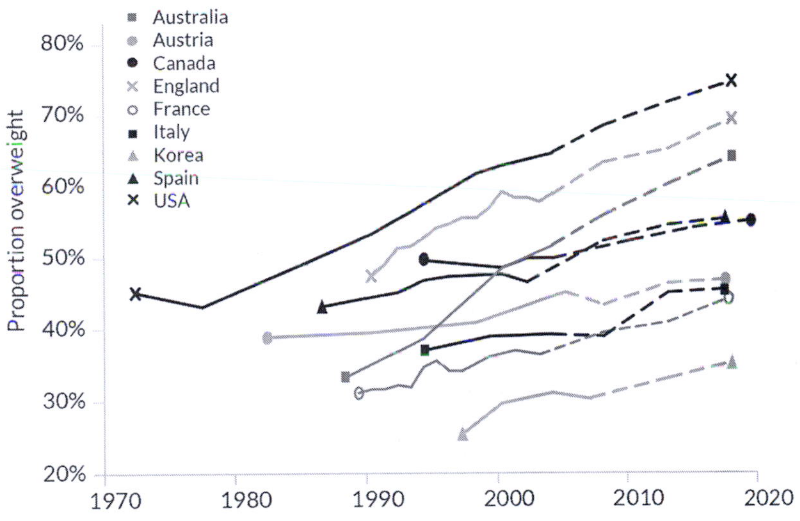

[50] G. Koritzky, E. Yechim, I. Bukay & U. Milman, "Obesity and Risk Taking. A Male Phenomenon," *Appetite* 59 (2) (2012): 289–97; C. Nederkoorn, F. T. Y. Smulders, R. C. Havermans, A. Roefs & A. Jansen *Appetite* 47 (2) (2006): 253–256; A. Drewnowski & S. E. Specter, "Poverty and Obesity: the Role of Energy Density and Energy Cost," *The American Journal of Clinical Nutrition* 79 (1) (2004): 6–16.

[51] *OECD*, "Obesity and the Economics of Prevention: Fit not Fat—France Key Facts," http://www.oecd.org/els/health-systems/obesityandtheeconomicsofpreventionfitnotfat-francekeyfacts.htm (accessed September 7, 2014).

Early marriage and sexual activity

Late age of marriage is also an indication of high C, being part of the biological pattern which delays breeding when food is limited. The minimum allowed age of marriage in England was 12 in 1275, rising to 16 by 1885, reflecting a rise in the age of puberty.[52] Similarly, as we saw in chapter six, the average age of marriage rose steadily from medieval times until the eighteenth and nineteenth centuries.

Age of first marriage has not fallen in the past century but age of sexual activity has, and biologically speaking that is what matters.[53] This change has been especially marked in the second half of the twentieth century. In France, the age of first sexual activity for women has fallen from 20.6 in the 1950s to 17.5 today, and for men from 18.8 to 17.2.[54]

In Britain, 27% of young women are sexually active before the age of consent, compared with just 4% for those born in the 1950s.[55] Similar patterns can be found in the US, where the percentage of females who first had sex by the age of 17 almost doubled from 20% to a little under 40% between 1972 and 1987. The average age of first sex in the US was 16.9 in 2005, although there are signs that it may have risen in recent years.[56] The reasons probably have to do with the recession cycle as indicated in

---

[52] S. Roberston, "Age of Consent Laws,"*Children and Youth in History* http://chnm.gmu.edu/cyh/teaching-modules/230 (accessed September 7, 2014).

[53] Biologically speaking, lower C increases sexual activity and thus raises the birth rate, but in humans it has the opposite effect on births.

[54] *INED*, "Age at first sexual intercourse in France," http://www.ined.fr/en/teaching_kits/population_of_france/age_first_intercourse_france (accessed September 7, 2014).

[55] S. Buckler, "New Report Reveals Sexual Behaviour across the Different Age Groups," *GovToday* (December 15, 2011), http://www.govtoday.co.uk/health/44-public-health/9286-new-report-reveals-sexual-behaviour-across-the-different-age-groups (accessed September 7, 2014).

[56] Figures cover the entire U.S. population, including minority groups, Ranking America, “The U.S. Ranks 13th in Age of First Sex,” 2009, http://rankingamerica.wordpress.com/2009/01/28/the-us-ranks-13th-in-age-of-first-sex/ (accessed September 7, 2014);
Real Clear Politics, “Virginity Rising, (2011), http://www.realclearpolitics.com/articles/2011/03/10/virginity_rising_109173.html (accessed September 7, 2014); Facts on American Teens' Sexual and Reproductive Health, *In Brief: Fact Sheet*, Guttmacher Institute June 2013, http://www.guttmacher.org/pubs/FB-ATSRH.html (accessed September 7, 2014).

chapter ten, which also indicates that the change is likely to be temporary.

The age of first sexual activity has lagged behind age of puberty as an indication of falling C, presumably because cultural norms are slower to break down than the change in temperament that is driving the change. There are also class differences, as shown by the *Kinsey Report* in the 1940s, with delayed sexual activity associated with economic success. A recent study found that the average age of first sex in low-income families in Boston, Chicago and San Antonio was 12 for boys and 13 for girls, with a handful having sex as early as 8 or 9.[57] Raising C by delaying sexual activity can have substantial benefits in terms of occupational success.

Sex: more activity, less control

Overall levels of sexual activity, and its escape from the confines of marriage, show a similar trend. As we saw in chapter six, standards of sexual behavior became increasingly strict up to the nineteenth century, at which time "respectable" women were believed to have little interest in sex.

According to Kinsey, women born in the first decade of the twentieth century were more sexually active than those born earlier. People coming of age in the 1920s were twice as likely to engage in premarital intercourse as those born before 1900, which is consistent with the popular image of that age—flappers, speakeasies, and the throwing-off of old constraints. Those born in the next two decades, and thus coming of age in the 1930s and 1940s, showed similar levels of sexual activity.[58] In other words, there was no restoration of traditional behavior, but no further decline in it either. The behavior of different cohorts can be explained in terms of the recession cycle, as discussed in chapter 10.

Then, from the 1960s onward, the generation born after the Great Depression of the 1930s launched what is commonly termed the "sexual revolution." This was undoubtedly eased by the ready availability of contraceptives such as the Pill, but reflects even more strongly the underlying fall in C. In line with the effect feedback cycle, of course, the

---

[57] *Futurity*, "Poverty Linked to Early Sexual Activity in Kids," (2009), http://www.futurity.org/society-culture/sex-starts-early-for-low-income-youth/

[58] Kaplowitz & Oberfield, "Reexamination of the Age Limit for Defining when Puberty is Precocious in Girls in the United States"; Kaplowitz et al. "Earlier Onset of Puberty in Girls"; Susman et al. "Puberty, Sexuality, and Health."

sexualization of society must also accelerate the decline in C.

This loosening of sexual restrictions resulted in a surge in the number of extramarital biths from the 1960s onward, as traditional courtship patterns broke down. Rates of extramarital birth in England and Wales rose during the twentieth century, dropped slightly after the world wars but then rose continuously from 1960 as C steadily declined (see Fig. 16.10 below).

Fig. 16.10. Rates of extramarital birth, England and Wales, 1900–2002—percentage of live births outside marriage.[59]

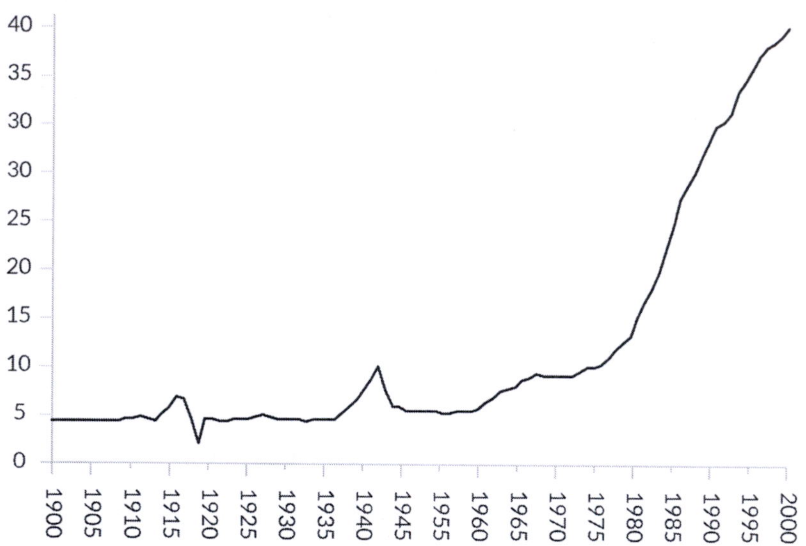

Similar patterns can be seen in the United States and other Western countries, and the rising incidence of divorce has a similar meaning. The West is moving away from the strong and relatively exclusive pair bonds of a high-C society towards greater sexual activity with multiple partners.

Tolerance of homosexuality has a similar meaning, as both an indication and a cause of declining C. In early-nineteenth-century England homosexuality was punishable by death. In 1861 the penalty was reduced to a sentence of hard labor for ten years to life, and then in 1885 to no less

[59] C. G. Brown, *Religion and Society in 20th-Century Britain* (London: Longman, 2006), 32.

than two years.[60] Today, homosexuals serve openly in the US military and an increasing number of jurisdictions recognize gay marriage.

These trends reflect a decline in C, as well as the abandonment of the religious and cultural technologies that support it. A recent study found that even among people who consider themselves "very religious," 35% of women and 39% of men consider it acceptable for unmarried 18-year-olds to have sexual relations. For the non-religious the figures were 74% and 79% respectively.[61] These are very similar to the cultural changes of the late Roman Republic, though taken to a far greater extent. It is worth noting that, in common with most indications of C, the decline has been most evident since the 1960s.

Decline of the nuclear family

As indicated in chapter 6, from medieval times to the nineteenth century the monogamous nuclear family became increasingly important as a social unit. This is the family and breeding pattern most characteristic of food-restricted animal societies, and also of high C human societies.

The rate of divorce has been rising since the peak of C in the mid-nineteenth century (see Fig. 16.11 below). The absolute rise in the divorce rate is not dramatic, but this is because of the rise in non-regular unions which by their nature tend to be more often short-term. One important result has been an increase in one-parent families. Between 1960 and 2000 the proportion of children raised by single parents grew from 9% to 28%. In Britain the proportion of people living as married couples with dependent children has fallen from 52% in 1971 to 37% today.[62] Once again, indications of falling C are most evident from the 1960s onward.

---

[60] M. Cook, *London and the Culture of Homosexuality, 1885–1914* (Cambridge: Cambridge University Press, 2008); R. Chipchase, "Attitudes towards Homosexuality in Victorian England," http://www.helium.com/items/2172125-attitudes-towards-homosexuality-in-victorian-england (accessed September 7, 2014);
Hamish, "Reflections on BNA, part 6: British Law," *The Drummer's Revenge*, http://thedrummersrevenge.wordpress.com/2007/07/25/reflections-on-bna-part-6-british-law/ (accessed September 7, 2014).

[61] New Strategist Publications, *American Sexual Behavior* (Ithaca, New York: New Strategist Publications, 2006), 80.

[62] T. Dalrymple, "The Paradoxes of Cultural Confidence:
Is Western culture in decline?" http://www.aims.ca/site/media/aims/Paradoxes.pdf (accessed September 7, 2014).

Fig. 16.11. Marriage and divorce in the United States, 1860–2000.[63] Increasing divorce is a sign of the breakdown in lifelong monogamous relationships—a clear indication of falling C.

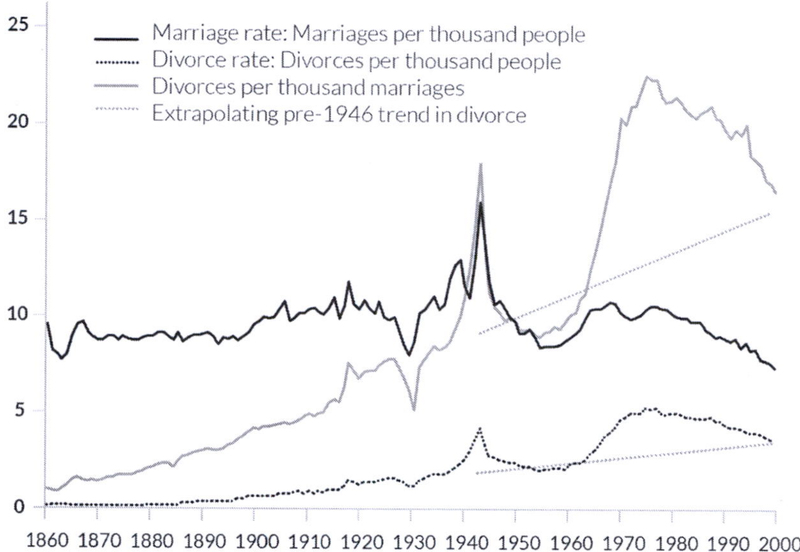

Less control of children

As detailed in chapter 6, control of children in Europe reached a peak in the nineteenth century, along with other indications of very high C. Books such as The Care and Feeding of Children (1894) by Dr Luther Emmett Holt reflected popular attitudes by encouraging the rigid scheduling of feeding, bathing and sleeping, and an avoidance of all undue stimulation of toddlers.[64] Similar patterns continued into the early twentieth century. Government pamphlets encouraged the strict discipline of children and tight feeding schedules, even warning parents that playing with their children could be harmful.[65]

---

[63] B. Stevenson & J. Wolfers, "Marriage and Divorce: Changes and their Driving Forces," *Journal of Economic Perspectives* 21 (2) (2007): 27–52.

[64] Internet FAQ Archive, "Parenting," (2008), http://www.faqs.org/childhood/Me-Pa/Parenting.html (accessed September 7, 2014).

[65] Internet FAQ Archive, "Discipline," (2008), http://www.faqs.org/childhood/Co-Fa/Discipline.html (accessed September 7, 2014).

The first signs of change can be seen in the 1920s. Amidst the growing prosperity of the post-war recovery, emphasis on rigidity was reduced, with the well-adjusted adult being viewed as more easygoing. This corresponded with the gradual rejection of the mechanistic and behaviorist idea of childrearing, and an increasing emphasis on the importance of meeting a child's emotional needs. For example, in 1936 pediatrician Charles Aldrich published Babies Are Human Beings, which advocates a move towards greater leniency.[66]

This went much further in mid-century, an attitude most strongly expressed by Dr Spock. He implored mothers to enjoy their children and reject traditional notions of childrearing. In recent decades this "revolution against patriarchy" has led to a minimum of discipline or control in American families.[67] Control of children is one of the most reliable and consistent measures of C, and here again there is a dramatic change in a very short period. Once more, the most rapid change is in the late twentieth century.

Declining sperm count and testosterone

As yet unpublished research suggests that sperm count is significantly higher in rats experiencing calorie restriction in early life. When female rats were 25% calorie restricted during the first 11 days after their litters were born, the adult sperm count of their male offspring was significantly higher. The same higher sperm count was found when a juvenile rat experienced 25% CR between 22 and 28 days.[68] This implies that declining C could result in a loss of fertility. As mentioned in chapter twelve, this may have been a problem in the Roman Empire, and declining Chinese gentry families also seem to have been relatively infertile.[69]

---

[66] Internet FAQ Archive, "Parenting."

[67] J. E. Block, *The Crucible of Consent: American Child Rearing and the Forging of Liberal Society* (Cambridge Mass.: Harvard University Press, 2012).

[68] J. Penman et al., Report series on studies on rats conducted at Latrobe University, 2012, (forthcoming).

[69] H.-T. Fei, "Peasantry and Gentry: An Interpretation of Chinese Social Structure and its Changes," in *Social Structure and Personality*, edited by Y. A. Cohen, 24–35 (London: Holt, Rinehart and Winston, 1961); P.-t. Ho, *The Ladder of Success in Imperial China* (New York: Columbia University Press, 1962), 129–145, 157–160, 166; E. O. Reischauer & J. K. Fairbank, *East Asia: The Great Tradition* (Boston:

Declining C could thus account for evidence of declining sperm count in Western men over the past two generations, something commonly attributed to estrogenic chemicals in the environment. These results are controversial and other studies have found no change, but a recent study confirmed a 32.2% decline in French sperm count between 1989 and 2005, with the proportion of properly formed sperm falling from 60.9% to 52.8%. This is especially convincing because of the large number of participants: over twenty-six thousand.[70] Declining C could provide an explanation for such a change, although further testing will be required to confirm it.

There is also evidence that the testosterone levels of American males have been declining over the past two decades, something to be expected as infant C (with its testosterone boosting effect as described in chapter two) falls to a lower level.[71]

Changes in moral teaching and the rise of passive welfare

In tracing the rise of the West we saw how the focus of moral attitudes changed from personal charity to values such as integrity, self-discipline,

---

Houghton Mifflin, 1960), 187, 223–228; F. Hsiao-Tung, *China's Gentry: Essays in Rural-Urban Relations*, edited by M. Park, , 205–206, 246–247, 270–272 (Chicago: University of Chicago Press, 1953).

[70] E. Carlsen, A. Giwercman, N. Keiding & N. E. Skakkebæk, "Evidence for Decreasing Quality of Semen During past 50 Years," *British Medical Journal* 305 (6854) (1992): 609.

J. Raloff, "That Feminine Touch: Are Men Suffering from Prenatal or Childhood Exposure to 'Hormonal' Toxicants?" *Science News* 145 (1994): 56–8; G. R. Bentley, "Environmental Pollutants and Fertility," in *Infertility in the Modern World: Present and Future Prospects*, edited by G. R. Bentley & C. G. N. Mascle-Taylor (New York: Cambridge University Press, 2000); J. A. Saidi, D. T. Chang, E. Goluboff, T., E. Bagiella, G. Olsen & H. Fisch, "Declining Sperm Counts in the United States? A Critical Review," *The Journal of Urology* 161 (2) (1999): 460–2; J. Auger, J. M. Kunstmann, F. Czyglik & P. Jouannet, "Decline in Semen Quality among Fertile Men in Paris during the Past 20 Years," *New England Journal of Medicine* 332 (5) (1995): 281–5; M. Rolland, J. Le Moal, V. Wagner, D. Royère & J. De Mouzon, "Decline in Semen Concentration and Morphology in a Sample of 26,609 Men Close to General Population between 1989 and 2005 in France," *Human Reproduction* 28 (2) (2013): 462–70.

[71] T. G. Travison, A. B. Araujo, A. B. O'Donnell, V. Kupelian & J. B. McKinlay, "A Population-Level Decline in Serum Testosterone Levels in American Men," *The Journal of Clinical Endocrinology & Metabolism* 92 (1) (2007): 196–202.

industriousness and principled philanthropy. These attitudes reached a peak in the nineteenth century, and were especially associated with the Victorian era. The prevalence of such attitudes is a strong indication of high C.

A recent study of 18–23-year-olds from a broad range of backgrounds found that they had few absolute standards of any kind. The only "traditional" values retained were that murder, rape and robbery were seen as wrong. Otherwise, morality was discussed in terms of what their peers will think and the chance of getting caught, and standards of right and wrong were a matter for personal choice.[72] Such attitudes were observed in the early-nineteenth century, to the evident shock of the middle-class people making the observations, but only in the very poorest urban areas.[73] By the early twenty-first century they had become mainstream.

As standards such as self-discipline and integrity have declined, so government handouts have increased, free from the stringent moral requirements of the Victorian era. As C declines, governments spend more and more of the national wealth on subsidizing low-income earners, paying state pensions, and generally supporting the needy, as can be seen from the increase in welfare spending over the past sixty years (see Fig. 16.12 below).

Far from being unique to our age, this is a return to the pattern of free and relatively unconditional charity of the Middle Ages, and similar to the corn doles of the late Roman Republic and early Empire. Though lacking the stern integrity of earlier times, the Roman government was increasingly generous in its subsidies to the urban poor. This began with the provision of subsidized grain by the popular leader Gaius Gracchus in 123 BC, and in 53 BC Clodius Pulcher provided free grain to the poor in a bid for political support. Later emperors offered free or subsidized grain along with popular entertainment, the well-known "bread and circuses." Their motivation was largely to buy political support, or at least political calm, but this applies just as much to our own era. In America today nearly half the population is receiving some form of government benefit, which

[72] C. Smith, H. D. Christoffersen & P. S. Herzog, *Lost in Translation: The Dark Side of Emerging Adulthood* (New York: Oxford University Press USA, 2011).
[73] Report of the Commissioners (1841), 7, http://www.forgottenbooks.org/books /Report_of_the_Commissioners_1841_1000269580 (accessed September 7, 2014).

makes it politically difficult to reduce welfare.[74]

Fig. 16.12. Welfare and pension spending as percent of GDP in the US[75]

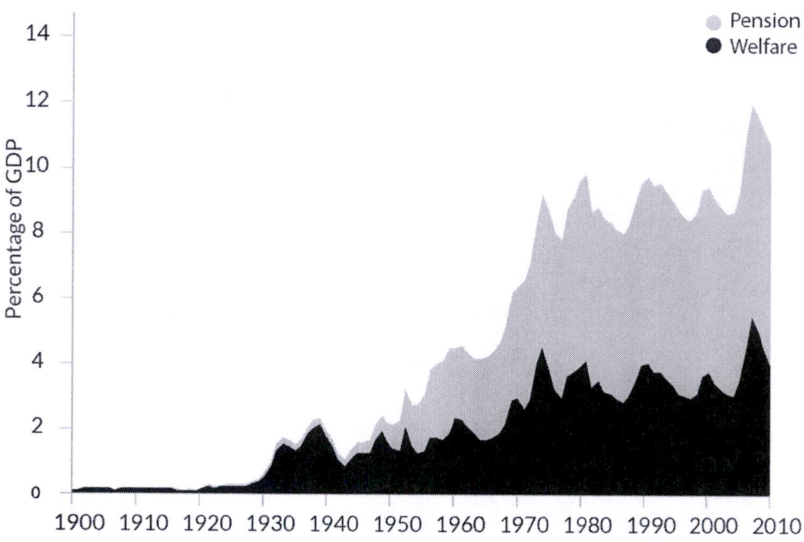

Decline of the market; rise of the state

A reliable indication of high C is that goods tend to be transferred by the action of the market, rather than through mutual sharing as in hunter-gatherer societies, or direct transfer by political authorities.

A steady rise in government spending is thus a strong indication of falling C, and especially the individualistic infant C. In most European countries governments now spend well over 50% of their GDPs.[76] Even in America, behind the trend in so many ways, the government's share has risen from 7% in the early 1900s to more than 42% in a little over a century. As Fig. 16.13 below suggests, in terms of expenditure the New Deal of the 1930s was no more than a slight fluctuation in this ongoing trend.

---

[74] *Wall Street Journal*, (June 12, 2013); W. Voegeli, *Never Enough: America's Limitless Welfare State* (New York: Encounter Books, 2012).

[75] US Government Spending, usgovernmentspending.com (accessed September 7, 2014).

[76] US Government Spending, http://www.usgovernmentspending.com /us_20th_century_chart.html (accessed September 7, 2014).

Fig. 16.13. Total (Federal, State, and Local) US government Spending.[77] Lower C people want governments to redistribute income and are less self-reliant and market-oriented. The late Roman Republic demonstrates similar trends.

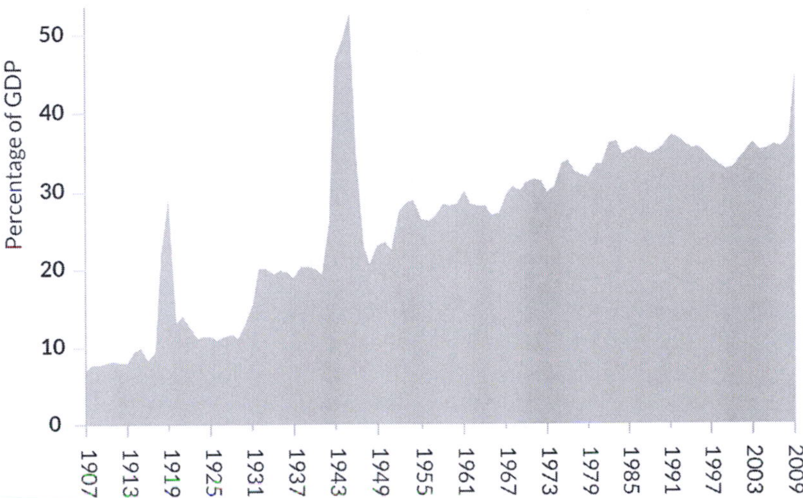

Accompanying this rise in spending is a massive increase in government regulation, with a multitude of laws covering wages, hiring and firing, licensing, planning permissions, rent controls, environmental issues and much more. While some of this (such as limits on pollution) is fair and necessary, much of it reflects the comfort that lower C people feel in having economic decisions made by government rather than individuals. As a result, they are ready to believe that government spending and controls promote growth, even when practical experience (and most economists) suggest otherwise.

Increased state control of the economy is not new to history— the Roman Empire demonstrates similar changes. By the third century AD local initiative and independence had been gradually smothered by the Empire's ever expanding bureaucracy, leading to crushing demands on the local populace. The Emperor Diocletian in the late third century made laws to fix prices and tie people to their ancestral professions. Trade declined and expenses soared as soldiers' wages strained an already stagnant economy. Failure to pay harsh taxes meant severe punishments, and many fled urban

[77] Ibid.

centers to avoid financial ruin and military service.[78] People with lower infant C are more accepting of government control, at least until C falls so far that governments are no longer viable.

In this area as in so many others, the twentieth and twenty-first centuries have shown unmistakable signs of falling C. America is not immune to the trend but somewhat behind other countries, especially in the "red states" that vote Republican. The reason this does not always correlate with economic performance is that "blue states" such as California have benefited the the most from falls in child V and stress.

Declining nationalism

Another characteristic of high infant C is loyalty to the nation state, especially the kind of loyalty that makes people adamantly resist control by alien rulers. In one sense, nationalism was most fully expressed in the aftermath of the First World War, when the defeat of the Central Powers plus a decline in child V allowed subjected nations such as Poland, Czechoslovakia and Ireland to gain independence.

But by other measures, nationalism has declined steeply since the early-twentieth century. Surveys in a number of European countries have shown a decline in national pride and confidence in the army.[79]

One expression is the growth of the European Union. Not only trade but immigration policies and many rules and regulations are now decided by nations acting in concert, a remarkable change from the fierce national rivalries of less than a century ago. The motivation has been economic rather than cultural, and there remains strong opposition from some sections of the population such as represented by the UK Independence Party, but it is a remarkable change in a very short time.

The rise of the United Nations also reflects declining nationalism. The vote of an international body is now regarded as an important determinant as to whether a nation should make war, even though some nations

---

[78] J. B. Bury, H. M. Gwatkin, J. P. Whitney, J. R. Tanner, C.W. Previte-Orton & Z. N. Brooke, *The Cambridge Medieval History. The Rise of Sacarens and the foundation of the Western Empire* Vol. 2 (New York: Macmillan, 1991) 40–41, 548–51.

[79] D. Mattei, "The Decline of Traditional Values in Western Europe," *International Journal of Comparative Sociology* 39 (1) (1998): 77–90.

(especially the US) are willing to act without such a sanction.

Also related is the ideology and practice of multiculturalism. People with high infant C show a strong preference for their own ethnic group, language and culture. As C declines other cultures are seen as equally valid, with the effect of allowing in large numbers of immigrants from poorer countries, contributing to the loss of national identity.[80]

With regard to local loyalties within nation states, trends have been more mixed. Federal countries such as the US and Australia have seen a marked increase in the relative power and importance of central government. For example, in 1900 more than half of all government spending in the US was at the local level, and most of the rest was by states. By the 2010s federal government accounted for more than 20% of GDP, with states accounting for 8–9% and local spending just over 10%.[81]

On the other hand, Europe in recent decades has seen growing regionalism in areas such as Scotland, Wales and Catalonia. The reasons for this are probably more economic than temperamental. The loss of the British Empire and growth of the EEC has made national unions less economically necessary. Also, in the case of Catalonia, democratic government and the absence of any credible external threat has made secession more feasible than it might once have been.

Declining impersonal loyalties

One outstanding characteristic of societies with high infant C is the strength of their impersonal loyalties. Very low-C societies show loyalty to individuals who are personally known, such as in hunter-gatherer bands. Moderate C involves loyalty to leaders who are identifiable but not personally known, such as monarchs. Very high-C people are loyal to impersonal institutions such as parliaments, political parties, codes of law and nations. But unions and business corporations are also impersonal institutions, reaching over regional and even national boundaries. It thus highly significant that all such loyalties are in obvious decline. Europeans

---

[80] *The Decline of Britishness: A Research Study*, Ethnos Research and Consultancy, Commission for Racial Equality, (May 2006), http://www.ethnos.co.uk/pdfs/10_decline_of_britishness.pdf (accessed September 7, 2014).

[81] US Government Spending, http://www.usgovernmentspending.com /us_20th_century_chart.html (accessed September 7, 2014).

in recent decades show less belief in parliaments, administration systems and political parties (as indicated earlier), but also in corporations and unions.[82] This is reflected not just in attitudes but in the continued and precipitous decline in membership of both unions and political parties (see Fig. 16.14 below).[83]

Fig. 16.14. Members of UK political parties 1975–2010.[84] Declining membership shows a lack of engagement with democratic politics.

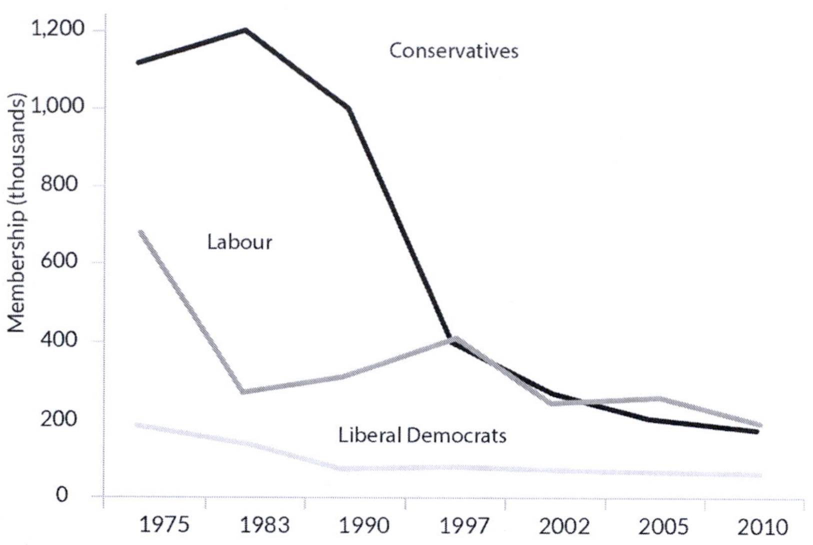

[82] Mattei, "The Decline of Traditional Values in Western Europe," 77–90.

[83] R. S. Katz, P. Mair, L. Bardi, L. Billc, K. Deschouwer, D. Farrell, R. Koole, L. Morlino, W. Muller, J. Pierre, T. Poguntke, J. Sundberg, L. Svasand, H. VandeVelde, P. Webb & A. Widfeldt "The membership of political parties in European democracies, 1960-1990," *European Journal of Political Research* 22 (1992): 329–345; P. Mair & L. Van Biezen, "Party membership in twenty European democracies, 1980-2000," *Party Politics* 7 (1) (2001): 5–21;
W. T. Dickens & J. S. Leonard, "Accounting for the Decline in Union Membership", The National Bureau of Economic Research, Working paper 1275, (1985); S. H. Farber & A. B. Krueger, "Union Membership in the United States: The Decline Continues" The National Bureau of Economic Research, Working Paper 4216, (1992).

[84] House of Commons Library, 2012.

There has also been a considerable loss of trust in government in general, one obvious measure being the decline in voter turnout in national elections (see Fig. 16.15 below).[85]

Fig. 16.15. Voter turnout in elections 1950–2010—France, Germany, UK, Japan, US.[86] Declining voter turnout reflects a declining engagement with the state.

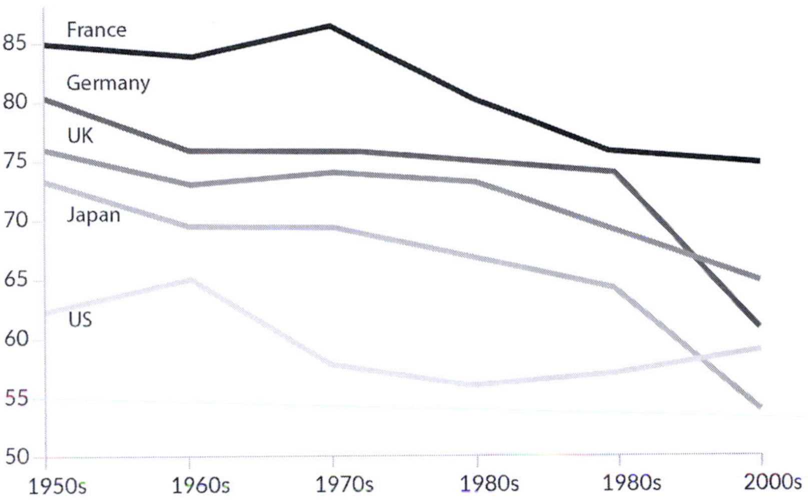

Significantly, US statistics show that voter turnout is lowest among groups with the lowest levels of C—people with less education, and the young. Barely 40% of eligible 18–24 year olds vote in Presidential elections, and only 20% of people without a high school diploma.[87] A continuation of such trends must eventually undermine democracy.

---

[85] R. Hardin, "Government without Trust," *Journal of Trust Research* 3 (1) (2013): 32–52

[86] R. J. Dalton, *Citizen Politics: Public Opinion and Political Parties in Advanced Industrial Democracies*, 5th edition (Washington DC: CQ Press, 2008), 37; *International Institute for Democracy and Electoral Assistance*, "Voter Turnout Database", International IDEA website, http://www.idea.int/vt/viewdata.cfm (accessed September 7, 2014).

[87] Peter Levine and Mark Hugo Lopez, "Youth Voter Turnout has Declined, by Any Measure " *CIRCLE: The Center for Information & Research on Civic Learning & Engagement*, September 2002, http://civicyouth.org/research/products/Measuring_Youth_Voter_Turnout.pdf (accessed September 7, 2014).

Declining work ethic

As indicated in chapter six, the rise of C to the nineteenth century was reflected in an increasingly powerful work ethic, and a willingness to work long hours at routine jobs in factories and farms. But since then the work ethic in the West has been in decline. For example, average working hours in the United States have dropped considerably in the past century (see Fig. 16.16 below).

Fig. 16.16. Average weekly hours worked per US male, all racial groups, aged 25–54, 1900–2005.[88] Lower C people are less inclined to work hard, so working hours drop with the fall in C.

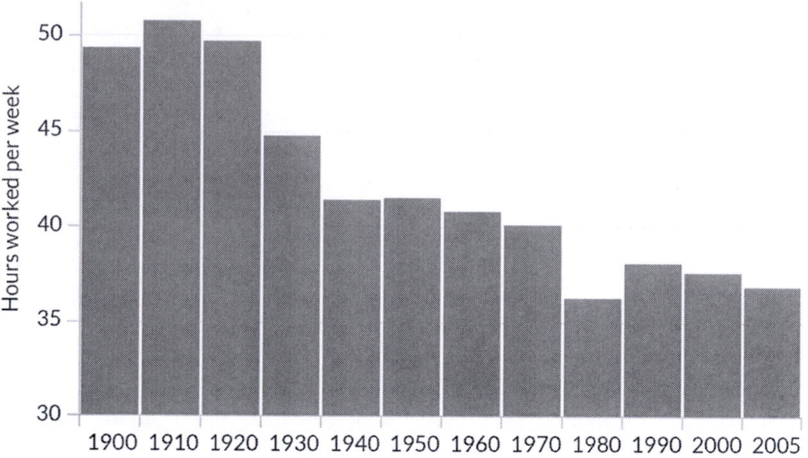

This change is only partially reflected in unemployment statistics because of the growing numbers on disability payments. Between 1960 and 2010 the percentage of Americans qualifying for Federal disability benefits rose from 0.7% of the labor force to 5.3%. The proportion of working-class men who were unemployed also rose steadily through these years, despite a large number of available jobs taken up by immigrants.[89]

Changes in labor laws reflect the growing demand for less strenuous working hours. In the United States at the turn of the twentieth century the

---

[88] W. A. Sundstrom, (2006), tables Ba4568 and Ba4589; Kendrick 1961, table D-10, http://weber.ucsd.edu/~vramey/research/Century_Published.pdf (accessed September 7, 2014).

[89] Murray, *Coming Apart*, 170–9.

average working week in manufacturing was between 53 and 59 hours, but by 2005 it stood at around 40. Similarly, the number of hours spent at college has declined from 40 hours a week from the 1920s –1960s to 27 hours a week in 2004, reflecting less strenuous demands on students.[90]

The affluence argument does not explain the reduced number of hours worked, since unskilled wages had actually fallen since the 1960s, meaning that people should logically need to work more hours to earn the same income. A better explanation is that people have simply become less industrious.

We have seen that higher C people have a more positive attitude to work and are more likely to enjoy it. The General Social Survey is the most widely used database for tracking American social trends, and one question it asked prime-aged white men was which factors they found most important when choosing a job. The choices were:

- High income
- No danger of being fired
- Chances for advancement
- Working hours are short; lots of free time
- Work important and gives a feeling of accomplishment

Until 1994, an average of 58% said the most important thing they looked for was a sense that their work was important and gave a feeling of accomplishment. By 2006, the proportion putting this in first place had dropped to 43%. Until 1994 the reasons least likely to be given as first priority were short working hours (4%) and no danger of being fired (6%). But in 2006, 9% gave short working hours as their first priority, and 12% gave job security.[91] This is a clear indication that work has become less valued for its own sake.

---

[90] Sundstrom, (2006); P. S. Babcock & M. Marks, "The Falling Time Cost of College: Evidence from Half a Century of Time," Use Data NBER Working Paper No. 15954, April 2010, https://mail.google.com/mail/u/0/?tab=wm#inbox/145280cff0a40755 (accessed September 7, 2014).

[91] C. Murray, *Coming Apart: The State of White America, 1960–2010* (New York: Crown Forum, 2012), 169.

Employers have noticed the change. A survey of American professionals by Ernst & Young found that baby boomers, born between 1946 and the mid-1960s, were more than twice as likely to be seen as cost effective as the generation born from the early 1980s. They were also far more likely to be considered hard working. On the other hand, they were judged to be slightly less entrepreneurial.[92]

A study of unemployed men in 2003–5 found that, by comparison with the unemployed in 1985, they spent less time in job searching, education and training. They did less useful work around the house and spent less time in civil and religious activities. They even spent less time on active pastimes such as exercise, sports, hobbies or reading.[93]

A comparison with other cultures shows that affluent societies do not necessarily work shorter hours. An perfect example of a relatively high-C society today is South Korea. In 1960, following the devastation of the war, South Korea was one of the poorest countries in the world, with an average income roughly on par with the poorest parts of Africa. As of 2012 it was richer than the European Union average, with a per capita GDP of $31,753, compared with $31,607 for the EU.[94]

Despite having achieved a First World standard of living, South Koreans still work 2,193 hours per year, compared to the OECD average of 1,718 hours.[95] Having industrialized later than the West, they have maintained a higher level of C than those in the Western world. But as C has begun to fall there are signs that the South Korean work ethic is weakening, as in other industrialized nations. Today, South Koreans typically work 206

---

[92] "Winning the Generation Game," *The Economist* (September 28, 2013).

[93] Ibid., 180–1.

[94] "What Do You Do when You Reach the Top?," *The Economist*, (September 28, 2013);
Central Intelligence Agency, "The World Factbook,"
https://www.cia.gov/library/publications/the-world-factbook/geos/ja.html
(accessed September 2014); B. Mason, "Britain: Income Inequality at Record High," World Socialist Web Site,
http://www.wsws.org/articles/2009/jun2009/inco-j04.shtml (accessed September 7, 2014).

[95] *BBC News Magazine*, "Who Works the Longest Hours?," (May 23, 2012),
http://www.bbc.co.uk/news/magazine-18144320 (accessed September 7, 2014).

hours a month on average, down from an average of 226 in the 1980s.[96] This gradual reduction in hours worked has been assisted by government regulation, including the introduction of a mandatory 40-hour working week (172 hours per month).

Differences in C can help explain the difference in work ethic of ethnic groups within a society. In the United States, both whites and Asian Americans stem from traditionally high-C societies. But Asians have been losing C for only a few decades while whites have been in decline for a century and a half. The hours spent studying by college students, averaged over a year, reflect these differences. Asians (highest C) studied an average of 15 hours a week, whites (moderate C) a little over 10. African Americans, descended mainly from the moderate C people of sub-Saharan Africa, have the lowest C, so their hours of study are the lowest (see Fig. 16.17 below).

Increased debt and reducing savings

People with low time preference, which chapter six linked to high C, are willing to sacrifice current consumption to build future wealth. The essence of this is thrift, and we have seen thrifty habits rose from the Middle Ages to the nineteenth century and played a crucial role in the Industrial Revolution. Falling C should have the opposite effect of a preference for present consumption over future benefit. At a moderate level this simply means living on current income and not saving or investing for the future. At an extreme level it implies debt, which is consuming now at the expense of future wealth.

Thus it is no surprise that debt levels have grown dramatically in recent decades as C has fallen. As can be seen from Fig. 16.18 below, levels of saving have dropped dramatically in the US since the 1970s, a trend only slightly and briefly reversed by the Global Financial Crisis. Meanwhile, in the past six decades, debt has grown from 20% to 120% of personal disposable income.

---

[96] M. Baker, "In land of longest hours, workers get a break," *The Christian Science Monitor*, http://www.csmonitor.com/2001/0821/p1s3-woap.html (accessed September 7, 2014).

Fig. 16.17. Hours per week spent on homework by US college students.[97] Like the provincials who kept the Roman Empire afloat for four centuries, Asians have higher C. Theirs will drop also as they are influenced by the declining C culture of the West.

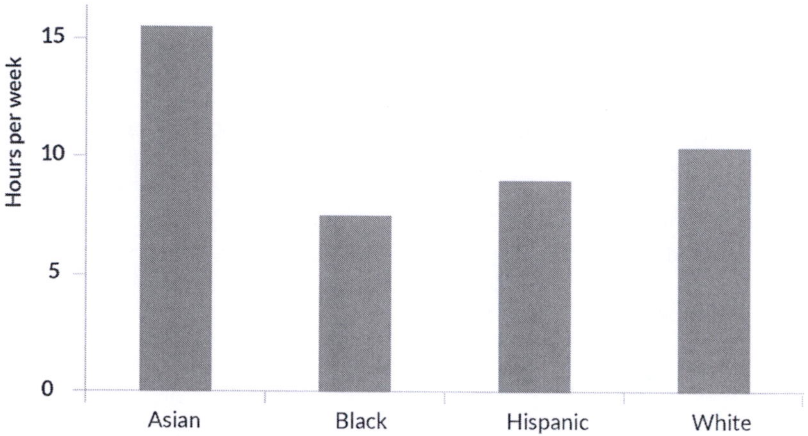

The same attitude impacts at the national level, with citizens increasing their consumption of government services without being willing to pay for them through taxes. The result is spiraling levels of government debt (see Fig. 16.19 below). This behavior is encouraged by economic theorists such as Keynes, who blame recessions on excessive saving and inadequate spending.[98]

---

[97] V. A. Ramey, "Is there a "Tiger Mother" Effect? Time Use Across Ethnic Groups," (2011), http://weber.ucsd.edu/~vramey/research/Tiger_Mothers.pdf (accessed September 7, 2014).

[98] J. Maynard Keynes, *The General Theory of Employment, Interest and Money* (Create Space Independent Publishing Platform, 2011).

Fig. 16.18. US Market Debt Outstanding as percentages of Personal Disposable Income.[99] (Household debt includes *consumer debt* and *mortgage loans*). As C declines, people save less and borrow more as they favor current consumption over future benefit—the reverse of what happened in the lead up to the Industrial Revolution.

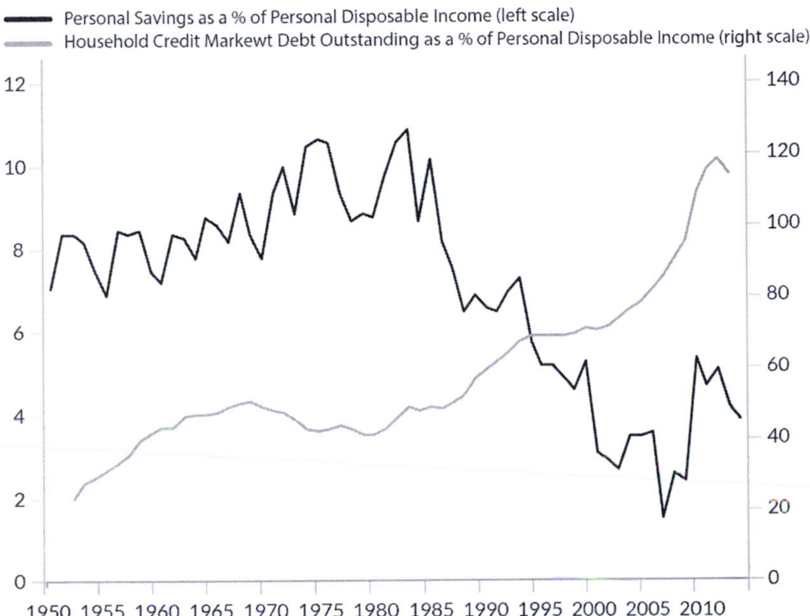

[99] US Department of Commerce, Bureau of Economic Analysis, http://www.bea.gov/histdata/NIyearAPFFiles.asp?docDir=Releases/GDP_and_PI/2012/Q4/Third_March-28-2013&year=2012&quarter=Q4 (accessed September 7, 2014).

Fig. 16.19. Gross Debt as percentage of GDP.[100] As C declines and people prefer current consumption to future benefit, they also pressure governments to plunge into debt.

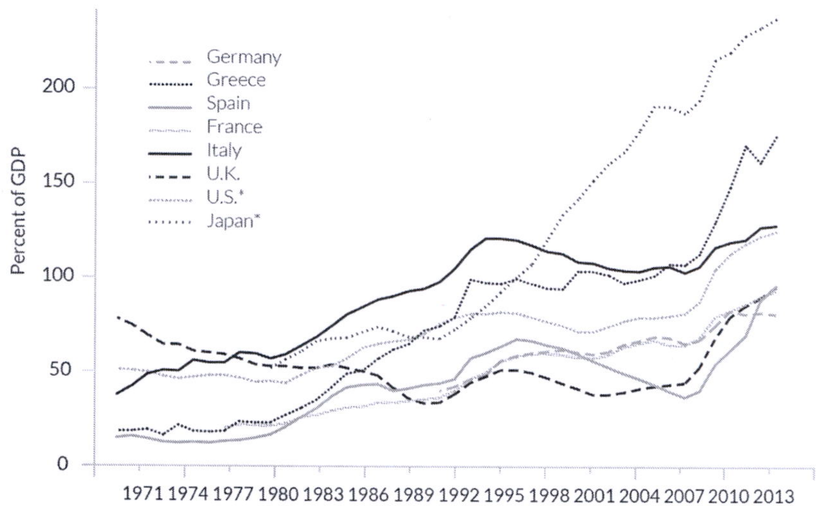

Decline in machine skills, engineering and science

People with moderate C can be competent at jobs requiring personal skills, such as government bureaucracy or retail, especially given the mental flexibility of low child V. But machine skills and engineering are especially the province of people with high infant C, which is the result of parental control in infancy or early childhood. Such peoples include the northern Europeans who launched the Industrial Revolution, the Japanese who were their earliest and most successful disciples, and the machine-minded Manus of the Admiralty Islands. Based on the theory developed in previous chapters, any decline in C should be expressed very early in a lack of interest in and aptitude for these areas.

As discussed in chapter 6, the peak of infant C in nineteenth-century Europe and America was associated with an unprecedented burst of technological progress. During this period pioneering figures such as Edison and Siemens drove industrial growth, and there were major new developments including the widespread use of steam trains, telegraph

[100] AMECO, US Fed Budget; IMF.

cables, radio communications, telephones, automobiles and more. Public attitudes were also extremely favorable toward technology, with engineers and scientists often presented as heroes, a popular fashion for international exhibitions of trade and industry. Technology-related novelists such as Jules Verne in France and Hans Dominick in Germany acquiring millions of readers.[101] But such enthusiasm has ebbed in recent years.

Continuing the graph of innovations in science and technology given in chapter six, and compared against population growth, the dramatic rise to the nineteenth century has been followed by an equally dramatic fall (see Fig. 16.20 below).

The same pattern can be seen for science, a reflection of high C which brought about massive changes in society in the nineteenth and early twentieth centuries. The decline of science, in particular physics and chemistry, has been particularly notable in Britain over the past fifty years. Britain was the first country to become wealthy and has seen a more obvious decline in C over this period. One sign is that falling applications have led to the closure of physics departments across the country. Since 1982, the number of A-level exam entries has halved, and as a consequence a quarter of universities that previously offered the subject have stopped teaching it since 1994.[102]

Newspapers have cited a "sustained national decline in demand for physics education," and the Institute of Physics, the professional body for physicists in the United Kingdom, described the trends as "enormously worrying," with serious implications for national prosperity.[103]

---

[101] F. S. Becker, "Why Don't Young People want to Become Engineers?," *European Journal of Engineering Education* 35 (4) (2010): 349–366.

[102] Materials World Magazine, "The Decline in Students Studying Physics," (November 1, 2006), http://www.iom3.org/material-matters /the-decline-students-studying-physics?c=574 (accessed September 7, 2014).

[103] L. Ward, "Physics Degree Courses Axed as Demand Slumps," *The Independent*, (January 23, 1997), http://www.independent.co.uk/news/uk/politics/physics-degree-courses-axed-as-demand-slumps-1284606.html (accessed September 7, 2007).

Fig. 16.20. Key innovations in science and technology plotted against population growth.[104] This is based on 8,583 key scientific and technological innovations, plotted against world population. Science and technology skills were strongest when the West reached its peak of infant C in the nineteenth century, and have since declined.

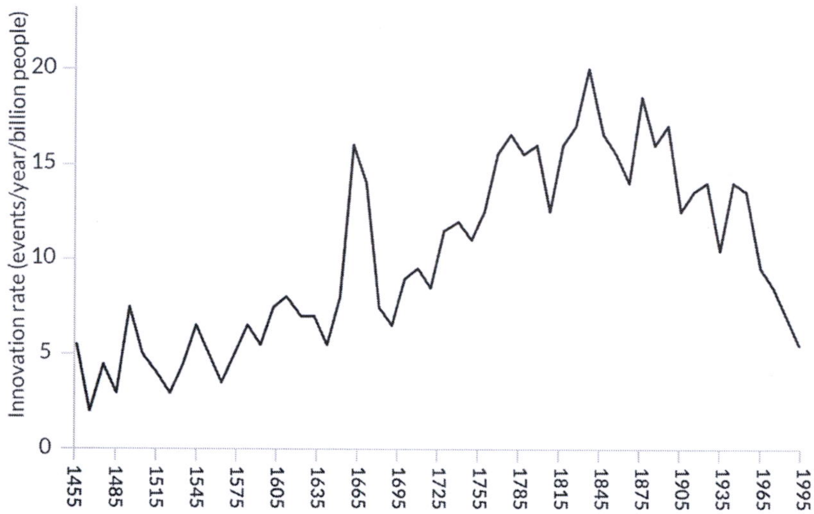

The academic decline in physics and chemistry in Britain has affected the nation's global standing in both these disciplines. Between 1940 and 1975, compared with the US, Britain won 37.5 % as many Nobel prizes in physics and 93.3% as many in chemistry—a remarkable achievement for a nation with one fifth of the population. British scientists were the most prolific winners of Nobel prizes of any country in Europe, but this level of achievement soon changed. Between 1976 and 2005 Britons won only 4.5% as many Nobel prizes in physics as the US, and 16.7% as many in chemistry. The decline can also be seen in the total number of British Nobel prizes, which fell over this period from 9 to 2 for physics, and from 14 to 6 for chemistry.[105] The US has done better not only because it is somewhat behind Britain in the trend to falling C, but because it has

[104] B. H. Bunch & A. Hellemans, *The History of Science and Technology: A Browser's Guide to the Great Discoveries, Inventions, and the People who made them from the Dawn of Time to Today* (New York: Houghton Mifflin Harcourt, 2004); J. Huebner, "A Possible Declining Trend for Worldwide Innovation," *Technological Forecasting and Social Change* 72 (8) (2005): 980–6.

[105] T. Dalrymple, *The New Vichy Syndrome* (New York: Encounter Books, 2010), 5.

benefited from a large influx of higher C immigrants such as Germans, Ashkenazi Jews, and in recent times Asians.[106]

Japan demonstrates a similar trend, even though it is still less than a century from its own peak of C. The first signs that young Japanese were losing interest in science and engineering appeared in the 1980s. Since then, enrollment in science and engineering has consistently declined, with students opting for careers in finance or economics. In a bid to make engineering look more attractive, companies have begun advertising campaigns to try to make the discipline appear "cool," as well as offering generous incentives, headhunting and even offering career coaches to help young Japanese climb the corporate ladder.[107] South Korea has also experienced a declining interest in science and engineering careers.[108]

American predominance in science and engineering continued after Britain went into decline, but the same trend is obvious in recent years there as well. The number of doctorates in Engineering dropped from around 5,700 in 1993 to around 5,330 in 2000.[109] In that last year 27% of bachelor degrees around the world were in natural sciences and engineering, including 52% in China, but in the US only 17% were in these disciplines.[110] The shortage of technical expertise is beginning to show. According to a recent Manpower survey, the vacancies that US employers find hardest to fill are the skilled trades such as welders, electricians and machinists, all occupations linked to C and especially infant C. The situation is about to become considerably worse, since more than half of

---

[106] C. Murray, *Human Accomplishment: The Pursuit of Excellence in the Arts and Sciences, 800 B.C. to 1950* (HarperCollins e-books, 2009).

[107] M. Fackler, "High-Tech Japan Running Out of Engineers," *The New York Times*, http://www.nytimes.com/2008/05/17/business/worldbusiness/17engineers.html?r=1&hp=&pagewanted=all (accessed September 7, 2014).

[108] S.-G. Kang, T.-C. Rho, S.-Y. Hahm & C.-S. Kim, "The present situations of engineering education and accreditation system in Korea," *Journal of Japanese Society for Engineering Education* 54 (6) (2006), https://www.jstage.jst.go.jp/article/jsee/54/6/54_6_6_50/_pdf (accessed September 7, 2014).

[109] National Science Foundation/Division of Science Resources Statistics, Science and Engineering Doctorate Awards: 2000, Detailed Statistical Tables, NSF 02-305 (Arlington, VA, 2001).

[110] R. B. Freeman, Does globalization of the scientific/engineering workforce threaten US economic leadership?, Working Paper 11457, National Bureau of Economic Research working paper series, (2005), http://www.nber.org/papers/w11457 (accessed September 7, 2014).

skilled tradesmen are 45 and older.[111]

Yet, despite this gradual decline, the US remains the world leader in innovation. One reason, as for Nobel prizes, is the number of foreign specialists it recruits. As of 2008 there were 7.8 million foreigners working in highly skilled professions, lured by high wages and other incentives to help fuel America's technology industry.[112] There was also a very large number of Asian doctoral students.

The proportion of foreign students receiving doctorates in these fields is also increasing. Between 1998 and 2001, the proportion of US citizens receiving doctorates in engineering fell from 43.3% to 41.1%, respectively.[113] In 2006, NSF data revealed that foreign students earned approximately 36.2% of the doctoral degrees in the sciences, with approximately 63.6% of those in engineering.[114]

This shift towards foreign students in science and engineering began as early as the 1980s and this share rose 25% between 1990 and 2005. The increasing reliance on foreign talent to fill gaps in science and engineering is having an impact on the background of technology entrepreneurs. Twenty-five percent of high tech companies founded between 1995 and 2005 had at least one foreign born founder, and 75% of companies launched by American venture capital had one key leader born overseas. In Silicon Valley, the proportion was more than one half. The same trend can be seen in the launch of firms backed by venture capital, many of which involve high levels of technology. In the 1970s only 7% of such firms were founded by immigrants. In the 1980s the proportion rose to

---

[111] J. Wright, "America's Skilled Trades Dilemma: Shortages Loom As Most-In-Demand Group Of Workers Ages," Forbes.com (March 7, 2013) http://www.forbes.com/sites/emsi/2013/03/07/americas-skilled-trades-dilemma-shortages-loom-as-most-in-demand-group-of-workers-ages (accessed September 12, 2014).

[112] Fackler, "High-Tech Japan Running Out of Engineers."

[113] E. Frauenheim, "Tech doctorates decline 7 percent," *CNet News*, (January 6, 2003), http://news.cnet.com/2100-1001-979385.html (accessed September 7, 2014).

[114] Christine M. Matthews, "Foreign Science and Engineering Presence in U.S. Institutions and the Labor Force," Congressional Research Service, (October 28, 2010), http://www.fas.org/sgp/crs/misc/97-746.pdf (accessed September 7, 2014).

20%.[115] Meanwhile, American high school students do poorly in mathematical tests, by international standards. [116]

The main reason for this reversal is that the native born are less interested in the science and technology jobs which are most dependent on high infant C. It is even more striking given that these engineers and scientists are largely from nations with relatively low infant C but traditionally high C, such as China, India and Vietnam. As discussed in chapter fourteen, the Chinese very likely have higher infant C than certain other societies such as Arabs and sub-Saharan Africans, but Europeans were once world technological leaders by a huge margin. This advantage has now been reversed.

Of all the areas of human endeavor, technology is the one most dependent on infant C. This makes it the area where the native born first begin to lose their advantage as C declines. Commerce follows closely behind, and after that academia, bureaucracy, politics and the arts. The more personal the skill, the longer the native born retain their advantage.

Decline in Productivity

Progress in science and technology is one factor that has made large gains in productivity possible over the past century. To this can be added improvements in business processes, and the development of a worldwide trade network involving literally billions of people, allowing vast economies of scale. Widespread education has also made it possible to train any number of skilled and specialized people, and to have them move to the areas where they can be most usefully employed. IQ scores have risen and people are increasingly flexible in their thinking. The US has a special advantage because of its ability to attract skilled and talented people, aided by the spread of English as a world language of business and science.

In recent times the computer revolution has opened up great possibilities for gains in efficiency, surely as much as anything that has been possible

---

[115] R. Lenzer, "40% of the Largest U.S. Companies Founded by Immigrants or Their Children," Forbes.com, (April 25, 2013), http://www.forbes.com/sites/robertlenzner/2013/04/25/40-largest-u-s-companies-founded-by-immigrants-or-their-children/ (accessed September 12, 2014).

[116] National Science Board, *Science and Engineering Indicators 2006*, Vols. 1 and 2. (Arlington, VA: National Science Foundation, 2006).

in the past. The operation of Moore's law, according to which computer capacity doubles every two years, is in itself a remarkable expression of this potential. A child's portable device now contains more processing power than that used to send men to the moon in 1969. In short, there is no physical or technological reason why productivity gains in the twenty-first century should not be equal to anything achieved in the past, allowing Western economies to continue their growth while others catch up.

But the outlook changes if we see economic development as primarily driven by epigenetically determined personality traits. As wealth and the turning of the civilization cycle undermine C, people become less economically capable. It is thus of particular interest to note that productivity growth seems to be slowing in Western nations. For example, US labor productivity in the nonfarm business sector declined from an average annual rate of 2.8% in 1948–73 to 1.8% in the years since then. Even a surge driven by the IT revolution between 1990 and 2007 could not achieve the average productivity gains of the post-war decades (see Fig. 16.21 below).

Fig. 16.21. Productivity change in the US nonfarm business sector, 1947–2011.[117] Productivity growth slowed in the late-twentieth century, despite some benefit from the computer revolution. Lower C people are less creative, less productive and less hard working.

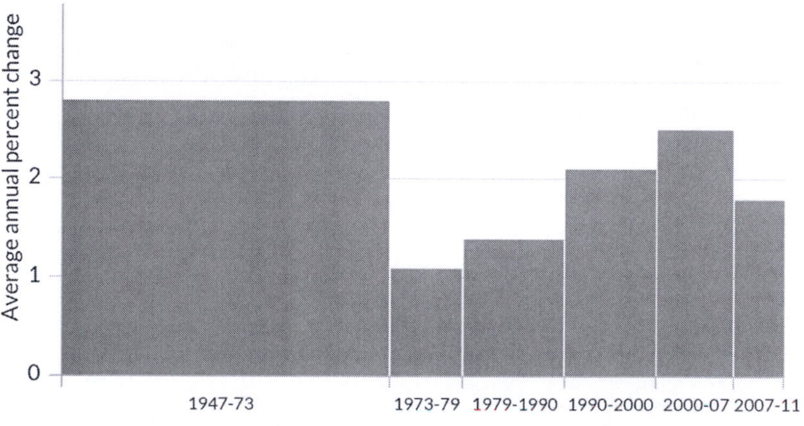

[117] United States Department of Labor, Bureau of Labor Statistics, "Productivity Change in the Nonfarm Business Sector, 1947–2013," http://www.bls.gov/lpc/prodybar.htm (accessed September 7, 2014).

Declining Innovation

A decline in innovation seems to account for a part of this slowdown. For example, in January 2013 *The Economist* noted that despite a revival in business activity in Silicon Valley since 2010, there is a widespread feeling that the pace of innovation has been slackening for decades. It also suggested there was a general slowdown of innovation since the 1970s in areas as diverse as speed of travel and medicine.

If biohistory is correct then this trend will continue. People will become steadily less capable of productive and innovative behavior, and the same temperamental change will make them steadily more favorable to government regulation and control. This indicates that economic growth in Western nations will cease in the relatively near future, to be followed by economic decline. The same will happen, though perhaps with a certain delay, in "emerging" economies such as China.

In the Roman Empire, economic growth ceased first in Italy during the first century BC. Only later, as Roman culture and wealth undermined the C of provinces such as Spain and Gaul, did the economy of those areas start to decline. The relative economic strength of Western nations can be understood in much the same way. The nations with highest infant C, notably Germany and Scandinavia, appear to be somewhat behind the trend, though this advantage can only be temporary. As discussed earlier, America has been helped economically by large-scale elite immigration of higher C people from Asia.

But it is not necessary to wait several decades to test this aspect of the theory. Physiological measures of C should significantly correlate with measures of economic efficiency and productivity, determining not only the success of individuals but that of ethnic groups and nations.

The peak of economic growth

Economic growth reflects the level of C and especially infant C, but C has been falling for 150 years in Europe and America and 70 years in Japan. Despite this, economic growth continued during the twentieth century, and may even have accelerated for a time. The reason can be found in our understanding of the way in which V and C decline, as discussed in chapter twelve. The first stage is a decline in V and especially in the "conservative" child V, with the result that people become more flexible and entrepreneurial. Initially, this more than compensates for a moderate fall in C.

Earlier in this chapter it was suggested that the decline in V and stress was especially rapid in the late nineteenth and early twentieth centuries. Women's rights were extended, capital punishment largely eliminated, and punishment of children banned in most schools. By comparison, changes in these areas since the 1960s have been relatively minor.

In terms of C, there is some evidence of decline in the early twentieth century including a fall in the age of puberty and declining work hours, but the most significant changes have occurred since the 1960s. These include the loss of almost all limits on sexual activity and the decline of the nuclear family. In other words, while both V and C have declined in the past 150 years, the decline of V was more rapid up to the 1960s and the fall of C more rapid since then. In fact, this is exactly what we should expect from our model of the civilization cycle (see Fig. 16.22 below).

Fig. 16.22. Civilization cycle of Western Europe and United States, 1850–2010. The civilization cycle suggests that the fall of V is faster in the early stage of civilization decline, and the fall of C faster in the later stage. In the modern West this changeover occurred in the 1960s.

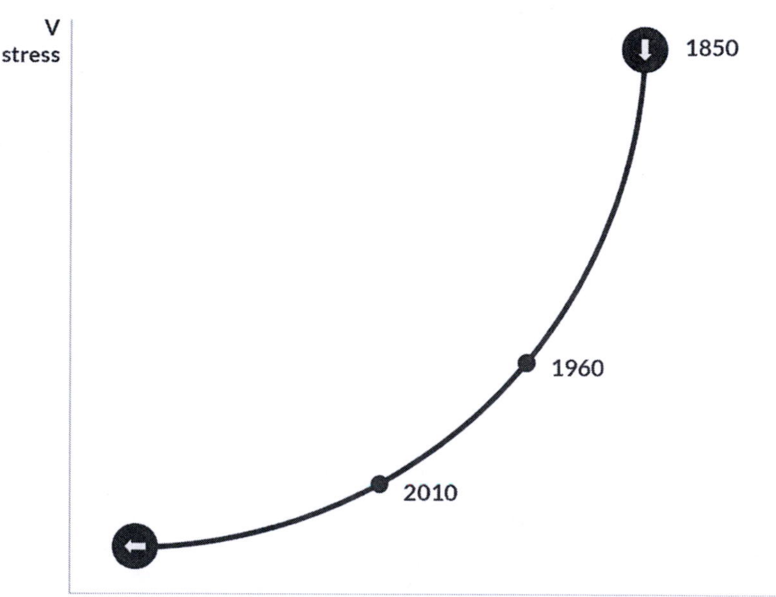

If economic growth is promoted by C but inhibited by child V, then the optimum condition for growth would be one where V has fallen rapidly and C is still largely in place—in other words, the 1960s. During the twentieth century, the only time for which good information is available, the rate of economic growth per capita was in fact greater between 1940 and 1970 than before or after, as Fig. 16.23 below shows for the US.

Fig. 16.23. US Growth Rates 1900-2013—average annual growth in real GDP per capita. The most rapid economic growth was in 1940–70, bracketing 1960 when C was highest relative to V.[118]

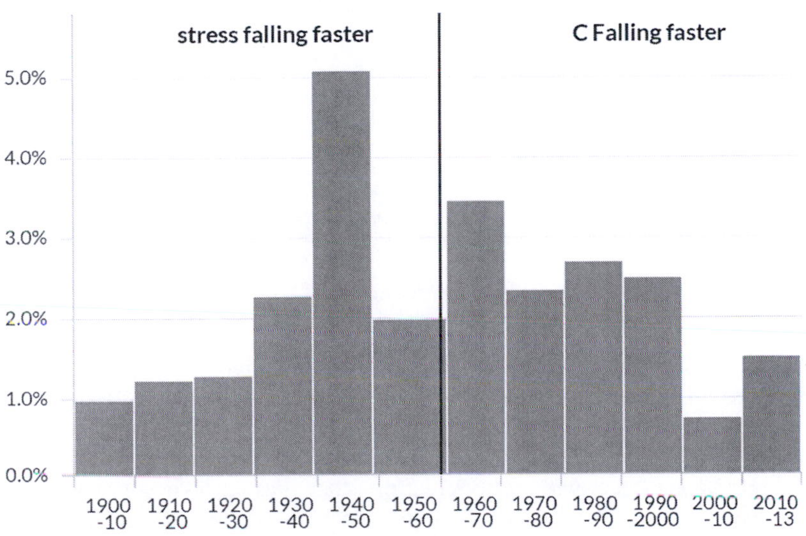

When seeking the reasons for economic growth, economists tend to focus on areas of government policy such as efficient legal systems, lack of bureaucratic regulation, and free labor markets including the right to hire and fire. These certainly have a major impact, but they are not the whole story. In the 1980s the Thatcher government in the UK launched a bold series of pro-market reforms, the effects of which can be seen in Fig. 16.24 below.

Fig. 16.24. UK Growth Rates 1900–2013—average annual growth in real GDP

[118] S. H. Williamson, "What Was the U.S. GDP Then?" *Measuring Worth*, (2014), http://www.measuringworth.com/usgdp/ (accessed September 7, 2014).

per capita. Economic growth was strong between 1950 and 1970, bracketing the year 1960 when there was an optimum combination of high C and low V. Growth surged briefly in 1980–2000 following pro-market reforms by the Thatcher government, but then went back to the pattern of decline.

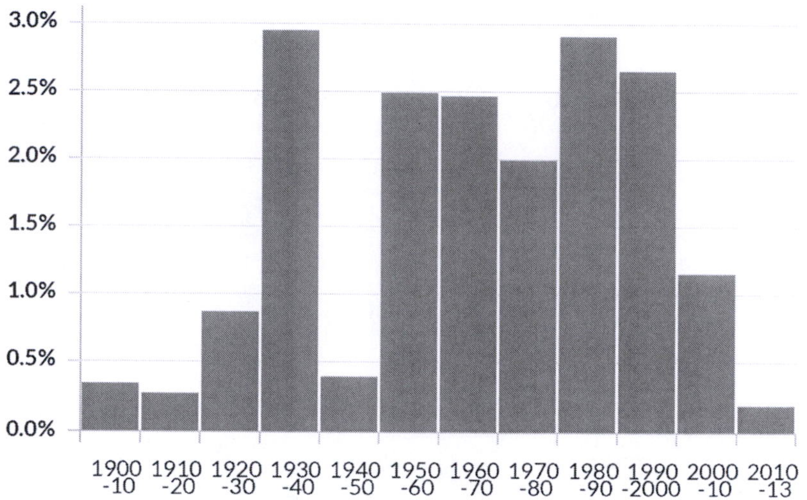

Leaving aside the 1930s when the economy recovered from the Great Depression, growth was especially strong in the 1950s and 1960s, dropped slightly in the 1970s and then surged again in the 1980s and 1990s. Since 2000 it has fallen once again in a trend very similar to that of the US, and entirely consistent with a population that has grown steadily less productive since the 1960s. The pro-market policies of the Thatcher government gave a temporary boost to the economy in the 1980s and 1990s but could not affect the underlying trend to lower productivity based on temperament. And pro-market policies become less politically feasible since populations with falling infant C are less and less favorable to hands-off governments and the free market.

This surge of economic growth around the 1960s is exactly the same pattern as observed in the late Roman Republic, when growing distaste for farming coincided with a commercial boom in the late third and second centuries BC. For example, Italian agriculture changed from a system based on smallholders growing their own food to one largely characterized by slave estates producing wine and olives for the market, with most grain imported from North Africa. Roman traders and moneylenders also became active throughout the newly conquered provinces.

Men with flexible minds took advantage of new opportunities. Marcus Licinius Crassus, for instance, who was born of a senatorial family but in quite modest circumstances, rose to become probably the richest man in Roman history. His wealth was built through political connections, trafficking in slaves, silver mines, and most notoriously real estate. One of his practices was to come to the site of a burning building and buy it and its neighbors at a knockdown price, using his army of clients to put out the fire. Moneylending was another common road to wealth at this time. Such men benefited from the decline in child V which accompanied the early stages of declining C. People may become less suited to the hard and monotonous work involved in farming, but they are better at tasks that require an agile mind.

The Immigrant Advantage

The advantages of falling V only last for a generation or two, before the continuing loss of C saps economic productivity. What saved Rome and allowed it to flourish for several centuries, as discussed in chapter twelve, was immigration. The Romans benefited from a constant influx of higher C newcomers—ex-slaves and provincials, with new elites rising to take over from the declining ones.

Much the same has happened in the modern West, with immigrants such as Eastern European Jews outstandingly successful in the postwar years but now giving way to new elites from higher C cultures, especially Asians. The Japanese, unable to accept large-scale immigration by the peculiarities of their culture, have experienced a lengthy stagnation.

Economic Decline

Of course, even for the West this is only a short-term solution, as the immigrants are acculturated and the C of the majority population continues to fall. All indications are that Western economies must stagnate and then, within a very short time, start sliding backwards. The decline of the work ethic and of innovation, loss of machine skills, declining productivity, spiraling debt and increasingly intrusive governments all contribute to decline. A particular problem will arise from the retirement of the large baby-boom generation, whose infant C reflects the higher level of C prior to 1960. Generations born from 1960 onwards should prove increasingly less capable.

As with the birth rate, there is little governments can do about this trend. Pro-market reforms of the Thatcher variety are less and less possible, and in any case do not affect the basic problem of changing temperament. Increased investment in education can have little effect, especially since it is largely in "soft" subjects irrelevant to the needs of the marketplace. Its effect on temperament may even be negative, since university environments tend to undermine C-promoting values. All projections show that economies in Western countries are headed towards stagnation and decline.

Disease

Another factor that should cause concern is resistnce to disease. At first sight this is a major positive, since epidemic disease has declined drastically in the West over the last century and a half. A curious point is that most of this decline happened *before* the invention of antibiotics.

To take one example, tuberculosis was a virulent disease in Western countries in the eighteenth and early nineteenth centuries, responsible for almost 25% of all deaths.[119] But between 1851 and 1935, tuberculosis mortality in England and Wales dropped from more than 300 per 100,000 people to fewer than 100, with a similar decline in the virulence of other infectious diseases, and before any effective medical treatments were in place.[120] This is not because the disease was eliminated, like smallpox or polio, since even today an estimated one third of the world's population has been infected by TB.[121] The invention of antibiotics has further reduced mortality, but in historical terms they have had far less impact.

As in so many areas, laboratory studies provide answers. We have seen from rat experiments that food restriction (high C) increases resistance to disease, a relationship born out in European history. The Black Death of the late fourteenth and early fifteenth centuries can be attributed to a lemming cycle G-150 period, but it was *not* repeated in the next G-150 period around 1700. By this time C was higher, contributing to immunity.

---

[119] B. R. Bloom, *Tuberculosis : Pathogenesis, Protection, and Control* (Washington, D.C.: ASM Press, 1994).

[120] L. G. Wilson, "The Historical Decline of Tuberculosis in Europe and America: its Causes and Significance," *Journal of the History of Medical and Allied Science* 45 (3) (1990): 366–396.

[121] "The Sixteenth Global Report on Tuberculosis," World Health Organizatio, (2011).

However, disease resistance is also reduced by stress. The unique virulence of Plague around 1400 can be explained by a maximum of the combined level of low C and high stress, which civilization cycle theory suggests peaked around that time.

The 1960s mark the opposite phase of the civilization cycle—the maximum of the combined level of high C and low stress. As noted earlier, it is the time when temperament is most favorable to economic growth. For the same reason, it should also be the time when people are most resistant to disease (see Fig. 16.25 below).

Fig. 16.25. Civilization cycle related to disease. Around 1960 the level of C was highest relative to V and thus stress. The combination of high C and low stress made people unusually resistant to disease. The opposite point of minimum resistance can be found around 1400, the period of the Black Death.

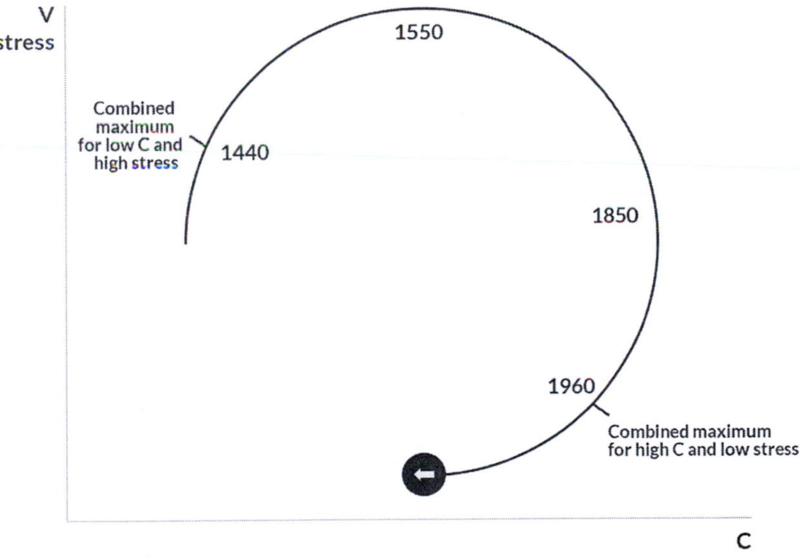

If this theory is correct then infectious disease should currently be on the rise, and this is exactly what we find. From the 1980s there has been a resurgence of disease in many countries—not only old killers such as tuberculosis and cholera but new ones including Legionnaire's disease,

HIV, hepatitis, SARS, and avian influenza.[122]

Once again, Roman Empire provides a similar pattern. The Romans also experienced a massive resurgence of disease in the late Empire, starting with the great plague of 165–180 AD and with another of extraordinarily virulence in the sixth century. Both, as noted in chapter twelve, were in lemming cycle troughs.

Growing gap between rich and poor

Still another variable following the same pattern is equality. We saw in chapter twelve how declining C in the late Roman Republic led to independent farmers, who had been the backbone of the Roman state, giving up their land and flocking to the city to live on subsidized corn. As discussed earlier, the late Roman Republic enjoyed a commercial boom because increasing flexibility of thought, as a result of declining child V, more than compensated for the early loss of C.

However, in a pre-industrial society only a minority of occupations require such flexibility. For peasant farmers whose success largely depended on a grinding regimen of work, loss of C was a serious problem. They were in effect the middle class of Roman society, and their disappearance led to a growing gap between rich and poor since these displaced farmers had little economic value.

In an industrial society there is a far greater premium on flexibility of thought. The strong work ethic and tolerance of routine provided by high C may be sufficient for fruit pickers and some assembly line workers, but most modern occupations demand more. Farmers need to be shrewd businessmen, juggling debt and the purchase of capital equipment, all the while watching market conditions. For the rest we require doctors, lawyers, bureaucrats, teachers, entrepreneurs, computer technicians, engineers, carpenters,

---

[122] N. Bhatti, M. R. Law, J. K. Morris, R. Halliday & J. Moore-Gillon, "Increasing Incidence of Tuberculosis in England and Wales: a Study of the Likely Causes," *British Medical Journal* 310 (6985) (1995): 967–989; P. Davies, "The Worldwide Increase in Tuberculosis: how Demographic Changes, HIV Infection and Increasing Numbers in Poverty are Increasing Tuberculosis," *Annals of Medicine* 35 (4) (2003): 235–243; R.A. Weiss & A. J. McMichael, "Social and Environmental Risk Factors in the Emergence of Infectious Disease," *Nature Medicine* 10 (2004): 570–576.

plumbers, architects, accountants and more. People with moderately high C but low child V are well suited to such jobs, which is why economic growth peaked in the decades around the 1960s. This also means the economic value of ordinary citizens peaked at the same time so that, contrary to the Roman experience, there was an increase in *equality*. One measure of this is that unemployment fell to a very low level.

But as the fall of C accelerated in the late twentieth century, so the gap between rich and poor began to widen again (see Fig. 16.26 below).

Fig. 16.26. Share of total market income, including capital gains, going to the top 10%.[123] A fall in child V permitted greater equality in the first century after the peak of C. After that, rapidly declining C made much of the population less capable and so inequality increased.

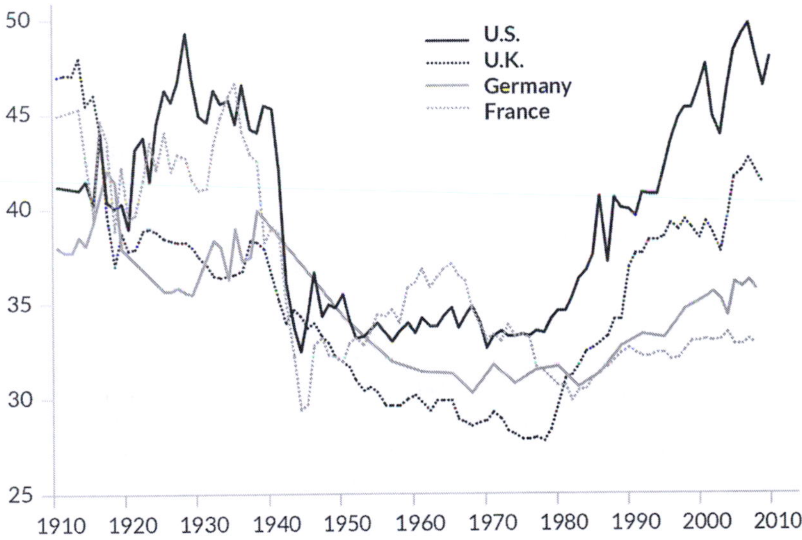

The US has experienced the greatest rise of the super-wealthy, with the top 1% earning almost 20% of the national income in 1928, a share which had fallen to around 7% by the 1970s, and then back up to a post-war high of more than 16%. But the same pattern applies to other Western nations, regardless of which measure of inequality is used. The OECD noted a growing gap between rich and poor in its 2008 report Growing Unequal.

[123] World Top Incomes Database, (2012). Missing values interpolated using 5% and top 1% series.

The report found that income inequality in OECD nations was at its highest for half a century, and that the average income of the wealthiest 10% of the population was about nine times that of the poorest 10% across the OECD, up from seven times 25 years ago.

Even relatively egalitarian countries such as Germany, Denmark and Sweden have experienced a rise in the income gap, with the richest 10% out-earning the poorest 10% by 6 to 1, compared to around 5 to 1 in the 1980s. On the same measure the wage gap in Italy, South Korea and Japan is currently 10 to 1, and in the US and Israel 14 to 1.[124]

The fall of C in East Asian societies started later than in Europe and America, so the change to greater inequality has been delayed, but even here there has been a marked change over the past decade or two. This can be measured by the Gini coefficient, a standard measure of inequality, where zero means everybody has the same income and one means the richest person has all the income. In 2000, South Korea measured an exceptional low Gini index of 0.286, which rose sharply to 0.325 by 2008. In Japan, the Gini rating rose slightly from 0.297 in 1993 to around 3.1 by 2005. By comparison, as of 2009 the Gini rating was 0.36 in Britain and 0.408 in the US.[125]

One important reason for increasing inequality is that the decline of C seems to have been more drastic in people with lower incomes. Consider the nuclear family, a strong indicator of C. In 1970 fewer than 10% of white American children were born to unmarried mothers. By 2010 this had risen to more than 60% for the children of mothers with less than twelve years of education, compared to under 5% for mothers with sixteen or more years.[126] Among lower income groups divorce rates are higher and married people less happy. According to the GSS, by 2010 lower income married people were far less likely to consider their marriage as very

[124] D. Flavelle, "Why the Gap between Rich and Poor in Canada keeps Growing," The Star, http://www.thestar.com/business/article/1097055--why-the-gap-between-rich-and-poor-in-canada-keeps-growing (accessed September 7, 2014).

[125] "What do you do when you reach the top?," The Economist, (November 9, 2011),
http://www.economist.com/node/21538104 (accessed September 7, 2014); Central Intelligence Agency, "The World Factbook"; Mason, "Britain: Income Inequality at Record High."

[126] Murray, *Coming Apart*, 161.

happy, compared to earlier times and to better off couples.[127]

Unmarried and divorced parents are especially likely to exercise minimal discipline. As quoted in one community study in a Pennsylvania town:

> Then you hear why the discipline was only minimal—"Well, you know, I talked to them and they said this, that, and the other, and I figure "Maybe he's right" … You want to be the cool parent, the friend parent, the great parent that the kid does whatever he wants, however he wants, dresses great.[128]

People on lower incomes work, on average, far fewer hours than the better off. At one level this reflects the lack of full time employment opportunities, but more fundamentally it is because of a lack of interest in the jobs available. A measure of this is that first generation migrants tend to work long hours in low wage jobs. Their children are less willing to do such work and consequently often have worse employment outcomes, despite presumably better English.[129] Third-generation immigrants tend to do more poorly still.[130]

Low-income people are also less likely to be religious. All of these trends are evident in better educated people over the last half century, but they are far more extreme in lower income groups. Such a divergence could account for the entire increase in inequality, largely a gap between skilled and educated people and the unskilled poor. As people lose C their economic effectiveness declines, and so also does their income.

Another aspect of this is increased "time preference," as indicated earlier. With falling C large segments of the population are less willing to sacrifice present consumption for future benefit, so fail to gain the education and occupational skills required for success in the modern economy. In discussing the rise of the West we saw that falling time preference actually reduced the premium for skilled wages compared with the Middle Ages.

---

[127] Kaplowitz & Oberfield, "Reexamination of the Age Limit for Defining when Puberty is Precocious in Girls in the United States"; Kaplowitz et al. "Earlier Onset of Puberty in Girls"; Susman et al. "Puberty, Sexuality, and Health."

[128] Murray, *Coming Apart*, 220.

[129] H. J. Gans, "Second-Generation Decline: Scenarios for the Economic and Ethnic Futures of the post-1965 American Immigrants," *Ethnic and Racial Studies* 15 (2) (1992): 173–192.

[130] G. Carliner, "Wages, Earnings and Hours of First, Second and Third Generation American Males," *Economic Inquiry* 18 (1) (1980): 87–102.

So it is that rising time preference among poorer people has the opposite effect, so that the premium for occupational skills starts to rise. It is the unskilled and unemployed who have shown the greatest increase in time preference. They are more likely to get into debt, indulge in drugs and overeat, resulting in obesity. Through all these forms of behavior runs the common thread of an excessive focus on short-term interests.

Political partisanship

Thus it is that the ‘sweet spot’ around 1970, when stress was lowest relative to C, had a number of positive effects. Economic growth was rapid, inequality low, and disease resistance high. Another consequence is what seems to have been a time of relatively bi-partisan politics, an example of which can be found in the voting records of American Senators and Congressmen.

In 1889-90 the ideological positions of Republicans and Democrats were far apart, with the most liberal Republicans far more conservative than the most conservative Democrats. By 1963-64 there was considerable overlap between the parties, making it possible for LBJ to negotiate through major changes such as the Civil Rights Act. By 1983=84 Congress was more divided, with very little overlap. This became more pronounced in 1997-98, and by 2011-12 the situation was almost back to the rigid ideological divisions of the 1880s.[131]

Higher stress periods such as the 1930s and 1970s, the product of the recession cycle, also showed high levels of division, but the general trend since the 1960s has been to an increasing rigidity and unwillingness to compromise. This trend can be expected to continue in coming decades, exacerbated (if biohistory is correct) by economic decline.

Economic collapse

In the nineteenth century, people of European descent had a number of characteristics that were favorable to economic development. These included a strong work ethic, a positive attitude to the market economy, a willingness to sacrifice present consumption for future benefit (including investment in education), machine and engineering skills, ethical attitudes (including a stress on integrity), and a considerable capacity for

[131] “Powering Down”, *The Economist*, November 8, 2014

innovation. All of these can be related to a high level of C, and especially infant C.

In the twentieth century C began to decline, a process that has accelerated since the 1960s. Though not favorable to long-term economic health there have been a number of short-term factors that countered this effect. The most important was a significant fall in child V which has made people more flexible, thus enhancing for a time equality and productivity.

Another is that economic capacity depends to a large extent on experience in infancy and thus reflects the C of the previous generation. The economically dominant baby boom generation was born in the period before the most rapid fall in C. Still another is mass immigration from higher C societies, especially Asia in recent years. To this can be added the increased outsourcing of manufacturing to these countries, especially China, which has brought down the cost of goods. A further benefit is a better understanding of economics, which has reduced trade barriers and brought inflation under control.

But within a few decades, given the continuing fall in C, Western economies must stagnate and then decline. The Japanese economy has already stalled, less than a century after its own peak of C. We have followed several civilizations through the process of collapse, but in none of them has the fall of C been as dramatic as that experienced by the West over the past fifty years. This is because none achieved our extraordinary level of wealth. Rome in the late Republic was a wealthy society by most historical standards, but mobs were still capable of rioting when the high price of bread left them hungry. By contrast, the biggest poverty-related health problem in the West is obesity. What this means is that economic collapse should be faster and more complete than anything seen in the ancient world.

We are accustomed to seeing government as the solution to economic ills, but government can do nothing about this problem. Lifting government controls can dramatically increase a nation's wealth in the short term, as shown most dramatically in recent years by the rise of China. But increased wealth only accelerates the fall of C, which leaves a society far worse off in the long run. The Sung dynasty in China was by historical standards a time of unusual prosperity. Barbarian invasion ended it in the north, as did the Mongol conquest in the south.

The triumph of capitalism in the Roman world is, ironically, one of the main reasons that the collapse of the Roman Empire was so catastrophic and the subsequent dark ages so extended. Chinese of the Qin and Han dynasties never abandoned peasant farmers as the cornerstones of the economy. By contrast with the mass depopulation of Europe, population seems actually to have grown in south China in the centuries after the collapse of empire.

Socialist systems are far less effective at creating wealth and so do less to reduce C in this way, but they tend to erode or deliberately eradicate the traditional beliefs and ideologies which support C. The Soviet Union undermined patriarchy (and thus V) by advancing the position of women. More fundamentally, it taught atheism to several generations of children, with Christianity barely tolerated and only in the approved Orthodox Church.

In China, Mao's indoctrination program almost obliterated a two and a half thousand-year-old Confucian heritage. Not only was Confucianism barred from schools, but a strident 1973 campaign destroyed temples and burned sacred texts. Attempts to revive Confucian thinking in recent years only illustrate how little of it remains, at least in the form of a coherent ideology.[132] The tight regimentation and control of Communist regimes had C-promoting effects, but these were lost with the end of Communism, leaving the population without effective C-promoters of any kind.

China has been the outstanding success story of the twenty-first century in economic terms, but biohistory predicts that its decline will be more rapid than most. As an ideology, nationalism may provide support for government, and materialism may create short-term wealth, but neither can support C. Other so-called "developing" nations will not be far behind as traditional beliefs and religions decline, along with birth rates.

The one major exception to this pattern is in parts of the Muslim world such as Afghanistan where the decline in the birth rate has been less drastic and traditional values have remained strong. But as we have seen, the personality structure of these high-V peoples makes them far less capable of economic development.

---

[132] "Confucius Makes a Comeback in China," *Bloomberg Business Week*, (November 1, 2012).

Political decline

Once economic collapse has begun, the end of democracy cannot be long delayed. Even now, most governments have plunged deeply into debt in an effort to buy support, just as demagogues bought support from the Roman mob with bread and circuses. Aging populations will demand ever more services as tax revenues decline, compelling governments to renege on debts or (more likely) inflate the currency. Hyper-inflation in turn will further depress economic activity. Then follows one man rule, or foreign conquest as nation states grow too weak to defend themselves and too demoralized to care.

As with the fall in birth rate and economic productivity, it is important to recognize that governments can do little. The Roman Republic did not fall because of the lack of clever political ideas. Politicians such as Cicero and Caesar were as brilliant as any in the history of the Republic. It fell because soldiers (and many civilians) felt more loyalty to individual leaders than to their own institutions and laws, thus paving the way for the Empire.

Biohistory predicts that the collapse of the modern West will be more rapid than that of ancient Rome, but should not result in anarchy. The Romans were able to assimilate immigrants and provincials into their own declining C culture, but Muslim immigrants into Europe are more resistant to the cultural influence of their adoptive nations, and their source populations far larger. The long-term result must be a stable and conservative Muslim peasant culture like that of the Ottoman Empire, with grindingly poor masses ruled by a privileged elite. North America might follow a similar path, or perhaps develop some rigidly conservative Christian version.

Knowledge of technology may not be completely lost, but books are of little use without the mechanical aptitude to make use of them. Africa remains in poverty despite billings in aid and trade, Western teachers and universities, and the most advanced technology on-line at the click of a mouse. Only a technological temperament can support a technological society.

And then, once the level of S has risen, there can be no more brilliant low-S civilizations such as those of ancient Greece or modern Europe. On current trends this, and not liberal democracy, is the true "end of history."

CHAPTER SEVENTEEN

ADVANCING BIOHISTORY

The book presents a model for major social changes such as economic growth and decline, wars, and the rise and fall of civilizations. It cannot be, and is not intended to be, a final and complete version. Much of the basic biological research remains to be done, and there will undoubtedly be significant changes in the years ahead. Important changes have been made even in the final weeks as this book has been prepared for print. There are, for example, major differences within the research team on whether C-promoters in late childhood have effects distinct from those on adults. Updated research results will be posted on our web site www.biohistory.org.

The Need for Testing

But if biohistory is correct, even in broad outline, the implications are serious for areas that include economic development, the treatment of anxiety, alcoholism and drug abuse, the wisdom of attempting to overthrow autocratic regimes, policies to deal with population decline, and much more. The first priority must be a program of scientific research to test key aspects of the theory. Fortunately this would be relatively easy, using equipment that is available in the biological science departments of most universities. As discussed in the Introduction, biohistory is an empirical, scientific theory that generates hypothesis for testing, and from the results the theory may be confirmed, modified or discarded. For those who object to our conclusions, and there will be many, we challenge you to do the science and prove us wrong.

This final chapter will present some initial thoughts on a research program, including areas of uncertainty that require further work. Readers with a background in the biological sciences will undoubtedly come up with more.

Measuring epigenetic changes in animals

A high priority is to measure as accurately as possible the epigenetic changes in the brain associated with conditions of calorie restriction and stress at different ages. This will confirm whether infant C and child V exist as discreet systems, distinct from the effects of environment at other ages. Important questions include whether calorie restriction in infancy increases C as well as infant C, something that has been assumed but that may not be correct. At what age do infant C effects cease? Are the effects of C-promoters in late childhood distinct from those in adult life, apart from child V effects? Do child V effects continue into adolescence? How much stronger are C effects in adolescents, compared with those in later life, and how permanent are they?

Measuring epigenetic changes in humans

The above experiments would be relatively simple because epigenetic changes as a result of V and C-promoters are mainly expressed in the brain. The next stage is the development of tests for V, child V, C and infant C from blood, saliva or urine.

Once such tests have been developed, studies of human volunteers could be done to test for correlations of epigenetic measurements with attitudes and behavior. For example, people with the epigenetic signatures for C should be more occupationally successful, those with infant C more innovative and successful at 'impersonal' occupations such as mechanics and engineering. People with higher C, especially compared with infant C, should be more religious. High V people should be more aggressive and more likely to join the military, and those with high child V should be more traditional in values and accepting of authority.

Cross generational C- and V-promoters

An important question to investigate is to what extent C and V are set before birth. Direct epigenetic inheritance can be measured by subjecting male rats to mild CR and looking for effects on their infant offspring, a project our research team is currently pursuing. Effects in the womb could also be assessed by cross-fostering rats at birth, thus separating pre-natal and post-natal influences.

Animals in a semi-natural environment

An especially useful experiment would be to set up a colony of rats or mice in an environment where they have ample room to breed, form territories and so forth. Different food conditions could be delivered by varying the calorie content of food, or by electronically restricting access to certain animals. The value of such a setup is that animals could be observed in a quasi-natural environment where there is more opportunity to show the full gamut of social behaviors.

Restricted animals should show the characteristics of C. They should spend more time with their young, be less tolerant of other adults and less social, and spend more time on patrol and less resting. They should be more discriminating in choice of mates, have more elaborate courtship rituals, and prefer mates similar in appearance and behavior. By varying the timing of food restriction it would be possible to examine the naturalistic behavior of animals restricted at certain ages such as infancy, the juvenile period and immediately after puberty.

Intermittent food restriction, or some other stress such as cortisol in drinking water, should lead to behavior characteristic of high V: aggression, males dominant over females, intensive mothering of infants with early weaning. If rats or some other group living animal were used, they should be more likely to form hierarchical and cohesive groups with joint defense against outsiders. Our research team is currently developing just such an experiment for mice.

Primate studies

Similar experiments with primate populations, if logistically and ethically possible, would be even more revealing. In particular it would be interesting to know whether primates subject to mild food restriction control their offspring's behavior to a greater extent.

Testing the C and V of primates in the wild

Levels of C and V of primates in the wild could be tested by collecting feces or blood testing anesthetized animals. Within the same species, animals with smaller troops should have higher levels of C, those in more hierarchical and aggressive groups higher V, and so forth. Systematic rating of behavior, such as the amount of time spent by mothers with youngsters of different ages, should correlate with physiological measures of C, troop size and other indications.

Studying and Creating Lemming cycles

Biohistory predicts that any group of mammals should show lemming cycles given severe, intermittent food shortages. The previously described setup could be used for this, if kept in place for long enough, though lemming cycles might be easier to observe in wild-caught populations which have higher C and V to begin with.

The lemming cycle theory could be tested by taking samples from animals at different stages of the cycle in the wild, and looking for the predicted epigenetic signatures. For example, animals born at the growth peak should show a maximum of V, and those in the next generation the maximum of child V. Once such cycles are established, attempts could be made to lengthen them by (for example) gradually ramping up the level of stress. This would help to identify exactly why lemming cycles in human populations lengthen as civilizations collapse.

It would be of particular interest to work out exactly what 'G' is. Lemming cycle theory suggests that it has important effects on attitudes and behavior, including maximizing the birth-rate, but is not V. Instead, the highest G generation gives birth to the highest V generation. G also seems to promote national unity and independent thought, but not for any obvious biological reasons. Is there a distinct epigenetic signature for high G?

Some more laboratory tests

CR animals could be tested for the intensive food-searching behavior of high C animals like gibbons, such as their willingness to do monotonous tasks for a food reward. This would have to be in a situation where CR animals were temporarily well fed so that hunger was not the driving force.

An attentive but stressed mother should produce offspring with a highly effective stress response: low resting CORT, a rapid increase in response to stress, and an equally rapid fall thereafter.

Wild animals given plentiful food for several generations should show behavioral changes indicative of lower C and V, even without genetic selection. They should become less aggressive, more tolerant of crowding, more promiscuous, and with reduced care of young.

Early CR and testosterone

One of the earliest findings of the research project was that early CR seems to increase testosterone, just as CR in later life reduces it. This was a key finding that made possible the development of the theory on war, ideas on double standards of sexual behavior, and more. An urgent need is to work out the source of this effect. Is it a C effect or stress effect or both? Is it a result of maternal behavior or a direct epigenetic effect?

Some tests on humans

Severely restricting the sexual outlet of people (especially adolescents) for a few weeks should produce physiological changes associated with C and changes in the attitudes and behaviors associated with high C. The latter would include religious sentiment, the capacity for work and self-discipline, reduced sociability, and a stronger sense of ethics in areas such as honesty.

Another experiment would be to subject a group of people to arbitrary social ‘rules’ requiring them to act in a certain manner on a regular basis. Compared with controls, they should also show physiological and behavioral changes indicative of high C.

People in military units undergoing basic training, especially of the more rigorous forms such as commando training, should show an increase in V. Those who started with higher V should make more effective soldiers.

People who join fundamentalist or otherwise very strict religious groups, such as monasteries, should have higher anxiety to begin with but less with time.

Identifying men who gained most of their adolescent sexual outlet through nocturnal emissions would make a useful study, by giving a clear picture of individuals with high C. They should be more introverted, more occupationally successful, have more children, and be more likely to show intense religious involvement.

Testing age cohorts

People born at the end of major wars should have physiological and behavioral indications of higher V than cohorts born before or after. These include Europeans and Japanese born in 1944-45, mainland Chinese born in 1948-49, and Iranians born in the mid-1980s. They should also, on average, be more aggressive than people born earlier or later, though in a form related more to group than individual (e.g. criminal) violence. In other words, this is an aggression that causes people to group together to fight against a common enemy, rather than the all-on-all hostility of a Stalinist terror regime. This test would also make it possible to identify exactly which are the crucial ages for the stress effect. The current model assumes the main effect is in the first two years or life, but it could be earlier or later or both.

People experiencing recessions in late childhood (age 6-12) or adolescence should have higher C and child V. This should also make clear which of these two effects is stronger, or even if both effects can be found.

Ethnographic studies

One useful project would be to re-do the cross-cultural survey, using a wider range of societies and a more rigorous experimental design, something currently being undertaken by our research team. At the time the initial survey was done the theory was at an earlier stage, so there was no attempt to measure parental control between the ages of two and five.

Even better would be to measure precisely the degree of control and punishment of children at different ages in different societies, perhaps through the use of surveillance cameras set up in the homes of volunteers. Individuals and cultures with higher C will be expected to control their children more tightly at all ages. Those with higher child V (for example, Middle Eastern Arabs) should show indulgence of infants and harsher treatment in late childhood. Use of physiological tests would help to determine exactly what treatment at which ages affects infant versus child C, etc.

Measures of C and V in populations world-wide should reflect the patterns identified in the text, such as indigenous Amazonian peoples having low C, or north Chinese higher V than south Chinese.

People from high S societies, such as Arabs or Chinese, could be tested for greater interest in babies versus older children through such measures as pupil dilation. It is predicted that women from high S societies would show greater pupil dilation than women from low S societies when presented with pictures of babies, indicating a greater level of engagement and interest.

In societies where women are more tightly controlled than men, the disparity between male and female testosterone should be greater.

Predicting economic downturns

Ongoing regular measurement of testosterone and stress hormones in human populations, such as through saliva tests, should be able to help predict the timing and severity of economic recessions, especially if conducted in a number of nations. An increase first in testosterone and then in stress hormones should precede severe recessions.

C-promoting supplements

An effective confirmation of at least some aspects of the theory would be the development of a C-promoting supplement, which would have the potential to treat alcoholism, drug addiction, obesity, poor maternal behavior and other conditions associated with low C. Our research team has achieved promising results and is working to develop them, with the details kept confidential for commercial reasons.[1]

Scientific Collaboration

Scientists and other scholars interested in collaborating on any aspects of the research may contact the author through www.biohistory.org. A number of PhD scholarships are available for dedicated students with good academic results and an interest in pursuing aspects of the research.

---

[1] Any rights will be owned by the non-profit Biohistory Foundation and used to further the research program.

GLOSSARY OF TERMS

C

A physiological system that adjusts behavior to conditions of limited food. Characteristics include delayed breeding, social intolerance (at the extreme, monogamous territories), stronger parental care, and activity in exploration and in the search for food, even when not hungry. Increased by chronic mild food shortage or (in humans) controls on behavior and especially sexual activity. Characteristics of C in humans include hard work, discipline, and willingness to sacrifice present consumption for future benefit. C-promoters in childhood and adolescence have more lasting effects than those in later life.

C-promoters

Factors such as food shortage and limits on sexual behavior that increase C.

Infant C

That form of C arising from parental control or other C-promoters in infancy and (to a lesser extent) early childhood. Characteristics include independent thinking, machine skills, hereditary loyalties, preference for rulers who are similar in language and culture, impersonal loyalties, honesty. Ultra-high levels seen as responsible for the Industrial Revolution.

Adult C?

That form of C arising from sexual restraint and other influences after puberty. Has general C characteristics such as promoting hard work and self-discipline, but with a particular emphasis on religious commitment and love of children. Not clear if this is distinct from the effects of C-promoters in late childhood.

V

A physiological system that adjusts behavior to conditions of generally plentiful food but with occasional severe stresses from famine or predator attack. Aspects include aggression, strong group cohesion, confidence and morale, intolerance of crowding, migration, patriarchy, indulgence of infants by anxious mothers combined with early weaning and exposure to severe stresses in late childhood (age 6-12)

V-promoters

Factors such as sexual restraint, patriarchy and occasional stresses that increase V.

Child V

That form of V arising from V-promoters such as parental control and especially punishment in late childhood. Associated with traditional thinking and acceptance of authority.

Stress

A physiological system that involves chronically high levels of cortisol and other stress hormones. Increased by crowding and threat and by higher levels of V.

Civilization cycle

The links between C and V that explain the rise and fall of civilizations, when combined with C- and V-promoters contained in advanced religions. Low C permits V (and thus stress) to rise. High V and stress increases C. High C (especially infant C) reduces V. Lower V allows C to fall. The most visible signs of this cycle is that a rise in C is accompanied by a rise and then fall in V and stress, and a fall in C is accompanied by a fall and then rise in V and stress.

S

A factor, presumably genetic, that causes people to be more indulgent of younger children and especially infants, compared with older children. Increases C relative to infant C, and also strengthens child V. This in turn makes societies more stable but less innovative.

Lemming cycle

A fluctuation in a trait known as ‘G’, believed to be a mammalian mechanism triggered by occasional food shortages which causes an alternation between V and C. Reflected by changes in population growth, political cohesion, and habits of thought. Normally around 300 years in humans but lengthens in 'Dark Ages' following the collapse of civilizations.

G

The central year of the time of the population growth phase of the lemming cycle, though it may not be the actual year of fastest growth.

Recession Cycle

The reason for shorter-term fluctuations in population growth. Behavioral characteristics include hedonistic behavior in times of declining birth-rate such as the 1920s and 1960s, a recession and/or period of political extremism when birth rates are low as in the 1930s, and more conservative attitudes when birth rates are rising such as in the 1940s and 1950s.

The driving force of the recession cycle is that people spending their late childhood and adolescence in prosperous time have lower C and perhaps child V, causing hedonistic behavior and impelling people to seek out crowds and other stimulus. Because tolerance for stimulus is set by V, the result is a gradual rise in stress until high enough that a recession and/or political extremism occur. The generation passing their late childhood and adolescence in such hard times have higher C and perhaps child V, causing a rise in the birth rate and more stable and conservative times.

INDEX